TRENCHLESS TECHNOLOGY

TRENCHLESS TECHNOLOGY

Pipeline and Utility Design, Construction, and Renewal

Mohammad Najafi, Ph.D., P.E.

Director of Center for Underground Infrastructure
Research and Education (CUIRE)
School of Planning, Design, and Construction
Michigan State University
East Lansing, Michigan

With Contributions by

Sanjiv Gokhale, Ph.D., P.E.

Associate Professor of Practice
Director of Construction Management Program
Vanderbilt University
Nashville, Tennessee

McGRAW-HILL

New York Chicago San Francisco Lisbon London Madrid
Mexico City Milan New Delhi San Juan Seoul
Singapore Sydney Toronto

The McGraw-Hill Companies

Library of Congress Cataloging-in-Publication Data

Najafi, Mohammad.
Trenchless technology : pipeline and utility design, construction, and renewal / Mohammad Najafi.
p. cm.
Includes bibliographical references and index.
ISBN 0-07-142266-8 (alk. paper)
1. Pipelines—Design and construction. 2. Trenchless construction. I. Title.

TA660.P55N35 2004
621.8′672—dc22 2004062405

2 3 4 5 6 7 8 9 0 DOC/DOC 0 1 0 9 8 7 6 5

ISBN 0-07-142266-8

The sponsoring editor for this book was Larry S. Hager and the production supervisor was Sherri Souffrance. It was set in Times Roman by International Typesetting and Composition. The art director for the cover was Margaret Webster-Shapiro.

Printed and bound by RR Donnelley.

This book was printed on recycled, acid-free paper containing a minimum of 50% recycled, de-inked fiber.

I would like to acknowledge the support and love of my wife, Homa and my daughters Maryam and Zinat and would like to dedicate this book to my brother and my professor Mr. Madi Najafi who dedicated his life to serve others.

Mohammad Najafi

I am grateful for the love and support of family, and wish to dedicate this book to my wife, Latha.

Sanjiv Gokhale

CONTENTS

Chapter 17. Lateral Renewal 347

Chapter 18. Localized Repair 355

Chpater 19. Modified Sliplining 371

Chapter 20. Underground Coatings and Linings 385

Chapter 21. Thermoformed Pipe 401

FOREWORD

It is with great pleasure that I take the opportunity offered by Dr. Najafi to welcome a new reference book to the field of trenchless technology. Dr. Najafi, as a student of Dr. Tom Iseley, was one of the early members of the Trenchless Technology Center and has since been involved in education and research in the field of trenchless technology at several universities and with projects for a variety of agencies and organizations. In this book, he has assembled a wealth of information on the inspection, installation, and rehabilitation of underground utilities using trenchless technologies. The book organizes and provides the information on various technologies in a very useful way and a nice feature of the book for teachers is the provision of sample questions at the end of each chapter. The variety of sources from which the material in the book has been assembled leads to some stylistic variations in the text but it will be very useful for the practitioner and student alike to have all the information in one place. The book is primarily descriptive in nature but, where appropriate, the standard design approaches for such topics as pipe and liner design, pipe friction during jacking, thrust block design, and so on are presented.

I believe that this book will find good use by the student, teacher, and practitioner alike. It is more comprehensive than any previous North American book on trenchless technology and covers the whole field in one text. Over the next decade, we expect to see continued advances in the understanding and reliability of trenchless techniques, better developed design approaches, and a continued effort to reduce costs while improving quality control. Such efforts will lead to a continual expansion of knowledge in the field. This book will provide an excellent baseline against which to measure our progress in the coming decade.

RAY STERLING
Contractors' Educational Trust Fund Professor of Civil Engineering and Director, Trenchless Technology Center, Louisiana Tech University
2002–2005 Chairman, International Society for Trenchless Technology

PREFACE

This book is a collection of the existing literature and methods successfully used to install and renew underground pipelines and utility systems. It is intended as a reference book for design and consulting engineers, utility owners, pipeline professionals, government agencies, municipalities, contractors, manufacturers, and professionals who are involved with planning, design, construction, operation, and maintenance of underground pipeline and utility systems. The book is also intended as a textbook for undergraduate and graduate students from all engineering and technology disciplines. The preferred prerequisites for this course would be fluid mechanics, statics, and mechanics of materials, although instructors can select the different sections for a semester or quarter course based on the background of students.

The field of trenchless technology is so vast that no single individual can master it all. Therefore, to prepare this book a team of pipeline professionals and industry leaders came together in three different capacities of advisory board, contributors, and reviewers. In addition, members of the board of the Center for Underground Infrastructure Research and Education (CUIRE) provided support and advice throughout the preparation of this book. The names of these individuals are presented in the Acknowledgments section following the Preface or at the end of each chapter.

No book can be complete and include all related topics. There will be new capabilities for current trenchless technology methods, and new methods will be developed in future. Design engineers, utility owners, and contractors should consider the existing and proven methods and those viable methods that will be available. To complement the materials presented in this book, a Web site will be dedicated by the Center for Underground Infrastructure Research and Education (CUIRE), Michigan State University at www.msucuire.org and Department of Civil and Environmental Engineering, Vanderbilt University at http://www.cee.vanderbilt.edu/grad_constmgmt.html to present case studies, design examples, test questions, and homework problems. Those faculty members and industry professionals who would like to contribute or update the materials presented need to contact the authors. All contributions will be acknowledged in the future revisions and Web site presentations.

Underground pipeline construction and renewal projects present many risks. There is no guarantee that the methods presented in this book will be successful at all times and at all project and site (surface and subsurface) conditions. Moreover, there is no endorsement and/or recommendation of the proprietary methods and brand names mentioned in this book. The specified method characteristics, diameter range, maximum installation, typical application, and accuracy is based on project and site conditions, type of equipment used, and experience and training of the crew and the operator. Design engineers, project owners, contractors, government agencies, and all other parties involved in trenchless technology projects should consider the risks involved and assume appropriate contingencies in the contract documents. Methods successfully used in some applications may not be applicable in some other conditions due to change in project, site, and/or soil conditions. Design engineers and pipeline owners need to be involved in the selection of appropriate trenchless methods for their specific project conditions and do not leave the trenchless method selection entirely to the contractor.

Additionally, a differing site condition clause needs to be included in the contract to cover contingencies, in case the trenchless project is stopped and cannot continue after one or more attempts, due to no fault of the contractor.

This book consists of 22 chapters, which are divided into three parts. Part one includes Chapters 1 through 6. Chapter 1 introduces the reader to trenchless technology and gives an overview of these methods. Chapter 2 presents an overview of the benefits of these technologies. In order to execute successful trenchless renewal projects, an assessment of current pipe conditions is necessary. Also the pipe must be cleaned and principles of asset management applied for a cost-effective and efficient pipeline renewal project. Chapter 3 provides information on pipeline asset management, inspection, and cleaning. Chapters 4 and 5 cover some important topics in the design of trenchless construction (TCM) and renewal (TRM) methods. No trenchless construction method can be used without an appropriate quality pipe. Chapter 6 covers an overview of pipe materials specific for use in trenchless technology. Parts two and three of this book divide trenchless technology methods into two broad divisions of trenchless construction methods (TCM) and trenchless renewal methods (TRM). Chapters 7 through 12 are devoted to individual trenchless construction methods. Chapters 13 through 22 are focused on different trenchless renewal methods. These chapters provide detailed information for design and construction of each of these technologies. Such information as introduction to the method, range of applications, project requirements and risk assessment, method description, soil and/or site compatibility, main characteristics, capabilities, and limitations for each method are presented in these chapters.

MOHAMMAD NAJAFI, PH.D., P.E.
Michigan State University

SANJIV GOKHALE, PH.D., P.E.
Vanderbilt University

ACKNOWLEDGMENTS

This book would not be possible without the help, support, and patience of Mr. Larry Hager, senior editor, McGraw-Hill Professional. The authors would like to thank McGraw-Hill Professional and the editorial staff for their support of this publication. The cooperation and hard work of Neha Rathor and Ben Kolstad, International Typesetting and Composition and their colleagues are acknowledged. The efforts of Alhad Panwalkar, CUIRE Research Associate, who spent many long hours coordinating the review process and writing the manuscript, are appreciated. The authors would like to acknowledge support of their colleagues at their respected universities during the preparation of this book.

Advisory Board

David G. Abbott *Jason Consultants Group*

John Capocci *Barbco Inc.*

Sggi Finnsson *Digital Control Inc.*

Mark Gallucci *Marketing and Training Manager, Digital Control Inc.*

Joanne Hughes *Raven Lining Systems Inc.*

Tom Iseley *BAMI and BAMI International*

Bernie Krzys *Benjamin Media Inc.*

Jerry Norton *Norton Construction Company*

Lynn E. Osborn *Insituform Technologies Inc.*

Steven K. Rose *Marketing Manager, Prime Resins Inc.*

Bill E. Shook *AP/M Permaform*

Richard O. Thomasson *Parsons Brinckerhoff*

Mark G. Wade *Wade & Associates Inc.*

Mark Wallbom *Miller Pipeline Corp.,*

Grant Whittle *Ultraliner Inc.*

Contributors

Frank Canon *Baroid IDP, Sec. 4.6 on Drilling Fluids in Chap. 5*

Dennis J. Doherty *Jacobs Civil Inc., Case Study in Chap. 14, Sliplining*

Dec Downey *Jason Consultants Group, Case Study in Chap. 2, Social costs of utility Construction: A life cycle cost approach*

Sggi Finnsson *Digital Control Inc., Sec. 4.7 on Locating and Tracking for Horizontal Directional Drilling in Chap. 5*

Bhavani Sripathi Gangavarapu *Portions of Chap. 2*

Joanne Hughes *Chap. 20, Underground Coatings and Linings*

Bill E. Shook *AP/M Permaform, Chap. 22, Manhole Renewal Methods*

Jadranka Simicevic *Trenchless Technology Center, Portions of Chap. 16 on Pipe Bursting*

Raymond L. Sterling *Trenchless Technology Center, Portions of Chap. 16 on Pipe Bursting*

Sacha Tetzlaff *Strand Associates Inc., Wisconsin, Case Study for Chap. 13, Cured-in-Place-Pipe*

Steven L. Woodman *City of Beloit, Wisconsin, Case Study for Chap. 13, Cured-in-Place-Pipe*

Ken Weeks *KRW Associates, History of Directional Drilling, Chap. 10*

Grant Whittle *Chap. 21, Thermoformed Pipe*

Reviewers

Lori Burgett *Kokosing Construction Co.*

Dec Downey *Jason Consultants Group*

Joanne Hughes *Raven Lining Systems Inc., Chap. 22*

G. Alan Johnson *Wade and Associates, Chap. 3*

Ray Huchinson *City of Atlanta, Chap. 3*

Tom Iseley *BAMI, Chaps. 1 through 3*

Tim Mahan *Consultant, Chap. 2*

Stewart Nance *Raven Lining Systems Inc., Chap. 22*

Paul Nicholas *Wirth Soltau, Chap. 11 and 12*

Lynn E. Osborn *Insituform Technologies Inc., Chaps. 1 and 13*

Mark Wallbom *Miller Pipeline Corporation, Chaps. 1 and 2*

Grant Whittle *Ultraliner Inc.*

Current Board Members of the Center for Underground Infrastructure Research and Education (CUIRE)

Jim Barbera and John Capocci *Barbco*

Jeff Bistodeau *Premier Events*

Lori Burgett *Kokosing Construction Co., Membership Committee Chair*

Robert Carpenter *Underground Construction Magazine*

Eugenia Chusid *City of Santa Monica*

Ben Cocogliato *TT Technologies Inc.*

Mark Dionise *Michigan Department of Transportation, Chair*

Troy Freed *SOS Service Group Inc. Membership Committee Vice Chair*

Peter Funkhouser *Michigan Department of Transportation*

Mark Gallucci *Digital Control Inc.*

Irvin Gemora *National Association of Sewer Service Companies (NASSCO)*

Daniel Hanson *Hanson Engineering*

John Heiberger *Consumers Energy*

Cliff Henderson *Hobas Pipe USA*

Mark J. Holbrook *Hamburg Township DPW Admin.*

Larry W. Johnson *Hobas Pipe USA*

G. Alan Johnson *Wade & Associates Inc.*

David Kozman *American Water Services, Technical Committee Chair*

Steve Kramer *Jacobs Civil Inc.*

Bruce Kuffer *Fishbeck Thompson Carr & Huber, Vice Chair*

Dan Liotti *Midwest Mole Inc.*

William (Tim) Mahon *Consultant*

Mike Moore *McLaughlin Manufacturing Company*

Carl M Pearson *Kenny Construction Company*

Gaylord Richey *Astec Underground/American Augers*

Jim Rush *Benjamin Media Inc. (Trenchless Technology Magazine)*

Jim Scott *Solution Resource*

Gary M. Soper *Benton Charter Township*

Jeff Sowers *TBE Group Inc.*

Don Spencer *City of Grand Rapids*

Jamie Taggert *Peninsular Technologies*

Rob Tumbleson *Akkerman Inc.*

CHAPTER 1
OVERVIEW AND COMPARISON OF TRENCHLESS TECHNOLOGIES

1.1 INTRODUCTION

The conventional method for construction, replacement, and repair of underground utilities has been trenching or open-cut. Based on the type of work, this method is also called dig-and-install, dig-and-repair, or dig-and-replace. This method includes direct installation of utility systems into open-cut trenches. Open-cut methods involve digging a trench along the length of the proposed pipeline, placing the pipe in the trench on suitable bedding materials and then backfilling. Most of the times, the construction effort is concentrated on such activities as detour roads, managing the traffic flow, trench excavation and shoring, dewatering (if needed), backfilling and compaction operations, bypass pumping systems, and reinstatement of the surface. This results in a small part of the construction effort actually being focused on the final product, which is the pipe installation itself. In some cases, the backfilling and compaction and reinstatement of ground and pavement alone amount to 70 percent of the total cost of the project. As such, considering all the project parameters, the open-cut method is more time-consuming and does not always yield the most cost-effective method of pipe installation and renewal. In recent times, due to the understanding of the various *social costs* involved with open-cut, this method of installation is being discouraged. Social costs include cost to general public, environmental impacts, and damage to pavement, existing utilities, and structures. A breakdown of social costs is presented in Chapter 2 of this book.

Advancements in technology and improvements in obtaining geotechnical data and development of new equipment have led to improvements in utility-pipe installation work. These alternative means of installing and renewing of underground utility pipes have been developed to facilitate utility construction and renewal with minimal surface disruption and social costs. These techniques are called *trenchless* technology (TT) or *NO-DIG* installation and renewal of pipelines and utilities. The expression *trenchless* or *NO-DIG* is actually a misnomer, because many trenchless methods especially for new installations require a pit, shaft, or some type of surface excavation. A better definition for TT techniques would include methods of pipeline and utility installations with *minimum* amount of surface excavation. Therefore, these methods would offer *less trench, less footprint, environmentally friendly*, and *less impact* construction and inherently enhance safety, productivity, and cost-effectiveness in the construction process. TT usually provide a safer environment for the workforce compared to open-cut technologies, and at the same time improve performance by eliminating the double handling of soil. Additionally, there is no need to excavate and

FIGURE 1.1 Hand mining. (*Courtesy of Akkerman Corporation.*)

then to backfill. Although many TT methods, such as tunneling, pipe jacking, and hand mining existed for more than 100 years (Fig. 1.1), the TT industry was officially established in the United States shortly after its creation in United Kingdom in 1986. Table 1.1 presents some historical development of trenchless industry.

1.1.1 Summary of Trenchless Technology Benefits

Trenchless methods have many advantages over conventional open-cut methods, such as they:

- Minimize the need to disturb existing environment, traffic, or congested living and working areas.
- Use predetermined paths provided by existing piping, reducing the steering and control problems associated with new routing.
- Require less space underground, minimizing chances of interfering with existing utilities or abandoned pipes.
- Provide the opportunity to upsize a pipe (within the technology limits) without open-trench construction.
- Require less-exposed working area, and therefore, are safer for both workers and the community.
- Eliminate the need for spoil removal and minimize damage to the pavement (the life expectancy of pavements have been observed to be reduced by up to 60 percent with dig-up repairs), and disturbance to other utilities.

1.1.2 Trenchless Technology Market Share

More than 300,000 miles of underground utilities, including water, sewer, and gas, electrical power, cable television, and telephone, are constructed around the world each year, with an estimated market value of greater than $35 billion. The U.S. sanitary sewer market is

TABLE 1.1 Some Historical Development of Trenchless Technology Industry

Year	Activity
1946	Publication of *Underground Construction Magazine*, with coverage of trenchless technologies beginning in the 1970s.
1976	Establishment of the National Association of Sewer Service Companies (NASSCO). NASSCO assisted in preparation of the EPA Sewer Evaluation & Rehabilitation Manual (EPA-600/2-77-017d).
1985	First NO-DIG Conference in London
1986	Establishment of the International Society for Trenchless Technology (ISTT).
1989	Establishment of the Trenchless Technology Center (TTC) at Louisiana Tech University.
1990	Formation of the North American Society for Trenchless Technology (NASTT).
1990	Trenchless Technology Center (TTC) conducts the very first international trenchless technology seminar in Houston
1991	1991–First NO-DIG Conference in the U.S. (Kansas City).
1991	Establishment of Gulf Coast Trenchless Association (CGTA) in Houston
1992	First International NO-DIG Conference in the U.S. (Washington, DC).
1992	Publication of *Trenchless Technology* magazine.
1994	Publication of NO-DIG *Engineering* technical journal.
1994	Establishment of Center for advancement of Trenchless Technologies (CATT) at the University of Waterloo.
1994	Establishment of Pipeline Rehabilitation Council (PRC).
1995	First Underground Construction Technology Conference (UCT) in Houston.
1995	Establishment of the Center for Innovative Grouting Materials and Technology (CIGMAT) at University of Houston.
2001	Publication of ASCE Standard Construction Guidelines for Microtunneling (CI/ASCE 36-01).
2002	Establishment of the Center for Underground Infrastructure Research and Education (CUIRE) at Michigan State University
2002	Establishment of NASSCO Pipeline Assessment and Certification Program (PACP).
2003	Establishment of Buried Asset Management Institute in Atlanta.
2004	Establishment of Pipeline Infrastructure Research Center (PIRC) at Pennsylvania State University.
2004	Establishment of BAMI-International
2004	Publication of ASCE Manual of Practice No. 106 on Horizontal Auger Boring.
2005	Publication of the book, *Trenchless Technology: Pipeline and Utility Design, Construction, & Renewal*, by McGraw-Hill.

estimated to include 5,200,000,000 ft of pipe and 21,000,000 manholes. An estimate of the total U.S. pipeline-renewal market is approximately $330 billion. According to Water Infrastructure Network (WIN), the 20-year need for both water and wastewater industries is about $1 trillion. According to North American cost indices for 1988 to1998, trenchless construction methods (TCM) have gained a market share of around 20 percent by cost in pipe installation and renewal for utility services.

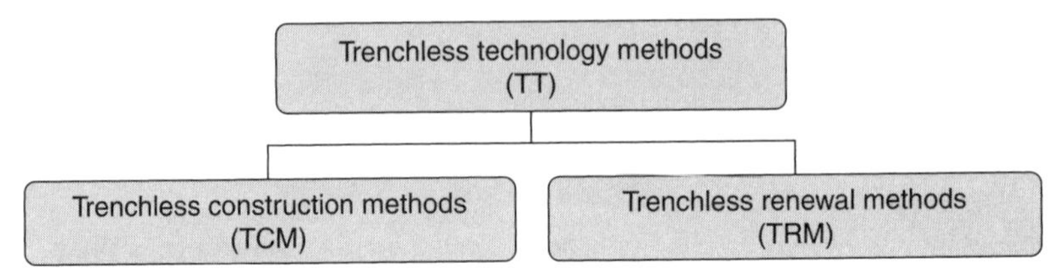

FIGURE 1.2 Main divisions of trenchless technology methods.

1.2 TWO MAIN DIVISIONS OF TRENCHLESS TECHNOLOGY METHODS

TT methods are divided into two main areas: TCM and trenchless renewal methods (TRM) as shown in Fig. 1.2. TCM include all the methods for new utility and pipeline installations, where a *new* pipeline or utility is installed. TRM include all the methods of renewing, rehabilitating, and/or renovating, an *existing, old*, or *host* (collectively called existing in this book) pipeline or utility system. Each of these main categories is further divided into subcategories as defined in the following sections.

An important note needs to be provided here. As for any type of construction project, a trenchless construction and renewal project is site specific and successful completion of a trenchless project is dependent on many factors. Such parameters as *accuracy, maximum installation length, diameter range*, and *type of application*, as presented in this chapter, are actually dependent on specific project conditions. Equipment operators' skills and contractors' experiences greatly influence the degree of accuracy and maximum installation length that can be obtained on a trenchless project. Also, type of equipment used, site surface and subsurface conditions, existing pipe conditions, and other project conditions will determine the applicability and capability of a trenchless method which must be carefully investigated by a competent professional engineer during the planning and design stages of a project. Therefore, main characteristics of different trenchless methods, as presented in the following sections, should only be used as a guide. Additionally, TT are advancing rapidly and engineers, project owners, as well as contractors need to keep abreast of latest developments in the industry in order to select the most cost-effective and appropriate method for their project conditions.

1.3 TRENCHLESS CONSTRUCTION METHODS (TCM)

TCM for installation of new pipelines and conduits include all methods of installing new utility systems below grade without direct installation into an open-cut trench. TCM are divided into two broad categories of worker entry required and worker entry not required as shown in Fig. 1.3.

Figure 1.3 also divides TCM into three major categories: horizontal earth boring (HEB), pipe jacking (PJ), and utility tunneling (UT). HEB includes methods in which borehole excavation is accomplished through mechanical means *without* workers being inside the borehole. As Fig. 1.3 indicates, both PJ and UT techniques require workers inside the borehole during

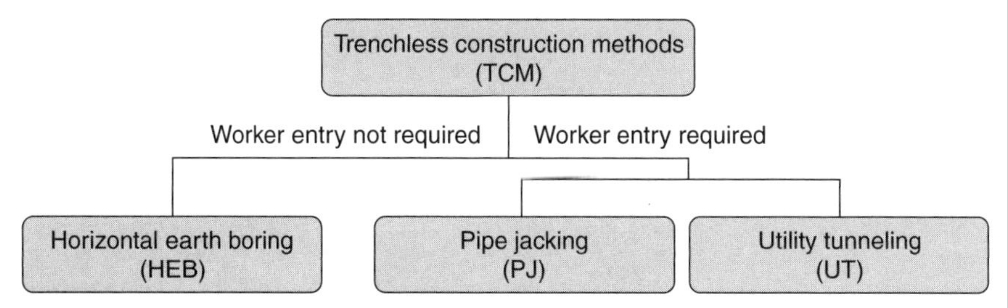

FIGURE 1.3 Trenchless construction methods.

excavation and pipe and/or temporary-support installation process. However, PJ is differentiated from UT by the support structure. PJ methods use installation of prefabricated pipe sections during the tunnel excavation. In this method, new pipe sections are jacked from a drive shaft or pit so that the complete string of pipe is installed simultaneously with the excavation of the tunnel. Although UT techniques may use the same excavation equipment, a temporary support structure is constructed from the excavation face. Normally, the support structure is traditional tunnel liner plates (TLP) or steel ribs with wooden lagging (SRw/WL). Table 1.2 presents main characteristics of PJ and UT methods. Figure 1.4 illustrates an inside view of a UT project.

The HEB is further divided into a number of methods as shown in Fig. 1.5. These methods are briefly described in the following sections and presented in detail in later chapters.

1.3.1 Horizontal Auger Boring (HAB) Methods

HAB is a cost-effective method of installing a steel casing pipe where it crosses a road, highway, or railroad track. This process simultaneously jacks a steel casing from a drive pit through the earth while removing the spoil inside the encasement by means of a rotating flight auger. The auger is a flighted tube having couplings at each end that transmit torque to the cutting head from the power source located in the bore pit and transfers spoil back to the machine. The casing supports the soil around it as spoil is being removed. Usually, after installation of casing, a product pipe is installed and the annular space is filled with grout. Table 1.3 presents the main characteristics of HAB methods.

TABLE 1.2 Main Characteristics of Pipe Jacking and Utility Tunneling

Method	Diameter range (in)	Maximum installation (ft)	Pipe material	Typical application	Accuracy
Pipe jacking and utility tunneling	42 & up	1500	RCP, GRP, steel	Pressure and gravity pipe	±1 in

FIGURE 1.4 View of a utility tunneling process.

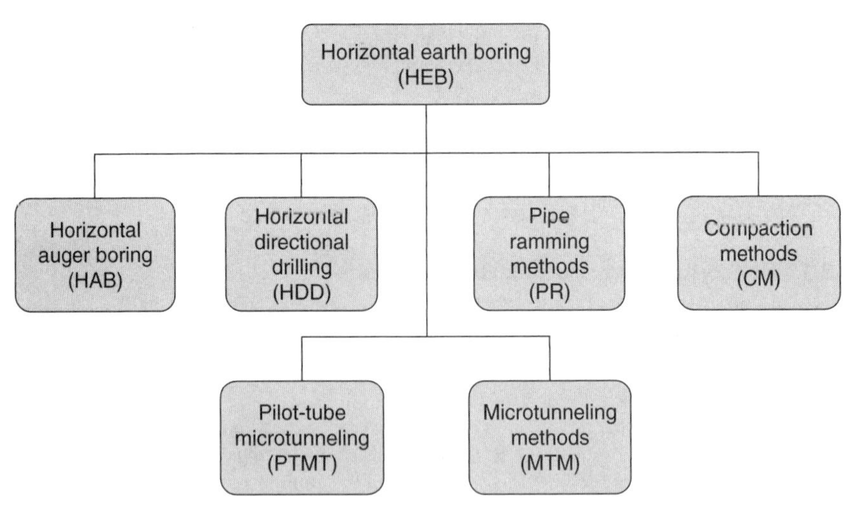

FIGURE 1.5 HEB categories.

TABLE 1.3 Main Characteristics of Horizontal Auger Boring (HAB) Methods

Method	Diameter range (in)	Maximum installation (ft)	Pipe material	Typical application	Accuracy
Auger boring	4–60	600	Steel	Road and rail crossing	±1% of the bore length
Auger boring steered on grade	4–60	600	Steel	Pressure and gravity pipe	±12 in
Auger boring steered on line grade	4–60	600	Steel	Pressure and gravity pipe	±12 in

TABLE 1.4 Main Characteristics of Microtunneling Methods

Method	Diameter range (in)	Maximum installation (ft)	Pipe material	Typical application	Accuracy
Microtunneling	10–136	500–1500	RCP, GRP, VCP DIP, Steel, PCP	Gravity pipe	±1 in

1.3.2 Microtunneling Methods (MTM)

Microtunneling boring machines (MTBM) are mainly used for installation of a gravity pipeline such as a sanitary or storm sewer. MTBM are laser-guided, remotely controlled, and permit accurate monitoring and adjusting of the alignment and grade as the work proceeds so that the pipe can be installed on a precise line and grade. Initially, microtunneling methods were developed for pipes 36 in or less, whereas currently the same technology is used for larger pipes where a remote-controlled technology is required. These methods require a drive shaft for jacking the pipe and an exit shaft for retrieving the MTBM. Table 1.4 presents the main characteristics of MTBM. Figure 1.6 presents MTBM.

1.3.3 Horizontal Directional Drilling (HDD) Methods

HDD methods are mainly used for installation of pressure pipelines and cable conduits. These methods involve steerable systems for installation of both small- and large-diameter pipelines. Directional drilling methods involve a two-stage process. The first stage consists of drilling a small-diameter pilot hole along the desired centerline of a proposed line. The second stage consists of enlarging the pilot hole to the desired diameter to accommodate the utility line and pulling the utility line through the enlarged hole. Sometimes the enlargement process may involve several steps where the desired diameter is obtained gradually. These methods are so named because of their unique ability to track the location of the drill bit and steer it during the drilling process. The result is a greater capability in placing the utilities in difficult underground conditions. The directional drilling methods can be classified into the following three broad categories of small (mini-HDD), medium (midi-HDD), and large (maxi-HDD). Table 1.5 presents the main characteristics of HDD methods. Figure 1.7 illustrates boring operation in HDD method.

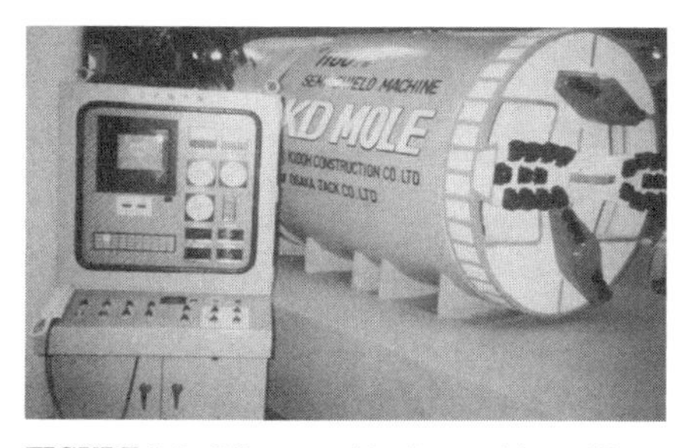

FIGURE 1.6 Microtunnel boring machines. (*Courtesy of Iseki Corp.*)

TABLE 1.5 Main Characteristics of Horizontal Directional Drilling Methods

Method	Diameter range (in)	Maximum installation (ft)	Pipe material	Typical application	Accuracy
Mini-HDD	2–12	Up to 600	PE, steel, PVC, clay, FRP	Pressure pipe/cable	Varies
Midi-HDD	12–24	Up to 1000 H	PE, steel, ductile iron	Pressure pipe	Varies
Maxi-HDD	24–48	Up to 6000 H	PE, steel	Pressure pipe	Varies

1.3.4 Pipe Ramming (PR) Methods

PR methods are mainly used for installation of utilities for road and railroad crossings. Using an air compressor, this process hammers a steel casing pipe inside the earth from a drive pit. The pipe might be hammered closed-end (for small diameters less than 8 in), or open-end (for diameters of 8 in or more). When using large diameters, the spoil is pushed out of the steel casing using air pressure. Table 1.6 presents the main characteristics of PR methods.

1.3.5 Compaction Methods (CM)

CM (also called impact moling) are mainly used for installation of small-diameter conduits (usually less than 8 in) under the roads and streets. As it is called, this method forms a borehole by compressing the earth while pushing or pulling a pipe or conduit. Table 1.7 presents main characteristics of compaction methods. CM are widely used on non-engineered jobs. Owing to this and their similarity with pipe ramming methods, this book does not dedicate a chapter to this technique.

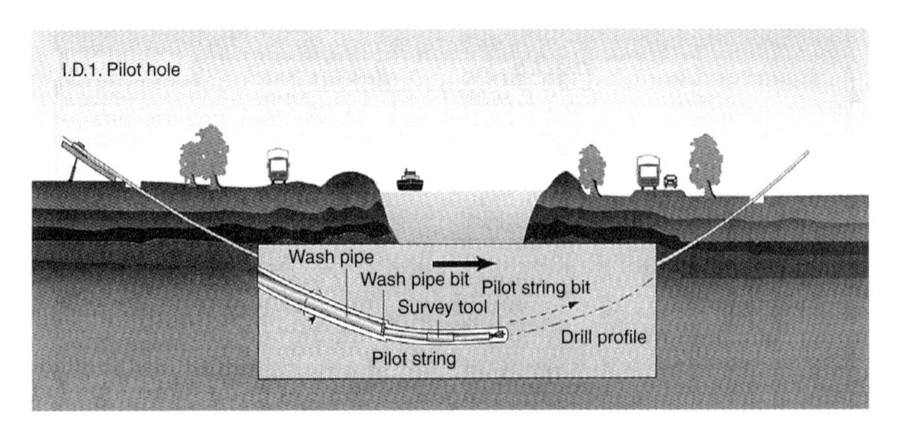

FIGURE 1.7 Boring operation in HDD method. (*Courtesy of DCCA.*)

TABLE 1.6 Main Characteristics of Pipe Ramming Methods

Method	Diameter range (in)	Maximum installation (ft)	Pipe material	Typical application	Accuracy
Pipe ramming	Up to 120	400	Steel	Road and rail crossing	Dependent on setup

1.3.6 Pilot-Tube Microtunneling (PTMT)

PTMT is an alternative to conventional microtunneling. PTMT combines the accuracy of microtunneling, the steering mechanism of a directional drill, and the spoil removal system of an auger boring machine. PTMT employs auger and a guidance system, using camera mounted theodolite and target with electric light emitting diodes (LEDs) to secure high accuracy in line and grade. When conditions are favorable, PTMT can be a cost-effective tool for the installation of small-diameter pipes of sewer lines or water lines. This technique can also be used for house connections, direct from the main-line sewers. Typically, pilot-tube machines can be used in soft soils and at relatively shallow depths. Jacking distances are limited to 300 ft or less. Table 1.8 presents the main characteristics of PTMT.

1.4 TRENCHLESS RENEWAL METHODS (TRM)

Many TRM are available or currently under development. These methods can be used to replace, rehabilitate, upgrade, or renovate (collectively called renewal methods) where a new design life is given to existing pipeline systems. The renewal methods can also be used to *replace* and *enlarge* existing pipelines. Throughout this book, the term *renewal* is used when trenchless methods are applied to extend the design life of pipelines. When the trenchless methods are used to repair pipelines without extending their design life, the term *repair* is used. The basic TRM can be categorized into the following types (Fig. 1.8):

1. Cured-in-place pipe (CIPP)
2. Underground coatings and linings (UCL)
3. Sliplining (SL)
4. Modified sliplining (MSL)
5. In-line replacement (ILR)
6. Close-fit pipe (CFP)

TABLE 1.7 Main Characteristics of Compaction Methods

Method	Diameter range (in)	Maximum installation (ft)	Pipe material	Typical application	Accuracy
Compaction methods	Less than 8 in	250	Any	Pipe or cable	±1% of bore length

TABLE 1.8 Main Characteristics of Pilot-Tube Microtunneling

Method	Diameter range (in)	Maximum installation (ft)	Pipe material	Typical application	Accuracy
PTMT	6–10 in	300	RCP, GRP, VCP DIP, Steel, PCP	Small diameter gravity pipes	±1 in

7. Localized repair (LOR) or point source repair (PSR)
8. Thermoformed pipe (ThP)
9. Lateral renewal (LR)
10. Sewer manhole renewal (SMR)

The choice of trenchless pipeline renewal methods depends on the physical conditions of the existing pipeline system, such as pipeline length, type, material, size, type and number of manholes, service connections, bends, and the nature of the problem or problems involved. The problems with an existing pipe may include structural or nonstructural, infiltration or inflow, exfiltration or outflow, pipe breakage, joint settlement, joint or pipe misalignment, capacity, corrosion and abrasion problems, and so on. Other features of the renewal systems such as applicability to a specific project, constructability, cost factors, availability of service providers, life expectancy of new pipe, and future use of pipe should also be considered. Figure 1.8 presents the main divisions of pipeline renewal methods. This chapter presents a brief description of these methods with detailed information in the following chapters. For detailed guidelines for selection of a specific TRM, refer to Chap. 4.

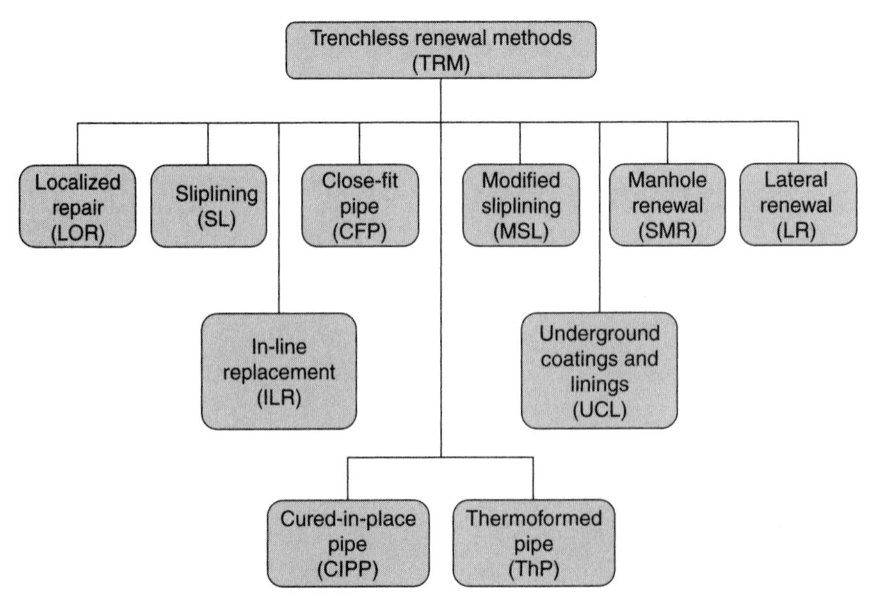

FIGURE 1.8 Basic trenchless renewal methods.

TABLE 1.9 Main Characteristics of Cured-In-Place Pipe Methods

Method	Diameter range (in)	Maximum installation (ft)	Liner material	Applications
Inverted in place	4–108	3000	Thermoset resin/fabric composite	Gravity and pressure pipelines
Winched in place	4–100	1500	Thermoset resin/fabric composite	Gravity and pressure pipelines

1.4.1 Cured-In-Place Pipe (CIPP)

The CIPP process involves the insertion of a resin-impregnated fabric tube into an existing pipe by the use of water or air inversion or winching. Usually the fabric is polyester felt material, fiberglass reinforced, or similar. Normally water or air is used for the inversion process with hot water or steam used for the curing process. The pliable nature of the resin-saturated fabric prior to curing allows installation around curves, filling of cracks, bridging of gaps, and maneuvering through pipe defects. CIPP can be applied for structural or non-structural purposes. Additionally, systems using felt impregnated polyester resin or ERC (E-glass corrosion resistant) fiberglass provide very good corrosion resistance. CIPP also has excellent strength, and can be designed as a standalone system to sustain entire loading on the existing pipe. Table 1.9 presents the main characteristics of CIPP methods. Figure 1.9 illustrates CIPP installation process.

FIGURE 1.9 CIPP installation process. (*Courtesy of Insituform Technologies.*)

FIGURE 1.10 A lining process (courtesy of Raven Lining Systems).

1.4.2 Underground Coatings and Linings (UCL)

The spraying of a thin mortar lining or a resin coating onto pipes is another method of pipeline renewal. For nonworker-entry pipes (usually for host pipe less than 48 in diameter), coatings and linings can provide improved hydraulic characteristics and corrosion protection. However, they may not enhance the structural integrity of the line and have little value in sealing joints or leaks. The lining materials may include concrete sealers, epoxy, polyester, silicone, vinyl ester, and polyurethane and is sprayed directly onto pipe walls using a remote-controlled traveling sprayer. For worker-entry pipes, sprayed mortars (shotcrete or gunite) are effective and widely used for renewing pressure pipes and gravity sewers and can be used for structural purposes. Figure 1.10 illustrates a coating process. The main characteristics of underground coatings and linings are presented in Table 1.10.

TABLE 1.10 Main Characteristics of Underground Coatings and Linings

Method	Diameter range (in)	Maximum installation (ft)	Liner material	Applications
Underground coatings and linings	3–180	1000	Epoxy, polyester, silicone, vinyl ester, polyurethane, and cementitious materials	Gravity and pressure pipelines

TABLE 1.11 Main Characteristics of Sliplining Methods

Method	Diameter range (in)	Maximum installation (ft)	Liner material	Applications
Segmental	24–160	1000	PE, PP, PVC, GRP (–EP and –UP)	Gravity and pressure pipelines
Continuous	4–63	1000	PE, PP, PE/EPDM, PVC	Gravity and pressure pipelines

1.4.3 Sliplining (SL)

SL is mainly used for structural applications when the existing pipe does not have joint settlements or misalignments. In this method, a new pipeline of smaller diameter is inserted into the existing pipeline and usually the annulus space between the existing pipe and new pipe is grouted. This installation method has the merit of simplicity and is relatively inexpensive. However, there can be a significant loss of hydraulic capacity. Table 1.11 presents the main characteristics of sliplining methods. Figure 1.11 illustrates a segmental SL method.

1.4.4 Modified Sliplining (MSL)

MSL includes methods in which pipe sections or plastic strips are installed in close-fit with the existing pipe and the annular space is grouted. There are three variations of the MSL method: panel lining (PL), spiral wound, and formed-in-place pipe.

FIGURE 1.11 Segmental sliplining method. *(Courtesy Midwest Society for Trenchless Technology.)*

TABLE 1.12 Main Characteristics of Modified Sliplining Methods

Method	Diameter range (in)	Maximum installation (ft)	Liner material	Applications
Panel lining	More than 48 in	Varies	GRP	Gravity pipelines
Spiral wound	6–108	1000	PE, PVC, PP, PVDM	Gravity pipelines
Formed-in-place	8–144	Varies	PVC, HDPE	Gravity pipelines

Panel linings can be used to structurally renew large diameter (more than 48 in or worker-entry) pipes. This method can accommodate different shapes, such as noncircular pipelines. The main type of material for this method is fiberglass.

The spiral-wound process uses a layered composite PVC liner and cementitious grout to renew the existing pipe. The combination of the ribbed profile on the PVC liner and the highly fluid nature of the grout produces a highly integrated structure with the PVC liner *tied* to the existing pipe through the grout. For worker-entry pipes, the structural strength of the renewed pipe is determined by the grout characteristics.

Formed-in-place pipe (FIPP) applications include renewal of wastewater, stormwater, and culverts for diameters ranging from 8 in to 12 ft regardless of the shape and material of the host pipe.

Table 1.12 presents main characteristics of MSL methods. Figure 1.12 illustrates a PL process.

FIGURE 1.12 Panel lining process. (*Courtesy of ChannelLine.*)

1.4.5 In-Line Replacement (ILR)

When the capacity of pipelines is found to be inadequate, then ILR should be considered. There are three categories representing ILR: pipe bursting, pipe removal (also called pipe eating) and pipe insertion. Pipe bursting, as the name implies, uses a hammer to break the old pipe and force particles in the earth whereas a new pipe is pulled and/or pushed in its place simultaneously. There are different variations of the pipe bursting method:

Pneumatic pipe bursting: Where a pneumatic hammer is used to break the existing pipe.

Static pipe bursting: Where the energy to break the old pipe is in the pulling machine itself and there is no active percussion going on in the head. This is a much quieter method and defiantly preferable in soggy soils or when attempting to break a ductile iron pipe where there is need of a brute force.

Hydraulic pipe bursting: Where the bursting head articulates to create the bursting action, without the noise of the pneumatic systems.

The pipe insertion method (also called pipe expansion) jacks a new rigid pipe (such as clay) into the existing pipe. Clay and ductile iron are the two most widely used segmental pipes with this method of pipe bursting or expansion.

Pipe removal, also known as pipe eating, can be performed by use of an HDD rig, an HAB machine or an MT machine. In this method, the old pipe is broken into small pieces and taken out by means of slurry (HDD or MTBM) or auger (HAB). Table 1.13 presents the main characteristics of ILR methods. Figure 1.13 illustrates a pneumatic pipe bursting method.

1.4.6 Close-Fit Pipe (CFP)

This type of trenchless pipeline renewal temporarily reduces the cross-sectional area of the new pipe before it is installed, then expands it to its original size and shape after placement to provide a close-fit with the existing pipe. This method can be used for both structural and nonstructural purposes. Lining pipe can be reduced on-site and reformed naturally or by heat and/or pressure. There are two versions of this approach: structural and nonstructural. Table 1.14 presents the main characteristics of close-fit lining methods. Figure 1.14 presents a nonstructural close-fit pipe process.

TABLE 1.13 Main Characteristics of In-Line Replacement Method

Method	Diameter range (in)	Maximum installation (ft)	Liner material	Applications
Pipe bursting	4–48	1500	PE, PP, PVC, GRP	Pressure and gravity pipelines
Pipe removal	Up to 36	300	PE, PVC, PP, GRP	Pressure and gravity pipelines
Pipe insertion method	Up to 24	500	Clay, ductile iron	Pressure and gravity pipelines

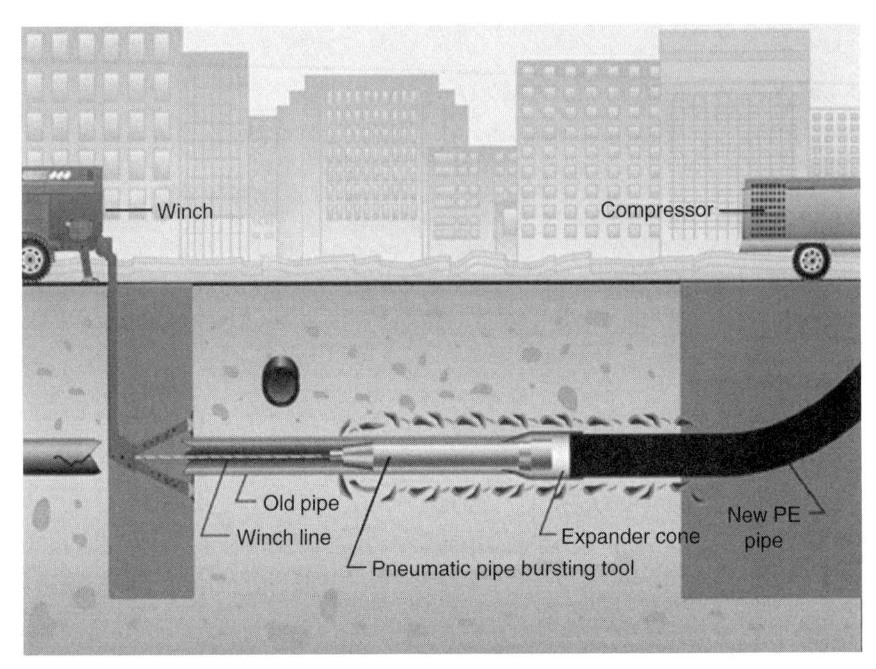

FIGURE 1.13 Pneumatic pipe bursting method. (*Courtesy of TT Technologies.*)

FIGURE 1.14 A nonstructural close-fit pipe going through diameter-reduction process.

TABLE 1.14 Main Characteristics of Close-Fit Methods

Method	Diameter range (in)	Maximum installation (ft)	Liner material	Applications
Close-fit pipe structural	3–24	1000	HDPE, MDPE	Pressure and gravity pipelines
Close-fit pipe nonstructural	3–63	1000	HDPE, MDPE	Pressure and gravity pipelines

TABLE 1.15 Main Characteristics of Point Source Repairs

Method	Diameter range (in)	Maximum installation (ft)	Liner material	Applications
Robotics	8–30	NA	Epoxy resins/ cement mortar	Gravity
Grouting	NA	NA	Chemical gel grouts, cement-based grouts	Any
Internal seal	4–24	NA	Special sleeves	Any
Point CIPP	4–48	50	Fiberglass, polyester, etc.	Gravity

1.4.7 Localized Repairs (LOR) or Point Source Repair (PSR)

When local defects are found in a structurally sound pipeline, localized or point source repairs are considered. Systems are available for remote-controlled resin injection to seal localized defects in pipes ranging from 4 to 24 in. in diameter. The point repair methods are generally used to address four basic problems. The first involves maintaining the loose and separated pieces of unreinforced existing pipe aligned to ensure the load-bearing equivalent of a masonry arch. The second is to provide added localized structural capacity or support to assist the damaged pipes to sustain structural loads. The third provides a seal against infiltration and exfiltration. Finally, the fourth is to replace missing pipe sections. Table 1.15 presents the main characteristics of PSRs. Figure 1.15 presents a LOR technique.

FIGURE 1.15 Internal seal point repair. (*Courtesy of Miller Pipeline Corporation.*)

TABLE 1.16 Main Characteristics of Thermoformed Pipe

Method	Diameter range (in)	Maximum installation (ft)	Liner material	Applications
Thermoformed pipe (ThP)	4–30	1500	HDPE, PVC	Gravity and pressure pipelines

Chemical grouting can be applied in cases where compression rings are used, but the action of stopping the leaking is actually performed by grout being *injected* or *forced* into the defective joint rather than the lining itself. Chemical grouting can also be applied from the surface above the repair area via probe grouting. This process not only seals the leak, but also stabilizes the surrounding soils behind the pipe. A new liner in a pipe cannot be considered a 100 percent repair if the soil surrounding the pipe still contains voids that create an unstable and shifting environment.

Another category of grouting is *Fill and Seal*, a patented flood-grouting technology, which in one continuous operation, seals from one entry point of sewer mains and simultaneously fills all the related joints, laterals, and manholes. By successive application of two innate chemical solutions, which automatically infiltrate to all the locations where the sewer pipe leaks, either by infiltration or exfiltration, this method repairs all cracks and voids in sewer pipes.

1.4.8 Thermoformed Pipe (ThP)

Thermoformed is the recognized terminology in North America for pipes that have a reduced cross section by folding for insertion and are subsequently heated to thermoform them to conform to the host-pipe dimensions. This type of trenchless pipeline renewal uses a new, folded PVC or PE pipe that is expanded by thermoforming to fit tightly inside the existing pipe. Both PVC and PE used for this method have a long performance history in pipe applications that verifies not only their structural capacity, but also other important long-term performance characteristics such as chemical and abrasion resistance. The main characteristics of ThPs are presented in Table 1.16.

1.4.9 Lateral Renewal (LR)

Sewer service laterals can be renewed using any of the methods used for renewal of main lines such as chemical grouting, cured-in-place pipe, close-fit pipe, pipe bursting, and spray-on lining. Table 1.17 presents the main characteristics of LR methods.

TABLE 1.17 Main Characteristics of Lateral Renewal Methods

Method	Diameter range (in)	Maximum installation (ft)	Liner material	Applications
Lateral renewal	4–8	100	Any	Gravity pipelines

TABLE 1.18 Main Characteristics of Sewer Manhole Renewal Methods

Method	Liner material	Applications
Coatings and linings–cementitious	Cementitious	
Coatings and linings–polymers	Epoxy, urethane	
Thermoplastic liners	PVC	
CIPP	Resin saturated polyester felt or fiberglass	Sewer manhole
Pressure grouting	Cementitious	
Chemical grouting	Polymers	
Inserts	Fiberglass	

1.4.10 Sewer Manhole Renewal (SMR)

SMR methods are provided to prevent surface water inflow and groundwater infiltration, repair structural damage, and protect surfaces from damage of corrosive substances. When renewal methods do not solve the problems cost-effectively, manhole replacement should be considered. Selection of a particular renewal method should consider the types of problems, physical characteristics of the structure, location, condition, age, and type of original construction. The extent of successful manhole-renewal experiences and cost should also be considered. SMR can be divided into the following methods: cementitious, cast-in-place, cured-in-place, and profile PVC. Another method is GRP inserts where the old corbel is removed and a new GRP manhole is installed inside the existing manhole. Table 1.18 presents the main characteristics of SMR methods. Figure 1.16 shows the manhole lining process.

In cases where the structural integrity of the manhole has not been jeopardized, chemical grout injection can be a cost-effective solution to stopping inflow and infiltration. Generally most of the manhole problems can be effectively solved by the use of chemical grouting alone, which is also economical. Even in situations where liners are being specified, many engineers still require chemical grouting before liner installation as a means of solidifying the surrounding soils behind the manhole to ensure a successful liner application.

1.5 SUMMARY

This chapter presents an overview of different TT methods with a summary of their capabilities. TT techniques have become a method of choice to replace the traditional open-cut pipeline installations and replacements owing to their many benefits. TCM offer solutions for new installation of pipelines and utilities. TRM stop existing pipe leaks, resist corrosion and

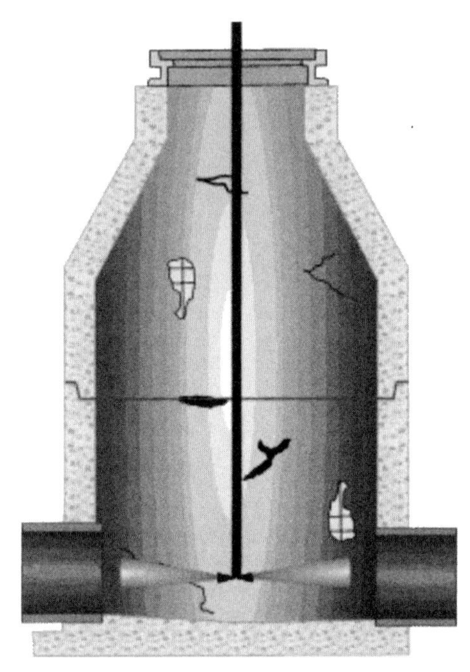

FIGURE 1.16 Manhole lining process. (*Courtesy of Perma Liner Industries.*)

abrasion, and install a new pipe in place of the existing and deteriorated pipes with a new design life.

REVIEW QUESTIONS

1. Define trenchless technology methods.
2. Describe the six benefits of trenchless technology methods.
3. Compare trenchless construction and trenchless renewal methods and give five examples of each method.
4. Describe the microtunneling method.
5. Describe the CIPP method.
6. Compare pipe jacking and utility tunneling.
7. List and describe the different methods of horizontal earth boring (HEB).
8. Recommend a pipeline construction method (TCM) for the following conditions. Explain why you recommend this method:
   a. 36-in diameter.
   b. 500-ft drive length crossing an Interstate highway.
   c. Gravity application.

9. Recommend a pipeline renewal method (TRM) for the following conditions. Explain why you recommend this method:

a. 24-in diameter.

b. Pressure pipe.

c. Minimum installation length of 2000 ft.

d. Structural application.

BIBLIOGRAPHY

Iseley, D. T., and M. Najafi (Eds.) (1995). *Trenchless Pipeline Rehabilitation*, The National Utility Contractors Association (NUCA), Arlington, Va.

McKim, R. A. (1997). Bidding strategies for conventional and trenchless technologies considering social costs, *Canadian Journal Civil Engineering/Revue Canadienne de Génie Civil*, 24(5):819–827.

Najafi, M. (1994). *Trenchless Pipeline Rehabilitation: State-of-the-Art Review,* Trenchless Technology Center, Ruston, La.

Thomson, J., et al. (1998). An overview of the economics of trenchless technology, *No-Dig Engineering*, 5(2):21–23.

U.S. Environmental Protection Agency (EPA) (1991). *Handbook of Sewer System Infrastructure Analysis and Rehabilitation*, Office of Research and Development, Cincinnati, Ohio.

Water Infrastructure Network (WIN), Available at http://www.win-water.org.

CHAPTER 2

SOCIAL COSTS OF UTILITY CONSTRUCTION: A LIFE CYCLE COST APPROACH

2.1 COST OF A PROJECT

The development of underground infrastructure, environmental concerns, and economic trends is influencing society, resulting in the advancement of technology for more efficient, environmentally friendly, and cost-effective utility and pipeline construction and renewal. Traditionally, construction has been concerned with infrastructure on relatively under-developed sites. However, because of increased above- and below-ground congestion in urban environments, the demands for construction methods have changed. Developed sites no longer allow trench or traditional open-cut construction. The demand for methods that minimize disruption and destruction of the surface is becoming more important. In the past, these methods have been limited and were often considered too expensive. With the latest developments in trenchless technology (TT) and the increasing competition, the total construction cost associated with the installation of a utility by trenchless methods may now be less expensive than by traditional trenching methods. Trenchless construction methods include all the methods of underground pipeline and utility construction and renewal with minimum or no surface or subsurface excavation.

The cost-effective construction of a pipeline project requires a clear understanding of all the cost factors associated with the specific conditions of the project. The designer must include all the cost items in the project budget. These cost categories include planning and engineering costs (also called preconstruction costs), direct, indirect, and social costs (construction costs), and postconstruction costs (also called operation and maintenance costs). Capital costs have traditionally been calculated only by considering the preconstruction and construction costs (without the social costs and postconstruction costs) of the project. However, the life cycle cost of a project (the total cost from inception to demolition after the useful design life) includes the three main parts of preconstruction cost, construction cost, and the postconstruction cost as indicated in Table 2.1.

The following sections present how the preconstruction, construction, and postconstruction costs show the great difference between trenchless and open-cut construction.

2.2 PRECONSTRUCTION COSTS

Table 2.2 presents a breakdown and comparison of cost factors for preconstruction costs. It should be recognized that decisions made during the preconstruction phase would have

TABLE 2.1 Life Cycle Cost of a Project

Preconstruction	Construction	Postconstruction
• Conceptual planning, risk, and impact analysis • Land acquisition • Surveying and documentation of existing site conditions • Easements • Permits • Design fees and preparation of contract drawings • Legal fees	• Direct construction costs (labor, material, and equipment) • Indirect (overhead) construction costs • Inspection and testing costs • Social costs	• Operation • Maintenance • Depreciation • Loss of revenue due to emergency repairs

a significant impact on the total life cycle cost (Fig. 2.1). As an example, a trenchless technique may provide a realignment advantage, which may shorten the overall length of the pipeline, and thereby reduce the number of manholes, and/or eliminate pump stations, which would significantly reduce the life cycle costs of the project. Figure 2.2 illustrates how a TT method may provide the most cost-effective route selection, and therefore, reduce total project costs.

Depending on the type of project, such as new construction or pipeline renewals, the engineering and design costs can be major (new installation such as a tunneling or pipe jacking project) to minor (pipe renewal). In addition, preparing the bid documents for an open-cut option would require substantial field survey and plan preparation. After construction, the new pipe may have the risk of built-in defects such as ground movements due to backfill and compaction. Trenchless methods have inherent benefits that provide higher construction productivity and involve less risk.

2.3 CONSTRUCTION COSTS

The construction costs include direct costs, indirect costs (overhead costs), and social costs.

TABLE 2.2 Cost Factors for Preconstruction Costs

Cost factor	Open-cut	Trenchless
Field survey work and plan preparation	Major	Minor
Engineering and design	Major	Major to minor
Legal issues (acquisition of easement and right-of-way, detour roads, and the like)	Major	Minor
Working area requirements	Major	Minor
Subsurface investigation requirements	Minor	Major
Preparation of bid documents	Major	Minor

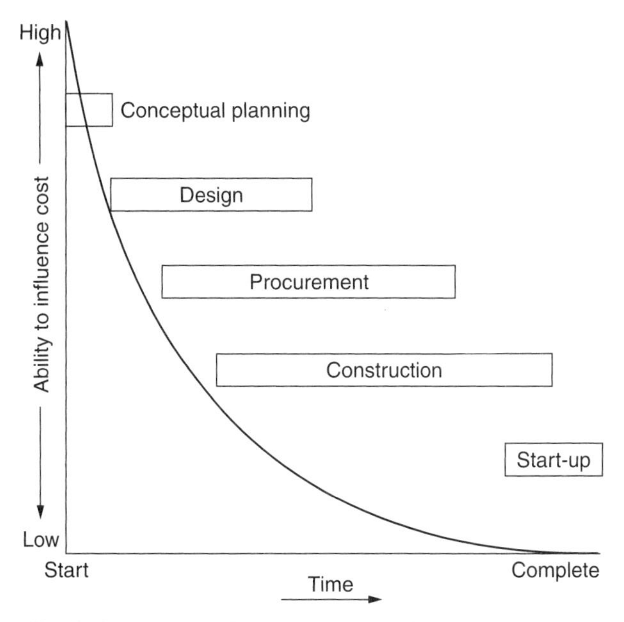

FIGURE 2.1 Relative ability to influence cost (CII, 1986).

2.3.1 Direct Costs

The direct costs of construction include cost of labor, materials, subcontractors, and equipment that are directly involved in the construction of the project. Shoring or sloping trench excavation walls, safety guardrails, dewatering, pipe material, labor costs, spoil removal, backfill and compaction, and so on, are direct costs of construction and should be included

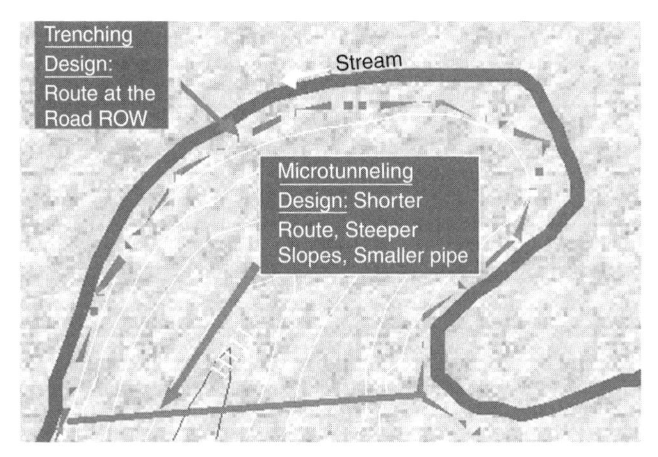

FIGURE 2.2 Trenchless technology may provide the most cost-effective route selection. (*Courtesy of Midwest Society for Trenchless Technology.*)

TABLE 2.3 Cost Factors for Direct Costs

Cost factor	Open-cut	Trenchless
Mobilization and demobilization	Major	Minor
Shoring and sloping trench walls	Major	Minor
Dewatering	Major	Minor
Spoil removal	Major	Minor
Cost of detour roads	Major	Minor
Backfill and compaction	Major	Minor
Reinstatement of surface	Major	Minor
Construction equipment costs (include both owning and operating costs)	Major	Minor
Labor costs	Major	Minor
Pipe material costs	Minor	Major to minor

with the takeoff quantities. Table 2.3 presents some of the direct cost factors for open-cut and trenchless alternatives.

Dependent on the type and size of the project, the construction equipment costs are usually higher for open-cut projects. For example, in a trenchless pipeline renewal project, the equipment costs are minor; however, if the same project were to be constructed by an open-cut method, the use of heavy trucks and excavators would be necessary and the equipment costs could be significant. In open-cut projects, the cost of fuel is significantly more than trenchless projects.

The pipe used for pipe jacking and microtunneling must have enough bearing capacity to take the jacking loads that are usually higher than the long-term dead and live loads exerted on the pipe. As a result, with pipe jacking and microtunneling operations, pipe material is more expensive than open-cut but a better quality end-product is achieved. Additionally, because of arching effects of soil, the soil and traffic loads on the pipes installed with TT are less than those installed with open-cut projects.

Studies have shown that owing to the cost of double handling of soil and required reinstatement of the surface for open-cut projects, these costs can add up to 70 percent of the total costs of an open-cut project. These reinstatement costs are usually minimal for TT installations.

2.3.2 Indirect (Overhead) Costs

The indirect or overhead costs of the construction basically include all the costs that are not *directly related* or applied to the actual construction operations. These costs are normally fixed and *spread out* over the entire project. Examples of indirect costs include, head office and job overhead costs, such as taxes, temporary utilities, field supervision, traffic control, and insurance. Indirect costs are usually calculated after direct costs are estimated and are most often added as a percentage of the direct costs. The determination of indirect costs requires considerable construction knowledge and includes the greatest variation in construction cost estimating and can be approximately 20 percent of the direct cost of a utility project. However, the indirect costs are dependent on the duration of the project and will increase as duration of the project increases. As trenchless construction methods usually involve higher productivity and reduced waste, the duration of these projects is normally

TABLE 2.4 Cost Factors for Indirect Costs

Cost factor	Open-cut	Trenchless
Head-office costs	Major	Minor
Field-office costs	Major	Minor
Field supervision costs	Major	Minor
Cost of temporary facilities	Major	Minor

less than open-cut projects. Trenchless methods, therefore, will have less indirect costs than open-cut projects. Table 2.4 presents some of the indirect cost factors for open-cut and trenchless alternatives. Figure 2.3 presents a *sample* breakdown of labor, materials, social, and overhead costs for open-cut and trenchless projects.

2.3.3 Social Costs

The social costs of construction include inconvenience to the general public and damage to the environment and existing structures. Social costs are becoming more important as the public awareness grows and the needs to conserve and protect our environment and quality of life are more understood. These needs have resulted in identification and evaluation of social costs of utility and pipeline installations. Social costs can be a major element in calculating the total life cycle cost of a project, which to a large extent, is a function of the method of installation adopted. Using trenchless methods can significantly reduce social costs. Social costs for open-cut projects can be as high as several times the value of construction, whereas for trenchless projects as low as 3 to 10 percent of the total cost of the project.

If social costs are evaluated and included in the overall cost of a project, TT methods can prove to be more cost-effective than open-cut method. This cost comparison for different construction methods needs to be performed during the design stage of a project and

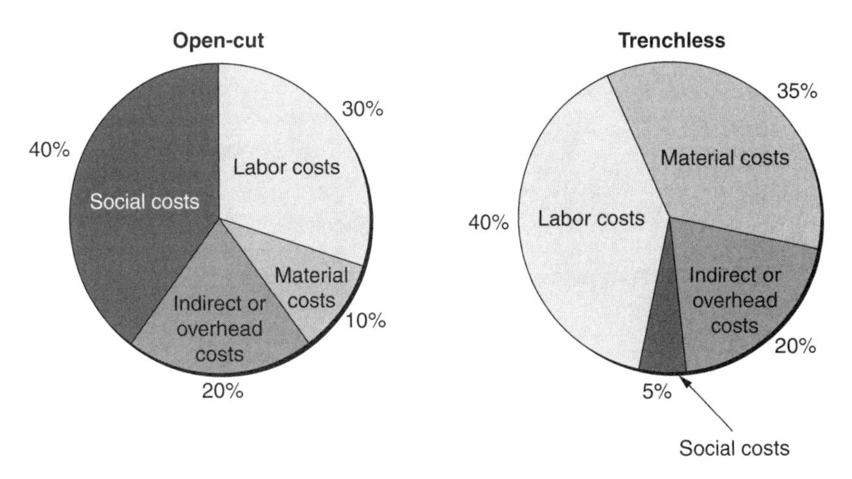

FIGURE 2.3 A *sample* breakdown of different cost categories for a utility project.

only *applicable* method(s) with *minimum* overall and social costs should be advertised for contractors' bids. As the contractors do not include social costs into their bids, as a general rule, the choice of construction method (between trenchless and open-cut) should *not* be left to the contractor.

The social costs of open-cut construction include the following major categories:

- Vehicular traffic disruption
- Road and pavement damage
- Damage to adjacent utilities
- Damage to adjacent structures
- Noise and vibration
- Heavy construction and air pollution
- Pedestrian safety
- Business and trade loss
- Damage to detour roads
- Site and public safety
- Citizen complaints
- Environmental impacts

Figure 2.4 presents an example of the social costs of open-cut construction that includes damage to pavement and roads, lane closures, safety issues, and damage to green areas and trees.

Vehicular Traffic Disruption. Use of open-cut method on roads and pavements causes traffic delays and congestions. The general public pays for the increased amount of time spent in waiting during traffic delays or going through the detoured roads. Surface excavations become more costly to commuters in areas of heavy traffic, such as business districts and major streets and roadways. In addition to the extra time, traffic disruption costs

FIGURE 2.4 Social costs of open-cut construction. (*Courtesy of InSituform Technologies*).

include cost of extra fuel, maintenance and repair, and depreciation of vehicles. Cost of traffic disruptions, delays, and diversions can be as great as the total construction costs or several times greater than the value of construction (Thomson et al., 1998). Considering the importance of traffic disruption costs, a detailed discussion of these costs is presented at the end of this chapter.

Road and Pavement Damage. Pavement cuts increase the roughness of a pavement structure in both immediate and surrounding areas of the cut and constitute a major portion of social costs. Pavement trenches introduce discontinuities in the pavement. A rough pavement can quickly lead to structural failures by allowing the vehicles to bounce, thus creating greater load on the pavement. Often because of the poor pavement restoration techniques, the same work has to be repeated within a few years. This not only increases the cost of the work, but also reduces the life span of the pavement by 30 to 50 percent.

Successive utility cuts, poor backfilling, and patching techniques also produce differential settlements and eventually require new pavement construction. Recent surveys conducted in Canada and the United States have shown that the restoration of road pavement following utility cuts, ranks as one of the main challenges facing municipalities and utility providers. Utility cuts are thought to contribute to the rapid deterioration of urban streets, causing a substantial loss of road serviceability and leading to increased maintenance demands, with huge cost implications for both municipalities and utilities. Several studies undertaken in North America have found that pavements in and around utility cuts deteriorate at an accelerated rate compared to uncut pavements. The resulting conditions of a utility cut influences pavement life, maintenance cost, aesthetics, vehicular damage, and motorist safety. Only a reduction in the number of utility cuts can preserve the pavement and maintain its design life. Figure 2.5 illustrates effects of multiple cuts on the pavement.

Damage to Adjacent Utilities. The possibility of damaging existing underground utilities is a major concern for a utility contractor. In case of damage to the adjacent utilities, the cost of repairing the service will be added to the cost of the project. Moreover, buried electrical cables, gas and oil pipelines, are a potential safety hazard to the crew working on the site and

FIGURE 2.5 Effects of utility cuts on pavement structure.

the general public. Using appropriate job planning, construction method, and equipment, as well as good utility locating, potholing, and safety procedure, will minimize possibility of damaging adjacent utilities.

Damage to Adjacent Structures. Underground utility construction can cause uneven settlements and distress in adjacent structures. Most of these settlement problems are caused by activities like dewatering, excess excavation, and improper techniques used in shoring and underpinning, mostly associated with open-cut construction.

Noise and Vibration. Noise and vibration are more associated with the open-cut method of utility construction. To construct underground pipelines and utilities using the open-cut technology, contractors have to cut open existing surfaces, such as pavements, using heavy machinery like jack hammers, drills, and heavy excavators. These equipment produce vibrations and noises that may lead to citizen inconvenience and complaints.

Heavy Construction and Air Pollution. Trenching and use of heavy excavating and trucking equipment produce dusty conditions. The serious health concerns associated with the dust result in an obvious public nuisance. This problem is even more complicated in critical areas such as schools and hospitals, and in areas of heavy urbanization such as downtown areas and major business areas in inner cities. Heavy construction equipment pollutes the air by generating three major pollutants: hydrocarbons, nitrogen oxides, and carbon monoxide. Motor vehicles also emit large amounts of carbon dioxide, which has the potential to trap the earth's heat and cause global warming. Additionally, the fuel costs of heavy construction equipment used in excavation, compaction, trucking, and backfilling are much more than those used in TT methods.

Pedestrian Safety. In some cases, during the open-cut construction, traffic is diverted to residential districts. The additional traffic in residential streets, when used as detour roads, is a potential risk to pedestrians and children. Also, open trenches and pavement cuts are potential safety hazards to pedestrians, especially for the elderly and children.

Business and Trade Loss. Construction in commercial districts is associated with loss to the local businesses sales. People try to avoid hard-to-get areas during open-cut construction that may result in closure of shops and other businesses for duration of the construction. Figure 2.6 illustrates a service station offering a discount to maintain sales during the utility and road construction.

FIGURE 2.6 Open-cut construction results in reduced business sales.

TABLE 2.5 Cost Factors for Social Costs

Cost factor	Open-cut	Trenchless
Vehicular traffic disruption	Major	Minor
Road damage	Major	Minor
Damage to adjacent utilities	Major	Minor
Damage to adjacent structures	Major	Minor
Noise and vibration	Major	Minor
Heavy construction and air pollution	Major	Minor
Pedestrian safety	Major	Minor
Business and trade loss	Major	Minor
Damage to detour roads	Major	Minor
Site and public safety	Major	Minor
Citizen complaints	Major	Minor
Environmental impact	Major	Minor

Damage to Detour Roads. During construction, use of detoured roads not suited or designed to take heavy traffic, such as residential streets, results in damage to pavement structure of these roads. The heavy traffic decreases the life span of detoured roads, which is an additional cost to municipalities and local governments.

Site and Public Safety. Site-related accidents to construction workers and the general public tend to increase in an open-cut utility construction. Collapse of trench walls and caving, and other fall accidents are common in open-trench utility construction.

Citizen Complaints. Disruption to the normal life of residents and businesses often generates public outcry and complaints to the authorities. Dust and noise are a major concern to the public during open-cut construction activity. Traffic delays and increased wait times due to lane closures frustrate people.

Environmental Impacts. Construction work in environmentally sensitive areas such as wetlands, green and wooded areas, developed neighborhoods, rivers, streams, natural habitats, public parks, protected areas, and historic places requires special effort. Sometimes, damage to these sensitive areas during the open-cut construction is not reversible. Damage to these sensitive areas is taken seriously by environmental advocates and the general public, resulting in an increase in administrative burden while working in these areas.

Table 2.5 summarizes major social cost factors for open-cut and trenchless alternatives. Social costs are project-specific and vary from project to project. More studies are needed to categorize and quantify these costs for different project conditions. For selection of a cost-effective method, design engineers need to quantify social costs based on specific conditions of their projects.

2.4 POSTCONSTRUCTION COSTS

The postconstruction costs of a project mainly include operation, maintenance, depreciation, and loss of income owing to emergency repairs. Table 2.6 presents some factors associated with postconstruction costs for open-cut and trenchless construction. The *out of*

TABLE 2.6 Cost Factors for Postconstruction Costs

Cost factor	Open-cut	Trenchless
Operation	Major	Minor
Maintenance	Major	Minor
Emergency repairs	Major	Minor
Loss of revenue due to emergency repairs	Major	Minor
Renewal costs	Major	Minor

sight, out of mind attitude, and lack of proper asset management and maintenance strategies, causes additional emergency repair costs that may be many times more than planned maintenance and renewal costs. The breakdown of these emergency costs and their total value are often underestimated because of improper allocation of costs and problems with budgeting, accounting, and job records. Figure 2.7 presents an emergency repair of a 21-in cast-iron force-main in the city of Atchison, Kansas. The pipe is used to transport sewage from a pump station to the wastewater treatment plant. During the 3-day emergency repair period, a 50 ft by 75 ft area was excavated and more than 1.5 million gallons of raw sewage was dumped into Missouri River. Also, during the emergency work an employee was injured. The *real* cost of emergency repair included the direct cost of crew and equipment working on the job (including excavation, repair, and restoration costs), the indirect or overhead costs including bypassing costs, and the social costs for environmental damage and worker injury costs. As the repair was not done in a developed area, there were no other social costs.

The operation and maintenance costs may be lower for trenchless construction owing to shorter routings and a better quality of pipe that is usually associated with trenchless construction. In many cases, the maintenance and repair costs are lower with trenchless methods when no surface excavation is required. Use of TT results in significant lowering

FIGURE 2.7 Real costs of emergency repairs. (*Atchison Daily Globe, Saturday, January 11, 2003.*)

of the live and dead loads on the pipe, compared to open-cut construction. The loss of revenue is also less with trenchless methods as these methods provide means for proper asset management and faster repair and maintenance of pipelines. The trenchless pipeline renewal methods provide means for replacement of old pipelines with a new design life. Implementing a trenchless renewal project will provide the following additional benefits:

- Reduce the risk of catastrophic structural failure of the pipeline.
- Facilitate the operation and maintenance by eliminating the need for unplanned emergency and unscheduled repairs.
- Reduce damage to the environment by eliminating leakage and bypassing during pipe breaks.
- Eliminate emergency unplanned repairs by providing means for inspection and spot repair or renewing the existing pipes.
- Improve hydraulic flow and reduce operation costs.
- Use less underground space by using the old pipe space to replace the existing pipe.

2.5 CALCULATING SOCIAL COSTS

The effect of social costs in increasing the total cost of a project is not very well understood. Some cities charge a utility-cut fee to recover some of the costs involved in the construction of the utility work, such as the cost of cutting and the cost of reinstating the pavement (NCMA TEK, 2003). Various state and federal laws were passed to compensate the utility cuts driven by construction (FHA, 2002). But these costs do not ease the money spent by an average person waiting in traffic due to delays and other social cost parameters as listed in previous sections. The average cost spent by a person due to traffic delay comes mostly from the money spent on fuel, vehicle depreciation, and time wasted while waiting in traffic or going through detour roads.

Studies have shown that the traditional open-cut method, although considered reliable, is a time-consuming and inefficient method of pipe installation and renewal. The costs and time associated with the traditional approach exceeds the estimated or bid amount due to the addition of various social and environmental costs involved. If unaccounted for in the project budget, some of these costs may show up during road maintenance by public agencies such as municipalities, which in turn look for tax money from the general public.

It is important to estimate the social costs of utility construction for using various methods of pipe installation. Government agencies and design and consulting engineers need to estimate the total life cycle costs for different project alternatives and select the best and most cost-effective method with the lowest social costs. Estimation of social costs is project specific and there is no universal method or model to calculate social costs for all the conditions. The following section is provided as a guide to quantify social costs for traffic disruptions. Similar logic can be used to estimate other types of social costs.

2.6 ESTIMATING COST OF TRAFFIC DISRUPTIONS

One of the most important portions of the social costs associated with open-trench utility construction is the traffic disruption (Bush, 2001). These costs are often ignored in preparing the initial estimate, owing to a poor understanding of their influence on the successful completion of the project. In most cases, the contractor is least concerned in considering and lowering these costs. The end-user of the road, such as a public agency and the general public is most affected by these costs. Public agencies when faced with unbudgeted and extra cost of utility construction may turn to the general public and increase taxes or issue special assessment fees. The following are the major categories under traffic and road disruption (Bush, 2001):

- Duration of the project
- Cost of fuel
- Cost of travel time
- Road damage
- Vehicular wear
- Loss of sales tax
- Loss of productivity due to noise and vibration
- Dust

In addition to the above, a direct influence of the construction activity can be estimated for loss of productivity and dust and dirt control (Boyce and Bried, 1994a). All of the above-mentioned costs have to be assessed to give a correct perspective of the traffic disruption costs.

2.6.1 Duration of the Project

The duration of the project plays an important role in the value of social costs involved in utility construction. For example, sometimes contractors need to close one or two lanes of traffic during open-trench construction. The lane-closure procedure often continues for the entire duration of the project, resulting in congestion and delays for daily commuters.

The cost of delay and congestion is significantly less for projects of short duration. But with an increase in time, the traffic disruption costs will increase. Also, the place and location of the lane closure affects the social costs of the project. A lane closure for a couple of hours on a city street would not be the same as a lane closure on a highway or a busy road. The time of the day for carrying out the work also affects the social costs. A lane closure during an urban peak hour can add significantly to the social costs, compared to a closure during off-peak hours. The time of the construction, such as winters, holiday seasons, and major events, also affects the social costs.

2.6.2 Cost of Fuel

Utility construction using open-trench method often results in lane closures and traffic congestion. The amount of time spent in traffic delays is directly related to the cost of fuel wasted. The cost of fuel is estimated based on the number of gallons wasted per car in waiting during traffic delays or going through detour roads. The average fuel consumption of a car is used to calculate the amount of fuel wasted in traffic. Costs of fuel for detour roads or delay per vehicle are calculated according to Eqs. 2.1 and 2.2.

$$\text{Cost of fuel for detour roads/vehicles} = (\text{average gal/mi}) \times (\text{average additional mile}) \times (\text{average cost of fuel/gal}) \tag{2.1}$$

$$\text{Cost of fuel for delay/vehicle} = (\text{average gal/unit time}) \times (\text{average time waiting}) \times (\text{average cost of fuel/gal}) \tag{2.2}$$

The average number of miles traveled per gallon can be used to calculate the number of gallons per mile and can be incorporated in Eq. 2.1. Data collected during case studies for traffic disruption represents a fraction of the total fuel consumption or the average miles traveled by vehicles can be used.

2.6.3 Cost of Travel Time

Travel time costs vary widely depending on factors such as the type of trip, distance of travel, traveler, and travel condition. For example, the delay cost during an emergency or crisis, rushing to a hospital or airport, may exceed a dollar per minute. On the other hand, a pleasant drive along a riverside might be considered a benefit rather than a cost. Per-minute travel time costs tend to be higher for passengers during uncomfortable and congested conditions. Based on earlier studies, Victoria Transport Policy Institute (VTPI) summarized the various factors affecting travel costs as follows:

- The costs for commercial vehicles, such as delivery trucks, include drivers' wages, vehicle costs, delay costs beyond a critical delivery time, and overhead costs.
- Personal travel time is usually estimated at one-quarter to one-half of prevailing wage rates.
- Travel time costs tend to be higher for driving under congested, crowded, or uncomfortable conditions.
- Travel-time costs per minute tend to increase for longer commutes of more than about 20 min.
- People with full-time jobs tend to have more demands on their time, than people who are retired or unemployed.

While calculating the cost of travel time, VTPI valued the drivers' time at $6.00/h (50 percent of $12.00 average wage) and passengers at $4.20/h (35 percent of $12.00). These values were used for automobiles and motorcycles. Urban peak speeds were estimated to average 30 mph with a 16.5 percent congestion-cost premium (VTPI, 2003), and off-peak speeds were estimated at 35 mph with no-congestion premium. Table 2.7 summarizes the travel-time cost per user.

TABLE 2.7 User Travel Time Costs per Passenger Mile in U.S. dollars

Vehicle class	Urban peak	Urban off-peak	Rural	Average
Average car	$0.230	$0.170	$0.150	$0.174
Compact car	$0.230	$0.170	$0.150	$0.174
Electric car	$0.230	$0.170	$0.150	$0.174
Van/light truck	$0.230	$0.170	$0.150	$0.174
Rideshare passenger	$0.180	$0.154	$0.135	$0.152
Diesel bus	$0.350	$0.280	$0.233	$0.275
Electric bus/trolley	$0.350	$0.280	$0.233	$0.275
Motorcycle	$0.230	$0.170	$0.150	$0.174
Bicycle	$0.350	$0.300	$0.300	$0.310
Walk	$1.000	$1.000	$1.000	$1.000
Telework	$0.000	$0.000	$0.000	$0.000

Source: VTPI (2003).

TABLE 2.8 Principal Means of Transportation to Work

Category	Number × 1000	Percent
All workers	118,041	100.0
Automobile	103,466	87.7
Drives self	92,363	78.2
Carpool	11,103	9.4
2 person	8705	7.4
3 person	1454	1.2
4+ person	945	0.8
Public transportation	5779	4.9
Taxicab	144	0.1
Bicycle or motorcycle	749	0.6
Walks	3627	3.1
Other means	987	0.8
Works at home	3288	2.8

Source: From NTS (2001).

As mentioned earlier, the cost of traffic delay due to utility cuts adds to the cost of travel time presented in Table 2.7. Boyce and Bried (1994b) and VTPI suggested using the following formulas in determining the cost of delay for detour.

$$\text{Cost of detour delay} = (\text{average time/mi}) \times (\text{additional miles to travel}) \times (\text{value of time in dollars}) \qquad (2.3)$$

$$\text{Cost of waiting delay} = (\text{average time waiting}) \times (\text{value of time in dollars}) \qquad (2.4)$$

Table 2.8 provides information about the means of transportation used by individuals. Most people prefer to use personal means of transportation. With the increase in the number of vehicles on the roads and the congestion thereafter, a few individuals are inclined to use other means of getting to work. Some of these individuals carpool to get to work, and some may use public transportation. There are also individuals who work at home, also called telework, and play an insignificant role in calculating the costs of travel time and congestion. In estimating the general cost of traffic delay or disruption, knowledge of the type or mode of transport preferred by a large section of the population has to be considered. Also, the other means of transport such as bus and train transit have to be considered to derive an accurate and consistent formula.

2.6.4 Road Damage

Road damage due to utility construction can be of two forms. One is the pavement damage due to utility cuts, trenching, and poor patching procedures. These damages show in the forms of potholes, surface roughness, and cracks. The second cost is the damage to detour roads, due to the additional heavy traffic during construction. These damages do not show up immediately and take a significant amount of time to develop.

Potholes and surface roughness form because of the differential settlement of the soil below the pavement. Also, poor drainage of storm water washes out the subgrade and eventually leads to the collapse of the pavement around the utility cut. Potholes and cracks in the pavement take time to develop and should be considered in the life cycle cost of the project for an effective estimate of the cost of the construction. Often, the local governments have to repair the pavement sometimes after completion of the project. The improper work of reinstating the pavement requires additional costs for repairing and become a burden to the local government agencies and eventually the general public.

Damage to detour roads also adds to the social cost for utility construction. Secondary roads are often used as detour roads during construction on a main road. As these roads are not designed specifically to take on heavy traffic, they may get damaged. This sort of damage can be in the form of wearing of asphalt, potholes, cracks, settlement and pavement collapse. For short duration projects, like a day or two, these costs can be insignificant. But for longer durations, these damages have to be added to the total estimate of the project costs.

The cost of restoring a pavement per foot has to be collected from the utility owner overseeing the work at the particular locality. In some cases, such costs cannot be easily recognized as pavement repair is done as an emergency sometimes after completion of utility construction, and would not be a part of a regular routine or periodic restoration of the pavement. The following formula can be used in estimating the cost of pavement restoration:

$$\text{Pavement restoration cost} = (\text{restoration cost/ ft}^2) \times (\text{number of ft}^2) \qquad (2.5)$$

2.6.5 Vehicular Wear

Additional social costs can be in the form of vehicle wear and tear, such as to the engine, transmission, and tire wear, owing to extra miles driven on detour roads. Improper techniques used in restoring the pavements often result in potholes, which damage the shocks, muffler, tires, axle, and chassis of the vehicles. According to a 1997 report by the Environmental Working Group and the Surface Transportation Policy Project, nearly $4.8 billion is spent each year to repair damage to cars resulting from run-ins with potholes, utility cuts, and other dangerous road conditions (Soucy Insurance, 2001).

A report by The Road Information Program, based on analysis of FHWA data, stated, "Driving on roads in need of repair or improvement, costs each American motorist an average of $222 annually in extra vehicle operating costs—$41.5 billion total. Pavement conditions are the single most important factor in determining extra vehicle operating costs. More than one-fourth, or 28 percent, of major roads in the country are in need of repair or improvement."

Cost of vehicle repairs due to potholes cannot be readily estimated, as they are not visible for a significant period of time. An analysis of pothole complaints may yield data for potholes or other related defects in the surface pavement. If such data were available these costs could be used in calculating the cost of damage and repair due to potholes.

$$\begin{aligned}\text{Cost of vehicular wear due to potholes} &= (\text{average amount of complains}) \\ &\quad \times (\text{number of complains})\end{aligned} \qquad (2.6)$$

Also the cost for repairing the vehicle damages because of potholes, reported for the duration of a project, can be accounted for in the estimate to get an approximate cost of vehicular wear. Vehicular wear can be estimated using the depreciation value of the additional miles traveled by the vehicle. Depreciation value of vehicles can be estimated using the following formula:

$$\text{Extra depreciation cost} = (\text{average depreciation cost/mi}) \times (\text{additional miles traveled}) \quad (2.7)$$

2.6.6 Loss of Sales Tax

Loss of tax revenue is incurred by businesses and shops affected by the utility construction. People try to avoid roads with lane closures due to utility construction. Loss of customers transforms to a loss in income for the shops. This cost should be added to the original construction cost to determine the life cycle cost of the project. The loss of business can be estimated using the following formula:

$$\text{Loss of sales tax} = (\text{average dollar loss/day}) \times (\text{duration of project in days}) \quad (2.8)$$

The loss of revenue by a business in a particular location may not be a loss of tax revenue. Often, the business profit or loss revolves within the community in the area of construction. Loss of business in one area due to construction would only transfer the potential clients or consumers to another shop or business within the area. In case of a specific shop or kind of business that is very specific to the locality with the lack of competition in the vicinity would mean a business loss for that particular community. In calculating the loss of business, care must be taken in locating a few other shops having the same kind of business within the vicinity. A coffee shop on a lane undergoing construction work might lose its business, but another shop on the parallel lane would gain. This is a transfer of consumers rather than a complete loss of tax revenue as explained earlier. There is also a risk of complete closures of shops and businesses in areas with successive utility construction work. Such construction would mean a complete loss of revenue for the shops as well as loss of tax revenue for the local government.

2.6.7 Loss of Productivity due to Noise and Vibration

Loss of productivity can be associated with the noise pollution generated during construction activity. Most of the time, the effect of noise on people is impossible to quantify. People react differently to noise; some can continue functioning with less productivity, whereas others are unable to put up with the noise. In residential neighborhoods, noise and vibration can disrupt the normal life of the residents. The following formula can be used to estimate the loss of productivity (Boyce and Bried, 1994b).

$$\text{Cost of productivity} = (\text{time lost/day}) \times (\text{number of persons}) \times (\text{value of time}) \times (\text{project duration in days}) \quad (2.9)$$

Time value is normally the hourly pay of the person, and the time lost would be estimated in hours lost per day due to the noise.

2.6.8 Dust

One way of estimating the cost of dust is to calculate the additional time spent in cleaning. The following formula can be used to estimate the cost of dust and dirt control:

$$\text{Cost of dust control} = (\text{increased cleaning time in hours/day}) \times (\text{hourly pay rate}) \times (\text{number of units impacted}) \times (\text{project duration in days}) + (\text{cost of cleaning materials}) \quad (2.10)$$

2.7 REDUCING SOCIAL COSTS

Municipalities and local governments need to take measures to reduce social costs. A project impact analysis must be prepared during the design phase. Some suggestions to reduce social costs are as follows:

- Plan realistic project durations. Unnecessary, too-short, or too-long project durations will increase social costs significantly.
- Select and advertise for bids with only the method having least social costs.
- Timing and scheduling of a project is important. Work during the weekends and holidays in business districts and during the workdays in residential neighborhoods. For residential neighborhoods, avoid early daily start-up times (before 8:00 A.M.) and daily late finish times (after 6:00 P.M.).
- Take measures to reduce noise, dust, and vibrations.
- Consider special needs of the community surrounding the work area for planning the construction work.
- Inform the pubic about work progress with periodic updates before, during, and after the work is completed.
- After project completion, take a survey to find problem areas and devise corrective measures for future work.

2.8 CASE STUDY: HISTORIC TREES PRESERVED BY TRENCHLESS TECHNOLOGY

A recent major sewer replacement project at the Royal Botanic Gardens, at Kew, London, United Kingdom, avoided tree root damage and was virtually undetectable to the visiting public, because of the careful use of TT by Perco Engineering Services Ltd. Figure 2.8 presents views of the project site.

The installation of 525 ft (160 m) of 9 in (225 mm) internal diameter pipe was part of a project to expand public facilities in the internationally famous gardens. It was essential

FIGURE 2.8 Views of the project site.

to avoid damage to the root systems of trees with historic and scientific value. An open-cut method would have carried the risk of damage that Kew Gardens could not accept.

While guaranteeing tree safety, Perco's hydraulic pipe burster had the added advantage of using only existing manholes for access. Working from Kew Gardens' 4 ft × 3.25 ft (1.2 m × 1.0 m) brick manholes, Perco employed their *Expandit* pipebursting technique to upsize the existing 6 in (150 mm) clayware sewer, including one drive of 150 ft (45 m). This technique involves jacking a powerful expanding mole through the existing pipe that it bursts and compresses into the surrounding soil. The existing pipe OD can range from 7 to 25 in (180 mm to 630 mm) and its hydraulic operation allows it to burst through a range of pipe materials, such as clay, concrete, asbestos cement, cast iron, and PVC. New pipe material can be clay, concrete, or polyethylene.

In this case, *Snap-it®* segmental pipes, in 2.6-ft. (790-mm) sections, were jacked into the resulting annulus. These pipes are manufactured in a number of water industry-approved polyethylene grades, from 10 in (250 mm) up to 40 in (1000 mm) OD and in a choice of standard dimension ratios (SDR). An ethylene-propylene-diene-monomer (EPDM) rubber O-ring seals the mechanical jointing method of the pipes.

The upsized sewer runs adjacent to the boardwalk that is the main road in the gardens, and passes outside the main gates, where it joins the public sewer. Consequently, the work could have potentially disrupted local residents and visitors alike. Using the pipe bursting method, the disruption and resulting social costs were completely avoided, to the extent that the royal opening of a new conservatory took place undisturbed as work was underway.

The ground conditions at Kew Gardens consisted of a primarily ballast-type material that proved to be a most suitable type of strata for hydraulic pipe bursting; this is because as the ground is displaced by the bursting head, the combination of fines and larger particle sizes enable the bursting annulus to be self supporting while jacking takes place, thus reducing the amount of skin friction that the pipe has to overcome. This type of ground is also advantageous when bursting longer lengths of pipe. The project duration was 4 weeks that provided an achievable time for preparation, installation, and reinstatement works.

At all times Perco worked according to the guidelines issued by the National Joint Utilities Group (NJUG), concerning the installation and maintenance of services close to trees. This defines a precautionary zone around the tree, the diameter of which is proportional to trunk girth. Only trenchless work or manual excavation is recommended within the zone. As such, pipe bursting system was perfectly completed to achieve the desired high standards required by NJUG.

2.9 SUMMARY

TT methods would become more cost-effective if all the life cycle cost factors, including the social costs, are calculated and considered in the design phase. This chapter discusses a breakdown of life cycle costs and associated social costs for utility construction. Also, this chapter considers the various cost categories within the social costs and focuses on calculating the cost of traffic disruption due to open-cut utility construction. Each of the project parameters contributes significantly in calculating the total social costs. Among different factors, the project duration contributes the most to social costs. Engineers and project owners need to calculate social costs based on specific conditions of a project to select the best and most cost-effective method of utility construction. Once each of the social cost parameters is estimated the most cost-effective method should be advertised for bids. The contractors should not be given the method selection option between open-cut and TT because they are not usually concerned with social costs of the project. Proper planning, scheduling, choice of construction method, and timing of construction operations can reduce social costs.

REVIEW QUESTIONS

1. What are the social costs of utility construction?
2. What constitutes the life cycle cost of a project?
3. Why is it important to quantify social costs for utility construction? Explain.
4. Why choice of construction method between open-cut and TT should not be left to a contractor? Explain.
5. Estimate the traffic disruption costs for the following conditions:
   a. City street with 500 cars/day.
   b. Project duration of 10 days.
   c. Assume the traffic is routed to a detoured road, which adds 2 miles to the travel distance of commuters.
   d. Assume on an average each car consumes 20 mi/gal and the cost of fuel is $2.15/gal.
   e. Assume the cost of time for commuters as $12/h.
   f. Assume speed limit on the main road to be 45 mi/h and on the detour road to be 30 mi/h.

REFERENCES

Boyce, G. M., and E. M. Bried (1994a). Benefit-cost analysis of microtunneling in an urban area, *Proceedings of No-Dig 1994*, Dallas, Tex.

Boyce, G. M., and E. M. Bried (1994b). Estimating the social cost savings of trenchless technique, *No-Dig Engineering*, 1(2):12–14.

Bush, G., and J. Simonson (2001). *Rehabilitation of Underground Water and Sewer Lines . . . The Costs Beyond the Bid*, University of Houston.

FHA (2002). *Manual for Controlling and Reducing the Frequency of Pavement Utility Cuts*, Federal Highway Administration, Washington, DC.

NCMA TEK (2003). *Reinstatement of Interlocking Concrete Pavements*, Technical Notes, National Concrete Masonry Association (NCMA). Available at www.orco.com/TechnicalNotes/National ConcreteMasonryAssocTEK/NCMA8.HTM.

NTS (2001). *National Transportation Statistics*, Bureau of Transportation Statistics, Washington D.C. Available at http://www.bts.gov/publications/nts/.

NTS (2001). *National Transportation Statistics*, *Passenger Car and Motorcycle Fuel Consumption and Travel*, and *Bus Fuel Consumption and Travel*, Bureau of Transportation Statistics. Available at http://www.bts.gov/publications/nts/.

Soucy Insurance (2001). *Potluck with Potholes*, Monthly tips, Soucy Insurance Agency, Woonsocket, RI.

Thomson, J., T. Sangster, and S. Kramer (1998). An overview of the economics of trenchless technology, *No-Dig Engineering*, 5(2).

VTPI (2003). *Transport Cost and Benefit Analysis—Travel Time Costs*, Victoria Transport Policy Institute, Victoria, Canada. Available at www.vtpi.org/tca.

BIBLIOGRAPHY

AASHTO (1997). *A Manual on User Benefit Analysis of Highway and Bus-Transit Improvements*, American Association of State Highway and Transportation Officials (AASHTO), Washington, DC.

ASCE (1999). Pipeline crossings, *ASCE Manuals and Reports on Engineering Practice No. 89,* American Society of Civil Engineers (ASCE).

Bodocsi, A., P. D. Pant, and R. S. Arudi (1995). *Impact of Utility Cuts on Performance of Street Pavements,* the Cincinnati Infrastructure Institute Department of Civil & Environmental Engineering, University of Cincinnati, Cincinnati, Ohio.

Business Strategies (2002). Disposition of a car, *Vehicle Mileage & Maintenance Record Book,* Business Strategies, a Division of Cedar Marketing Group, Inc.

Construction Industry Institute (1986). *Constructability: A Primer.* Publication 3-1, Austin, Tex.

Dukart, J. R (2000). *New Ways to Go Underground,* Utility Business, September 1, 2000. Available at http://utilitybusiness.com/ar/power_new_ways_go/index.htm.

Gangavarapu, B. S (2003). *Analysis and Comparison of Traffic Disruption Using Open-Cut and Trenchless Methods of Pipe Installation,* Unpublished Masters' Thesis, Construction Management Program, Michigan State University.

Hughes, D. M (2002). *Assessing the Future: Water Utility Infrastructure Management,* American Water Works Association, Denver, Colo.

Iseley, T., M. Najafi, and R. Tanwani (1999). *Trenchless Construction Methods and Soil Compatibility Manual,* 3rd ed., National Utility Construction Association (NUCA), Arlington, VA.

Kramer R. S., J. W. McDonald, and C. J. Thomson (1992). *An Introduction to Trenchless Technology,* International Thomson Publishing, Florence, NY.

Lansing Gas Prices, *Unleaded Gasoline Average Prices,* Gas Buddy Organization Inc., Canada. Available at http://www.lansinggasprices.com.

Sterling, R. L (1994). *Indirect Costs of Utility Placement and Repair Beneath Streets,* Final report, Underground Space Center, University of Minnesota, March 1994, Minneapolis, Minn., p. 52.

Vickridge, I., D. J. Ling, G. F. Read (1992). Evaluating the social costs and setting the charges for road space occupation. *Proceedings of ISTT No-Dig 92,* April 5–8, 1992, Washington, DC.

CHAPTER 3

PIPELINE ASSET MANAGEMENT, INSPECTION, AND CLEANING

3.1 BACKGROUND

The underground infrastructure systems span thousands of miles and form a significant part of the total U.S. infrastructure. Utility and pipeline systems form one of the most capital-intensive infrastructure systems and they are aging, overused, possibly mismanaged, and neglected. Many of these systems are deteriorating and becoming more vulnerable to catastrophic failures often resulting in costly and disruptive replacements. In spite of recent increases in public infrastructure investments, municipal infrastructure is deteriorating faster than it is being renewed. Organizations such as the U.S. Environmental Protection Agency (EPA), the American Society of Civil Engineers (ASCE), and the American Water Works Association (AWWA) have estimated water and wastewater needs for the next 20 years to be from US$150 billion to US$2 trillion. The ASCE's 2003 *Report Card for America's Infrastructure* gave wastewater infrastructure a "D," estimating an annual US$12 billion shortfall in funding needs nationally. According to AWWA, by the year 2020, the average utility will spend three times as much on infrastructure replacement as it does today. The sewer infrastructure of the United States must be assessed and upgraded to meet the requirements of the EPA Sanitary Sewer Overflow Policy and the guidelines of the Government Accounting Standards Board Statement 34 (GASB 34). The next regulatory frontier addresses sanitary sewer overflows (SSOs) from separate sanitary sewer systems. CMOM (Capacity, Management, Operation, and Maintenance) is the foundation for new regulations to address SSOs. A typical CMOM program would comprise two major elements as follows:

- General Standards: These standards establish the general requirements of the CMOM program.
  - Proper management, operation, and maintenance of collection systems
  - Providing adequate capacity to convey base and peak flow
  - Taking all feasible steps to stop and mitigate the impact of SSOs
  - Notifying the public and others with potential exposure to pollutants
  - Developing a written CMOM plan
- Management Programs: The CMOM plan should describe the management programs the wastewater agency has established for implementing the CMOM plan. The following elements should be covered in the management program:

- Goals
- Organizational structure
- Legal authority
- Measures and activities
- Design, rehabilitation, inspection, and testing standards
- Monitoring, measurement, and program modifications

Factors such as poor quality control, inadequate inspection and maintenance, and lack of consistency and uniformity in design, construction, and operation practices have adversely impacted on municipal infrastructure and are significant challenges to our systems. North America's water and wastewater pipelines (some of which are more than a century old) have corrosion damage, moving soils, changing temperatures, infiltration or inflow, in-service stress, and the continuous process of structural deterioration. At the same time, an increased burden on infrastructure as a result of significant growth in some sectors tends to quicken the aging process while increasing the social and monetary cost of service disruptions as a result of pipeline failures. These environmental and operating stresses inevitably lead to a number of pipeline failures throughout the year. Increasing concerns over health, safety, and the environment have contributed significantly to raising the visibility of pipeline risk management. The rapidly deteriorating old pipes and the expansion of present network owing to increasing demands, require the municipalities to prioritize the renewal process and installation of new pipelines. However, predicting and monitoring the condition of pipelines generally remain difficult tasks.

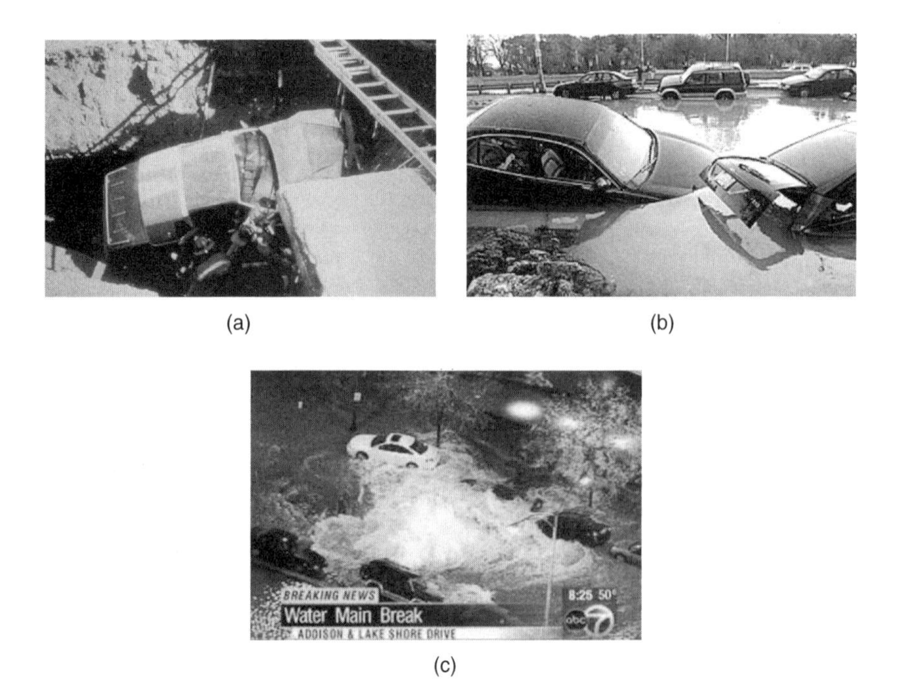

FIGURE 3.1 Examples of damage caused by pipeline failures. [*(a) is courtesy of Midwest Society for Trenchless Technology (b) and (c) are courtesy of Kenny Construction Company.*]

The final factor is that the low visibility components such as subsurface infrastructure become a low priority. Although nobody disagrees with the state of the system, the fallacy of GASB 34 or CMOM is that the liability numbers will be too large. Sewer pipes are not on anyone's radar screens. The buyers, system owners, and politicians do not want to think about how much money they will be required to spend. The users, homeowners, and businesses, do not want to pay any more (regardless of how inexpensive it is). Exacerbating this tendency to ignore the problem is an economic condition that limits the amount of funds everyone has, requiring dramatic priority shifting. A piece of pipe that is underground and that might not be failing too rapidly falls to the bottom of the list. Figure 3.1 illustrates some of the disastrous effects caused as a result of pipeline failures.

Maintaining and even enhancing wastewater collection systems are crucial, in order to have dependable transfer of wastewater to treatment facilities. When sewer systems deteriorate, water from excessive infiltration and inflow (I/I) enters the system, resulting in increased hydraulic loading at collection and treatment facilities. Consequently, increased hydraulic loading will intensify risk of SSOs and raise capital, operational, and maintenance costs. Therefore, it is necessary to apply principles of asset management to underground pipeline systems.

3.2 ASSET MANAGEMENT

Managing infrastructure assets presents several challenges and requires a wide array of resources. Any management strategy requires an estimation of the potential degradation of an asset over its life cycle and an analysis of the impact from asset failure and a set of actions (e.g., repair or renew). Infrastructure can be divided into two broad categories: aboveground and underground. The management of underground infrastructures such as pipelines and utilities has been traditionally overlooked because of their out-of-sight-out-of-mind nature. Considering failure to be a condition that prevents the designed use of the pipeline system, many times it is not the oldest pipe that has failed first. Factors such as pipe type and material, depth from ground surface, diameter, pipe section length, installation method, pipe environment and corrosiveness (soil and ground conditions), type of application, frequency of internal surcharge, joint type and configuration, external loadings, and extent of maintenance have a major role in pipeline failures.

The foundation of effective asset management is an accurate inventory. Buried infrastructure systems such as storm and sanitary sewers and water and gas pipes are particularly challenging because they are both hidden and distributed over relatively large geographic regions. Cost-effective management of these assets requires the right systems and processes to capture this information rapidly, maintain the information over time, and make decisions based on this information.

Utility owners and operators are faced with major expenditures to renew and extend the life of their buried infrastructure. Utility management, operations, and maintenance operators are faced with the task of allocating limited resources to their systems, with sparse data to back up their decisions. Given a set budget, these decisions include the following overall questions:

- What is our worst problem?
- What is the source of the problem and what is its extent?
- How do we correct the problem in a timely and cost-effective manner?

Some problems like sanitary overflows or low water pressure are obvious. Others, like pending sewer collapse or a water main break are hidden problems that can result in extremely expensive solutions. It is unlikely that all emergency repairs will be avoided, but by developing an asset management system (AMS), the number of emergency repairs can be minimized and budgets allocated for proper renewal work.

Development of an AMS requires preparation of necessary data. The immediate process includes determining the current information status and the goals to be achieved.

What Is the Goal of Pipeline Asset Management? The main goal of an asset management strategy is to meet the required level of service for present and future customers in the most cost-effective way through creation, acquisition, maintenance, operation, renewal, and disposal of assets. The key elements of pipeline asset management are

- Taking a life cycle cost approach
- Developing cost-effective management strategies for the long-term use
- Providing a defined level of service and monitoring performance
- Managing risks associated with asset failures
- Using physical resources in a sustainable way
- Continually improving asset management practices

An AMS enables pipeline operators to examine a pipeline in its totality, and focus to the specific pipe level data. It is a decision facilitation tool that combines different layers of information. The AMS centralizes, presents, and tracks diverse data sets collected from a variety of sources—including condition assessment inspection data. The asset management system has four main components:

1. Data centralization. A pipeline data audit is first conducted to catalogue the diverse data sets available to a pipeline operator. The enormous volume of pipeline information—including maps, pictures, inspection data, lay sheets, spreadsheets, and notes—is then incorporated into a logical data structure.

2. Data presentation. An AMS provides an accurate representation of in-service pipeline infrastructure. The system presents operators with single-source accessibility to pipeline related data sets between global positioning system (GPS)–accessible points illustrating:

- Pipe defects
- Pipe length
- Pipe joints
- Pipe position
- Pipe class
- Soil receptivity or chloride concentration
- Surface infrastructure
- Topography
- Land use
- Customer locations
- Results from condition assessment inspections
- Customer complaint records

3. Data tracking. The system records infrastructure characteristics, defects, repairs, and maintenance.

4. Capabilities. As a consequence of integrating this information into a logical data structure, detailed pipe level data is available to the pipe operator. This facilitates development of custom criticality indexes, compliance with legislative requirements, and development of accurate vulnerability assessments.

In addition, an AMS allows engineering staff to locate specific pipeline features with GPS equipment, simplifying the pipeline repair and renewal operations.

3.3 CAUSES OF PIPELINE DETERIORATION

Pipeline systems require constant maintenance and can deteriorate for a number of reasons. A comprehensive study performed by the Water Research Center (WRc), concludes that the concept of measuring the *rate of deterioration* of sewers is unrealistic, and deterioration is more influenced by random events in a sewer life span. Examples of random events include a heavy rain storm or an excavation nearby. Also, it maintains that severe pipeline defects do not always lead immediately to collapse. Sewer pipes are prone to certain types of failures, such as the type of material, physical design, age, functionality, and external and internal environment. Distress and collapse of a pipe are the results of the complex interactions of various mechanisms that occur within and around the pipeline. The impact of the deterioration of the pipeline system depends upon its size, complexity, topography, and service. Although it is almost impossible to predict when a pipe will collapse, it is feasible to estimate whether a pipe has deteriorated sufficiently for collapse to be likely. The mechanisms of pipeline deterioration are

- *Structural.* Cracks, fractures, breaks, and the like
- *Hydraulic.* Insufficient capacity, flooding, debris, encrustation, and grease
- *Corrosion.* Hydrogen sulfide, chemical corrosion, and external corrosion
- *Erosion.*
- *Operational problems.* Roots, blockages debris, maintenance procedures, and the like

Pipelines can have defects classified as built-in or long-term. Built-in defects are generated during pipeline construction and represent conditions that affect the performance of pipes after installation. Long-term defects are caused as a result of the deterioration process. Construction-related or built-in defects can be offsets in alignment, joints loosely fitted or loosened by vibrations, flattened or ovaled pipes, sags because of settlement, stresses caused by dynamic loadings of backfill, removal of trench sheathing and pilings, overburden compaction, and so on. Joints can experience construction defects, such as pinching of rubber gaskets, misalignment of gaskets, and squeezing because of *overshoving* of one pipe into another. A structural failure can be a crack, break, split, or cavitation of the pipe opening, or separation at a joint.

Examples of causes of long-term pipeline deterioration are sulfate corrosion as a result of sewer gases, excessive hydraulic flows, structural failures, leaks and infiltrations, and erosion. Bacteria in the wastewater stream convert the sulfates to hydrogen sulfides which, when released into the sewer air space, become oxidized into sulfuric acid. The sulphuric acid is reactive to some pipe materials corroding it. Severe corrosion can jeopardize the structural integrity of a pipe or manhole and lead to collapse. Any condition of pipeline deterioration, which occurs over an extended time period and is not a result of construction practice, is considered a long-term deterioration. Proper maintenance of pipelines is essential to keep the pipeline in good health.

The state of the surrounding soil is of fundamental importance in assessing the structural condition of a gravity pipe. The main factors that affect the rate of ground loss include gravity pipe defect size, hydraulic conditions (water table, and frequency and magnitude of surcharge), and soil properties (cohesive or noncohesive soil). Severe defects (larger than 4 in), high watertable (above pipe level), frequent and high magnitude of hydraulic surcharge, and soil types (silts, silty fine sands, and fine sands) can have serious effects on ground loss. Loss of side support will allow the sides of the pipe to move outward when loaded vertically, and collapse will likely occur once the pipe deformation exceeds 10 percent. Uneven loading of pipes because of joint displacement also accelerates the pipe deterioration process.

3.3.1 Modes of Pipeline Deterioration

Pipeline deterioration is a complex process; many factors are responsible for their deterioration and failure—structural, hydraulic, environmental, functional, age of the pipe, quality of initial construction, and so on. The intensity of structural failures depends on the size of the defect, soil type, interior hydraulic regime, groundwater level and fluctuation, corrosion, method of construction, and loading on the pipe. Hydraulic failures are caused by I/I problems. These I/I problems reduce the planned hydraulic capacity of pipes, increasing the potential for collapse. Figure 3.2 illustrates various kinds of internal and external forces acting on a pipe. The modes of failure depend on the type of environment and pipeline material.

Pipe breakage is likely to occur on pipes whose structural integrity has been compromised by corrosion, degradation, inadequate installation, or manufacturing defects. Pipe breakage types were classified by O'Day et al. (1986) into three categories: (1) circumferential breaks, caused by longitudinal stresses; (2) longitudinal breaks, caused by transverse stresses (hoop stresses); and (3) split bell, caused by transverse stresses on the pipe joint. This classification may be complemented by holes because of corrosion or impact. Circumferential breaks as a result of longitudinal stress are typically the result of one or more of the following occurrences: (1) thermal contraction (such as low temperature of the

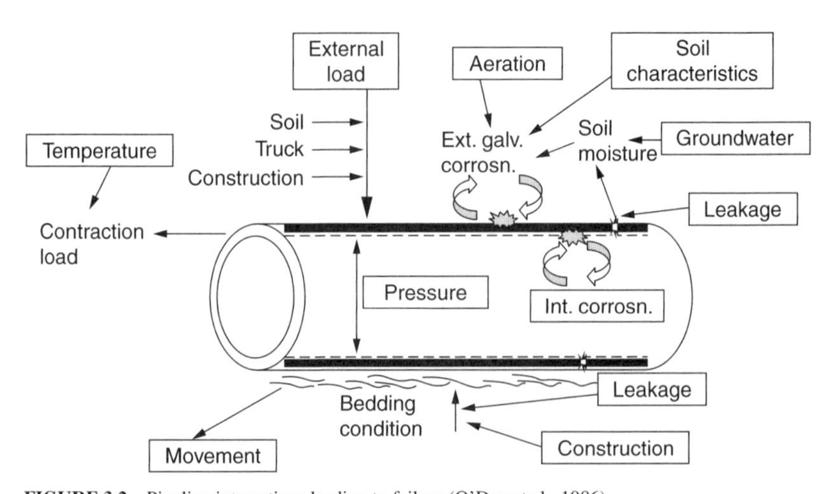

FIGURE 3.2 Pipeline interactions leading to failure (O'Day et al., 1986).

TABLE 3.1 Typical Defects in Pipes

Defect	Description
Longitudinal cracks and fractures	May occur at the springing level as well as at the crown and invert. A result of excessive *crushing* or *ring* stress
Tension cracks	Cracks are diagonal and spread from the point of overload, which is often a hard spot beneath the pipe
Circumferential cracks and fractures	Relative vertical movement of successive lengths of pipe causing cracks and/or fractures owing to excessive shear or bending stresses. Most likely to occur near joints
Broken pipes	Occurs when pieces of a cracked or fractured pipe visibly move from their original position. Normally represents a further stage in deterioration of a cracked or fractured pipe and is a very serious defect
Socket bursting	Excessive pressure inside the joint because of the expansion of the jointing material may cause a bursting failure of the socket
Deformed pipes	Occurs when a longitudinally cracked or fractured pipe loses the support of the surrounding ground

Source: From Davies et al. (2001).

sewage in the pipe and the pipe surroundings) acting on a restrained pipe, (2) bending stress (beam failure) owing to soil differential movement (especially clayey soils), or large voids in the bedding near the pipe (resulting from leaks, I/I, and so on), (3) inadequate trench and bedding practices, and (4) third-party interference (e.g., accidental breaks, and the like). Table 3.1 lists most typical types of defects found in pipes.

Structural Defects. Structural defect failure mechanisms include cracks and fractures in the pipeline material that are caused by a change in the forces around a pipeline or a change in the ability of the pipe material to resist existing forces The infiltration of groundwater through existing structural defects creates or increases the size of voids as the infiltrating water carries particles from the soil into the pipeline. The weakening of this soil makes the strata above the pipe vulnerable to surface collapse. The effects of infiltration on void formation are made worse by the process of exfiltration. Exfiltration occurs when water leaves the pipeline through structural defects during periods of hydraulic surcharge. Surcharge wastewater can scour or loosen more fines at the perimeter of the voids.

Dynamic forces that cause structural defects are large one-time events or smaller cyclic events that occur at a variety of frequencies (daily, seasonally, and so on). Large one-time events include periods of heavy surface construction, in-ground utility construction, or nonconstruction events such as earthquakes or landslides. These events are especially significant when coupled with a weakened material or voids in the soil. Many surface collapse failures are associated with degraded but functioning pipes that fail owing to a large one-time event (Delleur, 1989; WRc, 1986). Smaller cyclical dynamic loads include load transfer from above-ground activities, such as routine truck, machinery, and bus or train traffic, or in ground movements, such as those caused by expansive soils or frost heave.

Operational Defects. Operational defects failure mechanism originates from an increase in demand and a decrease in capacity. Infiltration and inflow are the two types of demand on a pipe system. Infiltration increases the demand as the number of structural defects grows. Inflow is the demand on the system from service connections or storm waters or both (ASCE, 1994; EPA, 1991). A decrease in capacity is the result of a decrease in the effective diameter of the pipeline and an increase in the roughness coefficient. The effective diameter is reduced by structural defects such as open and offset joints, broken pipe sections, root masses, grease, or collected debris.

Pipeline Deterioration Mechanisms. There are several theories that explain the deterioration mechanisms of buried pipelines. Many pipe system deteriorations are attributable to the following predominant mechanisms:

- Deterioration as a result of natural aging process—lack of maintenance exacerbates age-related deterioration
- Deterioration of pipes and joints as a result of soil-pipe interaction, operating conditions, and exposure to corrosive substances
- Freeze or thaw cycles, groundwater flow, and subsurface seismic activity that can result in pipe movement, warping, brittleness, misalignment, and breakage

3.3.2 Age of Pipe

A classical survival function relating the age of the pipeline to the failure rate is denoted by a bathtub curve as shown in Fig. 3.3. The early part of the curve shows *infantile failure*, which for pipes is representative of failure due to human factors in the actual laying of the pipe (manufacturing faults, tend to appear during that part). A period of time follows in which failure rate is generally low. When failure does occur it may depend on many factors, such as excessive loads not designed for, or settlement. As the pipes tend toward the end of their useful life the failure rate increases exponentially. This classic survival profile is known as the *bathtub* curve. The bathtub curve can be applied to an individual pipe, a group of pipes with similar characteristics, or the whole population of a pipe network.

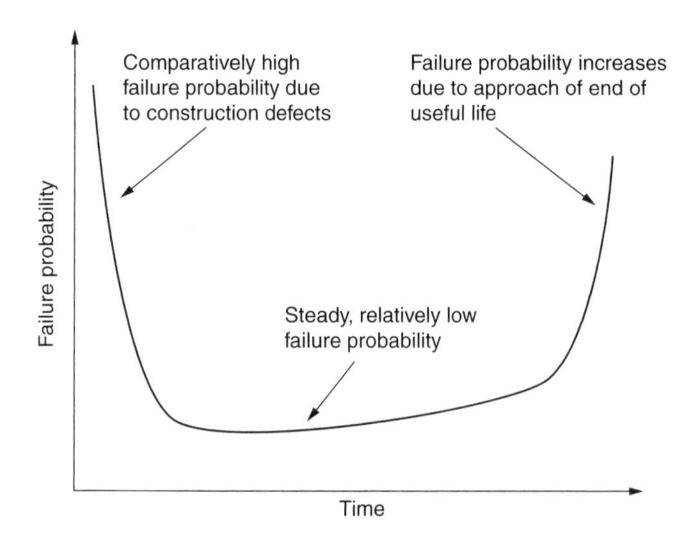

FIGURE 3.3 Bathtub curve of pipe performance with age.

The factors that accelerate the pipe aging process are discussed in detail in the following sections.

3.3.3 Pipe Size

A number of authors have investigated the relationship between pipe size and structural stability. Studies indicate that there is a decreasing trend in pipe failure rate with increasing diameter and is directly attributed to the increasing wall thickness and joint reliability with increase in pipe diameter. Larger wall thickness gives the pipe better structural integrity and improved resistance to corrosion failures. Many other studies have also shown that a larger proportion of failures have occurred on the smaller-diameter pipes.

Pipe size also affects the mode of failure. Smaller-diameter sewer mains (6 to 8 in) often experience beam (flexural) failure because of poor bedding conditions, however crushing failures (often longitudinal) are not likely to occur because of the relative length-to-diameter ratio. Conversely, larger sewer mains (10 in or greater) are likely to experience crushing failure, but are not likely to experience beam failure (O'Day, 1982). Although longitudinal bending stresses increase with increasing pipe diameter, they do so at a slower rate than the increase in the pipe's section modulus, hence pipes, which have high length-to-diameter ratios, are more likely to suffer from excessive bending stresses. Despite the fact that the above issue is well documented within the literature, there is little evidence of any numerical or statistical investigation of the effect of high pipe length-to-diameter ratios in regard to the mode of failure.

3.3.4 Pipe Section Length

Generally, longer pipe runs are less likely to deteriorate at a faster rate than the shorter ones. This may be because of the fact that longer pipe runs mean fewer or less severe bends in the pipe to accumulate debris, creating blockages or damage to the pipe from standing sewage. Another possible reason is that the longer pipe runs may be more of conveyance systems rather than collection systems, thus having fewer laterals connected to the pipes that can weaken a pipe system.

3.3.5 Gravity Pipe Gradient

The slope of a gravity pipe is found to have an impact on the condition of the pipe. For all conditions the same, the steeper the slope the higher is the possibility that pipe segments deteriorate. This may be owing to the fact that steeper pipe segments induce faster flow rates, resulting in greater possibility for erosion and surface wear to the inside walls or joints of pipe segments. Generation of deteriorating gases or movement of debris may also increase deterioration rates.

3.3.6 Gravity Pipe Joint Type

The main functions of a sewer pipe joints are to be watertight, durable, and resistant to root intrusion. Joint failures are due to leaks where pipe joints become separated or rubber ring is improperly seated. Joint type is an issue because the type of joint will influence the susceptibility of the pipe to specific failures. A large part of joint problems may be due to the amount of flexibility and lateral constraint the joint provides, as well as the pipe joint's actual strength and its ability to resist corrosion.

3.3.7 Pipe Depth

In investigating the effect of depth on pipe structural condition, found a steady decreasing defect rate to a depth of 18 ft, below which the defect rate began increasing with depth. It was suggested that the first occurrence probably reflected the decreasing influence of surface factors such as road traffic and utility or surface maintenance activity. The second occurrence or pattern was explained by the increasing effect of overburden pressure. Jones (1984) has suggested that, in shallow pipes, the effect of seasonal moisture variations in the soil surrounding might be significant. In an analysis of over 4400 pipe failures, Anderson and Cullen (1982) reported that 65 percent of all incidents occurred at a depth of 6.5 ft or less and 25 percent from 6.5 to 13 ft deep, although no indication is given of overall pipe depth distribution. Changes in cover depth may also be important in determining a pipe's structural stability.

3.3.8 Surface Loading and Surface Type

The location of a pipe will obviously affect the magnitude of surface loading to which it is subject; for pipes beneath roads the main component of such loading is likely to be that from traffic. Pocock et al. (1980) monitored the bending strain developed in a shallow buried pipeline owing to static and rolling wheel loads. The measured bending strains were found to increase linearly with axle load, the strains for any given load tending to decrease with increasing vehicle speed. Maximum strains were always associated with pipes that had been poorly bedded. Another possible cause of longitudinal failure is owing to a crushing load. This usually occurs in the large-diameter pipes (O'Day, 1982).

3.3.9 Frost Heave

Frost heave is defined as the vertical expansion of soils caused by soil freezing and ice lens formation. All underground structures require the consideration of frost heave effects, as they are capable of displacing portions or the entire underground structure.

Differential heave causes sections of pipe to experience nonuniform displacements, and this results in forceful flexural stresses. Uniform heaving may also prove to be a problem under certain circumstances where pipe joints are not subject to movement. Under this scenario, the pipe experiences stresses similar to a simple beam loading, in which case the pipe will experience bending stresses. Failure of pipe joints may be the result of the frost heave process. This may be a function of the type of connection and the type of fill material used between joints.

The conditions for frost heave require the presence of a frost-susceptible soil, the presence of a sufficient water source, and a ground temperature below the freezing point.

With all of the above factors present, there is the potential for damage owing to frost heave. The propensity for heave of a soil under freezing conditions is affected by properties such as grain size, rate of freezing, the availability of water, and by applied loads.

3.3.10 Frost Load

The failure of pipe pipes during winter could be attributed to increased earth loads on the buried pipes, that is, frost loads. In a trench, the frost load develops primarily as a consequence of different frost susceptibilities of the backfill and the sidewalls of the trench and the interaction at the trench backfill-sidewall interface. Trench width, differences in frost susceptibilities of backfill and trench sidewall materials, stiffness of the medium below the freezing front, and shear stiffness at backfill-sidewall interface play important roles in the generation of frost loads. Thus, it is preferable to use a backfill material that has matching or lower frost susceptibility than that of the sidewall in order to minimize the development of excessive differential frost loads.

3.3.11 Sewage Characteristics

Although domestic sewage is generally not aggressive to the fabric of the pipe system, the quality of the sewage varies from place to place and is dependent on several factors. It can vary from relatively weak domestic sewage, perhaps diluted with large quantities of storm water or infiltration, to strong and potentially aggressive sewage with a high proportion of trade effluent.

3.3.12 Soil-Pipe Interaction

Soil-pipe interactions are also a possible cause of pipe deterioration. The resistance of the soil-to-pipe union is important because the shear strength of the interaction can affect the degree of mobility of the pipeline and hence its ability to displace. In cold temperatures, the bond between the soil and pipe produces the amount of restraint in the pipe to shrink axially. A high soil-pipeline interaction will not allow the pipe to contract, and consequently the axial stress in the pipe will increase. It is also possible that a strong bond between the iron pipe and soil will cause excessive soil-pipe interface shear that may cause abrasion of the pipe coating. This abrasion may lead to premature corrosion of the pipe exterior.

3.3.13 Pipe Wall Temperature Gradients

For longitudinal failures, a failure mechanism might be when a high temperature gradient occurs across the pipe wall. If the temperature difference of the transported effluent and surrounding soil is significant, this temperature gradient can lead to unusually high hoop stresses, subsequently leading to failure.

3.3.14 Corrosion

Corrosion in metallic pipes essentially occurs by an electrochemical reaction between the outer surface of an exposed pipe and its surrounding soil environment. For corrosion to occur, there must be a potential difference between two points that are electrically connected in the presence of an electrolyte, in this case, the surrounding soil. With these conditions satisfied, a current will flow from an anodic area, through the soil to a cathodic area, and then back through the pipe wall to complete the circuit. The anodic area becomes corroded by the loss of metal ions to the electrolyte. Upon initiation, the corrosion process is self-sustaining, resulting in the formation of *pits* at the outer surface of the pipe, with a range of depths and widths. Different pipe materials have different characteristics in their reaction to corrosion. Factors like soil acidity, resistivity, pH content, oxidation-reduction, sulphides, moisture, and aeration levels have all been reported to influence corrosion rate (Romanoff, 1964), and correlations have been proposed between corrosion rates and soil electrochemical properties (Rossum, 1969). Longitudinal failures may also occur in combination with the weakening of the pipe wall owing to corrosion, at the weakest portion of the main wall.

The risk of *interior corrosion* of a sewer pipe interior depends upon the susceptibility of the pipe material to corrosion and the amount of corrosive chemicals in the wastewater. Interior pipe corrosion typically occurs from the formation and release of hydrogen sulfide. Hydrogen sulfide is formed in anaerobic conditions, such as those found in force mains, continually surcharged gravity pipes, debris piles, or pools caused by sagging lines. It is assumed that anaerobic conditions exist in open channel flow at the wetted perimeter, and therefore, a small amount of hydrogen sulfide is generated and some corrosion can occur if the conditions allow for release (ASCE, 1994). Hydrogen sulfide gas is released in turbulent conditions. Such conditions occur at siphon outlets, drop structures greater than 2 ft, discharge of force mains, interceptor intersections, a change in slope, and during high wastewater velocities.

Exterior corrosion depends upon the susceptibility of the pipe material to acidic ground substances and galvanic corrosion (Delleur, 1989). Acidic soils or groundwater attack unprotected cementitious or metal pipe materials, whereas stray currents in the ground cause a galvanic corrosion with metal or metal reinforced pipes.

3.3.15 Differential Pipe Temperature

Some literature speculates that a high differential temperature between the internal and external pipe wall can produce high temperature gradients. Under such conditions the inner and outer fibers will be subject to different temperature drops, resulting in differential strains and circumferential stresses.

3.3.16 Soil Type

The significance of the type of soil cannot be overlooked, as it is one of the most important factors, having effects on almost all of the above mechanisms. Its effects on frost heave, strength of soil-pipeline interaction, and external corrosion can be important for many failure mechanisms.

The type of soil the pipe is located in is also important for the aspect of differential heaving and thaw settlement. If a pipe is located at the interface of two different soil types, it has been shown that each soil will experience an uneven amount of frost heaving, and therefore have an influence on the amount of strain experienced by the pipe. In the same manner, thaw settlement will lead to differential stress distributions on the pipeline.

Soil corrosivity is a soil characteristic that must be considered for external corrosion predictions. Physical characteristics (particle size, friability, uniformity, organic content, color, and so on) have reflected corrosivity, based on observations and testing. Soil uniformity is important because of the possible development of localized corrosion cells. Corrosion cells may be caused by a difference in potential between unlike soil types, with both soils being in contact with the pipe. If it can be assumed that for a particular soil classification the approximate uniformity coefficient can be estimated, then the possibility of corrosion can be estimated.

3.3.17 Soil pH

In order to characterize external corrosion, it is necessary to find parameters which indicate the corrosivity of the soil. Soil pH is a good indicator of external corrosion because certain pH ranges allow for different corrosion mechanisms to occur. It has also been found that resistivity is a function of pH. For that reason, only one of the two may be required for characterization.

3.3.18 Groundwater Level

Use of the soil water content parameter is important from several aspects. As mentioned earlier, the rate of frost heave is controlled by the availability of free water. It is also important for external corrosion.

From the perspective of external corrosion, soil corrosion aggressiveness has been related to moisture content. Soils with moisture content above 20 percent are thought to be particularly corrosive (Jarvis and Hedges, 1994). Studies substantiate that moisture content is a factor contributing to soil aggressiveness.

3.3.19 Overburden Pressure

Overburden pressure is thought to be important because of its ability to help characterize frost heaving and soil-pipeline resistance. It can be characterized by the depth of cover and soil density. Literature indicates that the overburden pressure is important for the rate of heaving.

From the perspective of soil-pipeline interaction, it has been demonstrated that the frictional soil resistance is affected by pipe diameter and depth (Rajani et al., 1995). Also, from the perspective of mode of failure, larger pipes are more susceptible than smaller pipes to crushing failure. This is owing to depth, or the external loadings (i.e., roadways, large structures, and the like) to which the pipe is subjected (O'Day, 1982).

3.3.20 Temperature

The effects of temperature on pipe breakage rates have been observed and reported by many. Walski and Pelliccia (1982) suggested that pipe breakage rates might be correlated to the maximum frost penetration in a given year. To account for the lack of frost penetration data, they correlated annual breakage rates with air temperature of the coldest month, using a multiple regression analysis with age and air temperature as the covariates

$$N(t,T) = N(t_0)e^{At}\ e^{BT}$$

where t = pipe age, years
$N(t_0)$ = breaks per mile at t_0, years
T = average air temperature in the coldest month, °F
A, B = constants

Newport (1981) analyzed circumferential pipe breakage data and found that increased breakage rates coincided with cumulative degree-frost (usually referred to as freezing index in North America and expressed as degree-days) in the winter as well as with very dry weather in the summer. He attributed the increase in winter breakage rates to the increase in earth loads because of frost penetration, that is, frost loads, and the summer breakage rates to the increase in shear stress exerted on the pipe by soil shrinkage in a dry summer. He also observed that when two consecutive cold periods occurred, the breakage rates (in terms of breaks per degree-frost) in the first one exceeded those of the second one. He rationalized that the early frost *purged* the system of its weakest pipes, causing the later frost to encounter a more robust system.

3.3.21 Precipitation (Snow or Rain)

Snow cover is indicative of the insulating effect on ground temperature, as the snow will allow for the entrapment of heat into the ground. The amount of rainfall coupled with the soil type may be indicative of moisture content or hydraulic conductivity if these parameters are not measured regularly. Some literature indicates that corrosion resistance is enhanced during dry periods of the year (Smith, 1968). Therefore, inclusion of this parameter may be necessary to help characterize climatic changes as well as to infer adjustments to soil parameters.

3.4 PIPELINE INSPECTION: GRAVITY SEWER PIPE

Periodic pipeline inspection is very important to help define and prevent future problems. A study performed by the ASCE concluded that the most important maintenance activities are pipeline inspections and cleaning. Table 3.2 shows the average frequency of various maintenance activities. Maintenance of pipes is planned based on several factors such as:

Problems: Their frequency and location; 80 percent of problems occur in 25 percent of the system.

Age: Older systems have a greater risk of deterioration than newer ones.

Pipe material: Pipes constructed of materials that are susceptible to corrosion have a greater potential of deterioration and potential collapse. Nonreinforced concrete pipes, brick pipes, and asbestos cement pipes are examples of pipes susceptible to corrosion.

Pipe capacity: Pipes that carry larger volumes take precedence over pipes that carry a smaller volume.

Location: Pipes located on shallow slopes or in flood prone areas have a higher priority.

Pressure pipe or Gravity pipe: Pressure pipes have a higher priority than gravity, size for size, because of complexity of cleaning and repairs.

Subsurface conditions: Depth to groundwater, depth to bedrock, soil properties (classification, strength, porosity, compressibility, frost susceptibility, erodibility, and pH).

Corrosion: Hydrogen sulfide (H_2S) is responsible for corroding sewers, structures, and equipment used in wastewater collection systems. The interior conditions of the pipes need to be monitored and treatment needs to be implemented to prevent the growth of slime bacteria and the production of H_2S gases.

There are various condition assessment techniques that are used for sewer inspection and can be classified into three different groups. The first group, including conventional CCTV examinations, are techniques that determine the condition of the inside surface of

TABLE 3.2 Frequency of Maintenance Activities (EPA, 1999)

Activity	Average (% system/year)
Cleaning	29.9
Root removal	2.9
Manhole inspection	19.8
Inspection	6.8
Smoke testing	7.8

the sewer. The second group examines the overall condition of the sewer wall and, in some cases, the soil around the pipe. Finally, the third group detects specific problems within or outside of the sewer wall.

3.4.1 Inspecting within the Pipeline Walls

CCTV Inspection. The standard approach to sewer inspection is the examination of the inner surface of the pipe wall. The closed-circuit television (CCTV) technique is the most widely used sewer inspection technique and an effective method of inspecting the internal condition of sewers. This method is used generally where the flow depth is less than 25 to 30 percent of pipe diameter. Where flow depth is more than 25 to 30 percent, ultrasonic or sonar techniques can be used. Inspections using CCTV are beneficial as a permanent visual record is made and can be reviewed at a later time. There are two basic types of the CCTV system. Each uses a television camera in conjunction with a video monitor, videocassette recorders, and recording devices. In one case inspection is performed using a stationary or zoom camera mounted at a manhole so that it looks into the sewer, whereas in the other a mobile, robotic system moves through the sewer pipe itself. Operators using CCTV equipment must be trained to obtain the required information as effectively as possible as they must be able to translate the visual picture to a physical condition description while recording data on report forms. Figure 3.4 shows a typical CCTV inspection setup.

Stationary CCTV. These cameras mounted at a manhole are limited with respect to what they can see. Defects that are closer to the manhole are easier to locate and identify than the ones farther away. Stationary CCTV's usefulness in this respect will depend on whether the damage that can be detected by this type of system near a manhole in a sewer is representative of that throughout the entire section of the pipeline. In some cases, the major economic advantage of stationary CCTV is that it does not require the sewers to be cleaned so that a mobile inspection system can enter them. However, if the type of damage to be located is hidden under the debris in the sewer, cleaning will be necessary and mobile CCTV will be a preferred option compared to a stationary system. Stationary CCTV is less costly and may be used to identify defects in sewer lines and help prioritize pipelines for mobile CCTV. This method is useful for sewers upto 24 ft deep, and less than 36 in diameter.

Mobile CCTV. These systems are the most common means of inspecting sewer lines. This type of CCTV system uses a camera mounted on a robot that enters the sewer system. The camera generally looks forward as the robot system moves along the sewer axis, allowing the operator to examine and evaluate the entire length between a pair of manholes. Some CCTV systems have *pan and tilt* (since mid 80s) and zoom (since mid 90s) cameras attached to the

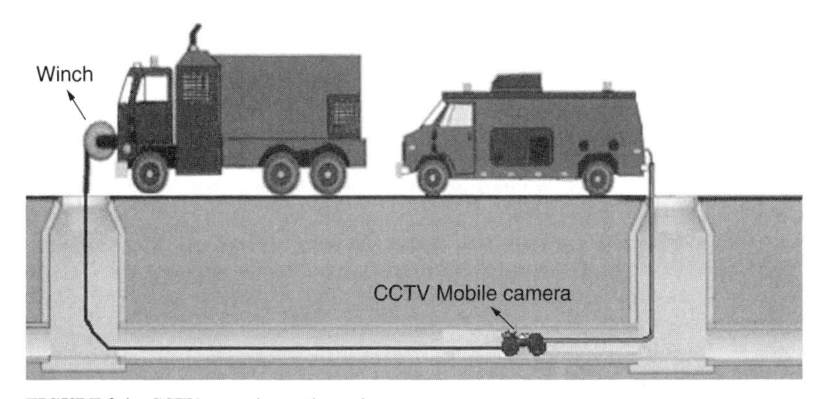

FIGURE 3.4 CCTV sewer inspection unit.

robot, which can find defects hidden from a forward-looking camera behind connections and other obstructions within the sewer line. Sonar or ultrasound systems are often attached to robots to examine the sewer below the waterline. It is also possible to obtain CCTV equipment with a *light line* attachment to assist in quantifying smaller sewer deformations. This system projects a line of light around the circumference of the sewer being examined to assist in assessing the shape of the sewer.

The CCTV inspection may miss certain type of defects, especially those that are hidden from the camera by obstructions as it looks down the sewer. Slight deformations of the sewer may go unnoticed, and any defect hidden beneath the water inside the sewer will definitely not be found.

The sewer faults and defects that can be identified through CCTV inspections might include longitudinal and circumferential cracks, collapse, displaced bricks, broken pipes, defective and displaced joints, evidence of abrasion or corrosion, siltation, encrustation, root penetration, loss of mortar, deformations, infiltration, and all lateral connections, and their degree of penetration (WRc, 1986).

Sewer Scanner and Evaluation Technology (SSET™). The SSET was developed in Japan in the mid-1990s as an advanced and innovative pipeline condition assessment technology. The development process was, to a large extent, financed by the Japanese government, which realized the limitations of CCTV pipeline inspection systems. SSET captures three streams of digital data without the need to stop the forward motion of the SSET probe within the pipeline. These data are forward view (FV), side scan (SS), and position data (PD).

The FV is nothing special as it is what is typically produced with any CCTV system. The SS is very special as it permits in real-time to open up the pipe and lay it out flat. This unique feature permits looking into 100 percent of the wall at the same angle and light intensity that minimizes operator error resulting from image skewness, light reflections, and shadowing. When the FV and SS are combined, it provides a three dimensional perspective. PD permits real-time mapping of the probe as it moves through the pipe by using an integrated inclinometer and gyroscopic network.

The SSET process consists of the field data acquisition (FDA) phase and the data analysis and interpretation (DAI) phase. The SSET FDA process allows the operator to capture complete and accurate pipeline condition assessment information automatically without requiring the operator to stop, code, and classify defects and features in the field. Taking this responsibility away from the field technicians, allows them to stay focused on what they are best qualified to do, which is to make sure that the equipment is being operated properly and the project is being managed safely. The DAI phase allows analysts to use advanced SSET analysis computer software to conduct a very thorough and accurate DAI report.

SSET was introduced in North America through a unique advanced technology evaluation process developed through the ASCE CERF (Civil Engineering Research Foundation) (CERF, 2001) that began in 1996. The CERF SSET evaluation program involved 13 sewer agencies throughout North America. Each sewer agency had a designated SSET evaluator. TTC served as a scientific consultant to CERF. Approximately, 10,000 ft of sewer lines were inspected by the SSET process during 1997 and 1998. At the end of this 2-year evaluation program, an official CERF report was developed that identified the strengths and weaknesses of the SSET. Based on this information, the second generation of SSET systems was developed.

Focused Electrode Leak Location (FELL) Technology. A new system, developed in Germany, the FELL system measures electrical current flow between a probe that travels in the pipe and a surface electrode. Pipe defects that allow liquids to flow into or out of the pipe cause a spike in the electrical signal, thereby, locating the sources of infiltration or exfiltration. The intensity of the measured current can be correlated to the magnitude of the leaks. The chief advantages of the FELL technology include: identifying leaking joints on mainline during dry weather; prioritizing leak repairs by intensity of leaks; determining

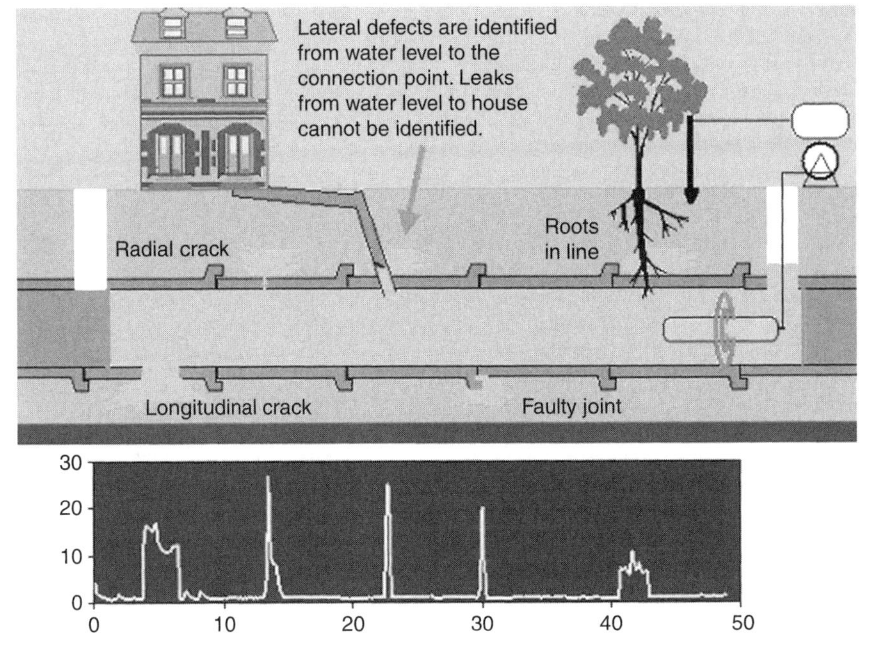

FIGURE 3.5 Working of FELL technology (Makar, 1999).

leaks in service lateral mainline connections; and an alternative to air pressure testing for acceptance of new and rehabilitated sanitary sewers. The setup is illustrated in Fig. 3.5.

The FELL test system uses a specially constructed electrode called a *sonde* that generates an electric field. The electric field is focused into a narrow disc 1-in wide set normal to the longitudinal axis of the sonde. A surface electrode (usually a metal stake) is inserted into the ground at the surface. When the sonde is placed in a *nonconducting* pipe that contains sewage (and/or water), the electric current flow between the sonde and the surface electrode is very small.

Defects in the pipe that would allow flow of fluid either *into* or *out* of the pipe provides an electrical pathway from the sonde, through the wall of the nonconducting pipe, and through the ground, to the surface electrode. When the sonde is close to such a defect, the current between the sonde and the surface electrode increases. Figure 3.5 illustrates the working of the FELL system.

Benefits and limitations of the FELL technology are as follows:

- FELL system is not a replacement for CCTV
- FELL system can be used to prioritize or eliminate sections that do not need to be televised; or to verify relative I/I removal after rehabilitation
- Use of the method is not cost-effective in pipe sizes larger than 24 in because of the time and cost of flooding the pipe
- Not recommended for testing of pipes at steep grade because of the increased possibility of flooding of house services
- Caution is needed in pipes with sharp bends because of the potential for damage to the electric cable
- Results appear to be independent of soil conditions and groundwater levels
- Results are reproducible

Laser-Based Scanning Systems. Laser-based scanning systems can be used to evaluate both the shapes of pipelines and the types of defects they contain. These systems are restricted to the part of the sewer above the waterline, but they can make more accurate inspections of sewer condition. An additional advantage with this technique is that the information from the laser scans is readily recorded and analyzed by computer, substantially reducing operator errors. This method is more effective because finer defects can be detected and the results of an examination and operator fatigue, which can lead to missed defects in a CCTV assessment, is reduced.

Ultrasonic Inspection (Sonar). Ultrasonic inspection (sonar) is performed using a beam of very high frequency coherent sound energy, with the frequency being many orders of magnitude higher than a human being can hear (Birks and Green, 1991). This method is best where the flow depth is greater than 75% of the diameter. Sound wave travels into the object being inspected and reflects whenever there is a change in the density of the material, with some of the energy in the wave returning to the surface and some passing on through the new material. The technique is capable of detecting pits, voids, and cracks, although certain crack orientations are much more difficult to detect than others. The ultrasonic wave reflects most easily when it crosses an interface between two materials that are perpendicular to the wave. The ultrasonic beams are well-suited to inspecting the inside surface of pipe walls. Typically they are used to examine the sewer below the waterline and therefore complement CCTV systems, which are confined to examining a sewer above the waterline.

Totally Integrated Sonar and CCTV System (TTSCT). Since 1986 a totally integrated system of CCTV mounted on top, and sonar mounted underneath the pontoon has been available. This method is used with flow depths from 25 to 75 percent of the pipe diameter, and pipe diameter greater than 24 in.

Wall Microdeflections and Natural Frequency of Vibration. Measurements of wall microdeflections and the natural frequencies of vibration of sewer lines are being developed specifically as a means of diagnosing brick sewer condition. The methods give information on the overall mechanical condition of the sewer line, rather than identifying specific defects (Makar, 1999).

A microdeflection in a pipe wall surface is created by applying pressure to the inside surface of the wall to deform it very slightly. In this case the intent is to measure the change in position versus the increase in pressure applied to the wall to indicate how well the grout between the bricks has been applied or whether the walls of a concrete or brick pipe have been damaged. It would be expected that a well-grouted brick wall would expand continuously (although not equally) in all directions as the pressure increases, provided the pressure is below that which would damage the grouting. A similar, equal increase would be seen in an undamaged concrete or vitrified clay pipe. Increasing in microdeflection in one direction while decreasing in another or a sudden change in the slope of a graph of applied force versus microdeflection would suggest that the wall was damaged. The major difficulty with this technique is determining the maximum safe pressure for use on a brick wall so that the inspection method does not damage it. This pressure will depend on the preexisting condition of the sewer. Although these pressures can be readily calculated for an undamaged sewer, the accuracy of such calculations is dependent on the knowledge of the strength of the mortar or concrete at any given point in the sewer. This will vary depending on the age and condition of each sewer section.

Therefore, care must be taken to avoid damaging sewer sections that have below normal strength but are still able to function properly. This safety consideration is not as important for concrete pipes, where the strength of the pipe material is more uniform around the pipe circumference. Microdeflections are restricted in use to rigid pipes where an entire pipe wall will be deflected by the applied force. Plastics such as PVC and HDPE cannot be inspected using this method, as local deformation of the pipe wall would tend to provide a false indication of the pipe condition. The restriction of the technique to materials such as brick, concrete, metal, and vitrified clay means that it is only sensitive to the wall condition, rather than that of the surrounding bedding.

Measuring the natural frequency of vibration also gives information about the mechanical behavior of a pipe wall, but in this case the process involves vibrating the wall at a range of frequencies and determining which frequency gives the largest vibrations (the natural frequencies). A section of good wall would be expected to have certain characteristic natural frequencies, whereas deviations from those frequencies would indicate that the wall or surrounding bedding was deficient in some manner. The application of this technique depends on the development of an understanding of exactly how the natural frequencies of different types of pipe wall would be expected to change with increasing damage. However, other factors can also affect the results of the natural frequency measurement, including changes in bedding material or quality, the amount of water in the pipe and the height of groundwater around the pipe. Considerable research is needed to determine if these effects can be separated from those produced by the actual damage to the pipe wall.

Ground Penetrating Radar. Although some of the inspection techniques described previously can give information about the conditions behind a pipe wall, their primary use is likely to be in determining the structural soundness of the wall itself. By contrast, ground-penetrating radar may occasionally give information about delaminations in concrete sewers, but its major use in sewer lines is in detecting potential problems behind the pipe walls.

Ground penetrating radar is also known by the name georadar or by the acronym GPR. Radar is well-known for its ability to detect airplanes or other flying objects, but with significant adaptations it can also penetrate rocks, sand, and other solid materials to detect buried objects and voids. GPR was initially developed by the U.S. military for use in detecting tunnels and mines. It has since been employed in the mining industry, in archeology and other areas. GPR is also currently in use as a means of nondestructively evaluating the condition of highways and concrete slabs.

The ability of GPR to detect subsurface voids has led to an interest in using it to evaluate the condition of sewers and other pipes. Although delaminations in concrete sewers could be detected by GPR systems, much of the interest in the technique is because of its ability to examine the bedding behind the pipe wall. Voids, rocks, and regions of water saturation produced by exfiltration should all be readily detectable by the technique. Recent research on this application has investigated its use in brick sewers, transport tunnels, and small-diameter sewer lines. Figure 3.6 illustrates a GPR output.

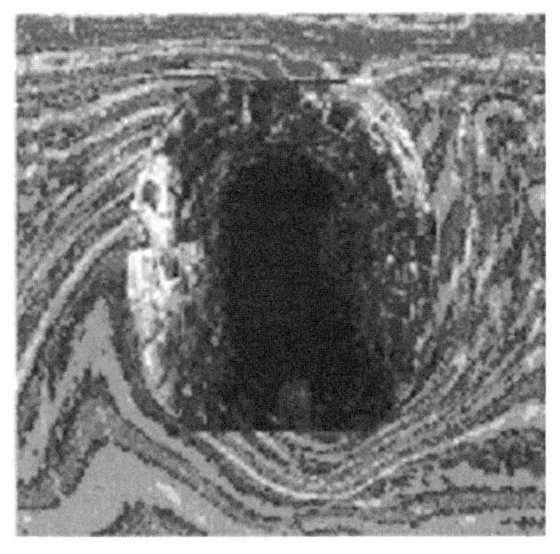

FIGURE 3.6 Ground penetrating radar (GPR) output.

Radar systems work by emitting a coherent beam of radio waves. These waves travel through space, air, or the ground until they reach an object with differing conductivity and dielectric constant, such as an airplane, a void in the ground, or an area saturated with water. Part of the radar wave is reflected off the interface between the two objects and propagated back to the transmitter. The rest of the wave passes into the new object and continues to travel in the original direction. Radar beams can also be attenuated by the nature of the material through which they travel.

Materials that are highly conductive, have high dielectric constants, or are magnetic will rapidly attenuate the radar beam. As a result, radar is attenuated very rapidly in metals, giving essentially zero penetration, but can travel very long distances in air and space. Sand, asphalt, clay, and ice fall between these two extremes, with the degree of attenuation depending on the amount of liquid water and salts present in the material. Ice is essentially transparent to GPR, allowing the technique to be used to map the bottoms of glaciers. It can also penetrate dry sand deeply. However, the depth of penetration in wet sand is much less, and in clays the penetration is further reduced (Benson, 1995). In these materials the presence of water increases the conductivity, whereas clays can also have significant dielectric constants. The presence of salt in the ground increases the soil conductivity and therefore further decreases the maximum penetration depth of a GPR system.

A summary of the current inspection techniques for sewers described in the previous sections is given in Table 3.3.

3.4.2 Structural Condition Rating of Sewers

The condition rating that follows sewer evaluation is used to objectively determine the current condition of sewers. A rating system that minimizes subjective evaluation and is repeatable can be effectively used to predict the future condition. It is acknowledged that it does not make sense to develop a sophisticated condition rating system if the deterioration process of a sewer structure is not fully understood; as is the case when new methodologies or materials are involved. However, comprehensive and objective rating systems can be developed for the most common sewer pipe materials and when adequate historical performance records are available.

Most rating systems are based on assessment of structural conditions with little consideration of hydraulics and I/I conditions, although these investigations can lead to long-term and cost-effective renewal measures. They require hydraulic modeling and simulations (which include comprehensive input data) and in-depth investigations of I/I, which can be expensive.

In the area of sewer management, there is no standard procedure to develop a condition rating of sewer pipes. Although a standard procedure for developing a comprehensive sewer condition rating does not exist, several methods of sewer condition rating (for brick and concrete or clay sewers) found in the literature have been reviewed to gauge the status of condition assessment methodologies for sewer systems.

Water Research Center (WRc). The Sewer Rehabilitation Manual (WRc, 1983) discusses the development of the structural rating system for concrete and brick sewers in the United Kingdom. The rating system involves three levels of structural condition. Each structural defect found in a concrete pipe is numerically scored based on the severity of the defect and the number of defects recorded in a pipe. These defects include: open joint, displaced joint, cracked, fractured, broken, deformed, and collapsed. For brick sewers, the defects are: mortar loss, displaced bricks, missing bricks, surface damage, fractured, and dropped invert. The inspector is provided with pictorial descriptions to determine the type and severity of each defect. For example, each *circumferential crack* found in a sewer pipe is assigned a score of 1. *Longitudinal* and *multiple* cracks are given

TABLE 3.3 Current Sewer Inspection Techniques: A Comparison (Makar, 1999)

Technique	Where to use	What will be found
	Inspection of the inner surface	
Conventional CCTV	Empty pipes, partially filled pipes above the water surface	Surface cracks, visible deformation, missing bricks, some erosion, visual indications of exfiltration/infiltration
Stationary CCTV	Pipes with less than 160 m distance between manholes	As CCTV
Light line CCTV	Pipes where deformation is an issue	Better deformation measurements and CCTV results
Computer-assisted CCTV	As CCTV, currently small-diameter pipes only	As CCTV, but with quantitative measurements of damage
SSET™	Pipes of diameter ranging from 8 to 24 in	As CCTV, but with higher sophistication and accuracy; can measure deformation of pipes
Laser scanning	Partially filled pipes, empty pipes	Surface cracks, deformations, missing bricks, erosion losses
Ultrasound	Flooded pipes, partially filled pipes, empty pipes	Deformation measurements; erosion losses; brick damage
	Inspection of pipe structure and bedding condition	
Microdeflections	Rigid sewer pipes	Overall mechanical strength
Natural vibrations	Empty sewer pipes	Combined pipe and soil condition, regions of cracking, regions of exfiltration
Impact echo	Larger diameter, rigid sewers	Combined pipe and soil condition, regions of wall cracking, regions of exfiltration
	Inspection of bedding	
Ground penetrating radar	Inside empty or partially filled pipes	Voids and objects behind pipe walls, wall delaminations, changes in water content in bedding material

scores of 2 and 5 (per crack), respectively. Each pipe is rated based on the number and score of defects within the pipe. A single collapsed pipe is scored 165, and so on.

The scores of all defects found in a pipe are then compiled to calculate *the peak score* accumulated in any 3.18 ft length. Additionally, *the total score* and *the mean score* for the entire length of sewer from upstream to downstream manholes are calculated. Based on these three scores, sewers are rated as grade 1, 2, or 3, where grade 3 represents the worst structural condition.

By considering the condition of the entire length of a sewer line from upstream to downstream manholes, sewer lines of different total lengths but similar scores are not equally rated. Consequently, shorter lines with more serious defects will not be rated below a longer line with less serious defects.

National Association of Sewer Service Company (NASSCO) with the collaboration of WRc developed the pipeline assessment codes in the United States. Table 3.4 describes the various structural condition distress terms given by NASSCO. Some defects encountered in sewers are listed in Fig. 3.7.

TABLE 3.4 Sewer Pipe Structural Condition Evaluation (NASSCO, 1996)

Pipe condition	Description
Collapsed pipe	Complete loss of structural integrity of the pipe as a result of fracturing and collapse of the pipe walls. Most of the cross-section area is lost to flow
Structural cracking with deflection	Pipe wall displacement plus cracks described by:
Longitudinal	Defect runs approximately along the axis of the sewer
Circumferential	Defect runs approximately at right angles to the axis of the sewer
Multiple	Combination of both longitudinal and circumferential defects
Slab-out	A large hole in the sewer wall with pieces missing
Sag	The pipeline invert drops below the downstream invert
Structural cracking without deflection	Sewer wall cracked longitudinally, circumferentially, or multiply, but not displaced
Cracked joints	The spigot and/or bell of a pipe is cracked or broken
Open joints	Adjacent pipes are longitudinally displaced at the joint
Holes	A piece of a pipe wall or joint is missing
Root intrusion	Tree or plant roots have entered the sewer through an opening in the pipe wall or joint
Protruding joint material	Joint sealing material or gasket is displaced into the sewer from its original location
Corrosion	When the cementitious pipe material shows evidence of deterioration illustrated by the following stages:
Condition 1	The pipe wall surface shows irregular smoothness, i.e., wall aggregate is exposed
Condition 2	The reinforcing steel is exposed
Condition 3	The reinforcing steel is gone and/or the pipe wall is no longer intact revealing the surrounding soil
Pulled joint	Adjacent pipe joints are deflected beyond allowable tolerances so that the joint is open
Protruding lateral	A service outlet or pipe section that protrudes or extends into the sewer varying in magnitude
Vertical displacement	The spigot of the pipe has dropped below the normal joint closure
Depth of cover	The amount of soil covering the top of the pipe

Water Environment Federation (WEF). American Society of Civil Engineers (ASCE). WEF-ASCE (1994) suggests assigning an importance factor to each condition evaluation criteria for the structural condition of brick and concrete or clay sewers. The structural condition of brick sewers involves the following aspects: sags, vertical deflection and cracks, missing bricks, lateral deflections, root intrusion, missing mortar, loose bricks, protruding lateral, soft mortar, and depth of cover. Concrete and clay sewer structural condition evaluation criteria include: collapsed pipe, structural cracking with deflection (longitudinal, circumferential, or both), slab-out sag, structural cracking without deflection, cracked joints, open joints, holes, root intrusion, protruding joint material, corrosion, pulled joint, protruding lateral, vertical displacement, and depth of cover.

The sewer degradation is broadly classified into five degradation sequences. All of the sequences started with an intact pipe and progressively degraded starting with the distinct

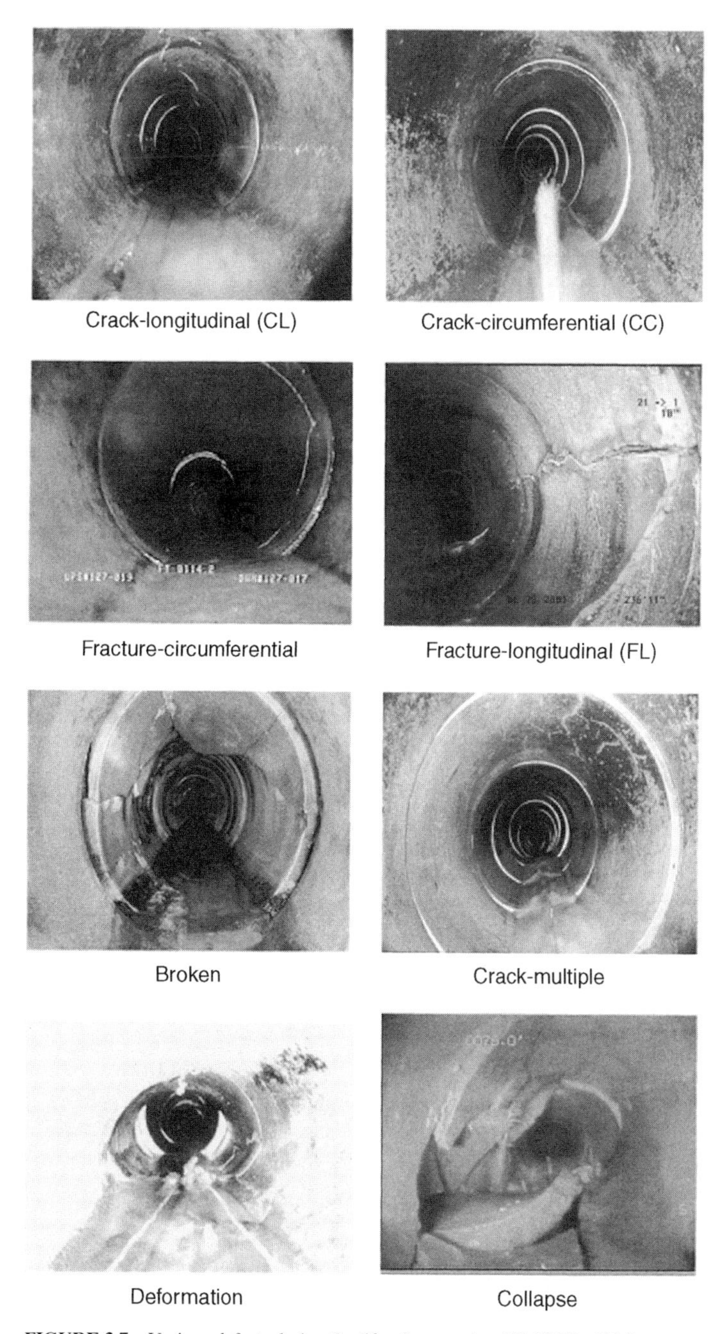

FIGURE 3.7 Various defects during the life of sewer pipe (NASSCO, 1996).

TABLE 3.5 Structural Distress Conditions Included in the Evaluation of Sewer Segments (WEF, 1994)

Structural condition	Description
Intact	Best possible sewer condition
Crack	Separation of pipe materials that run longitudinally or circumferentially along the sewer pipe
Open joint	Adjacent pipes are longitudinally displaced at the joints
Displaced joint	The pipe is not concentric with the adjacent pipe
Corrosion	The cementitious pipe material that shows evidence of deterioration from chemical action. The pipe wall surface shows irregular smoothness and aggregate on the cementitious material in the pipe is exposed
Deformation	Original cross section of the sewer is altered
Collapse	There is complete loss of structural integrity of the pipe. Most of the cross-sectional area is lost

sequences of (1) cracks, (2) open joints, (3) displaced joints, (4) corrosion, and (5) deformation, and ended in collapse. In each degradation sequence, there are various severity levels of each distress before it reaches the collapse from the initial intact condition. Table 3.5 contains a brief description of the degradation sequences.

Depending on the extent of the condition throughout a given sewer reach, a *minor*, *moderate*, or *severe*, a multiplier factor, such as 1, 2, or 3 is used, respectively. The overall numerical structural condition is then determined by calculating the total score. Based on how likely it is that the sewer will collapse, the internal condition-rating factor for overall structural condition can be determined. Sewers in rating 5 are in the most serious condition. This rating can be adjusted based on external factors such as soil types, surcharge, water table and fluctuation, and traffic condition.

3.4.3 Smoke Testing*

Smoke testing has been used for over 40 years for I&I studies and is an effective method for finding sources of storm water inflow. It can usually be carried out in dry period for sewer diameters 18 in or less. Other test methods that may be used for performing a complete evaluation of a sewer system are dye testing, manhole inspection, flow monitoring, and TV inspection. The principal sewer problems that will be identified by smoke testing include:

- Broken sewer pipes (provided there is a passageway for smoke to reach the surface and the breaks are above the groundwater table)
- Leaking manholes
- Illegal connections to sanitary sewers including building roof drains, sump pumps, and yard drains
- Cross-connections between storm and sanitary sewers
- Abandoned lines that were not correctly plugged

Figure 3.8 illustrates smoke testing of sewer lines.

Smoke Test Blowers. The objective of a smoke blower is to provide a high volume of air so that smoke will quickly arrive at the farthest points of the test section. The pressure

*This section was excerpted from USA BlueBook, available at www.usabluebook.com

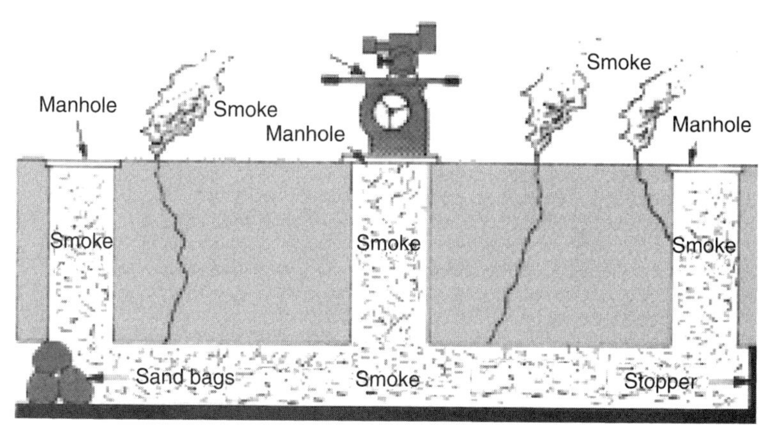

FIGURE 3.8 Smoke testing of sewers. (*USA BlueBook 2004.*)

developed by the blower must be high enough to force the smoke out through cracks, but not so high that it may force water out of drain traps in connected buildings. There are two basic types of blowers commonly used for smoke testing. *Squirrel cage* blowers have the disadvantage of weighing about 100 lb, but this is not a serious problem if you can back a truck up to each manhole to unload and install it. The advantage of this type of blower is that it can generally provide a higher static pressure in relation to cubic feet per minute output. In addition, if it should be found necessary, switching the size of the pulleys on the motor and the blower can change the blower output. *Direct-drive propeller* blowers are considerably more compact and their light weight makes them easier to carry and install. One disadvantage is that the output cannot be varied except to reduce the engine throttle.

Types of Smoke. There are two types of *smoke* available for smoke testing sewers. Classic smoke bombs (also called smoke candles) have been used since smoke testing first became popular. The smoke is generated from a chemical reaction and is very high in atmospheric moisture. It is very white and visible at relatively low concentrations. The smoke is free of oil and colored particles, which could leave residue and stains on clothing, furniture, or plants. The smoke bombs are either placed next to the fresh air intake of the blower or suspended beneath the blower in the manhole and the fuse is ignited with a match. Some blower models have a device for suspending the bombs under the blower. Many smoke test crews find it most convenient to lower the ignited candles into the manhole in a metal bucket. Bombs are available in 1 min and 2 to 3 min sizes. The 2 to 3 min size is also available with a double fuse so that two or more can be connected together for a longer burn. The size of the smoke bomb used depends on the length of the test section and the number of buildings connected to the section. Smoke candles have a limited shelf life. Once opened, a package of bombs should be kept in a dry location and used promptly. Bombs exposed to moisture or humidity may be difficult, if not impossible to ignite.

The other method of producing smoke is the *liquid smoke* system. This method involves injecting a petroleum-based product into the hot exhaust of the blower where it is heated and vaporized within the muffler or a heating chamber. One gallon of smoke fluid is generally considered to produce more smoke than 12 smoke candles. The primary advantage of using liquid smoke is that the production of smoke can be easily varied to a short or long *burn* as required for each test section. It is important when using liquid smoke that the muffler or heating chamber is as hot as possible before the liquid is turned on, otherwise the smoke will be rather thin. The production of smoke may also diminish after a minute or two if the liquid is being applied too fast and the heating chamber is cooling off. In general, smoke from a liquid system is considered to be not quite as consistently visible and does not travel as far as the smoke from candles. But if necessary, this can be compensated for by using smaller test sections. Many operators are finding these problems not as important as the convenience of using liquid smoke.

Smoke Testing Procedures

1. Advance notice is given to residents, fire, and police departments.
2. The manhole section or sections to be tested are isolated. One method is to use sandbags filled with medium-weight material (see Fig. 3.8), which can be put into place, by the use of a rope from the street surface. The incoming pipe of the upstream manhole and then the outgoing pipe of the downstream manhole are blocked. Care must be taken, when blocking the outgoing pipe of the downstream manhole, to prevent sewage flow from backing up and pushing the sandbag into the outgoing pipe.
3. The smoke blower is set up at an open manhole. Best results and control are achieved when only one manhole section is tested at a time. The blower is usually located at the upstream manhole because the outgoing pipe may be set up at a center manhole and both the upstream and downstream sections can be tested simultaneously. Testing more than two manhole sections at a time is less effective and gives the crew too much territory to observe during the relatively short life of the smoke bombs (3 or 5 min).
4. The smoke blower is started and given a brief warm-up to the side of the open manhole. A smoke bomb is ignited, placed in a bucket, and lowered into the manhole.
5. The blower is then positioned on the open manhole and set at full throttle. In less than a minute, smoke will surface from the roof vents of buildings and any directly connected points of inflow.
6. A smoke testing crew usually comprises three persons. One individual operates and maintains the blower whereas the other two walk the test area to locate and document the inflow sources indicated by the emerging smoke.
7. The crew documents significant and identifiable points of inflow. Documentation should be referenced to each manhole section tested. Sketches, field notes, photographs, and video recorders could be used to record test results.
8. It is recommended that those firms bidding on this type of work should have a documented history of 5 years of successful smoke testing. References and dates are to be submitted with bids.
9. All bidders will be licensed contractors and perform a minimum of 60 percent of the contract.

3.5 PIPELINE INSPECTION: PRESSURE PIPE

The rate of deterioration of a pressure pipe such as a water system is less a function of its age but a consequence of the cumulative effect of the external influences acting upon it. One of the challenges in maintaining these systems is that its problems are often not readily visible—and the consequences of its failure can often be severe. Water main breaks can lead to hazardous local conditions, public safety considerations, significant property damage, and extensive service interruption. In addition to costs for repair, replacement, or both, there may be revenue loss, water rationing, and investigative and legal costs.

Complicating the situation, water transmission pipelines face type-specific problems. The solution lies in understanding these problems—and understanding the condition assessment solutions that can provide an operator with the information needed to mitigate the inherent risks of operating water transmission systems.

Historically, condition assessment methodologies were somewhat primitive—usually comprising *walking the pipe* and performing a visual inspection. Moreover, broad-based infrastructure replacement decision factors, based on age, pipe size, pipe material, soil conditions, linings and so on, and provided information that was too general to result in an effective use of limited capital resources.

Although there is still no known solution for many of the challenges facing transmission pipeline operators today, an array of condition assessment techniques is available to help operators manage the risk. Several condition assessment techniques that can pinpoint individual pipes that are suffering from distress are commercially available. By using the information that these techniques provide, an operator can enact a selective and cost-effective rehabilitation program that targets individual pipes, thereby extending the safe and economic life of the pipeline. A brief overview of some of these condition assessment techniques is detailed in the following sections.

3.5.1 Remote Field Eddy Current/Transformer Coupling (RFEC/TC)

History. The RFEC/TC technique was developed by the Applied Magnetics Group in the department of Physics at Queen's University in Kingston, Ontario, Canada. The technique uses a combination of effects but originated from extensive research and development on the RFEC technique. The RFEC technique uses low frequency alternating current (AC) and through-wall transmission to inspect the electrically conducting tubes or pipes (such as ductile iron, cast iron, or steel) from the inside. It is able to detect not only the defects on the inside or outside of the pipe with approximately equal sensitivity, but also changes external to the pipe or, for concrete cylinder pipe (CCP), changes external to the cylinder. Conventional RFEC condition assessment techniques are limited to the surface nearest to the probe.

Theory. RFEC/TC is an internal manned inspection technique used to inspect large diameter (greater than 36-in diameter) prestressed concrete cylinder pipe (PCCP). A simple analogy would be to think of the RFEC/TC system as a radio receiver. The prestressing wire behaves like the radio antenna. With the antenna fully extended, the radio receives a clear signal from the transmitting station. However, if the antenna length is reduced (i.e., the continuity of the prestressing wire is broken), the clarity of the signal received is also reduced.

In more detail, the RFEC/TC technique comprises two distinct components. RFEC instruments have a transmitter and a receiver that are both induction coils. In a pipeline, as shown in Fig. 3.9, the transmitter is driven by an AC signal. This electric current in the transmitter generates two distinct magnetic fields. The first field is direct and inside the pipe. The circumferential eddy currents induced in the conducting pipe wall attenuates this field rapidly. A second, indirect field diffuses radially outward through the pipe wall and spreads along the pipe with little attenuation. This field rediffuses back through the pipe wall and is the dominant field inside the pipe where the detector, sufficiently spaced from the transmitter, is located. The returning field induces a voltage in the detector. Any disturbance in this indirect path causes a change in the magnitude and phase of the received signal. Figure 3.9 presents a pair of internal coils that are used in remote field-testing.

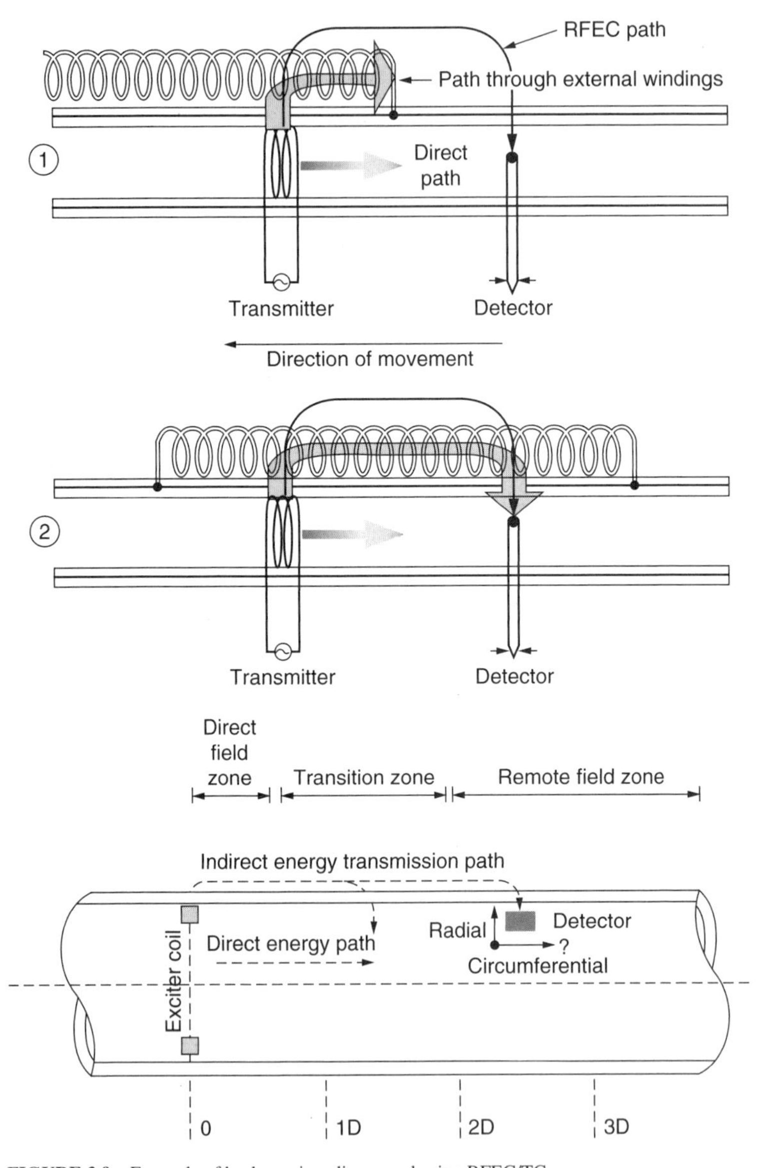

FIGURE 3.9 Example of broken wires discovered using RFEC/TC.

The signal from the transmitter is attenuated and slowed as it passes through the pipe wall. Wall thinning will allow the signal to be picked up by the detector sooner and stronger.

In Fig. 3.9, the transformer coupling, shown at the bottom, can be understood as an interaction between the indirect transmission path and the external winding in the concrete pressure pipe. The external winding is excited and will behave as a solenoid coupling flux between the exciter and detector. Thus, the signal received in a RFEC/TC system has two

FIGURE 3.10 RFEC/TC condition assessment inspection.

components. The remote field component will show a phase shift and attenuation consistent with the double, through-wall transmission of the field through the steel cylinder. The larger transformer component will dominate the response; however it will be reduced in the presence of broken wires.

During a RFEC/TC condition assessment inspection, the detected signal is recorded along the length of the pipeline. After the inspection, this signal is interpreted with respect to amplitude and phase. The amplitude represents the strength of the transmitted signal; the phase represents how long it takes for the signal to arrive at the detector. When both transmitter and detector are underneath the same set of prestressing windings, transformer coupling results in a very large amplification of the detector signal: disrupting the coupling by breaking a wire results in a decrease in the detector signal (Fig. 3.10).

During analysis, amplitude and phase logs are examined to identify the start and end positions for each pipe. Fixed references (i.e., outlets, steel sections, elbows, and manholes) within the data are identified and correlated with the information provided by the client. Each pipe is then examined for distress indications. A report is compiled wherein pipes are rank ordered according to level of distress, and the axial position and number of breaks are estimated. The report provides pipeline operators with the location, distribution, and number of wire breaks anywhere along the length or around the circumference of their PCCP pipe.

Applications. RFEC/TC establishes the baseline condition of each pipe in a PCCP pipeline. This technique can

- Detect broken prestressing wires in any size PCCP
- Find and quantify the number of breaks along the length and anywhere around the circumference of PCCP pipe
- Quantify wire breaks in embedded cylinder pipe (ECP), lined cylinder pipe (LCP), and noncylinder pipe (NCP)
- Spot wire breaks in pipes with or without shorting straps and in pipes with or without bonding straps

Replacing the pipeline is not the only option available to a pipeline operator, should a RFEC/TC inspection locate pipes suffering from distress. In fact, by using RFEC/TC to pinpoint the location of individual distressed pipes within a water distribution system, the cost of complete replacement can be avoided and the pipeline operator can make an

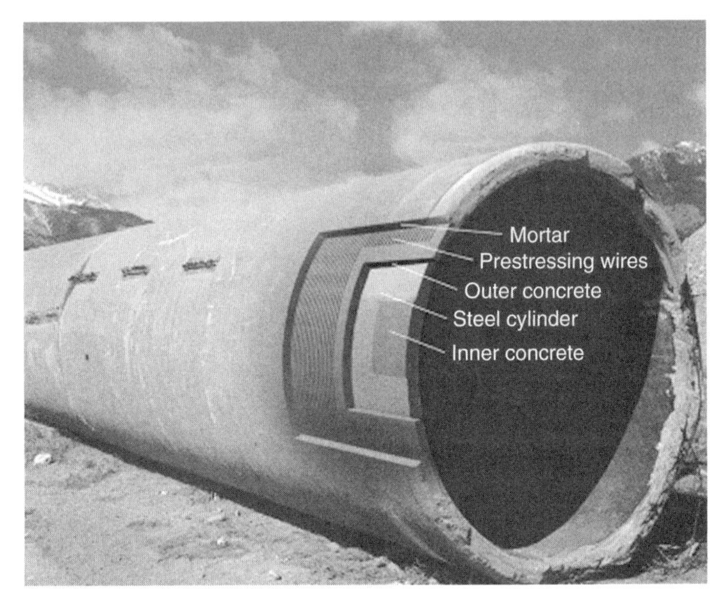

FIGURE 3.11 Schematic of a typical PCCP pipe.

informed decision to *selectively* renew, replace, or monitor *individual* pipes. Figure 3.11 shows a schematic of a typical PCCP pipe.

Should the RFEC/TC inspection locate pipes suffering from distress, there are three risk management options available:

Repair: There are a variety of repair methods available for extending the life of a PCCP pipe that is suffering from moderate amounts of damage.

Replace: Remove the distressed pipe or pipes from the pipeline and replace with new pipes.

Monitor: Once the baseline condition of the pipeline has been established, the operator may wish to monitor selected reaches of the pipe. Acoustic monitoring detects areas of the pipeline that are suffering from active distress. A monitoring program helps operators determine which specific pipes are degrading most rapidly, influencing decisions such as which pipes are to be rehabilitated or replaced.

3.5.2 Upper Homestake Pipeline Case Study

The Upper Homestake Pipeline, in Colorado, is a 66-in diameter noncylinder prestressed concrete pipe (PCP), manufactured by the International Pipe and Ceramics Corporation (Interpace) and jointly owned and operated by the cities of Aurora and Colorado Springs.

Black & Veatch Company was contracted by the Homestake Project to construct improvements to the existing pump station, which would increase the operating pressure in the pipeline. As part of the evaluation of the potential impacts to the pipeline, Black & Veatch

FIGURE 3.12 X-ray film placed around the pipe at the suspected distressed region.

signed a contract with the Pressure Pipe Inspection Company (PPIC) to perform a condition assessment of a portion of the pipeline.

Prior to the inspection, a 2-day calibration procedure was conducted. The calibration testing allowed PPIC to develop the calibration curves used to estimate the number of broken wires in each pipe section. The RFEC/TC inspection was then performed on 0.95 mi or 253 pipes. Analysis of the data obtained during this inspection determined that 10 pipes displayed evidence of wire breaks.

Black & Veatch signed a contract with Intermountain Testing to perform an x-ray examination of two of these pipes to verify the accuracy of the technology. One of the regions on a selected pipe was x-rayed and wire breaks were confirmed within 6 in of the location reported. As shown in Fig. 3.12 a pipe was excavated and x-rayed to confirm RFEC/TC inspection results. This photo shows the x-ray film placed around the pipe at the suspected distressed region.

The second pipe was then x-rayed. PPIC estimated the pipe to have a total of 15 to 20 wire breaks (one region of 10 breaks and one region of 5 to 10 breaks). The x-ray confirmed that there were 18 wire breaks in these two regions.

The results were consistent with the results obtained during the x-ray inspection and illustrate the accuracy of PPIC's RFEC/TC system and the importance of verification studies. PPIC continues to work with Black & Veatch and Homestake to evaluate other portions of the line.

3.5.3 Tarrant Regional Water District (TRWD) Case Study

The Tarrant Regional Water District in Ft. Worth, Texas, is a raw water supplier that serves 10 counties in north central Texas. Most of the 1.5 million people served by the district are in Tarrant County, which includes Fort Worth and surrounding communities. In 2001, the district delivered over 366,000 acre ft of water to its customers.

Both the Richland and Cedar Creek transmission pipelines are composed of PCCP. Beginning in 1981, 9 years after installation, the district began to suffer an increasing number of failures. The district explored corrosion as a possible cause by measuring cell-to-cell and pipe-to-soil potentials and found that its pipelines ran through extremely corrosive soil. As a result, TRWD enacted a successful cathodic protection program designed to minimize future corrosion. In addition, TRWD conducted a condition assessment of its

FIGURE 3.13 Corrosion along the shorting strap.

pipeline using PPIC's RFEC/TC pipeline evaluation system to determine the integrity of individual pipes.

Through the inspections, the district has been able to identify and replace a number of distressed segments. This has improved the integrity of the pipelines, and TRWD was able to happily reverse its earlier decision that it would have to replace them. The damage identified by the RFEC/TC system demonstrates that the Richland pipeline has fewer damaged segments than the Cedar Creek line, possibly because of the shorter time period it was in place before adding cathodic protection. Based on the number of pipe segments found with damage great enough to require replacement, the district expects that all major replacements will be completed in a few years—at a fraction of the cost of replacing the entire pipelines. Figure 3.13 illustrates corrosion along the shorting strap.

3.6 ACOUSTIC EMISSION TESTING (AET)

3.6.1 History

Acoustic emission testing (AET) is based on the detection of acoustic emissions. It is akin to the sonar technology used by the U.S. Navy for decades in antisubmarine warfare. In antisubmarine warfare, and in pipeline condition assessment, acoustic emissions of interest are detected by a series of acoustic sensors and screened for the known acoustic signatures that are emitted by an event of interest.

In 1991, the U.S. Bureau of Reclamation began experimentation to determine whether passive acoustic emission detection technology could be adapted to concrete pressure pipelines. These tests sought to take advantage of the fact that, during its manufacture, PCCP is reinforced by spirally wrapping high strength wire around a concrete cylinder. If the pipe is in a state of distress, the prestressing wire will be exposed and may corrode. When this occurs, the wire can break in a relatively brittle fashion, with an instant release of the tensile force in that strand of wire.

This testing confirmed that the breakage of a strand of prestressing wire, and the subsequent slippage of that wire, cause acoustic transient signals, and that these signals are propagated through the core of the pipe into the water column within the pipe.

3.6.2 Theory

AET is a passive on-line acoustic system that detects the frequency and number of distress-related acoustic events that occur over a defined period of time.

The origin of these acoustic events is determined through the precise identification of the arrival times of these signals at a series of acoustic sensors. Sound travels through water at constant speed of 4850 ft/s. The time it takes for sound to arrive at an acoustic sensor is directly related to the distance it travels. Therefore, the physical location of a wire break can be determined by comparing the arrival times of that event at both acoustic sensors by using the equation shown in Fig. 3.14. Acoustic sensors pick up these acoustic signals, which are then screened against a number of criteria (Fig. 3.14). The data are then analyzed to determine location and source.

Results are typically presented in visual form as both event logs and a pipeline overview. Event logs report each event in chronological order. The log indicates the precise time the event occurred and where it originated. The pipeline overview plots each

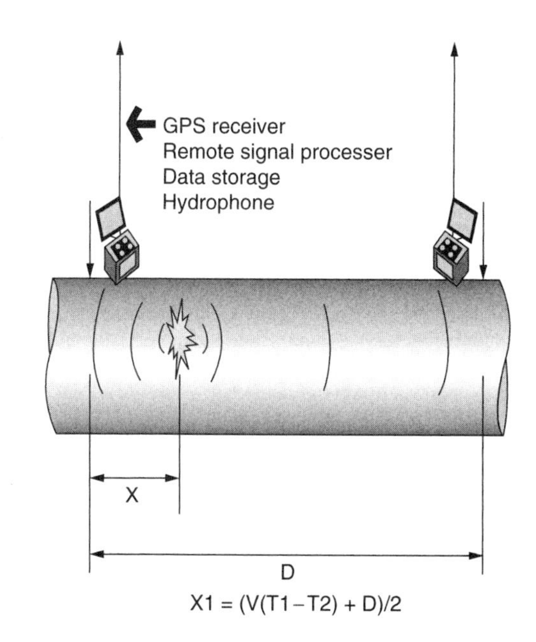

Where:

X1 = Distance of acoustic event from Acoustic Sensor #1
V = Velocity of sound in water
T1 = Time of arrival of acoustic signal at Acoustic Sensor #1
T2 = Time of arrival of acoustic signal at Acoustic Sensor #2
D = Distance between the two Acoustic Sensors

FIGURE 3.14 Detection of the origin of acoustic events.

event on a diagram of the pipe. The pipe is displayed on the X-axis across the top. Points of reference such as manholes are identified. Each day of monitoring is presented to the left on the Y-axis. Wire breaks are shown as circles. Vertical lines point to its origin on the pipe.

3.6.3 Applications

Acoustic monitoring is primarily used as a PCCP condition assessment tool. It is used preferentially when an internal pipeline inspection is just not possible (i.e., the pipe is too small for manned entry or cannot be dewatered). Advanced applications of the technology include:

Monitoring of PCCP during a pressure test. If an operator is considering raising the pressure in a PCCP pipeline, acoustic sensors can be used to determine if raising the pressure caused additional reinforcing wire breakage or slippage.

Monitoring a pipeline (of any construction type) for accidental construction damage. AET detects and records third-party damage events such as: inadvertent contact with the pipe cracks in the core caused by excessive external loading, or undermining of pipe bedding. This protects faultless parties from subsequent liability.

Monitoring of a pipeline after its baseline condition has been established. A monitoring program helps operators determine which specific pipes are degrading most rapidly and influencing decisions such as those pipes that are to be rehabilitated or replaced.

3.6.4 CMWD Case Study

Calleguas Municipal Water District (CMWD) in Thousand Oaks, California, annually conveys an excess of 120,000 acre-ft to local water agencies in southern Ventura County through a 130-mi system of pipelines. In 1999, following a failure in one of their major transmission mains, CMWD initiated an infrastructure renewal program with the twin goals of determining which PCCP pipelines were at the highest risk of failure and ensuring the integrity of any *at-risk* pipelines through system operation modifications and immediate capital projects. PPIC was commissioned to evaluate the baseline condition of the pipelines.

PPIC's RFEC/TC system conclusively identified that most of the pipes within CMWD's system were in pristine condition, though several individual pipes presented wire break evidence of distress. A few pipes presented inconclusive results. PPIC recommended an acoustic survey for the specific purpose of examining these pipe sections. Although the initial survey had not shown conclusive evidence of wire breaks, work in other pipelines had suggested potential reasons for concern. Figure 3.15 illustrates that PPIC's pipe rider tool is capable of inspecting a pipe as small as 36 in diameter.

As a result, and following the previous work, in May 2002, CMWD engaged PPIC to monitor a segment of their Conduit North Branch (CCNB) using PPIC's AH-4 autonomous acoustic sensors (Fig. 3.16). A 3965 ft segment of the CCNB was included in this test, which ran from May 6 to June 26.

The AH-4 system was installed in Calleguas with a solar panel to allow for extended periods of monitoring. As shown, each unit is typically stored in a secured lockbox beside the manhole. No distress-related acoustic events were picked up from the eight anomalous pipes. Based on this information, as well as other observations regarding the integral condition on these pipe sections, PPIC concluded that the eight anomalous pipes identified during the RFEC/TC inspection were not in a state of distress.

The information from this, and other condition assessment programs allowed CMWD to identify sections of its PCCP pipeline that were at the highest risk of failure. CMWD also prioritized capital expenditures by enacting a selective repair and replacement program.

FIGURE 3.15 PPIC's pipe rider tool is capable of inspecting pipes as small as 36 in diameter.

The reestablishment of the integrity of the pipelines, in part based on PPIC's condition assessments, have extended the safe and economic life of CMWD's pipeline.

3.6.5 Basin Electric Power Cooperative: Antelope Valley Station (AVS)

Antelope Valley Station (AVS), Beulah, North Dakota (Fig. 3.17) is a lignite-fired electric generating station located 7 mi northwest of Beulah, North Dakota. It has two units each rated at 450,000 KW. The raw water pipeline (RWPL) extends approximately 8 mi from Lake Sakakawea to the AVS. The pipeline was constructed in or around 1980. Since then, the pipeline has experienced failures and was earmarked for replacement. When AVS heard

FIGURE 3.16 AH-4 system.

FIGURE 3.17 Antelope valley station.

that PPIC could evaluate the condition of individual pipes within the RWPL, PPIC was contracted to assess the current condition of the pipeline and make recommendations on further strategies to continue the supply of raw water to AVS.

PPIC performed an AET inspection on the RWPL during the period of July 21, 2003, to August 29, 2003. A total of 30 locations, comprising combination valves (CVs) and air release valves (ARVs), on the RWPL were used to deploy the AH-4 acoustic sensors. PPIC's acoustic sensors can be installed through valves as small as 1 in diameter (Fig. 3.18). Ten AET monitoring systems were initially installed at sites 1 to 10 along the pipeline, and data were collected for a minimum of 100 h at these locations. The systems were moved as soon as the required number of monitoring hours was obtained. This procedure of moving and reinstalling the ten systems continued until all the 30 locations were tested. After testing all the locations for at least 100 h, specific sites were targeted for additional testing based on initial results.

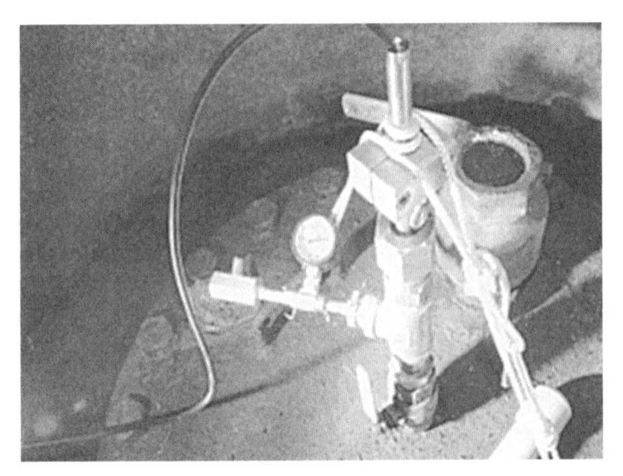

FIGURE 3.18 PPIC's acoustic sensors installed through valves as small as 1 in diameter.

The results of the AET test indicated that the RWPL was structurally sound throughout most of its length. However, as a number of localized events were recorded, testing indicated that the line was experiencing a significant degree of distress in several specific areas. The acoustic data, in conjunction with other related information sets, were presented to Basin Electric to support them in making preventive maintenance and renewal decisions.

3.7 TRANSMISSION MAIN LEAK LOCATION

3.7.1 History

Any leak in a pressurized water pipe will generate noise; the level and frequency of the noise will depend on a number of factors including the shape and size of the leak, the pipe material, and the pressure inside the pipe. Typically, leaks in water pipes will generate clearly audible sound.

The most widely used leak detection method is the leak noise correlator. This uses a pair of acoustic sensors or similar devices situated at a convenient location on the pipe some distance away from and on either side of the leak point. The leak noise correlator is a well-proven device for locating leaks on smaller pipes. However it becomes less reliable when used on larger-diameter pipes (12 in and larger) and on nonmetallic pipes (MDPE, PVC, asbestos, cement, or concrete); the leak noise is attenuated very rapidly as it travels away from the source in these types of pipes. At any significant distance from the leak point the leak noise signal is swamped by other background noise.

When the limitations of the correlator for larger pipes were recognized alternative methods of leak location in trunk mains were investigated. Based on these studies a group, led by Anthony Bond of the United Kingdom's WRc, decided in the early 1990s that the most reliable method of detecting leaks, and pinpointing their position accurately, would be to pass an acoustic sensor along the inside of the pipe and detect the point where the leak noise signal was greatest.

3.7.2 Theory

The Sahara® leak location system is an in-line condition assessment acoustic system that pinpoints the location and estimates the magnitude of leaks in transmission mains of all construction types.

The system is deployed through an opening of 2 in or greater with the mains still in service. An acoustic signal processor receives data from the acoustic sensor via an umbilical cable and feeds a display, shown at right, giving various characteristics of the noise as well as an audio signal. As the acoustic sensor passes the leak there is a peak in the sound level detected by it; processing of this signal in real time allows the precise leak position to be determined during the survey. Once a leak has been detected the locator head can be stopped at the leak point. The head can then be detected within the mains from the surface using the locator unit. The road surface can then be marked accurately for subsequent excavation and repair.

The system identifies leaks as small as 0.25 gal/h while pinpointing their location to within less than 3 ft. It can pinpoint multiple leaks even when many are to be found within one section and is unaffected by surface or pipeline noise. Depending on the pipe configuration, lengths up to 6500 ft can be surveyed in one pass. Surveys can be carried out in mains with a diameter of greater than 12 in; there is no limit on the maximum diameter. Figure 3.19 illustrates the unmistakable signature of a leak.

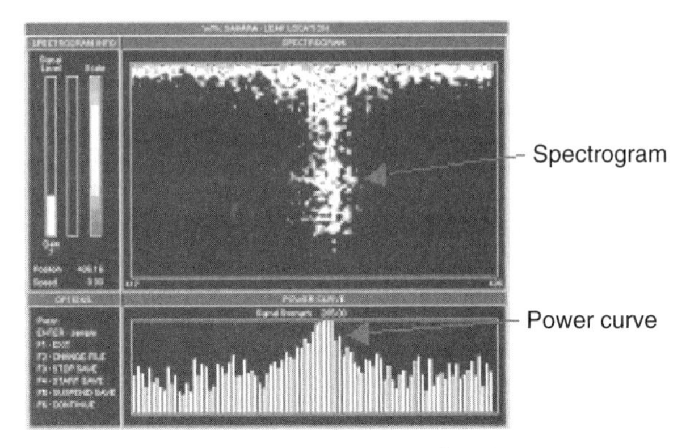

FIGURE 3.19 The unmistakable signature of a leak.

3.7.3 Applications

The Sahara® leak location system is primarily used as a condition assessment tool that pinpoints leaks in transmission mains of all construction types. The system is capable of detecting:

- Leaking joints in PCCP
- Weld leaks in steel
- Pinhole leaks in ductile iron, cast iron, PVC C905, asbestos, cement, or clay pipes

Advanced applications of the technique include:

- *Asset validation*: If the integrity of a given pipeline is critical, that is, it is found on an elevated slope where a failure could induce a mudslide, leak location can be used to validate the integrity of the pipe.
- *Legislative compliance*: Some municipalities mandate that water loss be restricted to a certain percentage, or absolute amount. Leak location can be used as a proof that the line is watertight.

3.7.4 Lydec Drinking Water Pipeline Repair Case Study

Since 1997, Lydec in Casablanca, Morocco, has delegated the management of the distribution of electricity, drinking water, and water purification services to Casablanca and Mohammédia.

As a result of expensive water import costs, Lydec wished to adopt a robust method for locating trunk main leaks. As conventional methods had proved ineffective, Lydec turned to the Sahara system. An initial 9-mi trial length was identified, with the surveys observed by Ondeo's Research Centre (NLTRC).

During the project, Sahara was deployed through existing on-line air valves (2.3 in to 3.9 in gate valves) at predetermined locations to allow for full traversal of the identified lengths. On an average, 2500 ft of pipe was surveyed from each insertion point as the mains were in a semiurban environment and were subject to a number of bends.

During the trial, 17 leaks were successfully located with pinpoint accuracy. The largest single leak identified exceeded 793 U.S. gal/day; no passive acoustic or visual cues were

FIGURE 3.20 Leaks identification.

evident. Subsequently, the system was retained and has afforded Lydec a significant reduction in imported water. Suez Lyonnaise Des Eaux now accepts clear surveys as confirmation of a leak-free line.

Leaks were successfully identified using the Sahara leak detection system (Fig. 3.20). "By using Sahara, we have been able to reduce our entire water importation by 3.5%. The savings afforded paid for the survey works in just eight months," El Hassane Benahmed, water resources manager, Lydec.

3.7.5 Thames Water, Puerto Rico, Case Study

Thames Water in Puerto Rico is responsible for the design, construction, and operation for water collection, treatment, and delivery of drinking water to the city of San Juan. It serves 1.6 million customers.

During the final construction phase of the Puerto Rico North Coast Super Aqueduct Project, prior to handover, leakage within the line was apparent. Static tests confirmed a small number of leaks, which though minor in terms of volume, were significant in that they may have led to commissioning delays. Conventional methods of leak detection were unable to detect all the leaks on-line and, with the commissioning deadline approaching, the client requested a Sahara survey (Fig. 3.21).

During the project, existing 8-in butterfly valves were adapted using a change plate to allow insertion of the equipment. In addition, three new saddle connections were fitted to allow access within sections where air valves were situated in excess of 6000 ft apart.

All leaks were successfully detected and located within the sections surveyed. The smallest leak located was just 14 gal/d. In each case, subsequent excavation was simplified by marking the leak location on the surface, ensuring minimal repair times. Following the repairs, the pipeline was proven to be leak-tight, passing all commissioning procedures and was successfully handed over.

FIGURE 3.21 Sahara survey for leak detection.

3.8 PIPELINE CLEANING METHODS*

Cleaning of sewers is prioritized based on the age of the pipe and the frequency of the problems it encounters. Cleaning is an important part of pipe maintenance. Sewer systems are inspected every 2 years for older pipes, and every 3 to 4 years for relatively new pipes. To maintain its proper function, a sewer system needs a cleaning schedule. Scheduled cleaning is proactive in that cleaning is done on a preventive basis to remove material prior to a stoppage occurring. Preventive cleaning activities can be supplemented by additional cleaning on an as-needed basis in cases where predictive information such as previous history, inspection data, pipe age and material, slope, or other information indicate a need for more frequent cleaning. Scheduled cleaning is usually coordinated with planned CCTV because televising requires a clean pipe for access and visually provides a much better picture of conditions.

Unscheduled cleaning is usually the result of a reported stoppage and is therefore reactive. When reactive maintenance is the primary form of maintenance (i.e., waiting until a failure occurs before performing maintenance), it will always result in poor system performance, especially as the system ages. Normally, this type of cleaning is done on an emergency basis to clear a stoppage, restore pipe capacity to full flow, and relieve a surcharging situation in the sewer that has caused a backup into homes, or an overflow, or both.

General requirements before cleaning include:

- Contractor must have means of trapping collected debris arising from cleaning.
- The disposal location for debris must be determined in advance and proper permits obtained.
- The contract must be awarded to a contractor with trained and certified crew with documented experience.
- It is very important to make sure that all safety requirements are met.
- The effectiveness of the cleaning operation is dependent on how well the equipment is maintained, as directed by the manufacturer.

---

* Portions of this section were excerpted from Collection Systems O&M Fact Sheet: Sewer Cleaning and Inspection, EPA (1999) and Chi-Yuan F. (2004)

3.8.1 Conventional Cleaning Methods

Table 3.6 summarizes some of the most commonly used methods used to clean sewer systems. These methods are used to clear blockages once they have formed, but also serve as preventative maintenance tools to reduce future problems. With the exception of flushing these methods are generally used in a *reactive* mode to prevent or clear up hydraulic restrictions. As a control concept, flushing of sewers is viewed as a means to reduce hydraulic restriction problems as well as a pollution prevention approach.

When cleaning sewer lines, local communities need to be aware of EPA regulations on solid and hazardous waste as defined in 40 CFR 261. They may also need to comply with the state guidelines on testing and disposal of hazardous waste. Although all the methods discussed in this section have proven effective in maintaining sewer systems, the ideal method of reducing and controlling the materials found in sewer lines is education and pollution prevention. The public needs to be informed that common household substances such as grease and oil need to be disposed in the garbage in closed containers, and not into the sewer lines. This approach will not only minimize a homeowner's plumbing problems, but will also help keep the sewer lines clear.

Rodding. The equipment may either be a power rodding machine (may be truck or trailer-mounted) or hand rods. Power rodding involves applying a torque to a steel rod as it is pushed through the line, rotating the cleaning device attached to the lead end of the rod. This method can be used for routine preventive maintenance, such as breaking up of grease deposits, cutting roots, loosening debris, threading cable for bucket machines or TV inspection equipment, and for emergency removal of blockages.

The method is fairly efficient in lines up to 12 in. in diameter but is less useful in larger lines. The method is ineffective for removing sand and grit accumulations, but may loosen the material so that it can be flushed out of the sewer. The rod has a tendency to coil and bend when used in large-diameter sewers. An electrically powered power-rodding machine is also available and can be used in smaller lines.

Balling. This equipment comprises an assortment of various sizes of sewer balls to fit different diameters of sewers, a tag line, winch, cable, reels, a water source, and a dump pick-up. The method uses the pressure of a water head to create high-velocity water flow around the ball. It is very effective in removing heavy concentrations of sand, grit, rock, and grease from the sewers. Balls are available in sizes from 5 to 24 in. This method is not recommended for locations with basement fixtures and in steep-grade hilly areas because of possible flooding of connected buildings. It cannot be used effectively when sewers have bad offset joints or protruding service connections because the ball can become lodged in the pipe, becoming distorted, thereby preventing it from doing an effective cleaning job.

Flushing. This is an inefficient hydraulic method that can be used at the upstream location of a collection system where low or sluggish flow results in deposition of solids. It is now rarely used since the introduction of the high-pressure water jet cleaners. A fire hydrant is normally used for this procedure. However, if a fire hydrant is not accessible, the equipment may comprise a water tank and a fire hose. The method is effective in removing floatables and some sand and grit. It is useful in combination with mechanical operations such as rodding or bucket machine cleaning. The method is not very effective in removing heavy debris and grit. It is not recommended for use in locations with basement fixtures and steep-grade hilly areas because of possible flooding of connected buildings.

Nearly all water utilities practice flushing, either conventional or unidirectional; this method is the oldest and costs the least. This method is very effective when poor quality water is present in the pipe, but cannot remove scaling. Unidirectional flushing (Fig. 3.22)

TABLE 3.6 Common Sewer Cleaning Methods (EPA, 1999)

Technology	Uses and applications
	Mechanical
Rodding	• Uses an engine and a drive unit with continuous or sectional rods • As blades rotate they break up grease deposits, cut roots, and loosen debris • They also help thread the cables used for TV inspections and bucket machines • Most effective in lines up to 12 in. in diameter
Bucket machine	• Cylindrical device, closed on one end with two opposing hinged jaws • Jaws open and scrape off the material and deposit it in the bucket • Partially removes large deposits of silt, sand, gravel, and certain solid waste
	Hydraulic
Balling	• A threaded rubber cleaning ball that spins and scrubs the pipe interior as flow increases in the sewer line • Removes deposits of settled inorganic material and grease buildup • Most effective in sewers ranging in size from 5 to 24 in
Flushing	• Introduces a heavy flow of water into the line at a manhole • Removes floatables and some sand and grit • Most effective when used in combination with other mechanical operations, such as rodding or bucket machine cleaning
Jetting	• Directs high velocities of water against pipe walls • Removes debris and grease buildup, clears blockages, and cuts roots within small-diameter pipes • Efficient for routine cleaning of small-diameter, low-flow sewers
Scooter	• Round, rubber-rimmed, hinged metal shield that is mounted on a steel framework on small wheels. The shield works as a plug to build a head of water • Scours the inner walls of the pipelines • Effective in removing heavy debris and cleaning grease from line
Kites, bags, and polypigs	• Similar in function to the ball • Rigid rims on bag and kite induce a scouring action • Effective in moving accumulations of decayed debris and grease downstream
Silt traps	• Collect sediments at convenient locations • Must be emptied on a regular basis as part of the maintenance program
Grease traps and sand or oil interceptors	• The ultimate solution to grease buildup is to trap and remove it • These devices are required by some uniform building codes and/or sewer-use ordinances. Typically sand or oil interceptors are required for automotive business discharge • Need to be thoroughly cleaned to function properly • Cleaning frequency varies from twice a month to once every 6 months, depending on the amount of grease in the discharge • Need to educate restaurant and automobile businesses about the need to maintain these traps and impose enforcement measures
	Chemicals*
General	• Used to control roots, grease, odors (H_2S gas), concrete corrosion, rodents, and insects • Root control: longer lasting effects than power rodder (around 2 to 5 years)

(*Continued*)

TABLE 3.6 Common Sewer Cleaning Methods (EPA, 1999) (*Continued*)

Technology	Uses and applications
	Chemicals*
	• H_2S gas: some common chemicals used are chlorine (Cl_2), hydrogen peroxide (H_2O_2), pure oxygen (O_2), air, lime $Ca(OH_2)$, sodium hydroxide (NaOH), and iron salts • Grease and soap problems: some common chemicals used are bioacids, digester, enzymes, bacteria cultures, catalysts, caustics, hydroxides, and neutralizers

*Before using these chemicals review the Material Safety Data Sheets (MSDS) and consult the local authorities on the proper use of chemicals as per local ordinance and the proper disposal of the chemicals used in the operation. If assistance or guidance is needed regarding the application of certain chemicals, contact the U.S. EPA or state water pollution control agency.

can remove most biofilm and sedimentation. It is widely recognized that flushing is the most useful and cost-effective method of pipeline cleaning. Conventional flushing clears the line of rust and sediment by opening the fire hydrants, though this method does not produce the water pressure or velocity to remove stubborn debris. Unidirectional flushing also uses fire hydrants, but opens them in sequence and only after a one-way flow has been established. Unidirectional flushing uses considerably less water than conventional flushing and can be used to locate hydrants in need of repair.

If the pressure in the system is too low or the diameter of the pipeline is too large to effectively create enough water pressure to flush the system, air scouring may be used. Air scouring can remove soft scaling as well as biofilm and sedimentation (awwarf.org).

Jetting. This is a hydraulic method of cleaning sewers, which directs high-velocity streams of water against the pipe walls at various angles (Fig. 3.23). The equipment comprises a truck-mounted high-velocity water machine, maintenance hole hose guide, debris traps, and a dump pick up or debris trailer. The method is very effective in cleaning flat,

FIGURE 3.22 Unidirectional flushing. (*Courtesy of AH Environmental Consultants, Inc.*)

FIGURE 3.23 Jetting of sewers (Atkinson, 2000).

slow flowing sewers. It is very efficient in removing grease, sand, gravel, and debris deposits in small sewers. It is also effective in breaking up solids in maintenance holes, and in washing structures. For more stubborn, strongly cemented silts, multi-pass iterations are necessary. The effectiveness in removing debris, however, decreases as the size of the pipe increases.

Scooter. This is a hydraulic method of cleaning a sewer line. The equipment comprises a scooter assembly, water tank truck, dump pick-up truck, tag line, and a power winch. The scooter itself is a steel framework on small wheels with a rubber-rimmed, round metal shield at one end. The top half of the shield is hinged and is controlled by a chain-and-spring system; the lower half is rigidly attached to the scooter frame. In operation, the shield acts as a plug to build a head of water. The pressure of the water behind the shield moves the scooter downstream, but the cable restrains it, allowing slow smooth progress downstream while the water forces past the shield rim and scours the pipe wall in a somewhat similar manner as the balling method. The high turbulence forces loosened debris to move downstream to be caught by the trap at the next maintenance hole. This method is very effective in removing heavy debris. The method is used for large-diameter pipes usually storm drains. Caution must be used in locations with basement fixtures and steep-grade hilly areas because of possible flooding of connected buildings.

Kites, Bags, and Tires. Kites, bags, and tires are devices more suited for hydraulically cleaning larger sanitary sewers in a manner similar to the balling method. The basic equipment includes a water tank truck, dump pick-up truck, and a power drum machine.

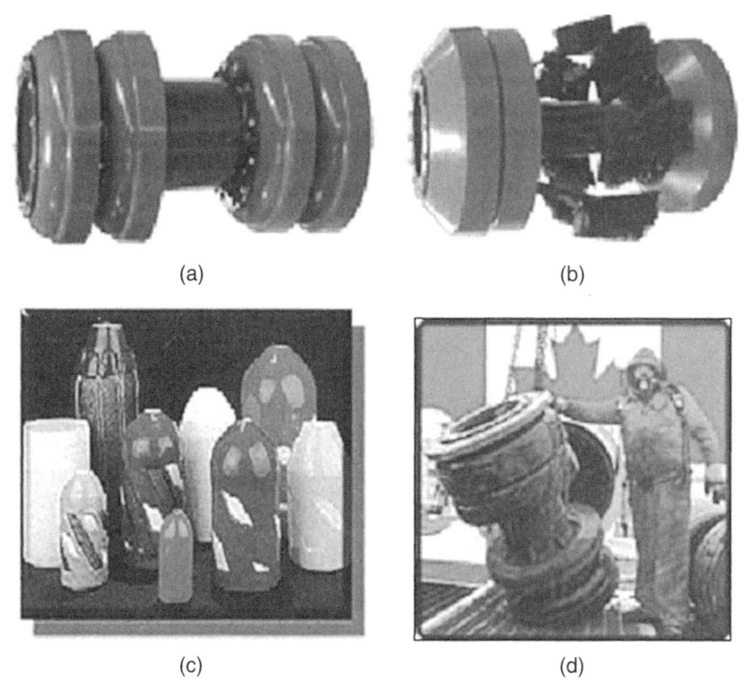

FIGURE 3.24 (*a*) Scraper cup pig. (*Courtesy of Apache Pipeline Products.*) (*b*) Conical cup pig with brushes. (*Courtesy of Apache Pipeline Products.*) (*c*) Foam pigs. (*Courtesy of Pigging Products & Services Association.*) (*d*) Big pig. (*Courtesy of Apache Pipeline Products.*)

The rigid rims of bags and kites cause the scouring action. The kite's shape creates a forward jet of water that scours the pipe wall. A tire, approximately 2 in smaller in diameter than the pipe, rigged to a restraining line, will respond to 2 ft or more head of water like a sewer ball. The poly pig is used for very large sanitary sewers and is not restrained by a line, but moves through the pipe segment with the water pressure built up behind it.

Poly Pigs. Pigging is the most readily available system of pipeline cleaning, and many types of pig are available. Cleaning pigs are designed to scrub or scrape sediment and adherents from the system. These are frequently used for cleaning pressure pipes. The devices are very effective in moving accumulations of decayed debris and grease downstream. They are also capable of washing ahead of it a full pipe of deposits, including roots. Caution must be employed in locations with basement fixtures and steep-grade hill areas because of potential flooding of connected buildings. Pigging can remove heavy sedimentation, hard scale, and biofilm. Different types of pigging vary in result, and more than one run down the line may be required (Fig. 3.24).

Launching and receiving pits are used to launch the pig into the pipe at the appropriate velocity and catch the pig as it finishes the run. Figure 3.25 illustrates a launching and receiving system.

Bucket Machine. The power bucket machine is a mechanical cleaning device effective in partially removing large deposits of silt, sand, gravel, and grit. These machines are used

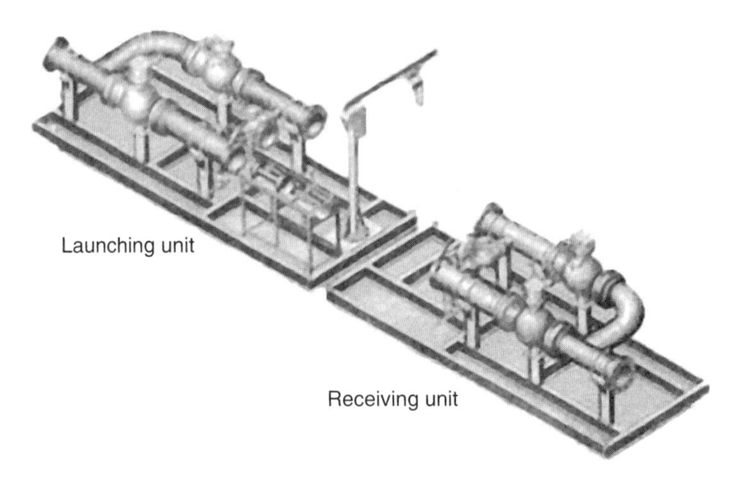

FIGURE 3.25 Pig launching-receiving system. (*Courtesy of Girard Industries.*)

mainly to remove debris from a break or an accumulation that cannot be cleared by hydraulic methods. In cases where the line is so completely plugged that a cable cannot be threaded between manholes, the bucket machine cannot be used. The bucket machine is usually trailer or truck mounted and mainly consists of a cable storage drum coupled with an engine with controllable drive train, up to 1000 ft of 0.5-in steel cable and various sized buckets and tools. The cable drum and engine are mounted on a framework that includes a 36-in vertical A-frame high enough to permit lifting of the cleaning bucket above the ground level. Typically, two machines of the same design are required. One machine at the upstream manhole is used to thread the cable from manhole to manhole. The other machine used at the downstream manhole has a small swing boom or arm attached to the top of the A-frame for emptying the cylindrical buckets. The bottom of the bucket has two opposing hinged jaws. When the bucket is plugged through the material obstructing the line, these jaws open and dig into and scrape off the material and fill the bucket. When the bucket is pulled in the reverse direction, the jaws are forced closed by a slide action. Any material in the bucket is retained as the bucket is pulled out through the manhole.

Silt Traps. Silt traps (or grit sumps) have successfully been used to collect sewer sediments at convenient locations within the system with the traps being periodically emptied as part of a planned maintenance program. The design and operational performance of two experimental rectangular (plan) silt traps in French sewer systems was reported (Bertrand-Krajewsk et al., 1996). Information on design procedures and methodology for silt traps is scarce.

Chemicals. Chemicals can be very helpful aids for cleaning and maintaining the wastewater collection system. Proper application of the right chemicals can be very effective to control root intrusion, odors, corrosion, and rodent and insect infestations. A chemical dosing program must be developed for this purpose. There should be a thorough evaluation and planning in preparing a chemical dosing program. The planner should be aware of the following facts:

- Chemicals cannot clear stoppages or blockages in sanitary sewer lines
- Chemical costs are high and increasing; hence, cost-effectiveness must be considered

TABLE 3.7 Effectiveness of Cleaning Methods (EPA, 1999)

	Effectiveness with each type of problem				
Solution to problem	Emergency stoppages	Grease	Roots	Sand, grit, scaling, and debris	Odors
Balling	–	High	–	High	Moderate
High velocity cleaning	Lowest	Highest	–	High	Moderate
Flushing	–	–	–		Moderate
Sewer scooters	–	Moderate	–	Moderate	–
Bucket machine scrapers	–	–	–	Moderate	–
Power rodders	High	Lowest	Moderate	–	–
Hand rods	High	Lowest	Low	–	–
Chemicals	–	Low	Moderate	–	Moderate

- Chemicals may be hazardous to employees, treatment process, and the environment
- Some vendors make elaborate claims for chemicals unproven in actual sewer cleaning situations

Hence, it is important that field demonstrations be required from prospective suppliers.

3.8.2 Effectiveness of the Conventional Cleaning Methods

Table 3.7 presents the effectiveness of different cleaning methods with respect to the different types of problems encountered in a sewer system. Bucketing and pigging can be effective in removing large deposits of silt, sand, gravel, and some type of solid waste. Pigging is also effective in removing accumulations of sedimentation, hard scale, biofilm, and grease; whereas scooters can effectively remove heavy debris and grease.

Cleaning with chemicals or water jetting can effectively clear the pipeline; although at times it can create new problems. These methods are very aggressive, and may damage the pipe that may need to be relined or coated with an anticorrosion layer. Chemical and water jetting can be used in combination with pigging, and pigs have been developed for specific use with these techniques. For example, the V-jet pig can be used after very aggressive treatments to spray corrosion inhibitor fluid to the top quadrant of the pipe. It also acts as a reservoir for the fluid.

High-pressure water jetting can remove the roots in the pipeline, and in the event of extremely large or tenacious root systems, a cutting tool can mount on the jetting unit to remove mass root intrusions (Fig. 3.26).

N-SPEC has developed a chemical system designed to work with pigs, during or prior to a pigging run. This system effectively "removes tenacious solids such as paraffin/asphaltine and compressor oil deposits." There are many benefits to a clean pipeline, such as higher operating efficiency, lower maintenance cost, and less internal corrosion. The most compelling reason to keep a pipeline clean is to extend the lifetime of that pipeline.

3.8.3 Limitations of the Conventional Cleaning Methods

Some of the limitations of the conventional cleaning methods are discussed in Table 3.8.

TABLE 3.8 Limitations of the Conventional Cleaning Methods (EPA, 1999)

Cleaning method	Limitation
Balling, jetting, scooter	In general, these methods are only successful when necessary water pressure or head is maintained without flooding basements or houses at low elevations. Jetting: The main limitation of this technique is that caution needs to be used in areas with basement fixtures and in steep-grade hill areas. Balling: Balling cannot be used effectively in pipes with bad offset joints or protruding service connections because the ball can become distorted. Scooter: When cleaning larger lines, the manholes need to be designed to a larger size to receive and retrieve the equipment. Otherwise, the scooter needs to be assembled in the manhole. Caution also needs to be used in areas with basement fixtures and in steep-grade hill areas.
Bucket machine	This device has been known to damage sewers and manholes. The bucket machine cannot be used when the line is completely plugged because this prevents the cable from being threaded from one manhole to the next. Setup of this equipment is time-consuming. Flushing: This method is not very effective in removing heavy solids. Flushing does not remedy this problem because it only achieves temporary movement of debris from one section to another in the system.
High-velocity cleaner	The efficiency and effectiveness of removing debris by this method decreases as the cross-sectional areas of the pipe increase. Backups into residences have been known to occur when inexperienced operators have used this method. Even experienced operators require extra time to clear pipes of roots and grease.
Kite or bag	When using this method, use caution in locations with basement fixtures and steep-grade hill areas.
Rodding	Continuous rods are harder to retrieve and repair if broken and they are not useful in lines with a diameter of greater than 300 mm (0.984 ft) because the rods have a tendency to coil and bend. This device also does not effectively remove sand or grit, but may only loosen the material to be flushed out at a later time.

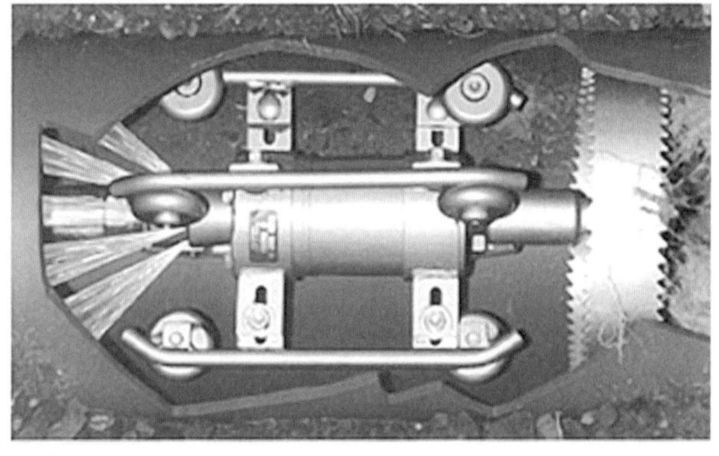

FIGURE 3.26 Root cutter device with high-velocity jetters. (*Courtesy of Sewer Equipment Company of America.*)

3.9 CHEMICAL CLEANING CASE STUDY

The Village of Westfield, New York is located in the grape growing area along Lake Erie between Erie, Pennsylvania and Buffalo, New York. It is a community with an aging distribution system that has 4-in diameter cast iron pipe that was installed pre-1930 in many of the older neighborhoods. Surface water is pumped to a reservoir and serves as a source for the treatment plant.

Consumer complaints were on the increase, with the vast majority occurring during the summer months. The consumers complained about problems with discolored water, taste, odor, and turbidity at their tap, even though a corrosion inhibitor was being fed at the treatment plant. Another problem that occurred was, reduced flows at the fire hydrants resulting in the fire marshal color-coding some of them for yields of less than 500 gpm. Several theories had been postulated as to the source of the problems but the utility personnel were not sure whether they stemmed from the reservoir, the distribution piping, or a combination of both. With the size of the reservoir and the variations in water temperature during the course of a year, it was thought that stratification was occurring that might cause the problems of discoloration, turbidity, bad taste, and odor. It was also thought that the warmer water temperature was the cause of deposits in the distribution line to slough off and thus cause the complaints. Several samples had been collected during break outages of the 4 in line and indications were that the pipe was reduced to as little as 2 in diameter in many sections.

In an effort to reduce the source of the problem at the reservoir, the water department personnel decided to install an aerator to provide a source of mixing to keep it oxygenated and prevent turnover in the summer. They also decided to clean sections of the distribution system lines on a planned schedule over a period of 5 years. They had a couple of bad experiences in the past with pigging that resulted in digging up lines owing to pigs getting stuck. As an alternative, they investigated using chemical cleaning as the method to rehabilitate the pipes. About 6900 ft of a 4-in diameter pipe in one of the oldest neighborhoods was given the first priority, to be cleaned.

The first step was to send a pipe sample to the HERC Products Incorporated laboratory for analysis. The deposit analysis indicated that the majority of the scale was iron oxide. The bench test and loop assembly test indicated that it would require a 30 percent solution of PIPE-KLEAN® Preblend to be circulated for 12 to 14 h. A proposal was then submitted and accepted to complete the project with as little disruption to the neighborhood and the system as possible.

The water department scheduled a series of meetings with the consumers to make them aware of what their planned course of action was going to be and what to expect during the cleaning process. They had decided to shut the water off for, approximately, 24 h and offer to provide bottled water for drinking. They had also suggested that the residents fill up sinks, tubs, and other containers for normal water use while the service was shut off.

To prepare for the cleaning, the water department personnel completed an examination of the service connections and replaced 20 percent of them because they were in poor condition. In addition, to set up recirculation loops for the pipe sections to be cleaned, it was necessary for the water department to install five flush hydrants.

The first phase project was divided into six individual sections ranging anywhere from 700 to 1795 ft in length. The goal was to clean each section in a 24-h time frame. At the start of cleaning in each section, water department personnel shut off the valves at the street, disconnected the meters in the homes, and plugged the service lines on the water main line to be cleaned. Hoses from the mobile recirculation unit (MRU) were connected to the hydrants to establish a recirculation loop. Water was pumped by the MRU through the loop to determine if there were any leaks (losing water) or cross-connections (gaining water). Once it was determined that the line was intact, the cleaning chemistry was added and circulated. The solution was tested periodically for acidity and dissolved iron.

As it turned out, every section was dirtier than expected and each required two shots of chemistry.

As the dissolved iron reached 4 percent in the cleaning solution, the solution was neutralized with caustic soda, discharged to the sanitary sewer and fresh PIPE-KLEAN® Preblend was charged back into the recirculation system. Once it was determined that the cleaning was complete, the PIPE-KLEAN® Preblend was removed from the distribution line and the line flushed with water until clear. Circulation was then established and the pH was raised to 10. A corrosion inhibitor, used by the water treatment plant in normal operation, was added and the system allowed circulating for several hours to passivate the line and to prevent flash corrosion. The passivation chemical was then evacuated from the system and discarded to the sanitary sewer. The distribution header was flushed with water until clear. Water department personnel then flushed each consumer's service line with water to the meter before reconnection.

In several sections during the cleaning process, leaks developed that required the line to be purged into the tank on the MRU so that the water department personnel could repair the leaks. In every case the leaks involved stop but not for the same reasons. The first leak that was encountered was a new stop that had just been installed. It was concluded that it had not been seated entirely and the penetrating characteristics of the PIPE-KLEAN Preblend found the discrepancy. The other leaks appeared to be the result of the cleaning solution removing corrosion products (copper oxide) that had built up and plugged holes in those valves and the cleaning solution did as it was designed to do, dissolve the deposits thus creating the leaks. The amount of deposit removed from the six sections is presented in Table 3.9.

Although the same finished treated water over the years was flowing through the pipeline, each of the sections exhibited varying types of deposits. In some cases there was definitely calcium carbonate, in one other, there was a slime layer, but in all, there was varying degrees of iron oxide as measured by the pounds removed, listed in the chart above.

The chemical cleaning of the sections selected resulted in an immediate improvement of water quality at the tap with no discoloration, turbidity, or odor complaints. Shortly after the cleaning process had concluded and service reestablished, some of the consumers mentioned a metallic taste that went away after about 24 h, numerous consumers had noticed a significant increase in water pressure at the tap, and water department personnel mentioned a vast increase in water flow at the hydrants. Where water had been normally falling a few feet from the hydrant, it was now discharging across the street. Flow test measurements at several hydrants are presented in Table 3.10.

The entire cleaning project was completed in 7 working days working around the clock with two crews. Each section required a little more than 24 h to complete from the time the service line was disconnected to when it was reconnected.

With less deposit in the lines it is expected that lower chlorine and inhibitor can be fed at the treatment plant because it will not be consumed or absorbed by the deposits.

TABLE 3.9 The Amount of Deposit Removed from Pipe Sections

Section	Length (ft)	Iron oxide (lb)
Kent Street	785	1110
Beckman Street	1550	1400
Second Street	700	230
Crandall Street	790	1160
Jackson-Billsboro	1795	2125
Pleasant-Ash-Riley	1200	1225

TABLE 3.10 Flow Test Measurements at Several Hydrants

Hydrant	Precleaning flow (GPM)	Postcleaning flow (GPM)
Beckman & Patterson	290	710
Pleasant & Kent	240	400
Billsboro	240	530

The energy costs are also expected to be lower with the restrictions removed in this portion of the system. The total cost of chemical cleaning was significantly less than the cost of replacement. Including all of the cost incurred by the village for the chemical cleaning contract plus their extra labor, overtime, flush hydrants, parts, and passivation chemicals, the total cost for the project was US$22.91/ft.

3.10 SUMMARY

This chapter discussed topics related to pipline deteriorations, asset management, and pipeline inspection and cleaning. A comprehensive asset management strategy would reduce possibilities of pipeline failures and emergency repairs. Understanding the causes of pipeline failures would help in design and construction of new pipeline systems. A thorough pipeline inspection, enables a pipeline operator to carry out a cost-effective trenchless renewal program that targets individual pipe needs.

ACKNOWLEDGMENTS

The following people have contributed in the preparation of this chapter:

Dr. Brian Mergelas, Xiangjie Kong, Mike Garaci, and Dave Caughlin, the Pressure Pipe Inspection Company Ltd., Ontario, Canada

D. L. Atherton, Department of Applied Magnetics, Queen's University, Kingston, Ontario, Canada

Anthony Bond, Technical Manager, Water Research Council, Warwickshire, United Kingdom

Jim Farmerie, HERC Products Incorporated, Portsmouth, Virginia

Guru Kulandaivel, Peter Kiewit Sons', Inc., Omaha, Nebraska

REVIEW QUESTIONS

1. Explain built-in and long-term causes of pipeline deterioration.
2. Describe the factors that influence pipe deterioration process.
3. What is an asset management strategy and list some of its benefits?
4. What are the main objectives of pipeline condition assessment?
5. Name and explain three types of sewer pipe inspection methods.

6. What is SSET and how does it work?
7. Describe three methods of pressure pipe inspection.
8. Describe the basic types of pipe cleaning methods.
9. Explain the process of chemical pipe cleaning.
10. Explain how ground penetrating radar (GPR) is used for pipeline inspection.
11. What is acoustic emission testing (AET) and how is it used?
12. What are the benefits and the limitations of FELL technology?

REFERENCES

Hughes, D. M. (Ed.) (2002). Assessing the Future: Water Utility Infrastructure Management, American Water Works Association (AWWA), Denver, Colo.

Anderson, D., and Cullen, N. (1982). *Sewer Failures 1981, the Full Year.* WRc External Report No. 73E.

Association of Metropolitan Sewerage Agencies. (2004). *Managing Public Infrastructure Assets to Maximize Cost and Maximize Performance,* Washington, DC.

Atkinson, K. (2000). *Sewer Rehabilitation Techniques.* Available at: http://www.hrwallingford.co.uk/projects/SEWER/sewer_home/S2d18.pdf.

Benson, A. (1995). Applications of ground penetrating radar in assessing some geological hazards, *Journal of Applied Geophysics*, 33(1–3).

Birks, A., and R. Green. (1991). *Nondestructive Testing Handbook Second Edition,* Vol. 7, Ultrasonic Testing, American Society for Nondestructive Testing, Columbus, Ohio.

Bungey, J. (1995). Testing concrete by radar, *Concrete*, November/December, Concrete Society, London, England.

CERF (2001). Evaluation of SSET, The sewer scanner & evaluation technology, *Civil Engineering Research Foundation (CERF) Technical Evaluation Report,* CERF Report No. 40551, March.

Chae, M. J., T. Iseley, and D. M. Abraham. (2003). Computerized sewer pipe condition assessment, New Pipeline Technologies, Security, and Safety, *Proceedings of the ASCE International Conference on Pipeline Engineering and Construction,* Baltimore, Md., July 13–16, 2003, pp. 477–493.

Chi-Yuan F. (2004). *Sewer Sediment and Control: A Management Practices Reference Guide*, EPA, Edison, N.J.

Clemena, G., and M. Sprinkel. (1987). *Use of Ground Penetrating Radar for Detecting Voids under a Jointed Concrete Pavement,* Transportation Research Record, No. 1109.

Daniels, D., and D. Schmidt. (1995). The use of ground probing radar technologies for non-destructive testing of tubes, International Symposium of Nondestructive Testing in Civil Engineering (NDT-CE), pp. 429–436, German Nondestructive Testing Institute (DGZfP), Berlin, Germany.

Davies, J. P., B. A. Clarke, J. T. Whiter, and R. J. Cunningham. (2001). Factors influencing the structural deterioration and collapse of rigid sewer pipes, *Urban Water*, 3, pp. 73–89.

EPA 832-F-99-031. (1999). *Collection Systems O&M Fact Sheet: Sewer Cleaning and Inspection,* Office of Water, Washington DC.

Gilbert, J., G. Campbell, and K. Rogers. (1995). Pirat—A system for quantitative sewer assessment, *International No-Dig,* Hamburg, Germany, September, pp. 455–462.

Gokhale, S., and M. Hastak. (2003). Automated assessment technologies for sanitary sewer evaluation, *Proceedings of ASCE New Pipeline Technologies, Security, and Safety Conference,* Baltimore, Md.

Goodman, D. (1994). Ground penetrating radar simulation in engineering and archeology, *Geophysics*, 59(2).

Hibino, Y., T. Nomura, S. Ohta, and N. Yoshida. (1994). Laser scanner for tunnel inspections, *International Water Power and Dam Construction,* June, IPC Electrical-Electronic Press, London, England.

Makar, J. M. (1999). Diagnostic techniques for sewer systems, *Journal of Infrastructure Systems,* 5(2), 69–78.

NASSCO (1996). *Recommended Specification for Sewer Collection System Rehabilitation.* NASSCO publications, Chambersburg, Pa.

Newport, R. (1981). Factors influencing the occurrence of bursts in iron water mains. *Water Supply and Management,* 3:274–278.

O'Day, D. K., R., Weiss, S., Chiavari, and D. Blair. (1986). *Water Main Evaluation for Rehabilitation/Replacement.* American Water Works Association Research Foundation (90509), Denver, Colo.

Peters, L., J. Daniels, and J. Young. (1994). Ground penetrating radar as a subsurface environmental sensing tool, *Proceedings of the IEEE,* 82(12):1802–1822.

Pisano, W. N. G., and G. Novac. (1997). Automated Sewer Flushing Large Diameter Sewers, in *Collection Systems Rehabilitation and O&M: Solving Today's Problems and Meeting Tomorrow's Needs,* pp. 12:9–12:20.

Pocock, R. G., Lawrence, G. J. L., and Taylor, M. F. (1980). *Behavior of a Shallow Buried Pipeline Under Static and Rolling Wheel Loads*, TRRL Laboratory Report 954.

Price, T. (1995). Inspecting buried plastic pipe using a rotating sonic caliper, *Proceedings of the Second International Conference on Advances in Underground Pipeline Engineering,* Bellevue, Washington, June 25–28, 1995, American Society of Civil Engineers, New York.

Walski, T. M., and A. Pelliccia. (1982). Economic analysis of water main breaks. *Journal AWWA*, 74(3):140–147.

Water Research Center (1986). *Sewerage Rehabilitation Book,* 2nd ed., Water Research Center/Water Authorities Association, United Kingdom.

WEF-ASCE (1994). *Existing Sewer Evaluation and Rehabilitation,* ASCE Books and Reports on Engineering Practice, No. 62, Alexandria, Va.

CHAPTER 4

DESIGN CONSIDERATIONS FOR TRENCHLESS PIPELINE CONSTRUCTION METHODS

4.1 INTRODUCTION

A successful trenchless construction project requires surface and subsurface investigations. Trenchless installation methods require the design engineer to provide the contractor with sufficient information to reasonably anticipate the obstacles that might be encountered and how drilling or boring operations should be carried out. During the design phase, surface and subsurface survey information will assist in determining the suitability of utility installation by the specific trenchless process. Obtaining and providing accurate surface and subsurface information will result in reducing the possibility of installation problems and change orders during the work as well as minimizing the possibility of litigation and dispute.

4.2 SURFACE SURVEY

A surface survey is required prior to the designing of a trenchless construction project. Each trenchless construction project has specific site requirements. A survey should be conducted along the centerline of the proposed bore path for a width of 100 ft. Predesign surface survey should include the following steps, at the least:

- Work area requirements
- Existing grade elevation data
- Surface features such as roadways, sidewalks, utility poles
- Boring or test pit locations
- Waterways and wetlands
- Visible subsurface utility landmarks such as manholes or valve boxes
- Structures adjacent to the bore path

Review of existing geological or geotechnical reports, maps, aerial photographs, and review of depositional history are important in developing a preliminary surface survey. For example, if the area has been subjected to glaciation, then cobbles, boulders, and gravel can be expected. If the area has been subjected to large landslides, trees and other items may have been buried and could be encountered. If the area has been subjected to meandering of relatively low-energy streams and rivers, then fine-grained deposits may be expected. Conversely, high-energy, steeply sloped stream beds may be covered with cobbles and boulders from nearby mountains. Typically, more than one type of geological process could complicate the depositional history. Surface contours are useful, but not important, as trenchless projects on the surface are limited to entry and exit pits and work areas.

4.3 SUBSURFACE INVESTIGATIONS

Subsurface investigation is the next step to surface survey. Subsurface considerations that impact trenchless design, and therefore need proper investigation, include presence of existing utilities or other manmade obstructions, method of placement, and geotechnical conditions along the proposed trenchless construction alignment.

4.3.1 Existing Utilities

Trenchless construction projects require the contractor to install the pipe without seeing the excavation area. Therefore, the contractor should be given a record of potential conflicts and utility clearances.

The local *one-call* service should be contacted as the first step prior to design and construction of a trenchless project. In the absence of *one-call* service, municipalities and utility companies should be contacted to obtain the required information. Obtaining as-built record drawings is also important but should not be relied on. Additional research is necessary, and usually conducted by the contractor to positively locate and expose existing utilities where running parallel less than 10 ft (at specified intervals, dependent on proximity and type of existing utility) or where crossing the bore path of the trenchless project. Methods of confirming subsurface utility locations include surface applied pipe locators, ground penetrating radar (GPR), vacuum excavation equipment, and seismic survey. Use of subsurface utility engineering (SUE) is highly recommended. SUE is a branch of engineering practice that involves managing certain risks associated with utility mapping at appropriate quality levels. It can be applied to utility coordination, utility relocation, utility condition assessment, communication of utility data to concerned parties, utility relocation costs estimates, implementation of utility accommodation policies, and utility design. More information on these methods can be found from the references at the end of this chapter.

4.3.2 Geotechnical Investigations

A second phase of subsurface investigation for trenchless construction projects is the determination of soil conditions. Once the proposed alignment has been identified, a geotechnical investigation should be performed. Investigations for complex installations should comprise two phases, a general geotechnical review and a geotechnical survey. A geotechnical survey alone may be sufficient for simpler installations.

General Geotechnical Review. A general geotechnical review involves examining existing geotechnical data to determine what conditions might be encountered in the vicinity of the installation. Existing data may be available from construction project records in the location of the trenchless project (buildings, piers, bridges, levees, and so on). Such an overall review will provide information that may not be developed from a geotechnical survey comprising only exploratory borings. It also allows the geotechnical survey to be tailored to the anticipated conditions at the site, thus enhancing the effectiveness of the survey.

Geotechnical Survey. For underground construction, it is necessary to know the actual soil stratification at a given site, the laboratory test results of the soil samples obtained at various depths, and the observations made during the construction of other underground structures built under similar conditions. The steps for subsurface investigation include the following:

- Determining the nature of soil at the site and its stratification
- Obtaining disturbed and undisturbed soil samples for visual identification and appropriate laboratory tests
- Determining the depth and nature of bedrock, if encountered
- Performing in situ field tests
- Observing surface drainage conditions from and into the site
- Assessing any special construction problems with respect to the existing structures nearby
- Determining groundwater levels, sources of recharge, and drainage conditions

The main methods of geotechnical surveys are as follows:

- *Ground penetration radar (GPR).* Useful in gravels and sands.
- *Acoustic (sonar).* Useful for determining depth of rock, interfaces between soft and hard deposits, and buried objects.
- *Geophysical methods.* Variations in the speed of sound waves or in the electrical resistivity of various soils are useful indicators of the depth of water table and of the bedrock.
- *Test pits or trenches.* This method is suitable for shallow depths only but allows visual observation over a larger area than is possible with samples from borings.
- *Hand augers.* Suitable only for shallow depths; only disturbed or mixed samples of soil can be obtained in this method.
- *Boring test holes and sampling with drill rigs.* This is the principal method for detailed soil investigations. Sampling interval and technique should be set to accurately describe the subsurface material characteristics taking into account the site-specific conditions. Typically, split spoon samples will be taken in soft soil at 5-ft depth intervals in accordance with ASTM D-1586.

It is essential that a thorough subsurface investigation is carried out in the design phase to identify the geologic conditions along the pipeline alignment. The anticipated geologic conditions comprise the most important factor in the selection of an appropriate trenchless construction method for a specific project. Groundwater conditions will have an important influence on the behavior of the ground and the potential for loss-of-ground. Groundwater levels should be determined, and pumping tests or other field tests should be conducted to estimate the hydraulic conductivity if dewatering is necessary and feasible.

The use of a geotechnical baseline for trenchless construction project is highly recommended (ASCE, 1997). The purpose of ASCE Geotechnical Baseline Report (GBR) is to establish a contractual statement of the geotechnical conditions anticipated to be encountered during underground or subsurface construction. This contractual statement is referred to as the baseline. Risks associated with conditions consistent with or less adverse than the baseline are allocated to the contractor, and those significantly more adverse than the baseline are accepted by the owner. The overriding philosophy adopted in the ASCE publication is that the owner owns the ground. If conditions are more adverse than projected in the baseline, the owner should pay the additional cost of overcoming those conditions.

The following is a list of recommended information to be obtained for a trenchless construction project:

- Standard classification of soils
- Gradation curves on granular soils
- Standard penetration test (SPT) values where applicable (generally unconsolidated ground)
- Particle size distribution including presence of cobbles and boulders
- Shear strength
- Atterberg limits (liquid, plastic, and shrinkage limits)
- Moisture content
- Height and movement of water table
- Permeability
- Cored samples of rock with lithologic description, rock quality designation, and percent recovery
- Unconfined compressive strength for representative rock samples (frequency of testing should be proportionate to the degree of variation encountered in the rock core samples); and Mohs hardness for rock samples. Where rock is encountered, it should be cored, in accordance with ASTM D-2113 to the maximum depth of the boring
- Presence of contaminated soils (hydrocarbons and so on)

Settlement Potential. Surface settlement is mainly a result of loss-of-ground during tunneling and dewatering operations that cause subsidence. During a trenchless technology project, loss-of-ground may be associated with soil squeezing, running or flowing into the heading, losses owing to the size of overcut, and steering adjustments. The actual magnitudes of these losses are largely dependent on the type and strength of the ground, groundwater conditions, size and depth of the pipe, equipment capabilities, and the skill of the contractor in operating and steering the machine. If passive earth pressure is exceeded, heave of ground surface may occur, causing damage to nearby utilities and other structures.

4.4 ALIGNMENT CONSIDERATIONS

As in all pipeline projects, identifying feasible trenchless technology alignments involves evaluating available right-of-way and easement acquisition issues, and determining the location of the existing utilities. Occasionally, alignments not considered feasible or economical for open-cut methods may be possible if trenchless construction methods are used. Pipelines constructed using trenchless construction methods can be located deeper, sometimes with only a small increase in construction cost. This may be a significant advantage if a deeper alignment can avoid existing underground utilities, potential conflicts, and utility relocations.

Straight horizontal alignments are generally preferred for trenchless technology projects. Straight alignments provide for more accurate control of line-and-grade and for a more uniform stress distribution on the pipe and joints, reducing the risk of eccentric loads that could damage the pipe.

To be feasible, a prospective alignment must have adequate jacking and receiving pit locations available. In addition, prospective jacking and receiving pit sites must be spaced at distances that are compatible with trenchless technology techniques. The maximum distance a pipe can be installed with a trenchless construction method is dependent on variables such as pipe size, structural capacity of the pipe, thrust capacity of the thrust block and the main jacks, soil conditions, effectiveness of the lubrication system, and specific project conditions such as operator's skill in steering the pipe.

Providing adequate space for staging construction operations is important so that pipe installation can be completed in an efficient manner. Construction access to the jacking pit must be provided for transporting tunnel muck, pipe sections, and tunneling equipment. In urban areas, traffic control requirements must be evaluated in selecting and laying out jacking pit sites. A typical jacking pit site needs enough space for the jacking pit, slurry tanks, a crane, pipe storage, and support facilities (e.g., a generator, power pack, and lubrication unit). The jacking pit should be a sufficient distance from overhead electrical lines to avoid hazards in operating the crane although in some areas a gantry system can be used instead of a crane for smaller pipe sizes.

The equipment arrangement is quite flexible and space requirements can be reduced for smaller sites, if necessary. Frequently, jacking pits have been located in the parking strip along the edge of a street with the equipment set up in a linear arrangement. Similar linear arrangements have also been used to stage trenchless technology operations from the median of wider, more heavily traveled streets without significantly impacting traffic flow. Staging area requirements can be further reduced by using each jacking pit to install two drives, one in each direction. This approach further minimizes the environmental impacts of construction by reducing the number of jacking pit locations.

4.4.1 Jacking and Receiving Pits

Jacking and receiving pits are vertical excavations with shoring and bracing systems. Several shoring systems are commonly used: sheet-pile systems with internal bracing, soldier pile or circular steel rib systems with timber lagging and internal bracing, and liner plate systems with steel rib supports. Watertight caissons may be needed for water bearing soils and deep installations. An important factor in the design of jacking and receiving pits is groundwater control. Dewatering systems using deep wells or well points are frequently employed. Alternatively, a groundwater cutoff can be used if relatively impermeable soils are present below water-bearing soils. Sheet-piles, for instance, could be driven into the impervious soils to cut off groundwater inflows. Sometimes ground freezing methods have been used for groundwater control but these methods are usually expensive. Grouting or any other method of groundwater control is necessary when launching the tunneling machine and advancing out of jacking pits or advancing into receiving pits unless groundwater levels are temporarily lowered in these areas and soils are stabilized. Where penetration grouting is not feasible, jet grouting techniques have been used to control groundwater inflows.

Caisson sinking methods are frequently used to construct shafts where dewatering or grouting methods would be difficult or uneconomical. This approach involves constructing the shaft by stacking up circular precast concrete sections while excavating inside the caisson below the groundwater level with a clamshell. After the caisson is sunk to the design elevation, a tremie concrete base slab designed to withstand the hydrostatic uplift pressures, is placed. With the slab in place, the water inside the shaft can be safely pumped out, leaving a dry excavation.

4.5 CALCULATING JACKING FORCE IN PIPE JACKING AND MICROTUNNELING

In pipe jacking and microtunneling methods, the jacking force required to push a pipe forward theoretically comprises the penetration resistance of the boring and steering head and the frictional resistance of the product pipe. In accordance with Fig. 4.1, the jacking force can be expressed by Eq. 4.1:

$$\mathrm{JF} = \mathrm{FP} + \sum \mathrm{FR} \tag{4.1}$$

where JF = total jacking (thrusting) force, lb
FP = resistance of the tunnel boring machines (TBM) (penetration resistance), lb
FR = frictional resistance, lb

The jacking force in the conventional microtunneling systems must be axially applied to the pipe from the main jacking station. Moreover, it must be ensured that the resulting stresses stay below the allowable stresses in all jacking pipe sections. Proceeding from the main jacking station, the jacking force decreases pipe by pipe in proportion with the given frictional resistance. Therefore, the highest jacking force in conventional microtunneling methods theoretically occurs in the pipe closest to the main jacking station just before the far end of the pipe arrives at the receiving pit.

The amount of jacking force required is governed by the soil type and its characteristics such as soil density and water content (location of water table); project characteristics such as height of cover, size of overcut, lubrication, overburden loads, time and jacking distance; and pipe characteristics such as pipe size, dimensional consistency, weight, resiliency, absorbency, and smoothness of its outer surface. On the other hand, the amount of jacking force that can be applied in any microtunneling project is limited to a great extent by the strength of the pipe material, the area of the jacking pipe at the smallest cross section, extent of eccentricity of the resultant jacking force, capacity of jacking equipment, and load-bearing capacity of the thrust block. Design parameters—such as the jacking rate and the interaction of the soil, or water pressure, or both at the face—are also important in determining the design jacking force. Oversize cut or use of lubricants may decrease the jacking force

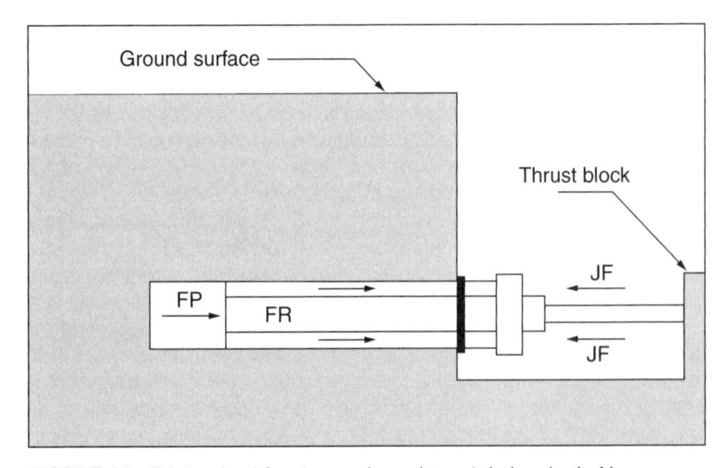

FIGURE 4.1 Frictional and face (penetration resistance) during pipe jacking.

up to 30 percent or more. On the other hand, occurrence of any unexpected obstructions, such as existence of boulders or restraint of pipes as a result of steering errors, can bring about a sudden increase in the jacking force. Consequently, it is desirable to install pressure relief valves at the drive pit and provide indicators on the control panel to ensure that the allowable jacking force is not exceeded.

Proper assessment of the skin friction, penetration resistance, and the required jacking force, is extremely important for (1) designing and selecting jacking pipes, (2) designing the thrust block, and (3) choosing the jacking method.

Calculation of Jacking Force. The jacking force is calculated according to Eq. 4.1:

$$\mathrm{JF} = \mathrm{FP} + \sum \mathrm{FR}$$

FR, the frictional resistance can be expressed as:

$$\mathrm{FR} = R \times S \times L \tag{4.2}$$

where R = circumferential frictional resistance (skin friction), lb/in^2
S = perimeter of pipe cross section = (outside diameter of pipe) $\times\ \pi$, in
L = jacking (thrusting) distance, in

Typical numerical values for circumferential frictional resistance (R) are shown in Table 4.1. These values are based on such project conditions as size of overcut, lubrication, pipe surface, pipe dimensional consistency, depth, installation idle time (soil clamping), soil density, and so on. It is difficult to accurately assess these values by using soil mechanics theory, therefore, pipe jacking and microtunneling contractors have developed empirical values after years of experience.

A rule of thumb used in the industry for estimating the maximum jacking force is to assume circumferential resistance of the pipe to be 1.5 psi per unit pipe or soil contact surface area. Lamb et al. (1993) reported the empirical value of circumferential frictional resistance of the pipe to be between 0.625 to 0.9 psi per unit contact surface area of the pipe. Also circumferential frictional resistance of the pipe can be calculated to be between 3 and 12 psi. Stein et al. (1989) reported an overall skin friction factor of 1.5 psi. It should be noted that the above values should be used only as a guide and may be considered very conservative depending upon project and soil conditions.

As stated previously, some of the main reasons for differences among the R values reported are owing to pipe material, type of soil, soil moisture content, and depth of cover, as well as, type of microtunneling equipment and details of the construction procedure (such as amount of overcut, use of lubricants, work stoppage, misalignment of pipes, and so on). Research carried out in Japan indicates that lubrication can result in

TABLE 4.1 Typical Values of Circumferential Frictional Resistance for Different Types of Soils

Soil	Clay	Silt	Sand	Clayey gravel	Swelling clay	Sandy gravel	Loamy sand
R (psi)	0.56	0.56	0.70	0.70	2.80	1.10	1.30

30 to 50 percent reduction for clayey soils and about 20 percent for sandy soil. In Germany, the reduction is generally observed to be about 30 percent for all soils (Klein, 1991). Klein reported that for a project in California the frictional resistance was observed to be between 0.75 and 2.5 psi surface of the pipe. The soil condition at the project reported was stiff clay and bentonite was used as lubrication.

Among the key parameters that affect the jacking force are, size of overcut, lubrication (including its quality, consistency, and both volume and pressure of injection), type of soil, pipe skin area and its quality, and jacking distance. For example, it has been shown that the relationship between height of cover and skin friction is nonlinear. In stable soils, the oversize cut can be maintained throughout the jacking distance and therefore the frictional force can only occur at the bottom of the pipe. In unstable soils, the oversize cut does not reduce skin friction; nonetheless this cut is necessary for steering operation of the tunneling machine. Increasing jacking distance, or the surface area of the pipe, or both has a linear relationship with increasing the skin friction. Lamb et al. (1993) report that the force owing to pipe weight is in the magnitude of 3 to 4 percent of the skin friction.

The penetration resistance opposes advancement of the boring machine throughout the microtunneling operation. The penetration resistance varies depending on the soil type and shape and steering action of the boring head, and therefore is controllable by the operator of the microtunneling machine.

For slurry shield microtunneling equipment, the value of the resistance of the leading pipe (FP) is usually calculated through Eq. 4.3:

$$FP = (P_e + P_w) \times \left(\frac{B_c}{2}\right)^2 \times \pi \tag{4.3}$$

where P_e = contact (point) pressure of the cutting head, psi
P_w = slurry pressure, psi
B_c = outside diameter of the shield (boring) machine, in

Geological conditions and existence of underground water table govern the contact pressure. The contact pressure of the boring head has to be higher than the active soil pressure and lower than the passive soil pressure to avoid settlement or heave on the surface.

The value of contact pressure of the cutting head (P_e) is generally assumed to be 20 psi. The value of the slurry pressure varies depending upon earth cover on top of the MTBM.

The Thrust Block or Abutment. The thrust block transmits the thrusting force to the soil support panels, such as sheeting wall, that on account of its natural stiffness directly transmit this force to the soil.

The allowable thrusting force (Q) of the thrust block is calculated by Rankine's passive soil pressure theory (Bowles, 1988). Figure 4.2 provides a schematic representation of the thrust block or soil interaction. The top of the thrust block is generally at some depth, h, below the existing ground surface. Resistance to failure in the soil behind the thrust block is assumed to occur on a plane extending at an angle of X equal to 45° – φ/2 from the horizontal starting at the base of the thrust block and extending upward through the existing ground surface. This thrust force is calculated in Eq. 4.4:

$$Q = \alpha B\left[\gamma H^2\left(\frac{K_p}{2}\right) + 2CH\sqrt{K_p} + \gamma HhK_p\right] \tag{4.4}$$

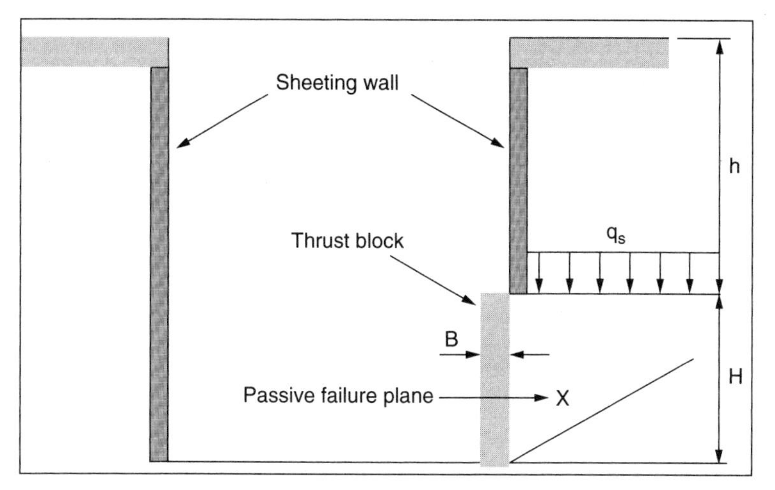

FIGURE 4.2 Calculation of loads on the throne.

where Q = allowable thrusting force, lb
α = coefficient = 2 (commonly set at 2)
B = thrust block or abutment width, ft
γ = soil density, lb/ft^3
K_p = coefficient of passive soil pressure = $\tan^2 (45° + \varphi/2)$
φ = angle of internal friction (shearing resistance of soil), degrees
C = soil cohesion, lb/ft^2
H = abutment height, ft
h = distance to top of abutment from ground surface, ft
$q_s = \gamma h$. (Fig. 4.2)

This equation provides the basic passive earth pressure for soil directly behind the thrust block that develops frictional resistance from both the angle of internal friction and cohesion components. The soil above the wall is converted to a uniform load applied to the top of the soil behind the wall, as shown in Fig. 4.2. When this transformation is made, the frictional resistance from this transformed soil can no longer be considered. Thus, the coefficient, α, is applied to account for the actual resistance conditions.

Evaluation of Uniform Vertical Load

Properties of a pipe. Although it is possible to reduce axial jacking loads on the pipe and install a low-bearing pipe with microtunneling, a pipe also must have enough strength, or stiffness, or both to perform its intended function. Strength for a pipe is its ability to resist stress. Stiffness is the ability of a pipe to resist deformation. Stresses in a pipe may be caused by loading such as internal pressure, soil loads, live loads, differential settlement, and longitudinal bending. Stiffness is directly related to the modulus of elasticity of the pipe material and the second moment of the area of the cross section of the pipe wall. Durability is a measure of the pipe's ability to withstand environmental effects with time. Such factors as corrosion resistance and abrasion resistance are durability factors.

A flexible pipe is defined as one that is capable of deflecting without cracking under soil loading (See Section 6.2.2). PVC pipe, for example, is considered to be a flexible pipe. Pipe materials that do not meet this criterion are considered to be rigid. Concrete and clay pipes fall in the rigid category of pipes. Classification of pipes to be either rigid or flexible is very important for designing a pipeline. For rigid pipes, strength to resist wall stresses as a result of the combined effects of internal pressure and external load is usually critical. For flexible pipes, stiffness is important in resisting ring deflection and possible buckling. Two components of vertical loading on the pipe are considered in Eq. 4.5:

$$q = w + p \tag{4.5}$$

where q = uniform load on the pipe, lb/ft^2
w = vertical soil (dead) load, lb/ft^2
p = vertical live load, lb/ft^2

Vertical Soil Load. The loads imposed on a pipe buried underground depend upon the stiffness properties of the pipe and the surrounding soil. This results in a statically indeterminate problem in which the pressure of the soil on the pipe produces deflections that, in turn, determine the soil pressure.

As an example, because PVC pipe is considered to be a flexible conduit, it is assumed to deflect under soil loadings. A flexible pipe derives its soil-load carrying capacity from its flexibility. Under soil load, the pipe tends to deflect, thereby developing passive soil support at the sides of the pipe. At the same time, the ring deflection relieves the pipe of the major portion of vertical soil load that is now taken by the surrounding soil in an arching action over the pipe.

Terzaghi's *trap-door theory* (Terzaghi, 1967) or the *silo theory* (Stein et al., 1989) can be applied to consider the arching effect that will occur in the soil when the flexible pipe begins to deflect. According to both the theories, the soil above the pipe with width *WS* (see Fig. 4.3) is assumed to move downward between two vertical planes. When the pipe deflects, the soil begins to shift downward. As it shifts, the friction and cohesion resistance along the potential failure planes begin to increase until the soil support above the pipe is redistributed. Basically, more of the soil weight in width *WS* is now supported by

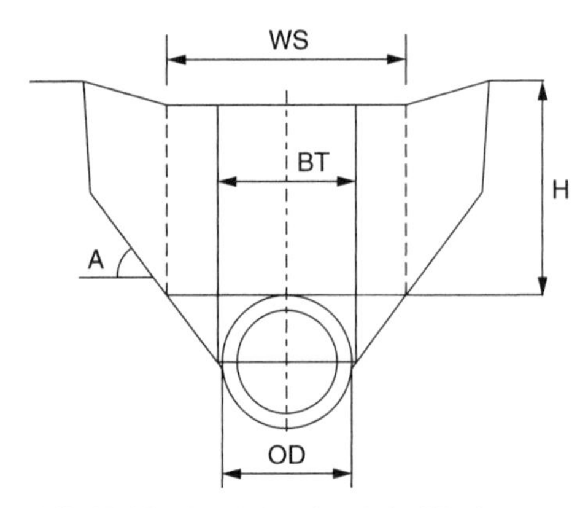

FIGURE 4.3 Calculation of vertical soil load.

the surrounding soil and less by the pipe itself. If the forces acting on a soil section of width *WS* in the vertical direction are summed and equated to zero, then we get Eqs. 4.6 and 4.7 as a result.

$$w = (\gamma - 2C/WS)C_e \tag{4.6}$$

$$C_e = \frac{WS}{(2K\mu)}[1 - e^{-2K\mu H/WS}] \tag{4.7}$$

where w = vertical soil dead load, lb/ft^2 (from eq. 4.5)
C_e = terzaghi coefficient of soil load
WS = width of soil above pipe affected by pipe deflection, ft
γ = soil density, lb/ft^3
C = soil cohesion, lb/ft^2
K = coefficient of static soil pressure
$\mu = \tan \varphi$
H = height of cover, ft
$A = 45° + \varphi/2$, (Fig. 4.3)

There are a variety of opinions on the proper value for *WS* based on the soil conditions and on the amount of shear movement that must occur in the soil to activate the full friction. Terzaghi's technique calculates a wider *WS* than other methods. This results in more soil load acting above the pipe requiring more frictional resistance to be mobilized along the vertical planes and more resistance by the pipe itself. Equation 4.8 defines *WS* as calculated by Terzaghi:

$$WS = BT\left[\frac{1 + \sin(45° - \varphi/2)}{\cos(45° - \varphi/2)}\right] \tag{4.8}$$

where $BT = OD + 0.1$
BT = point at which potential failure slope in soil becomes linear at angle of $45 + \varphi/2$ from horizontal, ft
OD = outside diameter of pipe, ft
φ = internal frictional angle of soil, degree

Vertical Live (Traffic) Load. Pipe placed under a pavement is also subject to traffic loading when the pipe is placed at a reasonably shallow depth. The primary variables expected to affect the critical stresses in pipeline are depth of burial, pipe diameter, pipe wall thickness, and the relative elastic moduli of the pipe and surrounding soil (O'Rourke et al., 1991). However, the live loads resulting from traffic are assumed to be point loads and are analyzed using either approximate methods or Boussinesq's theory of stress distribution. These analyses do not consider longitudinal bending in the pipe resulting from surface loads.

In an approximate method, the stress calculated is considered uniform over the horizontal area upon which it acts. At depths near the surface (about 3 ft), this will not be accurate, because of stress concentrations near the load. However, at the depth of most pipes, this uniform stress is a good approximation of the actual stress.

In the Boussinesq technique (Fig. 4.4), the pavement and underlying materials are assumed to be semiinfinite, isotropic, and elastic. This assumption, though a great simplification of the actual conditions, is used to model actual stresses.

The equation used for this method is as follows:

$$p = \frac{2P(2 + i)}{C(A + 2H \tan \theta)} \tag{4.9}$$

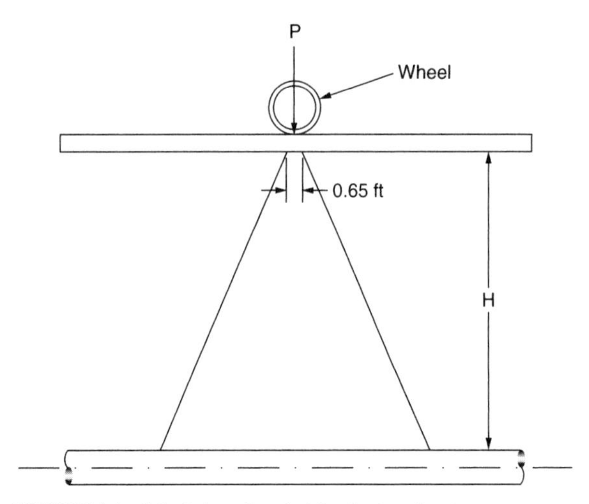

FIGURE 4.4 Calculation of vertical live load on the pipe.

where p = uniform distributed load at top of pipe, lb/ft^2
P = maximum wheel load, lb
i = impact factor
C = occupied width of a vehicle = 9 ft
A = contact length of tire = 0.65 ft
H = earth cover, ft
θ = angle of distributed load, degree

Note that the two P loads (Fig. 4.5) tend to overlap at some depth between the two wheels. Equation 4.9 assumes that the loads are spread uniformly on the area between the

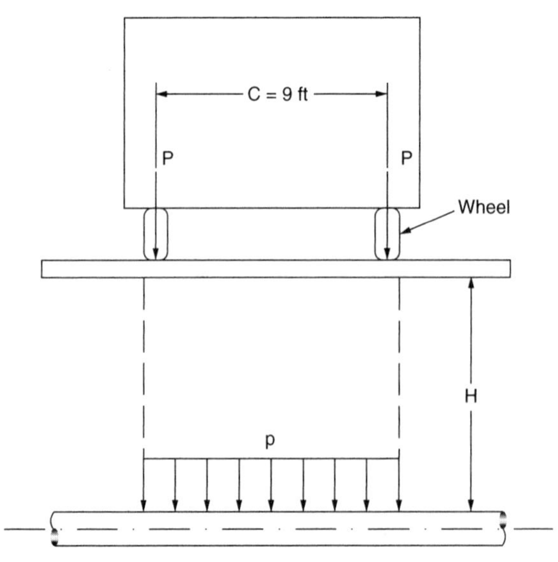

FIGURE 4.5 Details of the wheel loads on the pipe.

TABLE 4.2 Values of Impact Factor for Different Soil Depths Recommended by JRAS

Earth cover (H), ft	$H \leq 5$ ft	$5 < H < 20$ ft	$H \geq 20$ ft
Impact factor	0.65	0.5–0.1	0

two wheels and that there is no load distribution to the right or left of the wheel axle. This is a conservative approach giving higher stresses than actual.

The impact factor is calculated according to the values in Table 4.2. This impact factor accounts for the variability of the moving traffic load. The most severe loading condition occurs when the moving load is in the downward portion of a bounce as it traverses the pavement. If this bounce occurs over the pipe, the loading can be increased significantly. These values are according to Japan Road Association Standards (JRAS).

For comparison purposes, Table 4.3 presents impact factors recommended by the American Association of State Highway and Transportation Officials (AASHTO).

Two additional techniques for calculating live loads are available using the Boussinesq theory of stress distribution. The Hall method considers each tire load as a concentrated load, and the pipe load is calculated accordingly. The Newmark method considers each tire as a pressure spread over an area of pavement, and the pipe load is calculated. As with the JRAS, an impact factor is also included. All these methods obtain reasonable results and can be used to calculate effects of live load on the underground pipelines.

Live loads have little effect on pipe performance except at shallow depths. At extremely shallow depths of cover, a flexible pipe may deflect and rebound under dynamic loading. Therefore, special care should be taken for shallow burials in roadways to prevent surface breakup.

Bending Stress of the Pipe. The bending stress of the pipe is calculated according to Eq. 4.10:

$$\sigma = M/Z \tag{4.10}$$

where M = bending moment per unit length = $0.275 * 8 * r^2$, lb·ft/ft
$r = (D - T)/2$, ft
Z = section modulus per unit length = $BT^2/6 = T^2/6$, ft^3
q = vertical uniform load of the pipe, psi
D = outside diameter of pipe, ft
T = wall thickness, ft
B = unit length, ft

Deflection of the Pipe. Vertical deflection of the pipe is calculated according to Eq. 4.11 (Uni-Bell, p. 161):

$$\delta = 0.179 \times \frac{q * r^4}{EI} \tag{4.11}$$

TABLE 4.3 Values of Impact Factor for Different Soil Depths Recommended by AASHTO

Earth cover (H), ft	$H \leq 1$ ft	1 ft $< H \leq 2$ ft	2 ft $< H \leq 3$ ft	$H > 3$ ft
Impact factor	3	2	1	0

and

$$V = \frac{\delta}{2r} \times 100 \quad (4.12)$$

where V = rate (percent) of deflection, %
δ = deflection, in
q = vertical uniform load of the pipe, psi
r = (D – 2T)/2, in
E = modulus of elasticity, psi
I = moment of inertia per unit length = $BT^3/6 = T^3/6$, ft^4
B = unit length, ft

Longitudinal Loading on the Pipe. Only in truly ideal conditions a pipeline is subjected to only long-term vertical earth loading. There are other forces that in some way produce axial bending stresses in the pipe. These forces can be large, highly variable, and cannot be predicted with a high degree of confidence. Some examples of these forces are differential settlement, ground movement, and earthquake. A complete analysis of these forces are out of the scope of this book but can be found in some of the reference materials listed at the end of this chapter.

4.6 PRINCIPLES OF DRILLING FLUIDS FOR HORIZONTAL DIRECTIONAL DRILLING (HDD)

Bentonite-based drilling fluids are used in both vertical and horizontal drilling. The desirable drilling fluid properties, however, are different for horizontal and vertical applications. For example, a property of drilling fluid that is desirable in vertical drilling may be detrimental or used differently in HDD. It is important to remember that, although much the same materials is used in both vertical drilling and HDD, there are different needs as far as fluid properties are concerned. These properties are adjustable.

The main component of any drilling fluid is water. However, in most cases, water needs to be treated with drilling fluid additives to make it suitable for drilling operations. These additives are usually bentonite, polymers, or a combination of both. The quantity and type additive must be according to the soil type encountered at the job site.

Identification of soils is an important aspect of HDD projects. Problems usually arise with misidentification of soils, or change of soils, or both during the drilling and boring operations. Soil type determines all major decisions, such as selection of drilling fluid type and volume, and selection of bits and reamers. Sometimes a change of soils' moisture content, between the time the project is proposed and the time actual construction starts, may create problems. Therefore, an accurate identification of soils encountered at the job site just before start of drilling operation is necessary for successful and productive completion of the project.

Soils may be separated into two broad categories: *cohesive* and *cohesionless*. In cohesive soils (clay), the soil particles are very small and they tend to stick together owing to electrical charges and chemical bonding, and as a result, they are sticky and plastic. In the case of cohesionless soils (gravel and sand), the bonds are weak or absent and the soil particles do not tend to stick together. Silts typically have some characteristics of clays and sands, and can be some of the most difficult soils to work with in construction.

Soils can also be classified as course grained and fine grained. Course-grained soils are normally classified by particle size, distribution, and texture. Fine-grained soils are normally classified according to their properties as defined by the Atterberg limits. These properties can be determined in laboratory tests in a controlled environment. There is not unanimous agreement on the exact *size* division between the major soil types of clay, silt, sand, and gravel. Gravel and sand are usually considered coarse-grained, because the individual particles are large enough to be distinguished without magnification. Silts and clays are considered fine-grained soil because their small particles cannot be seen with unaided eyes. On a comparative basis, gravel sizes are between 0.08 to 8 in; whereas particle sizes for sand range from 0.003 to 0.08 in. Silt particle sizes range from about 8×10^{-5} to 4×10^{-3} in. Clay particle sizes are those less than about 8×10^{-5} in. It can also be said that coarse-grained soils are larger than No. 200 sieve and fine-grained soils are smaller (pass through) than No. 200 sieve. A brief description of different types of soil is given below.

Clay: Clay is an aggregate of microscopic and submicroscopic particles derived from chemical decomposition of rock constituents. It is a plastic within a moderate-to-wide range of water content. Dry specimens are very hard, and no powder can be detached by rubbing the surface of dried pats with the fingers. The permeability of clay is extremely low.

Sand and gravel: Sand and gravel are cohesionless aggregates of rounded subangular or angular fragments of more or less unaltered rocks or minerals. Particles with size up to 2 mm are referred to as sand and those with a size from 0.08 to 8 in (7.9 in) as gravel. Fragments with a diameter of more than 8 in are known as boulders.

Shale: Shale is a sedimentary rock mainly *composed of silt-size and clay-size particles.* Most shales are laminated and display fissility or splitting, because the rock has a tendency to split along relatively smooth and flat surfaces parallel to the bedding. When fissility is completely absent, the classic sedimentary deposit is called *mudstone* or *clay rock*. Depending on the mineralogy, void ratio, and degree of diagenetic (the process of chemical and physical change in deposited sediment during its conversion to rock) bonding or weathering, compressive strength of shales may range from less than 362 psi to more than 14,504 psi.

Silt: Silt is a fine-grained soil with little or no plasticity. Because of its smooth texture, inorganic silt is often mistaken for clay, but it can be distinguished without laboratory testing. If shaken in the palm of the hand, a pat of saturated inorganic silt expels enough water to make its surface appear glossy. If the pat is bent between the fingers, its surface again becomes dull. This procedure is known as shaking test. After the pat has dried, it is brittle and dust can be detached by rubbing it with the finger. Silt is relatively impervious, but if it is in a loose state, it may rise into a drill hole or shaft like thick viscous fluid.

Generally, for coarse-grained soils (sands and gravels), *bentonite* or a bentonite-polymer mix is used. For fine-grained soils (clays and silts), *polymer* or a polymer-bentonite mix. is used.

Coarse Soils and the Need for Bentonite. To drill into sand and gravel, it is desirable to have a fluid with two particular properties. First desirable property is that the fluid should not permeate through the coarse soils easily. Water by itself is of course not suitable for this. To solve this problem a substance called bentonite is usually added to water. Bentonite, when thoroughly mixed in water, breaks down into small particles called platelets. These flat plate-like particles are very small. In fact, one cubic inch of high-quality sodium bentonite, after mixing until it is broken down to its smallest dimension (the platelet), has enough surface area to cover 66 football fields. When this bentonite fluid is pumped into the borehole under pressure, the bentonite platelets start to plaster or shingle off the wall of the borehole and form a *filter cake* that cuts off the flow of fluid into the surrounding sand or gravel. The water phase of the fluid that does filter through this filter cake is referred to as *filtrate*. It is possible to improve the filter cake quality and *reduce* the amount of filtrate going into the surrounding soil by one of the two methods. More bentonite can be added (more available platelets) or it is possible to use certain polymers in conjunction with the bentonite to *tighten* the filter cake. In most cases, it is more desirable to use a bentonite-polymer system because the end result is a more pumpable fluid and more flowable slurry. It is important to remember, however, that in this case a polymer is used only to enhance the performance of the bentonite. Polymer by itself does not have the necessary beneficial solids to form a filter cake.

The second desirable property of drilling fluid is to provide suspension characteristics or *gel strengths*. Cutterheads and drilling tools such as bits or a reamers serve a very important secondary function. They are also responsible for mixing the soils, with the aid of the fluid, into flowable slurry. The drilling fluid must, therefore, be able to support, suspend, and carry these drilled spoils (cuttings). If the fluid does not have the ability to suspend the drilled material, that material will quickly pack off around the drilling rods, or more dangerously, around the product line (or lines) that are being pulled back into the borehole. Bentonite provides the carrying capacity (gel strength) needed to support the excavated material. Viscosity (the number of seconds that it takes for one quart of fluid to flow through a Marsh viscosity funnel) only gives an indication of the fluid thickness. It is possible to have a thick fluid (high viscosity) that has low carrying capacity (gel strength). This is why the gel strength is a much more important property to look for than viscosity. Water by itself has low viscosity and no gel strength. Polymers by themselves can give us rather high viscosity but low gel strengths. Bentonite, on the other hand, can give us both viscosity and gel strength.

Usually at least 30 to 35 lb of high-quality bentonite per 100 gal is required before any margin of safety is achieved in sand or gravel. Based on specific soil conditions at the job site, sometimes as much as 50 lb/100 gal might be needed. Sands and gravels can be tough and it is necessary to have enough margin of safety to be successful.

When slurried spoils flow out of the borehole either from the exit or entry side, they indicate an open bore path. Having a good slurried flow on both drilling the borehole and the backreaming operation is highly recommended. Before continuing the drilling fluids principles, the following section presents terminologies used with drilling fluids.

4.6.1 Drilling Fluids Terms

Viscosity. A primary objective in HDD is to maintain flow that is usually identified by viscosity. Viscosity is the measurement of the thickness of a fluid and determines the ease with which a fluid pours. It is defined as the property of a fluid that offers resistance to the relative motion of fluid molecules. Viscosity is measured with a Marsh viscosity funnel and is reported in seconds per quart. Viscosity measurements are important in HDD for different reasons than in vertical drilling. Gel strength and filter cake have more importance in horizontal drilling than viscosity but viscosity is a by-product of achieving these desirable properties. Excessive viscosity is undesirable because of higher pressures that can be generated by higher viscosity in the borehole when pumping horizontally. Adding more bentonite would improve gel strength and filter cake but it also raises viscosity level.

Gel Strength. Gel strength is the measurement of the suspension capability of a drilling fluid. Gel strength is measured with a rheometer or shearometer and is reported in pounds per 100 ft^2. Gel strength is of utmost importance, especially in coarse-grained soils (sands and gravels). In a vertical drilling, the solids are removed with viscosity and velocity. In HDD, the velocity is low because of type of pumps and the size of reamers used. In addition, high velocity rates that may be desirable in vertical applications may erode the less consolidated soils that are encountered at shallow depths as 3 to 4 ft often encountered in utility HDD. After a bit or a reamer cuts and mixes the soil with drilling fluid, the function of drilling fluid is to suspend these solids and keep them in suspension until they can be transported out of the borehole. This resulting slurry (solids + drilling fluid) acts like a conveyor belt to remove enough solids to make room for the product line. It is important to remember that, unlike vertical drilling, there is never an *empty* borehole in HDD. The slurry (solids + drilling fluid) aids in supporting horizontal bore paths until the product line is installed. Without gel strengths, the solids will not remain in suspension to maintain this support.

Filtration Control and Filter Cake. Although filtration control and filter cake come together (one cannot exist without the other), they are two separate properties of drilling fluid. In sand, for example, the filter cake quality is of extreme importance and it acts like a sealant, grout, or stabilizing property that maintains the integrity of the borehole. A good quality filter cake (wall cake) cannot be obtained without an acceptable filtrate (water loss). The filtrate amount in a sand situation, therefore, is secondary to the filter cake quality. In clay or shale, the opposite can be true. The low filtrate quantity is more important to prevent hydration, to prevent the water component of drilling fluid from mixing with clay particles and allow swelling to take place. In clay or shale, the filter cake is a secondary property, since a low filtrate volume cannot be obtained without good quality filter cake. Filter cake and filtrate are determined with a filtering process. Filter cake is reported in 32nd of an inch. Filtrate is measured in cubic centimeter (cc).

Fluid Density. Fluid density is not a major concern in HDD as it is in vertical drilling when related to hydrostatic head. Because of the relatively shallow depths at which most HDD installations are made, hydrostatic head does not come into play. This does not mean that drilling fluid (slurry) density is not important. In HDD, the density is used to measure the solid content of the fluid or slurry. A formula is used to convert the density of the fluid or slurry to solid content. This formula, $(\text{density} - 8.33) \times 8 = \%$ solids, is used in two ways. *First*, it can be used to determine the solid content of the return flow into the sump or catch pit. If the solid content is too high, it is an indication that the flow rate needs to be increased. If all available pump capacity is being used, it is an indication to slow down the speed of boring or backreaming operations. *Second*, density can be used to gauge the effectiveness of solid control equipment in a recycling system. The fluid density is measured with a mud balance and is reported in lb/gal.

Sand Content. Sand content determines the amount of sand that is in the fluid or slurry. Sand content should not to be confused with total solid content. The sand content is simply a determination of solids larger than 74 μm (*a millionth of meter or* number 200 sieve) that are entrained in the slurry. Sand content is measured with a special kit and is reported in percent of the total volume.

pH. pH is a measure of the acidity or alkalinity of a fluid, numerically equal to 7 for neutral solutions, increasing with increasing alkalinity and decreasing with increasing acidity. The pH scale commonly in use ranges from 0 to 14. pH is also an indicator of water quality. Low pH indicates the possible presence of calcium. pH amount can be adjusted using soda ash and, sodium carbonate. It is measured with pH indicator strips, papers, or meters.

Drilling Fluid Volume. Volume or volume flow rate is amount of drilling fluid pumped into the borehole. It has considerable importance in HDD as it relates to maintaining flow, reducing torque, and amount soil that will be transported out of borehole. Volume flow rate is expressed in terms of ft^3/sec or gal/min.

Solids. There are two types of solids—beneficial solids and detrimental solids. Beneficial solids are bentonite and other drilling fluid additives. Detrimental solids are those that become entrained in the fluid during drilling operations. There is a belief that sand is the detrimental solid because it may damage pumps. This is because when recirculating drilling fluids, usually too much emphasis is put on sand content. However, sand is probably the easiest solid to remove from slurry because of its size of 74 μm (*a millionth of meter or* number 200 sieve) and larger. Solids 73 μm in size and smaller fall into the category of fine solids or *fine soils*. It is the buildup of the fines that are, in many cases, being overlooked and troublesome. Because large amount of fine materials in slurry, can cause as much damage, if not more, to a pump.

Fine solids mainly come from drilling into reactive soils such as clays or shale. However, this is not always the source. They can also occur in rock drilling, especially when drilling harder rock. They can be the result of harder tungsten carbide insert bits that tend to make smaller cuttings, sometimes more like dust or flour than chip-like cuttings.

Simple tests can determine solid content of slurry. A mud balance can be used to determine the slurry unit weight. From this determination, the following formula is used to calculate the total solid content of the fluid:

$$(\text{Slurry unit weight, pounds per gallon} - 8.33) \times 8 = \text{solids contents as \% of total volume}$$

In the above formula, the average specific gravity of the solids is assumed to be 2.5. This calculation results in determination of total solid content that includes both fine soils and sand content. As stated previously, sand content is also easily determined using a sand content test kit. Sand content is also reported in percentage of total volume. Sand content and the total solid content are the two most important properties to know when operating solids removal equipment. Without knowing these values, there is no idea if solid removal equipment is working properly. Having a large amount of fines or solid content (larger than 3 percent) may result in wear on the impellers in centrifugal pumps that provide header or manifold pressure to hydro-cyclones.

4.6.2 Water: A Drilling Fluid's Very Special Ingredient

Such problems as, the bentonite does not mix well with water, the pump is not pumping right, when polymer is mixed, it gets all hard, when mixing equipment is turned off, bentonite starts to settle out and leaves clear water on top, the packing in this pump does not seem to last as long as it has in the past, it is taking a lot more bentonite to get our viscosity up, may relate to quality of water.

Low water pH, or calcium content, or both (hardness) is not friendly to the drilling fluid products. The bentonite and polymer products are all anionic (negatively charged). The calcium (Ca^{++}) in the *make-up* water not only carries a positive charge but is also a divalent ion that means that it has a strong positive charge.

What does this do to the bentonite? It was stated previously that when properly mixed, bentonite is sheared down to small individual platelets that carry a predominately negative charge. Because of this, each bentonite platelet is repulsed by other platelets, keeping the individual platelets separated. When calcium is present, the platelets tend to stay in little stacks like decks of cards and not shearing apart. With attractive forces present (instead of repulsive forces), platelets tend to clump into larger particles. This is why free water surfaces on top of tank when mixing equipment is turned off for a period of time. These larger particles want to settle out, indicating that the bentonite was not properly mixed in the first place.

The clumps of larger particles also stick to pump seats and valves, not letting the pump work freely. This is where the *clanging* in the pump can occur. The larger particles can also cut out pump packing system.

Polymers that also carry an anionic (negative) charge can be affected the same way. Calcium in the water can prevent the polymer from fully yielding or hydrating, keeping it from *blossoming* to work effectively.

What can be done about this problem? The first thing to do is to check water pH. On the pH scale, 7 is neutral. Both bentonite and polymers mix better with an elevated pH range of around 9. To raise the pH, a product called soda ash or sodium carbonate is used. This product both raises the pH and simultaneously treats out (precipitates) the calcium in the water. Soda ash can be easily found at drilling fluids supplier, or local fertilizer, or swimming pool supply store. It can be purchased in a 50- or 100-lb sack (the least expensive way to buy it) and keep it sealed in a 5-gal bucket. Soda ash should not be confused with sodium bicarbonate. Although the name sounds similar and will treat out calcium, sodium bicarbonate will lower the pH.

Using pH paper or strips, the results can be monitored until desired levels are obtained—adding soda ash until the pH is around 9. Usually a normal treatment requires about 1/3 to 1/2 lb per 100 gal of water. While using more soda ash cannot hurt, it is a waste of money, but it is better to use too much than not enough. Using city or tap water may still require use of soda ash, because of hardness (calcium) in city water supplies. It should be noted that quality of a specific water supply can change over time because of such reasons as drought, water main breaks, change in water treatment systems, and water source.

4.6.3 The Drilling Square

When a contractor has a drilling fluid problem, he or she is sometimes surprised when the cause of the problem might be other project factors than drilling fluid. Such project factors might be the soil, the bit, the reamer, and the volume of drilling fluid being pumped.

The factors that make a successful bore are too interrelated to point a finger at any one of them as being the cause of a problem in all cases. These factors can be shown in the form of a square (Fig. 4.6).

Notice that the base or foundation of this square is soil identification. Every decision should be based on properly identifying the soil we intend to drill. While for large diameter HDD projects (Maxi-HDD) a geotechnical investigation is required, for many small diameter and medium diameter (Mini- and Midi-HDD) it is not performed. In these situations contractors and HDD operators can take advantage of opportunities such as exposing existing utilities or digging a slurry catch pit, to obtain a sample of soil material. With using visual

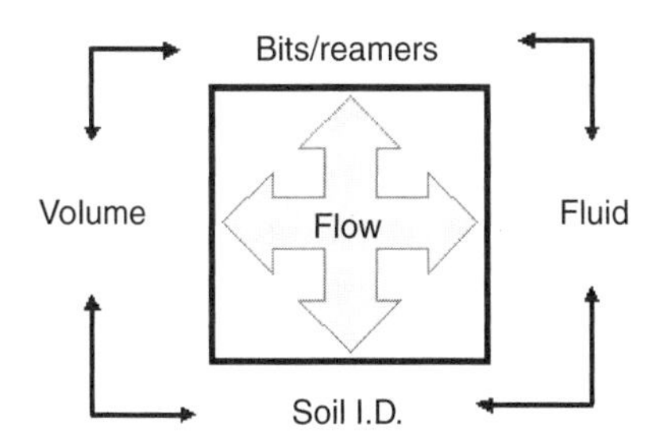

FIGURE 4.6 Drilling square.

and manual testing methods, a competent operator can identify the type of soil encountered with reasonable accuracy.

Several soil qualities are assessed in addition to the type. Those qualities are grain size, saturation, cohesiveness, and unconfined compressive strength. There are four types of grain sizes: gravel, sand, silt, and clay. If a grain of soil is larger than pencil lead, it is gravel, if it is smaller, it is sand. Your naked eye cannot see particles of clay and silt, but if soil clump when dug, it could be clay or silt.

Manual Testing. Manual soil tests are required before a protective system is selected. A sample taken form soil dug out into a spoil pile is tested as soon as possible to preserve its natural moisture. Manual tests include:

- **Sedimentation test:** Determines how much silt and clay are in sandy soil. Saturated sandy soil is placed in straight-sided jar with about 5 inches of water. After the sample is thoroughly mixed and allowed to settle, the percentage of sand is visible.
- **Wet Shaking test:** Wet shaking test is another way to determine the amount of sand versus clay and silt in a soil sample. In your hand, shake a saturated sample to gauge soil permeability based on the following facts: Shaken clay and silt resists water movement through it and water flows freely through sand.
- **Thread test:** Thread test determines cohesion. A representative soil sample is rolled between the palms of the hands to 1/8″ diameter and several inches in length. The rolled piece is placed on a flat surface, and then picked up. If a sample holds together for two inches, it is considered cohesive.
- **Ribbon test:** Ribbon test determines cohesion and is used as backup for the thread test. A representative soil sample is rolled out (using the palms of your hands) to 3/4″ in diameter, and several inches in length. The sample is then squeezed between thumb and forefinger into a flat unbroken ribbon 1/8″ to 1/4″ thick that is allowed to fall freely over the fingers. If the ribbon does not break off before several inches are squeezed out, the soil is considered cohesive.

4.6.4 Fluid Selection

Once the soil is identified, the type and amount of drilling fluid must be selected. As stated previously, generally bentonite is used for drilling sand and gravel and polymers are used for clay and shale. However, decision about product concentrations and additives need to be made in consultation with a drilling fluid engineer or representative of a drilling fluid manufacturer.

4.6.5 Bits and Reamers

Bits and reamers need to be matched with the soil type. If steering is a problem because the ground is too soft, we can go to a wider blade. If the ground is hard, going to one of the stepped or tapered carbide bits may be the answer. Bits are available with granular carbide in a bronze or steel matrix. Bits are designed with replaceable carbide inserts for rock (or soil that wants to act like rock). When we reach the exit point, we can observe the condition of the bit to see if we need to make any changes in our fluid or reamer selection. For example, if the bit is covered with clay, we probably need to add a wetting agent to the fluid to prevent stickiness. We may also need to choose a reamer that will chop up the clay to prevent large pieces from bridging off behind the reamer. If we have encountered any unexpected rocks, we may need a fluted or spiral reamer to press them into the side of the borehole. Again, bit or reamer selection and possible fluid changes are all dictated by the soil.

4.6.6 Volumes

We can make all the right decisions up to this point and still fail by not pumping enough fluid. In a non-reactive soil such as sand, we may be able to produce flowable slurry with as little as 1 to 1$^1/_2$ gal of fluid per gallon of soil, especially on short bores. We can do this because sand is inert and it does not swell and get sticky. Clay, on the other hand, can swell and get sticky. Because of this, 3 to 4 gal or more of fluid may be required per gallon of clay soil to maintain flow.

There is no universal soil. Because of this fact, there is no universal fluid, bit, or reamer. Nor is there a universal volume of fluid to be pumped per gallon of soil. Everything is interrelated.

4.6.7 Solid Removal Systems

As more utility contractors are moving up to larger drilling rigs, they are also moving toward the use of solid removal (solid control) systems. These solid control systems are not *stand-alone* units. They are a part of the drilling fluid system along with the drilling fluid and the fluid pump. Compromises and adjustments need to be made so that everything will work on an individual basis and will also work efficiently as a team.

Viscosity. High viscosity fluids do not pass through shaker screens as easily as lower viscosity fluids. They tend to create surface tension that can blind the screen openings and cause the fluid to pass off the end of the shakers. This can happen even with rather coarse-mesh screens. The viscosity problem continues on to the hydro-cyclones (desanders and desilters). Each size hydro-cyclone operates most efficiently within a certain pressure range. If the available pump pressure of the pump feeding the hydro-cyclone is being used up in moving the fluid, the efficiency of the hydro-cyclone is greatly reduced.

The viscosity problem may be coming from building too much viscosity with drilling fluid additives (bentonite and polymers). However, the problem can easily come from high solid content in the slurry that is being pumped from the sump up to the solid control unit. The solution is to pump more fluid during backreaming operation. The larger pumps that come with larger rigs are not used with full capacity during drilling operations, unless using mud motors. These larger pumps, however, are extremely necessary when backreaming. Because of the larger reamers, more soil is being cut and, *therefore*, more solids are being created in the slurry. If you do not have a larger pump, the backreaming operation must be slowed down to maintain lower solid content and, therefore, a lower slurry viscosity.

When adding water to the system, it needs to be added at the sump pit or at the shakers so that the thinning effect from any water additions will help the separation equipment be more efficient. A general rule is to add anything that thins the fluid as early as possible, at the sump for example. Anything that thickens the fluid should be added to the system as late as possible, at the pump suction pit for example. This is why the venturi mixers are always the last stage of a system.

When mixing fluid additives through a venturi or any other type of mixer, mix the material slowly over an entire circulation (tanks plus hole volume) or multiple circulations. This will give some assurance that the fluid and slurry is consistent throughout the system.

Polymer Additives. Certain polymer additives used routinely in HDD can cause problems, if mixed incorrectly or in high concentrations when using solid control equipment. Some polymers can cause a coating action on screens, making them blind off. This is not to say that we can not use this kind of polymer. We do, however, need to pay closer attention to the way we add the material to the system and where we add it. Drilling fluid additives of any type should never be added ahead of the solid control equipment: and always behind it in the suction section.

Material Additions While Drilling. Pump manufacturers are generally not in favor of mixing drilling fluid materials *on the fly* while drilling and, in many cases we can not blame them. When added rapidly, bentonite will ball up and cause damage to pump parts. Use of a *self-feeding hopper*, or a venturi hopper that is filled with bentonite and allows to flow quickly, is not recommended. Drilling fluid materials need to be added slowly! This will allow the venturi mixer to wet the material quickly and thoroughly and prevent these clumps from forming.

4.6.8 Problem Solving

Fluid Volumes: How Much Is Enough? There seems to be a certain amount of confusion about the role of drilling fluids in HDD, especially under roads, streets, and highways. There is concern about voids, washouts, and subsidence caused by drilling fluids and annular space (the space between the formation and the product line).

The first misconception is in the drilling process itself. The process by mistake is perceived by some to be a jetting action where the soil is cut literally by washing with water pressure. In HDD operations, however, we are involved in *fluid-assisted mechanical cutting*. We are cutting through the earth with a drill bit or reamer and the only real need for pressure is to keep the bit or reamer clean and maintain flow. The soil is blended with the fluid to form slurry and enough amount of this slurry should be allowed to flow out of the bored pathway to make room for the product line. Unlike vertical drilling, there is never an *empty* borehole. The slurry is what actually holds the bore path open and stable. The second misconception is that we are drilling with water. The drilling fluids have water as their base but certain additives are added to the water to make it suitable for drilling operation.

The drilling fluids are designed to have several functions. They work as soil stabilizers and lubricants. In sand and gravel, they work as grouts to hold the sand in place and prevent borehole caving. In clay materials, they retard swelling and reduce sticking of the soil to the bits, reamer, and the product line. These additives allow us to make the installation by allowing us to control the various soil types and keep the surrounding formation in as much of a native state as possible.

The next misconception is that we are leaving voids in the annular space between the formation and the product line being installed. An extremely popular question is "What happens to the annular space when the water goes out into the clay?" Studies have shown that by using a *designed* drilling fluid and proper installation procedures, the void in annular space will be gradually eliminated by redistribution of soil surrounding the product pipe. While this process is different for different soil and project conditions, the void in annular space should not cause any surface distress with a proper HDD installation.

As stated previously, there is no *universal* rule about how much fluid should be used for any type of project conditions. That depends on the soil type. Sand, for example, is inert. It does not swell and it does not get sticky. In sand, we can get by with using little more than 1 gal of fluid to each gallon of soil (see example below), especially for shorter bores. Longer bores require a higher percentage of fluid to maintain flowability of the slurry. Clay, however, is reactive and more fluid is needed to maintain flowability because of the swelling tendencies of the clay. In clay, several gallons of fluid may be necessary for each gallon of clay.

There is a simple formula used to calculate the soil volumes in the bore path:

$$(\text{Bit or reamer size in inches})^2 \div 24.5 = \text{gal of soil/ft of borehole (for English units only)}$$

For example for a 6-in bit or reamer:

$$6^2 \div 24.5 = 1.46 \text{ gal of soil per foot of borehole}$$

For a 10-in reamer:

$$10^2 \div 24.5 = 4.08 \text{ gal of soil per foot of reamed hole}$$

What happens if we use too little fluid and do not produce flowable slurry? If we are using an 8 in reamer, we are dealing with 2.6 gal of soil per foot. If we are pulling 5 in product line, we are pulling 1 gal of product line per foot. If we are in sand, we need *at least* 1 gal of fluid per gallon of soil. However, let us say that in this situation we are only pumping 1 gal of fluid per foot instead of the necessary minimum of 2.6 gal. There is a definite possibility of creating an undesirable speed bump across our highway. Why? We do not have flow because we have not created flowable slurry and we are now trying to force 4.6 gal of material (soil, product line, and fluid) into a space that originally only had enough room for 2.6 gal. Enough fluid *must* be pumped. Enough annular space *must* be created to allow for flow.

4.7 PRINCIPLES OF TRACKING AND LOCATING FOR HORIZONTAL DIRECTIONAL DRILLING

4.7.1 Introduction

It has been said that locating systems put the *directional* into directional drilling because these tools are responsible for allowing the drilling crew to make educated steering decisions based on real-time position and heading information. There are two main categories of these systems currently available, the walk-over locating system and the wireline guidance systems. There are variations within each of these categories as well as systems that overlap the categories. For the sake of simplicity, we will focus on the two categories but mention some of the variants as well. Majority of all HDD work is done with walk-over systems because these are much less expensive and simpler to use. The transmitters for use with these systems are typically battery powered (most often by C cell alkaline batteries) and maximum depth range of such systems is about 60 to 70 ft.

Wireline guidance as the name implies uses a transmitter that is powered by a wire that runs on the inside of the column of drill pipes. Generally, a section of wire needs to be spliced every time a new drill pipe is added. This is much more time consuming than using the battery-powered transmitters. These systems are most often required if the bore is either too deep or long for walk-over systems. An example would be a large river or multiple lane highway crossing. We will discuss both these systems in greater detail.

4.7.2 Overview of Walk-Over Locating Systems

A typical HDD walk-over locating system is shown in Fig. 4.7. In order to keep the terminology clear, let us start defining the following terms:

- *Receiver*. The locating or tracking device
- *Locator*. The person operating the receiver
- *Operator*. The person operating or controlling the drilling machine

A walk-over locating system typically comprises the following components:

- A transmitter (typically 15 in. in length by 1.25 in. in diameter)
- A receiver or locating device
- A remote display

FIGURE 4.7 HDD walk-over locating system.

The transmitter, beacon, sonde (transmitter), or probe as they are sometimes called, sends out a magnetic signal. A secondary signal, the data signal, is also emitted from the transmitter. This signal typically includes clock (roll), pitch, and transmitter temperature and battery status. The clock readout is typically in 12 to 24 distinct clock positions while the pitch or transmitter inclination is displayed in whole percent or degrees or in some cases $^1/_{10}$ of degrees or percent for gravity sewer installations that are increasingly becoming more common. Temperature information acts as a warning indication to the drilling crew and helps in avoiding damage to the transmitter's sensitive electronic components from the frictional heat generated during the drilling process. Battery status indicates the amount of battery life left in the transmitter.

The receiver picks up the transmitted information and translates the two signals into useful information for the locator. A copy of most of the pertinent locating information is sent via telemetry from the receiver to the remote display. This allows the machine operator to view most of the information the locator has access to.

4.7.3 Brief Theory Behind the Operation of Walk-Over Systems

The transmitter generates a magnetic dipole field. The frequency of this field is generally in the 25 to 40 kHz range although some systems employ much lower frequencies in certain situations. The more sophisticated receivers take advantage of the fact that the shape of the magnetic field is well understood. This allows the receiver to find the following locations within the magnetic field with great accuracy.

These are the forward and rear locate points as well as the actual over the transmitter position. The forward and rear locate points are the only two points within the magnetic field where the magnetic flux lines are vertical. Owing to the nature of the dipole field, these points also very accurately define an axis that runs along the center of the transmitter. Putting it differently, the forward and rear locate points indicate the direction of the transmitter. It should be noted that there is an *infinite* number of locate point pairs, depending on the depth of the transmitter. The deeper the transmitter is, the further apart the two points are, but they still define the axis of the transmitter. The distance between the locate points is, approximately, 1.4 (1.4142) times the transmitter depth when the transmitter is level.

As can be seen in Fig. 4.8, the magnetic flux lines are horizontal directly over the transmitter (assuming the transmitter is level) and using this fact, the position of the transmitter along the bore path can be quite accurately determined. This is often referred to as the

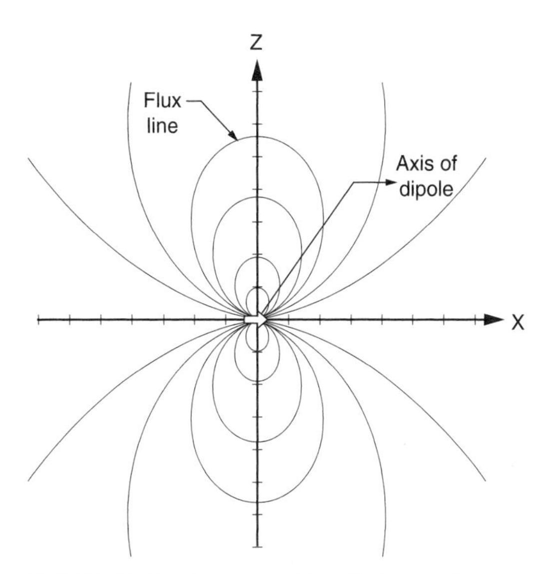

FIGURE 4.8 Flux lines produced by a dipole transmitter.

locate line as a contrast to the locate points (see Fig. 4.9). This line is perpendicular to the axis of the transmitter. Combining the information available from the two locate points (direction) and over-the-transmitter location (position along the transmitter axis), the direction and position of the transmitter below the surface is now known. Once this has been established, a depth reading can be taken.

4.7.4 Prebore Activities

Prior to doing any drilling there are a number of steps that need to be taken as far as the locating system is concerned. We assume that a proper bore plan has been developed,

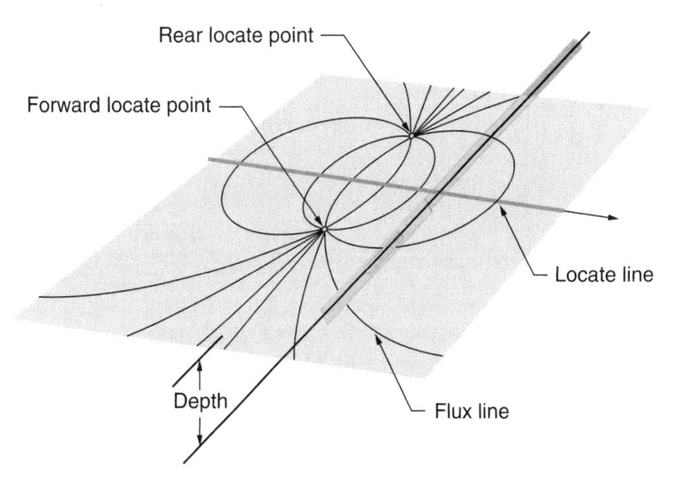

FIGURE 4.9 Locate points and the locate line.

that all existing utilities being crossed have been identified and exposed and that all the proper permits are in hand.

As most drilling takes place in urban areas, a common problem encountered in HDD locating is the effect of interference. Interference is often referred to as being either active or passive. Active interference can be defined as "anything that emits a signal or generates its own magnetic field." That being said, all things electrical emit a magnetic field. Some examples of active interference include power lines, traffic loops, fiber trace lines, invisible dog fences, and an unknown source. Do not assume that just because there is no evidence or markings on the ground that there is nothing there. Some of the possible effects of active interference on a receiver include erratic signal strength and depth readings, loss of pitch and roll data, and inaccurate calibration that may lead to depth errors.

Passive interference, as the name implies, does not generate a signal. It could be defined as "anything that blocks, absorbs or distorts a magnetic field." Examples include metal structures such as chain link fences, rebar, salt water, and other unknown sources. Anything that is conductive has the potential to act as a passive interference. Possible effects include depths appearing greater (or in some cases shallower) than they actually are, incorrect drill head location and direction, all information being blocked and incorrect calibration that may lead to depth errors.

Given that interference exists, it is very important to try to determine how much interference is present and how it can potentially affect the locating equipment. This is a two-step process. Start by walking the intended bore path without a transmitter turned on, looking for signal readings on the receiver. The higher the signal received, the greater is the interference. Walk the entire path making notes of the signal being read and identify potential problem areas.

Once the intended drill path has been walked, it should become clear which areas might be influenced by interference. It should be emphasized that this first walk through tests primarily for effects on depth readings. It however stands to reason that the amount of signal being registered by the receiver from the transmitter needs to exceed the interference by a significant amount to ensure adequate signal for locating. It is therefore important to be familiar with the locating equipment and the signal levels expected at various depths.

The second part of the test involves the transmitter. The purpose is now to see how reliable the pitch and roll signal will be. Let us assume that the planned depth for a particular bore is 12 ft. At this depth the receiver will see a pitch or roll signal from the transmitter in the ground of a given strength (signal magnitude). The issue to be determined is, is this signal powerful enough to overcome the interference? One way to find out is to simulate the bore. This requires two people, one carrying the receiver along the bore path and a second person carrying the transmitter. Insert batteries in the transmitter and move it away from the receiver one and a half times the anticipated drill depth. In this example, this would be 18 ft (1.5 × 12 ft). This is to take into account the signal loss that occurs when a transmitter is placed into the drill housing. Otherwise, one of the crew needs to carry a heavy housing with the transmitter in it at the 12-ft distance. The two walk the bore path keeping the required separation. If the pitch and roll information is not affected during this test, it is reasonable to assume clear sailing during the bore. Although this test is quite effective, it cannot always pick up all potential problems.

4.7.5 Walk-Over System Operation

Calibration and Test. Many drillers are in the habit (or are required by inspectors) to calibrate their receivers on a daily basis or before every shot. This is not necessary because most receivers will hold their calibration for an extended time. It is however necessary to check the receiver calibration before each bore.

Given what was discussed earlier about interference, in some cases it may be detrimental to calibrate too often as that can increase the likelihood of a bad calibration.

There are, however, situations when calibration is required.

Calibration is required when you change from one transmitter type to another, that is, from a standard to long range and also when you change from one receiver to another.

Calibration is required if you change from one drill head or transmitter housing to another.

Calibration is required if the receiver does not read the correct depth at the test distance after you have ensured that you are not in a high electrical interference area or near metal objects. Metal can weaken signal strength, thereby resulting in greater-than-expected depth readings. It is also normal for a transmitter to lose some signal strength as it ages.

The first thing to identify is an interference free area to perform either the calibration or the calibration test of the locating equipment. Calibration of the locating equipment takes place on top of the ground, with the transmitter in the transmitter housing. The housing should be at least 10 ft away from the drilling machine to minimize passive interference effects from the drilling machine's metal structure.

During the calibration process, the receiver measures the signal strength coming from the transmitter at a prescribed distance and assigns the signal strength a numerical value. In this country, this is most typically an inch measurement, whereas our colleagues overseas prefer the centimeter. Once the signal strength reading has been recorded, the receiver now has the proper information stored in its memory that allows it to calculate, based on any signal strength received, a depth measurement. It should be noted that the calibration procedure only involves the depth signal and the roll pitch readings are not in any way affected or *calibrated*. The next step is to test the calibration. This can be done by selecting a distance other than the calibration distance. A good rule would be to select a distance equal to the deepest part of the bore and compare the readings on the receiver to the measuring tape. If the discrepancy is minimal or acceptable, the drilling process can proceed.

Locating the Transmitter in the Ground. It is important to use a systematic approach when locating the transmitter. This not only speeds up the process but also reduces the guess work that otherwise might be involved. As mentioned in the theory section, the receiver is used to identify its location relative to the signal source, that is, the transmitter. It is recommended to first locate the forward locate point. This is because the first decision then becomes whether the transmitter is headed in the right direction. As the locate point is out ahead of the transmitter position, this allows the locator to make left and right steering decisions before the transmitter gets too far off the course. Once the forward point has been located and a steering determination has been made, the next step is to locate the head. As the forward locate point is directly in front of the transmitter, this is fairly easy. Walk back toward the drill machine and once the receiver indicates you are over the head, take a depth reading. Let us assume that 10-ft drill rods are being used and the head is located once each drill rod has been drilled in. The transmitter can of course be located as often as the locator wants. Given this, the next forward locate position will be about 10 ft (depends on the pitch angle of the transmitter) further out as will be the transmitter location. Find the forward locate point and determine whether any left or right course correction is needed. Walk back in the direction of the last known head location and once you are over the head, take a depth reading.

Some locating systems simplify the above procedure by showing the movement of the locate point and locate line in real time. This allows the locator to react very quickly to any course deviations and essentially locate the head in real time as opposed to once or twice per drill rod.

Remote or Target Steering. Most walk-over systems have the capability of remote steering. Using this method, the receiver is placed out ahead of the transmitter and is used as a target. Directional information is displayed on the remote display that allows the operator to steer the drill head toward the target (receiver). Some systems display current depth information as well in this mode. By using pitch information, the bore profile can be maintained and the directional indication is used to maintain the correct course. The distance that the receiver or target can be placed out in front of the transmitter, depends on the transmitter range and the depth the transmitter is at. As an example, using a standard transmitter at 10-ft depth, one can place the receiver about 25 ft out in front of the transmitter. This method is good for crossing small streams or single or dual lane roads without endangering the locator in traffic. The accuracy is however affected by any interference sources located between the transmitter and the receiver.

Magnetic Wireline Guidance Systems. These systems are generally employed when the depth of the proposed bore or the length is such that use of walk-over locating systems is not practical or feasible. There are two primary differences between the two categories of systems:

- No walk-over detection is required
- The wireline transmitters are typically powered by a wire (although exceptions do exist)

These systems most commonly rely on a wire connection inside each rod to power the transmitter and to relay all the data back to the operator station. This means that every time drill rod is added, a section of wire is spliced to continue the wire connection. In many cases, the drill rods themselves act as the ground path. The power source can either be the drilling machine battery, an external battery or any other DC power source that complies with the voltage and amperage requirements of the guidance system.

Wireline guidance (or steering tool systems as they are often called) systems rely on a magnetic compass within the transmitter to determine the transmitter heading (horizontal angle) in degrees in addition to the tool face orientation (clock) and inclination or pitch (vertical angle). Typically the clock has 360 positions and is indicated in degrees where 90° is the equivalent of 3 o'clock with the walk-over systems.

By integrating the magnetic heading as well as the inclination a depth and deviation from the planned course of the transmitter can be identified and displayed on a graph or in coordinates relative to the starting point of the bore. Using this system, all the steering information is available at the drilling machine operator station.

Owing to the reliance on a magnetic compass there are some special requirements for the use of such systems. Any metal in the vicinity of the compass will throw off the readings. As a result, the first requirement is that the transmitter needs to be in a nonmagnetic housing. The second is that the drill pipes themselves can affect these readings. Therefore, a nonmagnetic section of drill pipe, often 15-ft long, is required in between the nonmagnetic housing and the rest of the drill string. A third thing to be mentioned is a method to overcome magnetic disturbance caused by ferrous materials along the bore path. TruTracker®, which is sometimes referred to as TruTracking is a patented method that allows the wireline transmitter to be located independent of the magnetic readout. A coil of wire is laid out on the surface in accordance to a surveyed pattern. DC power source is used to send a current though the wire that generates a magnetic field received by the transmitter. In this manner, the position of the transmitter can be determined relative to the surveyed wire. In most cases, the actual guidance is performed by expert locating personnel (steering engineers) that accompany the wireline guidance system as an integral part of a rental package.

Hybrid Wireline Systems. There are locating systems available that could be called hybrid, that is, systems that overlap the walk-over and wireline system categories.

The first version would be the traditional walk-over systems used in conjunction with a cable transmitter. These transmitters have a depth range in excess of 100 ft and such range is not feasible using battery power. The alternative is to power the transmitter from an above-ground power source in the same manner as traditional wireline guidance systems. In this case however, the only data transmitted through the wire to the remote display are the clock, pitch, temperature, and battery status data. As long as there is a connection, these data are available to the operator. This system can also be very beneficial in situations of very high interference where the signal from a battery-powered transmitter is not strong enough to overcome site conditions. The locating operation using such systems is identical to the traditional walk-over systems where depth readings can only be achieved with a receiver positioned over the transmitter. These systems can however be used when traversing smaller rivers or highways where during the actual crossing the drilling crew relies on pitch and clock readings only from the remote display to maintain the predetermined bore profile. Depending on the width of the crossing, some amount of directional (left/right) control can be achieved toward the end of the crossing by using remote steering.

The second category of such systems is one that has the capabilities of both a traditional walk-over system as well as by using a special transmitter can function as a magnetic wireline guidance system. In this case, depending on the job requirements, the same system can be used for a small house connection in one project and then cross a large river in the next one. This system not only sends the entire traditional magnetic guidance information through the wire that is horizontal, but also sends out a magnetic field that can be tracked from above. In this manner, the location of the transmitter can be determined using the magnetic data and confirmed by walk-over locating methods.

Recent and Future Developments. Some of the more recent developments include a tension load and drilling fluid monitor for the pullback operation. This allows the drill rig operator to monitor, in real-time, the pull force on the product and the downhole mud pressure. The advantage of product tension information is that corrections can be made right away during the installation instead of afterward when the product fails the integrity test. The advantage of knowing the downhole mud pressure could result in avoiding frac-outs (inadvertent fluid returns) or worse, road buckling.

Another recent development is in the area of wire connections; each drill pipe is prefitted with a permanent connector. These connectors comprise inserts at both the pin and box end of every rod. Once installed, the electrical connection is secured when the drill stem is made up. What used to take anywhere between 5 and 15 min is now achieved in the time it takes to thread the rod ends together. Faulty splices and wires caught in the rod joints are a thing of the past, as are all the crimps, heat shrinks, heat shrink tools, and spools of wire. The connection system is expected to last the life of the drill pipe.

A large number of nonmetallic conduits have been installed in recent years without adequate documentation about their location and depth. Many of these are too deep for traditional utility locators. HDD locating manufacturers have responded by developing small, inexpensive transmitters that send out a locating (depth) signal. These can have a range of as much as 100 ft. The transmitter is pulled through the conduit and the location and depth can be determined using traditional HDD locating methods.

The developmental focus for many locating manufacturers is in the area of interference rejection, accuracy, versatility, and ease of use. Manufacturers are using increasingly more sophisticated digital signal processing (DSP) methods to improve the receiver's capability to filter out noise (interference) and thereby increase both range and accuracy. Versatility is being achieved by introducing as an example dual frequency systems that allow the locator to change frequency of both the transmitted and received signals for optimum performance in varying conditions. Ease of use is also important to minimize human error.

Obstacle detection is yet another field being studied. The premise is to develop methods that would allow the locator to receive a warning of some sort as they approach existing utilities in the ground. This would help minimize utility strikes and further enhance the overall HDD process.

4.8 SUMMARY

A successful trenchless construction method requires a thorough understanding of project surface and subsurface conditions. This chapter provided various parameters needed *during the planning and design stage* to select an appropriate trenchless construction method. Other design issues such as alignment considerations for pipe jacking and microtunneling and calculation of jacking forces were discussed. Principles of drilling fluids, an important aspect of successful HDD operation, was discussed in detail. Another important aspect of HDD process discussed in this chapter is the tracking system, using both walk-over and wireline technologies. Finally recent and future developments in tracking technologies were presented.

REVIEW QUESTIONS

1. Describe the elements of surface and subsurface investigation requirements.
2. List and describe methods of locating and identifying existing utilities.
3. Describe the importance of drilling fluids in HDD operations.
4. List and describe functions of drilling fluids.
5. How are the jacking forces calculated in pipe jacking operations? Explain.
6. Describe the different forces acting on the pipe in, during, and after installed in HDD projects.
7. Describe how the tracking and locating systems for HDD equipment work.
8. What are the different types of tracking systems? Explain.
9. What factors influence jacking forces? Explain.
10. Calculate the jacking force for a 24-in OD fiberglass reinforced polyester mortar pipe installed in a clayey soil for a distance of 500 ft. Use the circumferential frictional resistance values in Table 4.1.

REFERENCES

Ariaratnam, S. T. (2001). *Evaluation of the Annular Space Region in Horizontal Directional Drilling Installations*, Arizona State University.

ASCE (2005). *Manual of Practice for Pipeline Design for Installation by Horizontal Directional Drilling*, American Society of Civil Engineers, Reston, Va.

ASCE (2003). *Standard Guidelines for the Collection and Depiction of Existing Subsurface Utility Data*, American Society of Civil Engineers, Reston, Va.

ASCE (1997). *Geotechnical Baseline Reports for Underground Construction*, American Society of Civil Engineers, Reston, Va.

ASCE (2000). *Standard Practice for Direct Design of Precast Concrete Pipe for Jacking in Trenchless Construction*, American Society of Civil Engineers, Reston, Va.

Bennett, D., and P. A. Taylor (1993). Construction of microtunneling test facility at WES and preliminary test results, *Proceedings of Trenchless Technology Advanced Technical Seminar,* Vicksburg, Miss.

Blewitt, R. J. (1987). *Analytical Study of Stresses in Transmission and Distribution Pipelines Beneath Railroads*, Topical Report of Task 2, Department of Civil Engineering, Cornell University, Ithaca, NY.

Bowles, J. E. (1988). *Foundation Analysis and Design*, McGraw-Hill Inc., New York, NY.

Buczala, G. S., and M. J. Cassady (Eds.)(1990). *Buried Plastic Pipe Technology*, ASTM, Philadelphia, Pa.

Castronovo, J. P. (Ed.)(1991). *Pipeline Crossing, Proceedings of a Specialty Conference*, American Society of Civil Engineers (ASCE), New York, NY.

Concrete Pipe Design Manual (2000). American Concrete Pipe Association, Vienna, Va.

Das, B. M. (1990). *Principles of Foundation Engineering*, PWS-KENT Publishing Company, Boston, Mass.

Deflection—A Mark of Excellence (1979). Uni-Bell Plastic Pipe Association, Dallas, Tex.

Developments underground, NO-DIG (1989). *Proceedings of Fourth International Conference and Exhibition on Trenchless Construction for Utilities*, ISTT, London, United Kingdom.

Hair, C. W. III (1995). Site investigation requirements for large diameter HDD projects, *New Advances in Trenchless Technology: An Advanced Technical Seminar*, St. Louis, Mo., February 5–8.

Hair, J. D., and C. W. Hair III (1988). Considerations in the design and installation of horizontally drilled pipeline river crossings, *Proceedings of Pipeline Infrastructure Specialty Conference*, American Society of Civil Engineers.

Hair, J. D. (1995). Design and project management considerations involved with horizontal directional drilling, *New Advances In Trenchless Technology: An Advanced Technical Seminar*, St. Joseph, Mo., February 5–8.

Hair, J. D. (1991). Analysis of subsurface pressures involved with directionally controlled horizontal drilling. *Proceedings of the ASCE Pipelines Conference.*

Hancher, D. E., T. D. White, and D. T. Iseley (1989). *Construction Specifications for Highway Projects Requiring Horizontal Earth Boring and/or Pipe Jacking Techniques*, School of Civil Engineering, Purdue University, West Lafayette, Ind.

Handbook of PVC Pipe—Design and Construction (1993). Uni-Bell Plastic Pipe Association, Dallas, Tex.

HDD Consortium, *Horizontal Directional Drilling Good Practices Guidelines*, North American Society for Trenchless Technology, Arlington, Va.

Ingold, T. S., and J. C. Thomson (1989). Site investigations related to trenchless techniques, *Proceedings of the International Society for Trenchless Technology NO-DIG '89.*

Iseley, D. T., and D. H. Cowling (1994). *Obstacle Detection to Facilitate Horizontal Directional Drilling*, prepared by the Trenchless Technology Center for the American Gas Association.

Iseley, T., M. Najafi, and R. Tanwani (1999). *Trenchless Construction Methods and Soil Compatibility Manual*, 3rd ed., National Utility Construction Association (NUCA), Arlington, Va.

Khanfar, N. M. (1993). A comparison of loads and ground movements which occur from microtunneling methods, *A Thesis Presented in Partial Fulfillment of the Requirement for the Degree of Master of Science*, Department of Civil Engineering, Louisiana Tech University, Ruston, La.

Kienow, K. K. (Ed.) (1990). Pipeline design and installation, *Proceedings of the International Conference by American Society of Civil Engineers*, Las Vegas, Nev.

Klein, S. J. (1991). Geotechnical aspects of pipe jacking projects, *Pipeline Crossing, Proceedings of a Specialty Conference*, American Society of Civil Engineers (ASCE), New York, NY.

Kramer, S. R., W. J. McDonald, and J. C. Thomson (1992). *An Introduction to Trenchless Technology*, Van Nostrand Reinhold, New York, NY.

Lamb, E., R. Lys, and T. M., Garrett (1993). Specifying microtunneling pipe, *Proceedings of Trenchless Technology Advanced Technical Seminar*, Vicksburg, Miss.

Lawrence, J. R. (1992). Plastics: New materials of the century, *Engineering News Record*, 229(21): 3–18.

Liu, H. (2003). *Pipeline Engineering*, CRC Press, New York, NY.

Moser, A. P. (2001). *Buried Pipe Design*, 2nd ed. McGraw-Hill, Inc., New York, NY.

Najafi, M. (1993). Evaluation of a new microtunneling propulsion system, *Dissertation Presented in Partial Fulfillment of the Requirement for the Degree of Doctor of Engineering*, Department of Civil Engineering, Louisiana Tech University, Ruston, La.

National Utility Contractors Association (NUCA) (1993). *A Guide to Pipe Jacking & Microtunneling Design*, NUCA, Arlington, Va.

O'Rourke, T. D., and A. R. Ingraffea (1986 a). *State-of-Art Review: Practices for Pipeline Crossings at Railroads*, Topical Report (June 1985–June 1986), Department of Civil Engineering, Cornell University, Ithaca, NY.

O'Rourke, T. D., and A. R. Ingraffea (1986 b). *State-of-Art Review: Practices for Pipeline Crossings at Railroads, Executive Summary*, Topical Report (June 1985–June 1986), Department of Civil Engineering, Cornell University, Ithaca, NY.

O'Rourke, T. D., H. E. Stewart, A. R. Ingraffea, and S. L. El-Gharbawy (1991). Influence of soil-pipeline stiffness on bending stresses from surface loading, *Pipeline Crossing, Proceedings of a Specialty Conference*, American Society of Civil Engineers (ASCE), New York, NY.

Selig, E. T., and W. A. Nash (1988). *Buried Pipeline Research Needs*, Final Report on National Science Foundation Workshop, University of Massachusetts, Amherst, Mass.

Stein, D., K. Mollers, and R. Bielecki (1989). *Microtunneling*, Ernst & Sohn, Berlin, Germany.

Svetlik, H. E., (1995). Polyethylene pipe design for directional-drillings and river-crossings, *Proceedings of the North American Society for Trenchless Technology NO-DIG '95.*

Terzaghi, K., R. B. Peck, and G. Mesri (1996). *Soil Mechanics in Engineering Practice*, 3rd ed., John Wiley and Sons, New York, NY.

Trenchless cities, NO-DIG (1990). *Proceedings of Fifth International Conference and Exhibition on Trenchless Construction for Utilities*, Rotterdam, The Netherlands.

Trenchless construction for utilities, NO-DIG (1985). *Proceedings of the First International Conference*, The Institution of Public Health Engineers, London, United Kingdom.

CHAPTER 5
DESIGN CONSIDERATIONS FOR TRENCHLESS RENEWAL METHODS

5.1 INTRODUCTION

The most important step in the designing of a trenchless renewal technique is the selection of the most appropriate, cost-effective, and reliable method. Obviously, selection of a solution is based on the recognition of the problem or problems with the existing pipeline system. As there is no *universal problem* there is no *universal solution* and the method or methods selected must be suited for the individual project conditions. Sometimes the best method is found in the application of multiple systems to achieve the most cost-effective solution. The trenchless renewal method (TRM) can be applied for addressing structural problems, infiltration/inflow problems, capacity problems, corrosion problems, and so on. The problem area can be found through a complete evaluation of as-built drawings and other records, and inspection and monitoring of the pipeline systems. The design of a trenchless pipeline renewal system include (1) identification of pipe conditions and problem recognition and classification, (2) prioritization of problem considering strategies and long-term plans, (3) selection of an appropriate pipeline renewal method, (4) designing renewal methods based on project specific conditions, and (5) implementation and monitoring.

This chapter provides information on method selection and some design basics. As mentioned in Chap. 1, throughout this book the term *renewal* refers to all aspects of trenchless pipeline renewal methods, such as rehabilitating, upgrading, and renovating activities where the design life of the pipeline is extended. The term *repair* refers to a problem that is temporarily or permanently fixed without adding to the design life of the pipeline system.

5.2 TRM SELECTION PROCESS

The TRM selection criteria involve the following two main steps:

- Assessment of pipeline specific conditions including state of deterioration and identification of problems
- Selection of suitable renewal technologies

The selection of a renewal method requires knowledge of the pipeline interior conditions (corrosion, deposits, cracks, misalignments, settlements, joint problems, and so on) as well as the underground conditions around the exterior of the pipe. Pipe specific conditions, such as gravity or pressure, type and composition of fluid, workspace area and bypassing requirements, number and degree of bends, slope, depth, length, diameter, flow rate, internal deposit build up, and the like, must be evaluated. Accurate classification of defects is essential in selecting a suitable renewal solution. It is also important to estimate rates of deterioration and to predict pipeline failure. When the pipe and the surrounding soil is in equilibrium, the deterioration condition might remain stable for a longtime with minimum likelihood of catastrophic failure or collapse. Using an assessment technology that can accurately determine deterioration rates is an important component of the decision-making process. As technologies for pipeline assessment continue to expand, it is important to emphasize that appropriate trenchless renewal methods must be used to obtain a thorough understanding of the condition and the dynamics of the deterioration process.

In general, a trenchless pipeline renewal program can be summarized in four phases: (1) initial planning, (2) pipeline integrity assessment, (3) developing renewal solutions, (4) and implementing and monitoring.

5.2.1 Initial Planning

During this stage, to the extent possible, the pipeline operator or engineer should have information available on

- Pipeline inspection results
- Fluid (water, wastewater, gas) characteristic data (pH, temperature, chemical composition, and so on)
- Pipe specific conditions
- Job site conditions and availability of workspace

5.2.2 Pipeline Integrity Assessment

The focus in this task should be on

- Analysis of pipeline inspection data
- Pipeline hydraulic capacity
- Pipeline defects (leaks, corrosion, structural, capacity, infiltration or inflow, and so on)
- Pipeline environmental conditions (surface and subsurface surrounding the pipe)

5.2.3 Develop Renewal Solutions

Using the process described in this chapter as a guide, a suitable renewal method can be selected. The requirements for candidate method or methods selected must be evaluated with pipe specific conditions and other project needs and a final selection is to be made based on all project conditions. The guidelines in this chapter have been developed to allow the preliminary decisions. The final decision requires other considerations such as available budget, organizational policy and objectives, short-term and long-term plans for the pipe, history of previous applications of the candidate method, compatibility of the trenchless method with existing systems, availability of experienced and qualified contractors, and level of workability, ease, and confidence with a specific method.

5.2.4 Implement and Monitor

The pipeline owner or operator should assign a project manager to carry out the pipeline renewal task. A qualified contractor must carry out the pipeline renewal work under supervision of the assigned project manager. Post renewal monitoring activities should follow organizational operation, reliability, and maintenance (ORM) guidelines.

5.3 SELECTION OF A RENEWAL METHOD BASED ON EXISTING PIPE CONDITIONS

Among other factors mentioned previously, selection of the appropriate renewal technique depends on the type and application of the existing pipe being renewed and the type of defects that are being remedied. For pipes that are in poor structural conditions, but otherwise have sufficient hydraulic capacity, renewal methods can range from in-line replacement (ILR), thermoformed pipe (ThP), sliplining (SL), modified sliplining (MSL), panel lining (PL), close-fit pipe (CFP), and cured-in-place pipe (CIPP). More details on these methods are presented in later chapters.

For pipes that are in good structural conditions but are judged inadequate with regard to hydraulic capacity, renewal methods can comprise ILR and sometimes lining technologies (such as CIPP, CFP, MSL, and ThP); and in rare cases even open-cut (OC) method may be considered. For pipes that are in good shape, but otherwise have localized defects, any of the localized point repairs or point source repairs, described in Chap. 18, can be selected. For pipes that require corrosion and abrasion protection and stabilization, any of the underground coating and lining (UCL) methods (Chap. 20) can be used. For deteriorated manholes, any of the methods described in Chap. 22, Sewer Manhole Renewal can be considered.

According to ASTM F-1216, in general, trenchless pipeline renewal design encompasses two general categories of methods:

- Those in which the existing pipe continues to bear the soil and live loads, but has such problems as corrosion and I/I (*partially deteriorated*).
- Those in which a new pipe or other reinforcement is added to help or carry the loads (fully deteriorated). Recent studies have shown that the term *fully deteriorated* is fundamentally flawed because the existing pipe structure, even in its fully deteriorated state, sustains the soil and live loads and in reality is *not* fully collapsed. The research in this area is ongoing and more information can be found in the new ASCE Pamphlet on Design Guidelines for Pipeline Renewal Systems scheduled for publication in 2005 and references listed at the end of this chapter. New design criteria would enable engineers to design more cost-effective and reliable trenchless pipeline renewal systems, especially for larger diameters.

5.4 SELECTION OF A RENEWAL METHOD BASED ON SITE AND PROJECT CONDITIONS

Similar to any other construction project, a renewal project is site and project specific and many factors influence selection of a specific renewal method. Availability of work space, project surface and subsurface conditions, soil and watertable conditions (including fluctuation of watertable), number and condition of service laterals, diameter, depth, slope, age, and material and application of existing pipe, are examples of project conditions to be evaluated.

Other conditions might include local laws and regulations, such as restrictions on confined spaces and worker entry into manholes and pipelines. In this regard, before initiating a pipeline renewal program, local laws and regulations and organizational policies must be evaluated. Life expectancy and performance of the new pipe is another important design consideration. Not all trenchless renewal methods are proven to have the same design life of, say, 50 years or more. Each pipeline operator needs to make appropriate decisions based on its ORM strategy.

5.5 SELECTION OF A RENEWAL METHOD (A SIX-STEP PROCESS)

This section presents a six-step process for selection of an appropriate trenchless pipeline renewal method.

The six-step solution process summarized in Table 5.1 will guide the operator and the engineer to make rapid quality and cost-effective decisions. Using this process, the engineer can select methods to solve specific problems without having to develop a comprehensive background and knowledge of all renewal methods.

Step 1. The proper selection of renewal methods and materials depend on a comprehensive understanding of the specific problems to be corrected including external as well as internal pipeline conditions. In that regard, it is important to conduct a pipeline integrity assessment program to minimize the likelihood of an unexpected collapse, blockage, leakage, spill, and so on. This step includes initial planning and pipeline integrity assessment.

TABLE 5.1 How to Select a Pipeline Renewal Method

Step	Objective	Task
1	Define the problem.	Conduct complete and accurate pipeline condition assessment.
2	Identify candidate methods.	2A Review specific TRM applicability and technical compatibility parameters, and select methods likely to succeed. 2B Review the limitations of the selected candidate methods and confirm that the selected methods are compatible with project conditions. 2C Review the associated installation cost range. 2D Review the installation experience factors.
3	Make a final selection.	Select a candidate method.
4	Develop a better understanding of the selected method.	Locate the selected candidate method in Table 5.6 and develop an understanding of how it fits within the complete family of methods. For more detailed information on specific method refer to the appropriate chapter in this book.
5	Identify technology providers (contractors).	5A Review the vendors of the selected renewal method, using a trenchless technology directory. 5B Contact vendors if more information is needed.
6	Provide solution.	Develop the design and contract documents for the final renewal method selected.

As mentioned previously, during the initial planning, the operator should gather the available information (to the extent practical) on the pipeline, such as fluid characteristic data, inspection results, and job site conditions.

The initial planning information and data are shown in Table 5.2. A contractor usually does the pipeline integrity assessment. At the conclusion of this effort, the pipeline operator should be able to infer whether the pipeline problems are associated with:

- Cracks, fractures, and holes
- Corrosion
- Infiltration, inflow, and exfiltration
- Deformation
- Line deviation
- Capacity or hydraulic problems
- Joints
- Debris and obstructions
- Construction fracture (connections)

Table 5.3 lists some typical findings and conclusions from pipeline assessment effort.

TABLE 5.2 Pipeline Data and Information Gathered during Initial Planning

Category	Parameters
General	Pipeline maps, including detailed plan and profile view drawings
	Existing pipe shape (circular/noncircular)
	Existing pipe material
	Existing pipeline pipe diameter
	Soil type
	Ground cover depth
	Ground conditions
	Groundwater table
	Major obstacles (roots, joint settlements, straightness, bends, and so on)
	Existing liner (if any) material and thickness
	Joint types
Fluid characteristics data	pH levels
	Solid characterization
	Temperature
	High-pressure or high-velocity flow
Inspection results	Internal pipeline conditions
	Soil condition outside the pipeline
	Voids located around pipeline pipe
Jobsite constraints	Accessibility
	Flow interruption
	Personnel entry
	Excavation
	Operator service interruption
	Safety issues

TABLE 5.3 Information and Conclusions from Pipeline Integrity Assessment

Extent of the damage (localized or continuous)
In-line offsets, valves, other obstructions, and so on (number or sizes)
Circumferential breaks, if yes, maximum gap width (in)
Longitudinal breaks, if yes, maximum gap width (in)
Previous lining condition (structurally sound or breaking up or corroded)
Leaking joints
Separated joints, if yes, maximum gap width (in)
External and internal pipeline pipe corrosion
(maximum size hole in pipe now and allowed at the end of design life)
Possible sag in the pipeline
Need of pipeline cleaning prior to renewal
Need of enlargement of the discharge capacity
Is reduction of discharge capacity permissible?
Constructability issues involved with a specific renewal method
(Bypass requirements, access pit requirements, and the like)
Preferred new pipeline materials
Number of service laterals or connections

Step 2A. After the pipeline assessment has been completed and the problem has been defined, candidate renewal methods need to be identified. This step requires the reader to select one or more performance parameters that best represent the pipeline condition. Once the parameters are selected, using Table 5.4 will assist the selection of the preliminary candidate methods that are generally applicable and suitable for the performance factors selected. Table 5.4 also illustrates applicability of different renewal methods for specific pipe conditions. It should be noted that localized repair (LOR) or point source repair (PSR) solutions (also called spot repairs) should always be considered first, because they can be an important cost-saving alternative to renewing an entire manhole-to-manhole (MTM) section of a pipeline. Based on actual renewal costs, some decision makers have developed guidelines for determining when to use MTM solutions and when to use LOR solutions. One suggestion is that if three or more LOR solutions would be required between adjacent manholes, then it would be better to use MTM instead of LOR solution. This guideline is intended to be used only when the actual cost and effectiveness data are not easily available. Of course, if better data are available, they should always be used.

Step 2B. After selecting the potentially applicable methods (Step 2A), use Table 5.5 to learn about the limitations of the selected candidate methods.

Step 2C. Use Table 5.6 to check the candidate methods for cost feasibility and receive proposals.

Step 2D. Use Table 5.6 to check the candidate methods for overall and project specific installation experience.

Step 3. Select a final candidate method or, if there are still several candidates, make a list of the final candidates. To get to this point, the reader needs to know very little about the candidate methods but should have a good understanding of the overall pipeline condition.

Step 4A. Develop a further basic understanding of the installation process, performance characteristics, and limitations of the selected method. At this point, it is possible that the reader is not satisfied with any of the methods resulting from Step 3. The reader must go back to Step 1 and confirm that the pipeline condition assessment is thorough and accurate. Then the reader can repeat Step 2. It might be necessary to evaluate possible renewal

TABLE 5.4 Applicability of Various Renewal Methods

Renewal methods*	I/I/E†	Joint separation†	Corrosion†	Cracks/ holes†	Joint problems†	Structural problems†	Inadequate hydraulic capacity†
CIPP	Yes	Yes	Yes	Yes	Marginal	Yes	No
SL	Yes	Yes	Yes	Yes	Marginal	Yes	No
CFP	Yes	Yes	Yes	Yes	Marginal	Yes	No
PB	Yes	Yes	Yes	Yes	Marginal	Yes	Yes
PR	Yes	Yes	Yes	Yes	Yes	Yes	Yes
LOR-RR	Yes	Yes	No	Marginal	Marginal	Marginal	No
LOR-GR	Yes	Yes	No	Marginal	Marginal	No	No
LOR-IS	Yes	Yes	Yes	Yes	Yes	Yes	Yes
LOR-CIPP	Yes	Yes	Yes	Yes	Marginal	Yes	No
MSL-PL	Yes	Yes	Yes	Yes	Marginal	Yes	No
MSL-SW	Yes	Yes	Yes	Yes	Yes	Yes	No
UCL-CM	Marginal	Yes	Yes	Yes	Marginal	No	No
UCL-EL	Marginal	Yes	Yes	Yes	Marginal	No	No
UCL-SH	Marginal	Yes	Marginal	Yes	Marginal	Yes	No
UCL-GU	Marginal	Yes	Marginal	Yes	Marginal	Yes	No
ThP	Yes	Yes	Yes	Yes	Marginal	Yes	No

*For a list of abbreviations, see table 5.6; †Problematic pipeline conditions.

Corrosion. Pipeline corrosions.

Cracks or holes. Cracks owing to pipeline defects or lack of proper bedding or installation or operational overstress (built-in and long-term defects).

I/I/E (Infiltration/Inflow/Exfiltration). Water entering or exiting a pipeline system.

Inadequate hydraulic capacity. Hydraulic capacity insufficient to carry more flow.

Joint problems. Joint settlement, misalignment, sags, offsets owing to lack of proper bedding or installation or material.

Joint separation. Joints with infiltration or inflow problems that have caused a void in underlying soil.

Structural Problems. Inadequate structural capacity to carry soil and/or traffic loads.

TABLE 5.5 Limitations of Various Pipeline Renewal Methods

Renewal methods*	Must grout annular space	Must seal liner ends and/ or laterals	Open cut needed to reconnect laterals	Need entrance pit	Need bypass	More than 10% loss of diameter	Many joints in new pipe	Difficult to install if existing pipe is misaligned	Might damage nearby utilities	Not suit-able so all pipe mate-rials	Limited range of diameters (inches)	Personnel entry into pipeline is needed	No structural support
CIPP		X			X						4–108		
SL	X	X	X	X	Varies	X	Varies	X			4–160		
CFP		X		X				X			3–63		
PB		X	X	X	X			X	X	X	4–48		
PR		X	X	X	X		Varies		X	X	12–36		
LOR-RR		X		X	X	X					8–30		X
LOR-GR		X			X						3–180		X
LOR-IS					X						6–110		X
LOR-CIPP					X	X					4–48		X
MSL-PL	X	X		X	Varies			X			More than 48	X	
MSL-SW	X			X	X						6–108	Varies	
UCL-CM		X		X	X			X			3–180		X
UCL-EL		X		X	X			X			3–24		X
UCL-SH		X			X	X		X			48–180	X	
UCL-GU		X			X	X					48–180	X	
ThP		X		X	X			X			4–30		

*More information is provided in the individual chapters. For a list of chapters and method abbreviations, see Table 5.6.

TABLE 5.6 Cost and Experience Indicators for Various Renewal Methods

Renewal method		Reference chapter	Experience indicator	Cost indicator
CIPP (dia. >10 in)	Cured-in place pipe	13	High	High
CIPP (dia. <10 in)	Cured-in place pipe	13	High	Low
SL	Sliplining	14	High	Medium
CFP	Close-fit pipe	15	High	Low
PB	Pipe bursting	16	Medium	Medium
PR	Pipe removal	16	Low	High
LOR-RR	Localized repair–robotic repair	18	High	High
LOR-GR	Localized repair–grouting	18	High	Low
LOR-IS	Localized repair–internal seal	18	Medium	Medium
LOR-CIPP	Localized repair–cured-in-place pipe	18	High	High
MSL-PL	Modified sliplining–panel lining	19	Medium	Medium
MSL-SW	Modified sliplining–spiral wound	19	High	Medium
UCL-CM	Underground coatings and linings–cement mortar	20	Medium	Medium
UCL-EL	Underground coatings and linings–epoxy lining	20	Medium	Medium
UCL-SH	Underground coatings and linings–shotcrete	20	High	Low
UCL-GU	Underground coatings and linings–Gunite	20	High	Low
ThP	Thermoformed pipe	21	Medium	Low

Note: High cost means more than $10/ft per inch diameter; high experience means more than 20 years; medium cost means $5 to $10/ft per inch diameter; medium experience means 10 to 20 years; low cost means up to $5/ft per inch diameter; low experience means less than 10 years.

For other abbreviations, acronyms, and descriptions: See Glossary at the end of this book.

Many factors influence project cost, which is dependent on specific project conditions. The cost information in this table is the best judgment from municipal projects and includes the overall cost, including cleaning, inspection, bypassing, and sludge disposal. This cost information should be used with caution and for comparison purposes only. Typically, the cost of an industrial pipeline renewal project is 2 to 5 times the cost of a municipal project. Most of the experience has come from the municipal sector.

candidate methods that are rated as marginally suited (Table 5.4) to solve specific problems. When special design considerations are met, these marginal solutions might prove to be superior. After new methods are selected, the reader should repeat steps 2 and 3.

Step 4B. Use a trenchless technology directory to find local vendors or contractors who provide the selected method.

Step 5. Contact vendors of the selected method to obtain more detailed information and confirmation that the method is appropriate for the intended application.

Step 6. Develop the design and contract documents to procure the applicable method to solve the pipeline problem.

It is likely that in addition to the tables in this book, other sources of information will also be available to the decision makers. The decision process on the suitability of a specific method is a complex matter that can depend on many factors including the information provided in this book as well as the cost-benefit analysis, life cycle costs analysis, and long-term goals for a specific renewal method. When selecting a renewal method, all available data should be considered—not just the tables in this book. The final recommendation that is submitted for approval should include proper justification such as a summary of the pipeline condition survey as well as conclusions based on the tables in this book and other data used in the evaluation. It should be noted that trenchless industry is advancing rapidly and the user must keep abreast of new developments and capabilities of existing methods as well as any viable method which will be developed in future.

5.6 GENERAL DESIGN CONSIDERATIONS

Selection of the appropriate renewal technique depends, among other factors, on the type of pipe material being renewed and the type of defect that is being remedied. In cases involving installation of a slightly smaller pipe within an existing pipe, the question of hydraulic capacity of the new pipe needs to be considered. Fortunately, most of trenchless renewal techniques produce a pipe with a lower friction coefficient than the original existing pipe, thereby partially or totally compensating for the reduction in pipeline cross-sectional area. The issue of hydraulic capacity of the renewed pipe is not discussed in this book except to point out that the pipeline hydraulics need to be checked for those renewal techniques that reduce cross-sectional area of existing pipe.

5.6.1 Flow Bypassing Considerations

Most renewal technologies require that the flow in the line be temporarily rerouted during the time that the renewal work is being conducted. Depending on the circumstances, this can be a significant cost consideration, amounting to up to one-third of the total cost for a pipeline renewal job. In the municipal sector, there are a few renewal technologies, which do not as a rule require flow bypassing. These methods include the segmental sliplining method, internal seals method of localized repairs, and internally applied grouting in manholes. The key consideration here is simply whether the safety limitations applicable to a municipal or industrial pipeline would preclude personnel from entering the manhole or existing pipe while it is in service. For lines that carry mainly storm runoff this may be feasible in an industrial setting, although even here the safety considerations and reviews would be of paramount importance. Other than this exception, the general assumption for pipeline renewal is that the flow in the pipeline will have to be bypassed and rerouted during the time the line is being renewed.

5.6.2 Reinstatement of Laterals

For lines other than interceptors, the reinstatement of smaller diameter laterals is another task for most renewal techniques. For the majority of renewal methods this is done by cutting service laterals from the inside of the main pipe using a CCTV guided robotic device, and sealing the joint from within the main line using another set of robotic tools. The complete set of tools to conduct reinstatement and sealing of a lateral may all be mounted on a single robotic tractor or platform, requiring only one trip through the line to complete the reinstatement of laterals.

To be able to reestablish the laterals from inside the main, the laterals must first be located. Renewal techniques using flexible liner pipes are amenable to visual location of the laterals via CCTV by looking for a *dimple* on the renewed pipe wall. For methods that produce a stiffer wall, location of the laterals is either done by careful inspection using CCTV before renewal, or by using more sophisticated sensing techniques after renewal. Some techniques involve placing an electronic or radioactive source in the lateral that can be detected by an instrument inside the newly installed pipe.

When property ownership issues do not restrict access to the upstream end of the laterals, the laterals can sometimes be drilled from the lateral into the renewed pipe rather than from inside out. This option is probably more feasible for industrial pipelines than for municipal pipelines as the property ownership issue would generally not be a problem within an industrial operator.

Regardless of how the laterals are located or opened, sealing the connections is generally more difficult than sealing the line. If hydraulic integrity is the primary concern, this factor needs to be considered. In general, when conducting a hydrostatic or air pressure test on a new pipe, the allowable leakage rates are based on the assumption that the testing is done before the laterals are reconnected.

5.6.3 Open-Cut (OC)

Although not a trenchless method, open-cut (OC) always needs to be considered as an option during a renewal project. However, compared to a trenchless method, open-cut is often more costly and takes more time to complete. Most of the extra cost of OC over trenchless methods (about 70 percent of the open-cut cost) is for the restoration of the site. Also, the possibility of pipeline damage because of shoring removal, backfill, and compaction (built-in defects) during the OC should be considered. Open-cut involves disturbing the ground (both surface and subsurface) and requires potentially costly environmental disposal of any contaminated soil (e.g., from previous leaks) that might be discovered. Open-cut method can also disrupt surface traffic and other surface operations (refer to Chap. 2 for more information on social costs of OC method).

5.7 STRUCTURAL DESIGN OF CIPP

Cured-in-place pipe method is the most widely used of all the pipeline renewal methods (Fig. 5.1). For this reason, this section presents an overview of CIPP design procedure. This presentation also serves as an example for the design of other renewal methods in which same principles are usually applicable. CIPP should be designed in accordance with ASTM F-1216 entitled "Standard Practice for Renewal of Existing Pipelines and Conduits by the Inversion and Curing of a Resin-Impregnated Tube." This standard has a nonmandatory design appendix, which addresses the method used for calculating wall thickness. Virtually all CIPP suppliers and contractors recognize this standard. The standard for the CIPP is all-inclusive covering material requirements, construction methods, and design parameters. The equations and definitions used in this section are taken from ASTM F-1216.

As for any other renewal method, the first step in designing a CIPP project is identifying the condition of the existing pipe. ASTM F-1216 divides existing pipe conditions into two classes: partially deteriorated condition and fully deteriorated condition. The assignment of a partially or a fully deteriorated design procedure depends upon the existing condition of

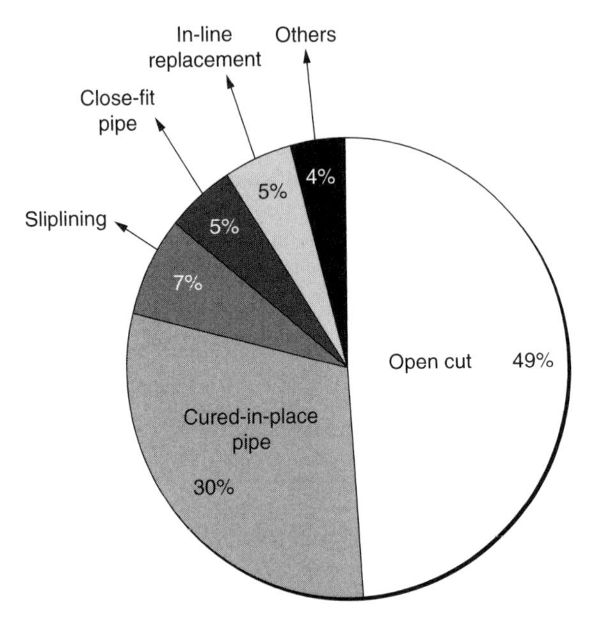

FIGURE 5.1 Usage statistics of pipeline renewal methods.

the existing pipe or its expected structural contribution over the CIPP design period. An engineer with experience in designing CIPP is most qualified in determination of a design procedure.

Alternative designs for CIPP have merits and potentially offer more accurate predictions of performance. However, it is not within the scope of this book to provide alternative design equations until they have been accepted and adopted by specific organizations such as ASTM and ASCE. Refer to references at the end of this chapter for more information on current research in this area.

5.7.1 Major Design Variables

As for any structural deign, there are many variables and parameters used in the design of the pipe wall thickness, which provides the required performance requirements (strength, stiffness, and so on) to resist the loads. For a proper design, it is important to use quantities that are representatives of the site conditions for each pipe section designed. If precise values are not known, good engineering judgment should be used to estimate the magnitudes of some of these variables. Design engineers with familiarity and experience with CIPP product and its design can offer required engineering judgments. Some of these parameters are specified below. Table 5.7 presents minimum recommended design properties for CIPP.

Flexural modulus of elasticity, E. The correct resin or tube selection for the effluent and exposure will determine the associated physicals used for that system. Fillers or reinforcements should be evaluated and used appropriately as needed. All of this must be

TABLE 5.7 Minimum recommended design properties for CIPP

Test property	Epoxy resin	Epoxy vinyl ester	Isophthalic polyester	Filled isophthalic polyester
Flexural modulus, ksi	250–300	350–450	250–350	400
Flexural strength, ksi	5.0	5.0	4.5	4.0
Tensile strength, ksi	3–5	3–5	2–3	2–3

considered prior to assigning a modulus value. Typical values for E are 250,000 psi (minimum accepted value to about 500,000 psi.

Long-term flexural modulus of elasticity, E_L. Independent analysis should be used to determine the 50-year retention modulus value used in the design calculations. Typically, a reduction of the short-term modulus of 50 percent is used for long-term design values.

Soil modulus, E'_S. The type of soil, depth of burial, and surface use will determine the range for soil modulus. If little is known about the conditions, a 700-psi value should be used. For example, if the pipe runs under a roadway at a depth of 10 ft or more, a minimum value of 1000 psi should be used.

Ovality, q. The existing pipe out-of-roundness significantly contributes to the wall thickness calculation of CIPP. Estimates of the percentage of ovality should be made using the best pipe dimension information available to ensure sufficient wall thickness. In existing pipes requiring renewal that have not been closely evaluated, 2 percent ovality is commonly used to estimate the required wall thickness. To quantify q, in the large-diameter existing pipes where personnel-entry is possible, actual in-pipe measurements should be used.

Factor of Safety, N. Usually the factor of safety for trenchless pipeline renewal methods (TRM) is 2. However, in large-diameter, worker-entry pipes where more accurate data and thorough existing pipe investigations data such as ovality, groundwater pressure, and so on can be quantified, a value of 1.5 is acceptable.

5.7.2 Design Background

The objective of underground pipe design revolves around the ability to develop a set of equations that can take loads such as groundwater, soil, traffic loads, and other loadings into consideration. Through practical experience and scientific study, it has been determined that underground cylindrical structures such as tubes or pipes usually fail by buckling when exposed to an external load. Timoshenko and others have published the earliest buckling theories in early 1900s. These equations were subsequently modified to take into account long tubes having practical thickness consistent with the pipe and structures available at the time. One of the first practical applications of this work was the successful development of the first submarines. The unrestrained buckling equation that was developed for long, thin tubes is given in Eq. 5.1.

$$P_W = \frac{2Et^3}{(1 - v^2)D_m^{\ 3}} \qquad (5.1)$$

where P_W = hydrostatic water pressure, psi
E = modulus of elasticity, psi
t = pipe wall thickness, in
ν = Poisson's ration, typically = 0.3
$D_m = (D_O - t)$ = mean pipe diameter, ft
D_O = mean outer CIPP diameter, ft

In the 1940s Spangler published work that was conducted on flexible piping systems. This work was the basis on which stiffness of flexible pipes was derived. The measurement of pipe stiffness has been standardized with ASTM D-2412 and is determined at a pipe deflection of 5 percent. This is a relatively simple test and is performed on a free-standing, unsupported pipe placed between two parallel plates that are pressed towards each other at a controlled rate. Spangler also developed a model for the deflection of buried flexible pipe that took into account factors such as dead load forces, pipe bedding, and soil modulus.

Work by these early pioneering engineers was extremely important in laying the foundation that is the basis of design equations used for CIPP today. However, it is important to understand that there is very little similarity between the loading experienced by installed CIPP and that of underground rigid or flexible pipe. CIPP is installed into existing pipe that has typically been buried for many years. As such, the soil has long since consolidated and the soil pipe system is typically stable. Therefore, installed CIPP is supported by the soil pipe system and the subsequent pipe deflections can be expected to be minimal. Also, when CIPP is installed into an existing pipe, the surrounding pipe provides constrained ring support to the CIPP under the influence of uniform hydrostatic water pressure.

Given enough pressure and long enough period of time, the CIPP can deform to the extent that it will produce catastrophic failure by buckling. In order to take the long-term effects of creep into account, the modulus of elasticity in buckling equation attributed to Timoshenko was modified to a long-term modulus. In addition, a safety factor and correction for pipe ovality was also added to obtain the restrained buckling equation. By substituting the dimension ratio (DR) for the mean diameter and rearranging, Eq. 5.1 will be:

$$P_W = \frac{2KE_L}{(1-\nu^2)} \frac{1}{(DR-1)^3} \frac{C}{N} \tag{5.2}$$

where E_L = long-term modulus of elasticity of pipe material, psi
K = enhancement factor, typically $K = 7$
$DR = D_O/t$, dimensionless
D_O = mean outside diameter of the CIPP, ft
N = safety factor
C = ovality correction factor (Table 5.8)

$$C = \left[\frac{D_{O\min}}{D_{O\max}{}^2}\right]^3 = \left[\left(1 - \frac{q}{100}\right) \Big/ \left(1 + \frac{q}{100}\right)^2\right]^3 \tag{5.3}$$

$$q = 100 \times \frac{(D - D_{\min})}{D}, \quad \text{or} \quad 100 \times \frac{(D_{\max} - D)}{D} \tag{5.4}$$

TABLE 5.8 Ovality Reduction Factor, C

$C = [D_{O\min}/(D_{O\max})^2]^3$										
Percent ovality	1	2	3	4	5	6	7	8	9	10
Reduction factor, C	0.91	0.84	0.76	0.70	0.64	0.59	0.54	0.49	0.45	0.41

Source: Lanzo Lining Services Design Manual

where C = ovality reduction factor (Table 5.8)
q = percentage of ovality of the original pipe
D = inside diameter of the original pipe, in
$D_{\min}$ = minimum inside diameter of the original pipe, in
$D_{\max}$ = maximum inside diameter of the original pipe, in

There is no single design equation that can be used for all different conditions that must be taken into account for the proper design of CIPP, so it is necessary to divide these conditions into different groups. For both gravity flow and internal pressure, design equations have been divided into categories of *partially deteriorated* and *fully deteriorated* conditions of the existing pipe. These piping conditions are defined in the following sections.

5.7.3 Partially Deteriorated Gravity Condition

Once again, a partially deteriorated gravity flow pipe is one in which the existing pipe may have displaced joints, cracks, or corrosion, but is structurally able to support all soil and surface loads. In this case, the existing pipe is intended to provide structural support over the full circumference of the CIPP. Assuming that a pipe is partially deteriorated, the CIPP will be designed to withstand uniform hydrostatic pressure over the full circumference of the CIPP. In addition, as a conservative approach, this design does not assume that the CIPP is attached or bonded to the existing pipe in any way. The hydrostatic pressure on the CIPP is calculated as given in Eq. 5.5.

$$P_W = \frac{H_W \text{ ft} \times 62.4 \text{ pcf (unit weight of water)}}{144 \text{ in}^2/\text{ft}^2} \tag{5.5}$$

where H_W = groundwater height above the top of the pipe and the other variables are defined earlier

A partially deteriorated pressure pipe may also have minor corrosion, leaking joints, and/or small holes, and should be free of any longitudinal cracks. In this case the existing pipe is assumed to be able to withstand the specified internal design pressure over the expected lifetime of the pipe. When assuming that a pressure pipe is partially deteriorated, it is assumed that the CIPP will conform tightly against the existing pipe for the full pipe section (i.e., including the bends or diameter changes, and so on) and uses the strength of the existing pipe to support the stresses. The thickness of the CIPP can be considered to span small holes but will not be of sufficient thickness to withstand design pressures. In addition, if the partially deteriorated pressure pipe is found to be leaking, the designer must consider external hydrostatic pressure to ensure that the minimum CIPP thickness is sufficient to withstand hydrostatic pressure over the design life of the new pipe.

5.7.4 Partially Deteriorated Gravity Flow CIPP Design

The restrained buckling condition must be considered when renewing a partially deteriorated existing pipe for gravity flow condition. In this case, the classical buckling equation that has been described previously is rearranged to solve for CIPP thickness as shown in Eq. 5.6.

$$t = \frac{D_O}{\left(\dfrac{2KE_LC}{P_W N(1-v^2)}\right)^{1/3} + 1} \tag{5.6}$$

where D_O = mean outer CIPP diameter, in
K = enhancement factor, typically $K = 7$
E_L = long-term modulus of elasticity of the pipe material, psi
C = ovality correction factor (Table 5.8)
P_W = external water pressure measured above the pipe invert, psi
N = safety factor, typically $N = 1.5$ to 2.0
v = Poisson's ratio, typically $v = 0.3$, dimensionless

For partially deteriorated gravity design conditions where the groundwater fluctuation is below the invert of the pipe, the hydrostatic pressure is equal to zero, and the restrained buckling equation cannot be used to calculate CIPP thickness. For this special design case, the calculated thickness of the CIPP must be equal to or greater than that which will produce a maximum DR of 100. ASTM requires a CIPP with stiffness in the range of 1 to 2 psi. When this special design condition exists, CIPP thickness is determined by Eq. 5.7.

$$t = D_O/100 \tag{5.7}$$

While designing for circular partially deteriorated pipe, the CIPP is under constant compressive hoop stresses. If the existing pipe is out of round or has localized ovalization, the predominate force on CIPP might be bending moment. For this special case, the CIPP must be checked to ensure that the bending moment do not exceed the long-term flexural strength of the CIPP. To determine this, the bending stresses on the CIPP are determined by Eq. 5.8.

$$\frac{S}{P_W N} = [1.5q/100\,(1+q/100)\,DR^2] - [0.5(1+q/100)DR] \tag{5.8}$$

where Eq. 5.4 defines q, and the other parameters have been defined previously.

Table 5.9 gives design thickness for different values of H_W (or P_W) using particular values for E_L, C, N, v, and K.

Partially Deteriorated Design Example. Determine the minimum CIPP wall thickness required for the following existing pipe conditions (Fig. 5.2):

- Existing pipe classification = partially deteriorated
- Mean outer CIPP diameter (D_O) = 24 in
- Minimum pipe diameter (D_{min}) = 23.1 in

TABLE 5.9 Partially Deteriorated Gravity Condition

Variables				
	E_L	175,000 psi	K	7
	C	0.84	$1-\nu^2$	0.91
	N		2	

Water depth (H_W), ft	Water pressure (P_W), psi	CIPP thickness (in) for each pipe diameter (in)																	
		6	8	10	12	15	18	21	24	27	30	33	36	42	48	54	60	66	72
1	0.4380	0.04	0.06	0.07	0.09	0.11	0.13	0.15	0.17	0.20	0.22	0.24	0.26	0.30	0.35	0.39	0.43	0.48	0.52
2	0.8700	0.06	0.07	0.09	0.11	0.14	0.17	0.19	0.22	0.24	0.27	0.30	0.33	0.38	0.44	0.49	0.54	0.60	0.65
3	1.3000	0.06	0.08	0.10	0.13	0.15	0.19	0.22	0.25	0.28	0.31	0.34	0.37	0.44	0.50	0.56	0.62	0.69	0.75
4	1.7300	0.07	0.09	0.11	0.14	0.17	0.20	0.24	0.27	0.31	0.34	0.37	0.41	0.48	0.55	0.61	0.69	0.75	0.82
5	2.1700	0.07	0.10	0.12	0.15	0.17	0.22	0.26	0.30	0.33	0.37	0.41	0.44	0.52	0.59	0.66	0.74	0.81	0.88
6	2.6000	0.08	0.10	0.13	0.16	0.20	0.24	0.27	0.31	0.35	0.39	0.43	0.47	0.55	0.63	0.70	0.78	0.86	0.94
7	3.0300	0.08	0.11	0.14	0.17	0.20	0.25	0.29	0.33	0.37	0.41	0.45	0.49	0.57	0.66	0.74	0.82	0.91	0.99
8	3.4600	0.09	0.11	0.14	0.17	0.22	0.26	0.30	0.34	0.39	0.43	0.47	0.52	0.60	0.69	0.77	0.86	0.94	1.03
9	3.9000	0.09	0.12	0.15	0.18	0.22	0.27	0.31	0.36	0.40	0.44	0.49	0.54	0.63	0.71	0.80	0.89	0.98	1.07
10	4.3300	0.09	0.12	0.15	0.19	0.23	0.28	0.32	0.37	0.42	0.46	0.51	0.56	0.65	0.74	0.83	0.93	1.02	1.11
11	4.7600	0.09	0.13	0.16	0.19	0.24	0.29	0.33	0.38	0.43	0.48	0.52	0.57	0.67	0.76	0.86	0.95	1.05	1.15
12	5.2000	0.10	0.13	0.17	0.20	0.24	0.30	0.34	0.39	0.44	0.49	0.54	0.59	0.69	0.78	0.88	0.98	1.08	1.18
13	5.6300	0.10	0.13	0.17	0.20	0.25	0.30	0.35	0.40	0.45	0.50	0.56	0.61	0.70	0.81	0.91	1.01	1.11	1.21
14	6.0600	0.10	0.14	0.17	0.20	0.26	0.31	0.36	0.41	0.46	0.52	0.57	0.62	0.72	0.83	0.93	1.03	1.13	1.24
15	6.5000	0.11	0.14	0.18	0.21	0.26	0.31	0.37	0.42	0.48	0.53	0.58	0.63	0.74	0.85	0.95	1.06	1.16	1.27
16	6.9300	0.11	0.15	0.18	0.22	0.27	0.32	0.38	0.43	0.48	0.54	0.59	0.65	0.76	0.86	0.97	1.08	1.19	1.30
17	7.7360	0.11	0.15	0.19	0.22	0.28	0.33	0.39	0.44	0.50	0.55	0.61	0.66	0.77	0.88	0.99	1.10	1.21	1.32
18	7.7900	0.11	0.15	0.19	0.22	0.28	0.33	0.39	0.45	0.50	0.56	0.62	0.67	0.78	0.90	1.01	1.12	1.23	1.34
19	8.2300	0.11	0.15	0.19	0.23	0.28	0.34	0.40	0.46	0.51	0.57	0.63	0.69	0.80	0.91	1.03	1.14	1.26	1.37
20	8.6600	0.11	0.15	0.19	0.23	0.29	0.35	0.41	0.46	0.52	0.58	0.64	0.70	0.81	0.93	1.04	1.16	1.28	1.39

Source: Lanzo Lining Services Design Manual

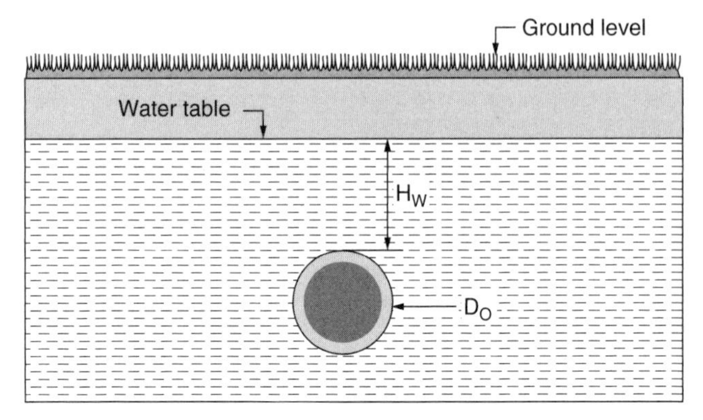

FIGURE 5.2 Partially deteriorated design example.

- External water above invert (H_w) = 8 ft
- Minimum CIPP flexural modulus (E) = 350,000 psi
- Minimum CIPP flexural strength (S) = 5500 psi
- Long-term modulus (E_L) = 175,000 psi
- Long-term strength (S_L) = 2750 psi

Step A. Determine hydrostatic pressure acting on CIPP (Eq. 5.5).

$$P_W = \frac{8 \text{ ft} \times 62.4 \text{ pcf (unit weight of water)}}{144 \text{ in}^2/\text{ft}^2} = 3.46 \text{ psi}$$

Step B. Calculate the pipe ovality.
Determine q using Eq. 5.4.

$$q = \frac{100(24 \text{ in} - 23.1 \text{ in})}{24 \text{ in}} = 3.75\%$$

Determine ovality reduction factor using Eq. 5.3.

$$C = \left[\frac{(1 - 3.75/100)}{(1 + 3.75/100)^2}\right]^3 = 0.715$$

Step C. Determine minimum CIPP design thickness using buckling Eq. 5.6.

$$t = \frac{24 \text{ in}}{\left(\frac{2(7) \times 175000 \text{ psi } (0.715)}{2(3.46 \text{ psi})(1 - 0.3^2)}\right)^{1/3} + 1} = 0.36 \text{ in}$$

Step D. Because the pipe is out of round, bending stresses must be calculated to ensure CIPP does not exceed the long-term flexural strength.

Determine S using Eq. 5.8.

$$DR = D_O/t = 24/0.36 = 66.67$$
$$S/(3.46 \text{ psi})2 = [1.5(3.75/100)(1+3.75/100)66.67^2] - [0.5(1+3.75/100)66.67]$$
$$S = 6.92\,[259.4 - 34.55]$$
$$S = 1556 \text{ psi.}$$

Step E. The minimum CIPP design thickness is 0.36 in because the bending stresses are less than the long-term CIPP flexural strength. However, had bending stresses exceeded the long-term flexural strength, then Eq. 5.6 would not have controlled the design. In this case, solving for the proper thickness can be accomplished by trial and error. In this case, DR that are smaller than those previously used can be selected until the bending stress is less than the long-term flexural strength of the CIPP.

5.7.5 Fully Deteriorated Gravity Condition

A fully deteriorated gravity flow pipe is one in which the existing pipe has insufficient strength to support all soil and live loads. A fully deteriorated pipe is characterized by severe corrosion, missing pipe, longitudinal cracks, and severely deformed pipe. Assuming the existing pipe to be fully deteriorated, the CIPP is designed as a new pipe that withstands all hydrostatic, soil, and live loads with adequate soil support.

An alternative design strategy for fully deteriorated gravity flow pipes is available to the designer in the areas where there are isolated sections of missing or severely offset pipe. In these areas, it may be possible to carry out localized repairs and renew the pipe as a partially deteriorated classification. However, each situation must be considered by a knowledgeable professional engineer.

A fully deteriorated pressure pipe is one in which the existing pipe has structurally failed, or has insufficient strength to operate at specified design pressures. A pipe may also be classified as fully deteriorated if it is determined that it will not be able to withstand design pressures at some point during the expected lifetime. A fully deteriorated pressure pipe is characterized by significant loss of wall thickness owing to severe corrosion, large holes, missing sections of pipe, and leaking longitudinal cracks. Assuming a pipe to be fully deteriorated, the CIPP is designed as a stand-alone pipe that will be able to withstand all internal pressure. In addition, the designer must consider that a fully deteriorated CIPP pressure pipe must be capable of withstanding external hydrostatic pressure.

5.7.6 Fully Deteriorated Gravity Flow CIPP Design

ASTM F-1216 and ASTM F-1743 specify use of a modified AWWA C950 design equation for a fully deteriorated gravity flow pipe. This equations adds an ovality reduction factor for consideration of long-term creep effects. Equation 5.9 presents the modified AWWA C950 equation that has been rearranged to solve for CIPP thickness.

$$t = 0.721\, D_O \left(\frac{(NP_t/C)^2}{E_L R_W B' E'} \right) \tag{5.9}$$

where P_t = total pressure due to water, soil, and live load acting on pipe, psi
R_W = buoyancy factor, dimensionless
B' = empirical coefficient of elastic support, dimensionless
E' = modulus of elasticity of adjacent soils or soil reaction, psi

Modified AWWA C950 formula specifies CIPP to have a minimum 50 percent of required stiffness (EI/D_O^3), which is 0.093 in. As shown in Eq. 5.10, to obtain a thickness of 0.093 in or more, a pipe designed with a flexural modulus of elasticity (E) of 350,000 psi would have a dimension ratio of 67. Therefore, if the CIPP stiffness is too low; the wall thickness must be increased accordingly to ensure design condition in Eq. 5.10 is met.

$$\frac{EI}{D_O^3} = \frac{E}{12(DR)^{33}} \geq 0.093 \tag{5.10}$$

where E = flexural modulus of elasticity of the CIPP, psi
I = moment of inertia, in^4, $I = t^3/12$

If the existing pipe is out of round or the CIPP may have localized ovalization, bending moment may predominate on the CIPP. For this special case, the CIPP must be checked to ensure that the bending moment does not exceed the long-term flexural strength of the CIPP. To satisfy this requirement, the bending stresses on the CIPP are determined by modifying Eq. 5.8 and substituting total pressure (P_t) to produce Eq. 5.11.

$$\frac{S_L}{P_t N} = [1.5\, q/100\, (1 + q/100)\, DR^2] - [0.5\, (1 + q/100)\, DR] \tag{5.11}$$

Where Eq. 5.4 defines q, and the other parameters have been defined previously.

Table 5.10 gives design thickness for different values of H_W and using particular values for E_L, C, N, R_W, and E'.

5.7.7 Total External Pressure on CIPP

For fully deteriorated design, all loads acting on the CIPP must be estimated to determine the total pressure (P_t). The total load is typically made up of hydrostatic pressure (P_W'), buoyancy corrected (see Sec. 5.9.8) soil load (P_S), superimposed or live loads (P_L), and other loads such as a vacuum load (P_V). Vacuum load is a special case and will not be discussed in this book. The total pressure on the pipe can be represented as shown in Eq. 5.12.

$$P = P_W' + P_S + P_L + P_V \tag{5.12}$$

5.7.8 Hydrostatic and Soil Loads

For the fully deteriorated design condition, watertable and pipe depths are determined from the top of the pipe. The hydrostatic pressure is determined with Eq. 5.5 (see Sec. 5.7.2) rewritten below.

$$P_W' = H_W \left(0.433\ \frac{\text{psi}}{\text{height (ft) of water}}\right)$$

TABLE 5.10 Fully Deteriorated Gravity Condition

Variables		
	E_L	175,000 psi
	C	0.84
	N	2
	R_W	0.67
	E'	1000, psi

Water depth, ft	CIPP thickness (in) for each pipe diameter (in)																	
	6	8	10	12	15	18	21	24	27	30	33	36	42	48	54	60	66	72
6	0.09	0.12	0.15	0.19	0.23	0.28	0.32	0.37	0.41	0.46	0.50	0.55	0.64	0.73	0.83	0.92	1.01	1.10
8	0.10	0.13	0.17	0.20	0.25	0.30	0.35	0.39	0.44	0.50	0.54	0.59	0.69	0.79	0.89	0.99	1.09	1.19
10	0.11	0.14	0.18	0.21	0.27	0.32	0.37	0.43	0.48	0.53	0.59	0.64	0.75	0.85	0.96	1.07	1.17	1.28
12	0.11	0.15	0.19	0.23	0.28	0.34	0.42	0.45	0.51	0.57	0.62	0.68	0.79	0.91	0.63	0.48	1.24	1.36
16	0.13	0.17	0.22	0.26	0.32	0.39	0.45	0.52	0.58	0.65	0.71	0.78	0.91	1.04	1.17	1.30	1.43	1.56
20	0.14	0.19	0.24	0.29	0.36	0.43	0.50	0.57	0.65	0.72	0.79	0.86	1.00	1.15	1.29	1.43	1.57	1.72
25	0.16	0.21	0.26	0.31	0.39	0.47	0.55	0.63	0.71	0.79	0.87	0.94	1.10	1.26	1.42	1.57	1.73	1.89
30	0.17	0.23	0.28	0.34	0.43	0.51	0.59	0.68	0.76	0.85	0.93	1.02	1.19	1.36	1.53	1.70	1.87	2.04
Practical minimum thickness, in	0.18	0.24	0.24	0.30	0.30	0.35	0.35	0.41	0.47	0.47	0.53	0.59	0.65	0.71	0.83	0.94	1.06	1.12

Source: Lanzo Lining Services Design Manual

TABLE 5.11 Soil Types and Unit Weight

Soil type	Unit weight, γ(lb/ft^3)
Sand and gravel	110
Saturated topsoil	115
Ordinary clay	120
Saturated clay	130

Source: Lanzo Lining Services Design Manual

The soil prism loading pressure is determined by Eq. 5.13.

$$P_S = \frac{w \cdot H_S \cdot R_W}{144\left(\frac{\text{in}^2}{\text{ft}^2}\right)} \tag{5.13}$$

where w = soil density, lb/ft^3 (see Table 5.11 for soil types and densities)
H_S = soil height above top of pipe, ft
R_W = water buoyancy factor, dimensionless

$$R_W = 1 - 0.33\ (H_W/H_S) \geq 0.67 \tag{5.14}$$

Other design parameters are the modulus of soil reaction or elastic support (E') and the coefficient of elastic support (B'). The modulus of soil reaction values used for CIPP design should typically represent stable undisturbed soils that would have E' values in the range of 700 to 3000 psi. Most typically a value of 700 psi is recommended for unknown soil conditions. Where the pipe is buried deep and the soil conditions are stable, values of 1000 to 1500 psi may be applicable. In areas known to have weak and unstable native soils, a value of 200 psi may be more appropriate. The coefficient of elastic support (B') is determined from Eq. 5.15. Table 5.11 presents elastic support coefficients at various soil depths.

$$B' = 1/(1 + 4 \cdot e^{-0.065 H_S}) \tag{5.15}$$

5.7.9 Superimposed or Live Loads

For a fully deteriorated design condition, dynamic live loads occur frequently and are a standard design condition for the parameter P_L. Live loads may be classified as either concentrated or distributed, depending on the soil pipe conditions and the depth the pipe is buried. In some cases the live load may be characterized by impact factors. Impact loading is generally only applicable for pipes that are relatively shallow (i.e., 2 to 5 ft). A maximum live load is recommended for a pipe buried beneath a highway, railway, or an airport runway. Guidelines for these live loads are provided by the American Association of State Highway and Transportation Officials (AASHTO), American Railway Engineers Association (AREA), and the Federal Aviation Agency (FAA) repectively. The most frequently encountered design condition is for pipes buried less than 7 ft from top of the pipe under active roads or highways. Live load pressures (P_L) associated with the surface load impacts are given in Table 5.12.

TABLE 5.12 Live Load Pressures and Impact Factors

	Highway, HS-20–44		Railway, E-80	
Soil height, H_S (ft)	Pressure, psi	Impact factor	Pressure, psi	Impact factor
0–1	>15.1	0.3	NA	0.4
1	15.1	0.3	NA	0.4
2	10.9	0.2	26.4	0.36
3	5.3	0	23.6	0.28
4	2.2	0	18.4	0.24
5	1.7	0	16.7	0.20
6	1.3	0	15.6	0.16
7	1.1	0	12.2	0.12
8	1.0	0	11.1	0.08
9	NA	0	9.4	0.04
10	NA	0	7.6	0
12	NA	0	5.6	0
15	NA	0	4.2	0

Source: Lanzo Lining Services Design Manual

Fully Deteriorated Gravity Design Example. Determine the minimum CIPP wall thickness required for the following existing pipe conditions (Fig. 5.3):

- Existing pipe classification = Fully deteriorated
- Mean outer CIPP diameter (D_O) = 48 in

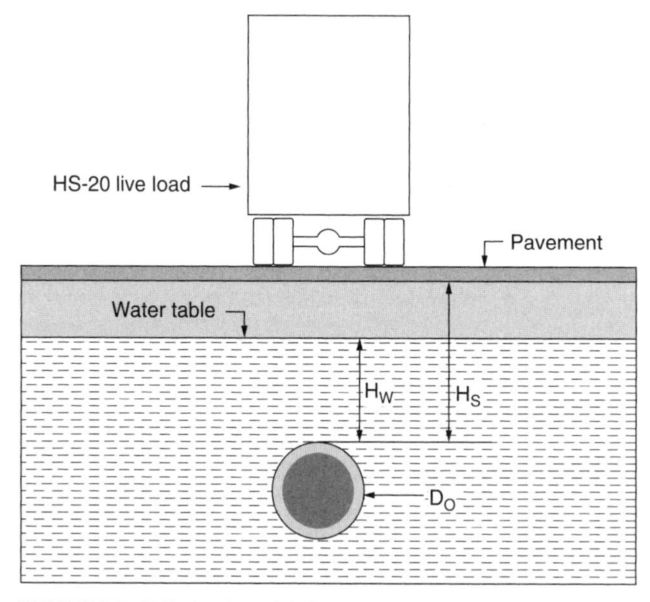

FIGURE 5.3 Fully deteriorated design example.

- Minimum pipe diameter (D_{min}) = 47.04 in
- External water above pipe (H_W) = 8 ft
- Depth of soil cover above pipe (H_S) = 15 ft
- Type of soil = Ordinary clay (120 lb/ft^3)
- Soil modulus (E_S') = 700 psi
- Live load = Live load HS-20
- Minimum CIPP flexural modulus (E) = 350,000 psi
- Minimum CIPP flexural strength (S) = 5500 psi
- Long-term modulus (E_L) = 175,000 psi
- Long-term strength (S_L) = 2750 psi

Step A. Determine the total load Eq. 5.12.

$$P_t = P_W{}^1 + P_S + P_L$$

Hydrostatic water pressure (Eq. 5.5)

$$P_W' = 0.433\ (H_W) = (0.433\ \text{psi/ft})\ (8\ \text{ft}) = 3.46\ \text{psi}$$

Soil load

$$P_S = wH_S R_W / 144\ \text{in}^2/\text{ft}^2$$
$$R_W = 1 - 0.33(H_W / H_S)^3\ \ 0.67 = 1 - \ 0.33\ (8\text{ft}/15\text{ft}) = 0.824,\ \text{which is} > 0.67$$

Ordinary clay soil density = 120 lb/ft^3 (Table 5.11)

$$P_S = 120(\text{lb/ft}^3) \times 15\ (\text{ft}) \times 0.824/144\ (\text{in}^2/\text{ft}^2) = 10.3\ \text{psi}$$

The soil pressure can also be determined by multiplying the soil prism pressure given in Table 5.13 and the buoyancy correction factor (R_W) given in Table 5.14.

$$P_S = 0.824(12.5) = 10.3\ \text{psi}$$

Live load (P_L) (Table 5.12) ~ 0.0 psi
Total load

$$P_t = 3.46\ (\text{psi}) + 10.3\ (\text{psi}) = 13.76\ \text{psi}$$

Step B. Calculate coefficient of elastic support.

$$B' = 1/(1 + 4e - 0.065Hs)$$
$$B' = 1/(1 + 4e - 0.065\ (15\text{ft})) = 0.40\ \ (\text{Table 5.15})$$

Step C. Calculate pipe ovality using Eq. 5.3.

$$q = \frac{100(48 - 47.04)}{48} = 2.0\%$$

TABLE 5.13 Soil Prism Pressure as Function of Water or Soil Height and Soil Density

Height of water, H_W' or soil, H_S, ft	Hydrostatic pressure P_W', psi	Soil prism pressure, psi Soil density, w, lb/ft^3				
		100	110	115	120	130
1	0.43	0.7	0.8	0.8	0.8	0.9
2	0.87	1.4	1.5	1.6	1.7	1.8
3	1.30	2.1	2.3	2.4	2.5	2.7
4	1.73	2.8	3.1	3.2	3.3	3.6
5	2.17	3.5	3.8	4.0	4.2	4.5
6	2.60	4.1	4.6	4.8	5.0	5.4
7	3.03	4.9	5.3	5.4	5.8	6.3
8	3.46	5.6	6.1	6.4	6.7	7.2
9	3.90	6.3	6.9	7.2	7.5	8.1
10	4.33	6.9	7.6	8.0	8.3	9.0
11	4.76	7.6	8.4	8.8	9.2	9.9
12	5.20	8.4	9.2	9.6	10.0	10.8
13	5.63	9.2	9.9	10.4	10.8	11.7
14	6.06	9.9	10.7	11.2	11.7	12.6
15	6.50	10.7	11.5	12.0	12.5	13.5
16	6.93	11.5	12.2	12.7	13.3	14.4
17	7.37	12.2	13.0	13.6	14.2	15.3
18	7.79	13.0	13.8	14.4	15.0	16.3
19	8.23	13.8	14.5	15.1	15.8	17.2
20	8.66	14.5	15.3	16.0	16.7	18.1
22	9.53	15.3	16.8	17.5	18.3	19.9
24	10.4	16.8	18.3	19.2	20.0	21.7
26	11.3	18.3	19.9	20.8	21.7	23.5
28	12.1	19.4	21.4	22.3	23.3	25.3
30	13.0	20.8	22.9	24.0	25.0	27.1

Source: Lanzo Lining Services Design Manual

Step D. Determine the minimum CIPP thickness for buckling.

$$t = 0.721 D_O \left(\frac{(NP_t/C)^2}{E_L R_W B' E'} \right)^{1/3} = 0.721(48) \left(\frac{[(2.0)(3.76)/(0.84)]^2}{175000(0.824)(0.4)(700)} \right)^{1/3}$$

$t = 1.0$ in

Step E. Check for minimum pipe stiffness.

$$DR = 48/1 = 48$$

$$350{,}000/12(48)^3 = 0.26 > 0.093$$

Check for pressure due to bending stress (Eq. 5.11)

$$S_L = 13.76\ (2)\ [(1.5(2)/100)(1+2/100)482] - [0.5(1+2/100)48]$$

$$S_L = 1920 \text{ psi}$$

TABLE 5.14 Water Buoyancy Factor, R_W

$R_W = 1 - 0.33\ (H_W/H_S) \geq 0.67$	
Ratio (H_W/H_S)	Factor (R_W)
0.00	1.00
0.05	0.98
0.10	0.97
0.15	0.95
0.20	0.93
0.25	0.92
0.30	0.90
0.35	0.88
0.40	0.87
0.45	0.85
0.50	0.84
0.55	0.82
0.60	0.82
0.65	0.79
0.70	0.77
0.75	0.75
0.80	0.74
0.85	0.72
0.90	0.70
0.95	0.69
1.00	0.67

Source: Lanzo Lining Services Design Manual

Step F. Therefore, the final design thickness for the CIPP is

$$t = 1.0 \text{ in}$$

The calculated bending stress (1920 psi) is less than the estimated long-term bending strength of the resin (i.e., 2700 psi) so bending stress does not control the design thickness.

5.7.10 Partially Deteriorated Pressure Flow CIPP Design

Important pressure pipe parameters include operating pressure(s), test pressures, surge pressures, and/or water hammer that may significantly exceed operating or test pressures. In addition, the design engineer and contractor must consider an existing pipe may change from partially to fully deteriorated condition after heavy cleaning. Therefore, it is recommended that the classified partially deteriorated existing pipe be tested for operating or test pressure to verify the condition of the pipe prior to CIPP. If the pipe is able to maintain the specified pressure then it can be classified as partially deteriorated. it should be noted that presence of small holes in the pipe may not allow maintaining pressure. When the condition of the pipe or the operating pressure cannot be well defined, it is recommended that the pipe be classified as fully deteriorated. Pressure pipe presents a higher CIPP risk; therefore, it is recommended that the contractor be experienced in this area.

TABLE 5.15 Coefficient of Elastic Support, B'

$B' = 1/(1 + 4e^{-0.065Hs})$

Soil height (H_S, ft)	Elastic support (B')
1	0.20
2	0.21
3	0.23
4	0.24
5	026
6	0.27
7	0.28
8	0.30
9	0.31
10	0.32
11	0.34
12	0.35
13	0.37
14	0.38
15	0.40
16	0.41
17	0.43
18	0.45
19	0.46
20	0.48
22	0.51
24	0.54
26	0.58
28	0.61
30	0.64

Source: Lanzo Lining Services Design Manual

The partially deteriorated design equation for internal pressure pipe given in ASTM F-1216 was derived with the assumption that the CIPP acts like a uniformly pressurized round flat plate with fixed edges covering an existing hole in the pipe. Thus CIPP is designed with the assumption that the aforementioned condition prevails and that bending stresses at and around the hole (if one exists), control the design thickness. This design assumption is more conservative than that of a square or rectangular plate. The ASTM F-1216 equation is rearranged to solve for CIPP thickness in Eq. 5.16.

$$t = \frac{D_O}{[5.33/P_i(D_O/D_h)^2(S_L/N)]^{0.5} + 1} \tag{5.16}$$

where D_O = mean outer CIPP diameter, in
P_i = internal pipe pressure, psi
D_h = hole diameter in existing pipe, in
S_L = long-term flexural bending strength for CIPP, psi
N = safety factor, $N = 2$ minimum

In order for the circular flat plate design condition to be valid, the criteria defined in Eq. 5.17 must be met. If this condition is not met, then the CIPP cannot be considered a circular flat plate and ring tension or hoop stress will dominate, and the CIPP will be designed as a fully deteriorated internal pressure pipe.

$$\frac{D_h}{D_O} \leq 1.83(t/D_O)^{0.5} \tag{5.17}$$

Once the CIPP thickness has been calculated, this value must be compared with the thickness calculated from Eq. 5.6 to confirm that the external hydrostatic pressure does not dominate the design condition. The larger thickness is then selected for the design.

5.7.11 Fully Deteriorated Pressure Flow CIPP Design

As discussed previously, it is critical to understand the physical conditions of the pipe and its operating parameters when designing for fully deteriorated pressure flow pipe. This design classification assumes that the existing pipe has no capability to hold the pressure and therefore, the CIPP must be designed to withstand all internal and external pressures as a standalone pipe. ASTM F-1216 assumes that pressure pipe is a *thick-wall cylinder* and internal pressure is calculated in Eq. 5.18.

$$P_i = \frac{2S_{tL}}{(DR-2)N} \tag{5.18}$$

However, Eq. 5.19 presents internal pressure for a *thin-wall cylinder* as follows.

$$P_i = \frac{2S_{tL}}{(DR-1)N} \tag{5.19}$$

Substituting Do/t for DR in Eq. 5.19 and after rearranging and simplifying, CIPP thickness is obtained from Eq. 5.20.

$$t = \frac{D_O}{(2S_{tL}/P_i N)+1} \tag{5.20}$$

where all the pipe design parameters have previously been defined.

Although the differences between a thin-wall and a thick-wall cylinder is relatively small, the thin-wall solution is a more conservative approach than what ASTM F-1216 recommends. The CIPP thickness for external pressure should also be checked against Eqs. 5.8 and 5.9 for fully deteriorated gravity flow pipe. The greatest thickness is chosen for the pressure pipe.

5.7.12 Hydraulic Design of CIPP

Gravity Flow. The installation of CIPP typically improves flow characteristics of the existing pipe. Flow in CIPP is improved because the inner surface of new pipe is extremely smooth and continuous, usually without any joints or discontinuities that create friction to flow. Typically the Manning's equation is used to predict flow rates in gravity or open channel conditions as given in Eq 5.21 below.

$$Q = \frac{1.486AR^{2/3}S^{1/2}}{n} \tag{5.21}$$

TABLE 5.16 Manning Coefficients for Typical Piping Materials

Pipe material	Range of Manning's coefficients (n)	Recommended Manning's coefficients (n)
Cured-in-place pipe (CIPP)	0.009–0.012	0.010
Vitrified clay	0.013–0.017	0.014
Concrete	0.013–0.017	0.015
Corrugated metal	0.015–0.037	0.020
Brick	0.015–0.017	0.016

Source: Lanzo Lining Services Design Manual

where Q = flow rate, ft^3/sec (cfs)
V = velocity, ft/sec
n = Manning's coefficient of roughness, dimensionless (see Table 5.16)
$R = A/P$ = hydraulic radius, ft
P = wetted perimeter of flow, ft
S = slope of grade line, ratio of vertical drop (h) to horizontal distance over which the drop occurs (L) usually expressed as a percentage (dimensionless)

When the pipe is circular and the flow is full as in a surcharged situation the Manning's equation may be modified to Eq. 5.22.

$$Q = \frac{0.463D^{8/3}S^{1/2}}{n} \tag{5.22}$$

where D = pipe internal diameter, ft, and other parameters are defined previously.

For circular pipe flowing full the Manning's equation can be abbreviated to obtain an easy comparison of flow capacity between CIPP and different the existing pipe materials as given in Eq. 5.23.

$$\%\ \text{Flow capacity} = \frac{Q_{\text{CIPP}} \times 100}{Q_{\text{exist}}} = \frac{n_{\text{exist}}}{n_{\text{CIPP}}}\left(\frac{D_{\text{CIPP}}}{D_{\text{exist}}}\right)^{8/3} \times 100 \tag{5.23}$$

The Manning's coefficients are dependent on many factors such as the type and condition of the existing pipe and the quality of CIPP installed. A conservative average of Manning's coefficient for CIPP in a relatively clean concrete, clay, or steel pipe has a 'n' of 0.010, as shown in Table 5.16. However, this coefficient might be subject to change over time as deposits may build up in an uncleaned pipe over time.

Gravity Flow Design Example. Determine the change in flow capacity when a circular 24-in concrete pipe is flowing full and is lined with a 0.47 in (12 mm) thick CIPP.

Step A. Select Manning's coefficients for the piping materials (Table 5.16).

(a) 'n' for CIPP = 0.010
(b) 'n' for concrete = 0.015

Step B. Determine inside pipe diameters.

(a) Existing concrete pipe D = 24 in
(b) New CIPP D = 24 in – 2(0.47 in) = 23.1 in

Step C. Determine increased flow capacity with Eq. 5.23.

$$\%\ \text{Flow capacity} = \frac{0.015}{0.10}\left(\frac{23.1}{24.0}\right)^{8/3} \times 100 = 135\ \%$$

Therefore, it is estimated that the CIPP increases the flow of concrete pipe by approximately 135 percent compared to that of the existing concrete pipe. This increase in flow was realized although the inside diameter of the CIPP was slightly smaller than the existing concrete pipe.

Pressure Flow. For determining the flow rate of pressure pipes, the Hazen-Williams equation is commonly used. Because of the inherent smoothness of CIPP inner surface, in this case, CIPP also increases the flow capacity of an existing pipe. The Hazen-Williams equation is given in Eq. 5.24.

$$Q = 1.318\, C \times R^{0.63} \times S^{0.54} \times A \tag{5.24}$$

where Q = flow rate, cfs
C = Hazen-Williams coefficient, dimensionless (see Table 5.17)
$R = A/P$ = hydraulic radius, ft
A = flow area, ft^2
P = wetted perimeter of the flow, ft
S = slope of the grade line, previously defined in Eq. 5.23

Hazen-Williams equation can also be simplified to provide a comparison of flow capacity between CIPP and the existing pipe as shown in Eq. 5.25 below:

$$\%\ \text{Flow capacity} = \frac{Q_{\text{CIPP}}}{Q_{\text{exist}}} \times 100 = \frac{C_{\text{CIPP}}}{C_{\text{exist}}}\left(\frac{D_{\text{CIPP}}}{D_{\text{exist}}}\right)^{8/3} \times 100 \tag{5.25}$$

Determination of flow capacities of CIPP relative to other existing piping materials is calculated in the same manner as given in gravity flow design example.

TABLE 5.17 Hazen-Williams Coefficients for Typical Pipe Materials

Pipe material/condition	Recommended Hazen-Williams coefficient '*C*'
Cured-in-place pipe (CIPP)	140
New steel or ductile iron (less than 1-year old)	120
Cement lined new steel or ductile iron	140
Steel (2-year old)	120
Steel (10-year old)	100
Cast iron (5-year old)	120
Cast iron (18-year old)	100
Tuberculated steel or cast iron	80

Source: Lanzo Lining Services Design Manual

ACKNOWLEDGMENTS

The CIPP design section presented in this chapter has been provided by Fred Tingberg, Jr., Lanzo Lining Services.

REVIEW QUESTIONS

1. Describe the different stages of TRM selection process.
2. Why pipe defect identification and inspection prior to renewal method selection is so important? Explain.
3. Define "partially deteriorated" and "fully deteriorated" design principles.
4. A 12-in diameter (D_O) sanitary sewer pipe with some cracking and offset joints is leaking and is in need of renewal (see Fig. 5.2). Through CCTV inspection, it appears that in at least one area, the pipe is no longer round and has deflected approximately 5 percent. The water table, through investigation is found to be normally 3 ft below the surface. The pipe is buried at a depth of 15 ft (H_S) to the top of the pipe. Assume a modulus of elasticity for the resin of 300,000 psi (long-term modulus, $E_L = 0.5(300{,}000) = 150{,}000$ psi) and a long-term flexural strength of 2500 psi. Determine the CIPP required wall thickness.
5. The 12-in diameter (D_O) sanitary sewer pipe (see Fig. 5.3) is cracked and leaking. The severity of the crack has progressed to the point that some pieces of the pipe wall have fallen into the pipeline and voids have begun to form around the pipe. The pipe is shallow and located under a busy city street subject to *HS*-20 loading. Assume that the pipe is buried in a clay soil with a unit weight of 120 pcf. Similar to Problem 4 above, the pipe is no longer round and has approximately 5 percent deflection. The pipe is buried at a depth of 5 ft (H_S). The water table is normally located 2 ft (H_w) above the top of the pipe. Assume a modulus of elasticity for the resin of 300,000 psi (long-term modulus, $EL = 0.5\ (300{,}000) = 150{,}000$ psi), a long-term flexural strength of 2500 psi, and a soil modulus of 1000 psi. Determine the required CIPP wall thickness.
6. Determine the change in flow capacity when a circular 18-in vitrified clay pipe is flowing full and is lined with a 0.36 in thick CIPP.

REFERENCES

Aggerwal, S. C., and M. J. Cooper. (1984). *External Pressure Testing of Insituform Lining*, Internal Report, Coventry Polytechnic.

ASTM 2412. (1992). *Standard Test Method for Determination of External Loading Characteristics of Plastic Pipe by Parallel-Plate Loading*, American Society for Testing and Materials, Conshohocken, Pa.

Engineering Design Manual for Rehabilitation of Cured-in-Place Pipe. (2004). Lamzo Lining Services.

Glock, D. (1997). *Uberkritisches verhalten eines ummauteltn kriesrohres bei wasserdruck von auBen und temperaturdehnung*, Der Stahblau.

Guice, L. K., and J. Y. Li. (1994). Buckling models and influencing factors for pipe renewal design, *Design Theory Workshop, North American NO-DIG '94 Conference Proceedings*, NASTT, Dallas, Tex.

Guice, L. K., T. Straughan, C. R. Norris, and D. R. Bennet. (1994). *Long-term Structural Behavior of Pipeline Renewal Systems,* Trenchless Technology Center, Louisiana Tech University, Ruston, La.

Iowa Statewide Urban Design and Specifications (SUDAS). (2004). *Design Manual: Chapter 14,* Ames, Iowa. Available at: http://www.iowasudas.org/.

Kleweno, D. G. (1997). Critical buckling behavior and life cycle predictions of CIPP, *NO-DIG '97 Conference Proceedings North American,* Seattle, Wash.

Lo, K. H., and J. Q. Zang. (1994). Collapse resistance modeling of encased pipes, *Buried Plastic Pipe Technology,* 2nd Vol., ASTM STP 1222, Dave Eckstein, Ed., American Society of Testing and Materials, Philadelphia, Pa.

McAlpine, G. A. (1996). Statistical analysis and implications of test data from long-term structural behavior of pipeline renewal systems, *Proceedings of the Water Environment Federation 69th Annual Conference & Expo,* Dallas, Tex.

McAlpine, G. A. (1996). TTC report on long-term structural behavior of pipeline renewal systems, *NO-DIG Engineering.*

Moore, I. D. (1998). Tests for pipe liner stability: What we can and cannot learn, *North American NO-DIG '98 Conference Proceedings,* NASTT, Albuquerque, N. Mex.

Omara, A. (1997). *Analysis of Cured-in-Place Pipe (CIPP) Installed in Circular and Oval Deteriorated Existing Pipes,* PhD Thesis, Louisiana Tech University, College of Engineering.

Schrock, J. B., and J. Gumbel. (1997). Pipeline renewal—1997, *North American NO-DIG '97 Conference Proceedings,* NASTT, Seattle, Wash.

Spangler, M. G. (1941). *The Structural Design of Flexible Pipe Culverts,* Bulletin 153, Iowa State College of Engineering Experiment Station, Ames, Iowa.

Standard Specifications for Highway Bridges, 14th ed., (1989). American Association of State Highway and Transportation Officials, Washington D.C.

Timeshenko, S. P., and J. M. Gere. (1988). *Theory of Elastic Stability*, 2nd ed., McGraw-Hill Publishing Co. New York. NY.

Tingberg Jr., F., and W. Cavalier. (1999). Sewage force main renewal using cured-in-place pipe technology, *NO-DIG '99 Conference, Proceedings* NASTT, Orlando, Fla.

CHAPTER 6
PIPE MATERIALS

6.1 INTRODUCTION

Trenchless construction methods cannot be successful without the use of a quality pipe. While the pipe must have certain qualities to be accepted as a final product, its properties and its interaction with soil and trenchless equipment will determine whether the trenchless project can commence and problems can be prevented. This chapter presents an overview of pipe material as they relate to trenchless construction and renewal methods.

The role of piping materials in water and sewer systems has traditionally received little attention in academia. At the undergraduate level, the vast majority of civil engineering curricula place a disproportionate emphasis on theoretical hydraulic design and piping materials, with the latter receiving little attention, if any. Although students are trained well in designing systems that incorporate various parameters such as maximum flows, surges, head losses, and so on, they rarely receive sufficient instruction on the importance of selecting the appropriate piping material for the design application. Many young engineers who may get involved in designing water and sewer systems for the majority of their careers are unsure of the differences between rigid and flexible conduits and the pipe-soil interaction mechanism for each type. *Traditional* piping materials mentioned in textbooks more than half a century ago continue to be the major focus in more recent textbooks, despite the numerous technological developments and changes in engineering and installation practices that have taken place in the last three decades in the field of pipe engineering. An understanding of the advantages and limitations of various materials for new construction as well as renewal is a necessary tool for pipeline engineers today.

6.1.1 Pipelines and Trenchless Technology

The emergence of trenchless technologies in recent years for construction and renewal of pipeline systems has resulted in the production and availability of various pipe materials with modified *smooth* joints. Traditional pipe construction projects use open trenching methods to install pipes of all types and sizes. Using open-cut methods, it is possible to install bell-and-spigot gasket-joint pipes because they are neither pushed (jacked) through, nor pulled into boreholes. Piping materials used in trenchless construction, on the other hand, must exhibit sufficient compressive or tensile strengths and have joints that will not be over-inserted or pulled apart if they are to be either jacked or pulled through underground bores.

In general, a pipe used for jacking must be round, have a smooth, uniform outer surface, and have watertight joints that also allow for easy connections between pipes. Pipe lengths must be within specified tolerances and pipe ends must be square and smooth so that jacking loads are evenly distributed around the entire pipe joint and point loads will not occur when the pipe is jacked in a reasonably straight alignment. Pipe used for pipe jacking must be capable of withstanding all the forces that will be imposed by the process of installation, as well as the final in-place loading conditions.

The driving ends of the pipe and intermediate joints must be protected against damage as specified by the manufacturer. The detailed method proposed to cushion and distribute the jacking forces is usually specified by applicable standards for each particular pipe material.

6.1.2 Traditional and Modern Piping Materials

Traditional piping materials for water systems include cast iron pipe (CIP), ductile iron pipe (DIP), asbestos-cement pipe (ACP), prestressed concrete cylinder pipe (PCCP), and reinforced concrete pipe (RCP). Asbestos-cement pipe, though no longer manufactured and installed in North America, accounts for large amounts of water pipe that are still in service today throughout the United States. In sanitary sewers, traditional piping materials include brick pipe (BP) vitrified clay pipe (VCP), concrete pipe (CP), and reinforced concrete pipe (RCP). Ductile iron pipe is also used in some gravity and pressure sanitary sewer applications (force-mains).

Use of pipelines has a long history. For example, more than 1000 years ago, the Romans used lead pipes for aqueduct system to supply water to Rome (Liu, 2003). The oldest clay pipe for water and sewer applications goes back to 4000 B.C. and was found in the ruins of the Palace of Knossos in Crete and was also used for drainage purposes in Egypt and Persia. The earliest known man-made concrete sewer, made in 800 B.C., is the Cloaca Maxima in Rome; it is made of stone and natural cement. The first cast iron pipeline was put into service in Siegerland, Germany, in the 1800s. The manufacturing of steel pipe with increased strength and availability of different pipe sizes in the 19th century revolutionized pipeline technology. In 1879, following the discovery of oil in Pennsylvania, the first long-distance oil pipeline was built in this state. This pipe was a 6-in diameter, 109-mi long steel pipe (Liu, 2003). Figure 6.1 presents a wood water pipe recently removed during a construction project in Mobile, Alabama.

FIGURE 6.1 An example of wood water pipe. (*Source: Dr. Tom Iseley.*)

Modern piping materials for water and sewer applications, now in use in North America for almost four decades, and much longer in Europe, include thermoplastic pipes such as polyvinyl chloride pipe (PVC), high density polyethylene pipe (HDPE) and medium density polyethylene pipe (MDPE), glass-reinforced plastic pipe (GRP), also called fiberglass pipe or fiberglass reinforced mortar pipe (FRM), or centrifugally cast fiberglass reinforced polymer mortar pipe (CCFRPM), and polymer concrete pipe (PC). The oldest-known PVC pipe was manufactured and installed in Germany in the early years of World War II, and continues to be in service today. Steel pipe, cast iron pipe, and ductile iron pipe are also used in various pressure and gravity applications.

All the above mentioned piping materials have commonly been used in open-trench construction. Until recently, the availability of joint-types for trenchless technology in some of these pipes has traditionally prevented them from being used in trenchless applications. Some others exhibit either low compressive or tensile strengths that prevent them from being jacked or pulled for installation. Detailed discussion of these properties for each piping material will be presented later in this chapter.

6.2 PIPE-SOIL INTERACTION

Pioneering work in the field of pipe engineering was done at the turn of the last century by Anson Marston and was continued in the subsequent years by others such as M. G. Spangler and R. Watkins. The theoretical foundation of piping systems was laid down by Marston in his first paper of 1913 titled *The Theory of Loads on Pipes in Ditches and Tests of Cement and Clay Drain Tile and Sewer Pipe.* The method of calculating the earth load to which a buried pipe is subjected in service, known to this day throughout the world as the Marston load theory, was outlined in this paper. In 1941, Spangler, then a student of Marston's, developed the theory of flexible pipe design, published in a paper titled *The Structural Design of Flexible Pipe Culverts.* In 1955, Spangler's student, Reynold Watkins, further refined Spangler's work of 1941, defining a new parameter, the modulus of soil reaction, E', and ultimately derived the present-day equation for predicting the deflection of a buried flexible pipe, known as the Modified Iowa Formula.

All the researches described above focused on one main theme: the behavior of a pipe in relation to the surrounding soils in which it is buried. Also referred to as the pipe-soil interaction mechanism, it is this property of a pipe that is used to place it into one of the two categories: rigid or flexible.

6.2.1 Rigid Conduits

The basic concept of Marston's load theory is that the load because of the weight of the column of soil above a buried pipe is modified by the response of the pipe and the relative movement of the side columns of soil to the central column. When the side columns of soil between the pipe and the trench wall (pipe zone) are more compressible than the pipe, this causes the pipe to assume load generated across the width of the trench. In other words, the pipe itself bears the brunt of the weight of the soil column directly above it. This is illustrated in Fig. 6.2(a) and is the typical pipe-soil interaction behavior displayed by rigid pipes. Concrete, asbestos-cement, and clay pipes fall into this category. When installing rigid pipes, it is important to have a good backfill with appropriate compaction so that the side columns do not experience an excessive settlement. If this occurs, the load on the rigid pipe may surpass its design strength capability and result in the pipe walls cracking.

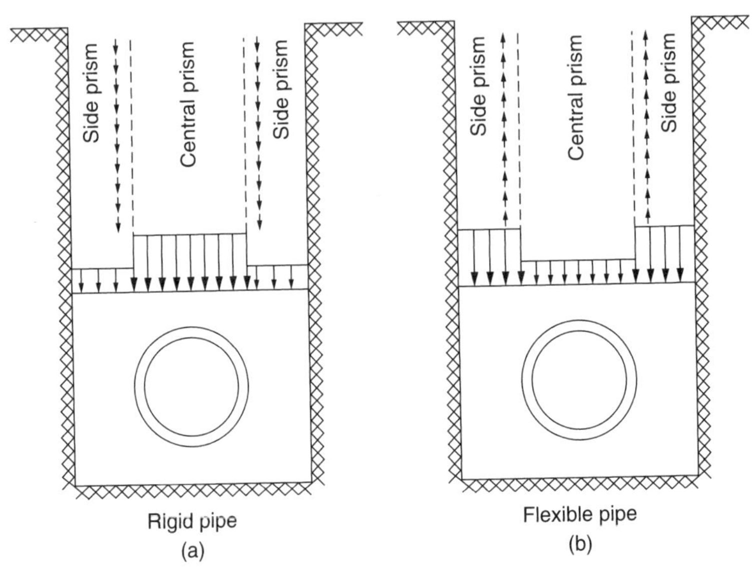

FIGURE 6.2 Trench load comparisons for (*a*) rigid (Marston load) and (*b*) flexible pipe (adjusted Marston load).

6.2.2 Flexible Conduits

The other scenario involves pipe that is capable of deflecting without cracking under soil loading. In this case, the central prism of the soil directly above the pipe settles more in relation to the adjacent soil columns (between the pipe and the trench wall). This settlement produces shearing forces that reduce the load on the pipe to an amount that is less than the weight of the prism directly above it. This is illustrated in Fig. 6.2(b). Pipes that exhibit these characteristics are defined as flexible conduits. Regardless of the pipe stiffness and the strength of the pipe, as the soil in the trench settles or moves downward compared to the trench sidewall, friction forces are generated that act to reduce the weight of the trench-wide soil column. Marston's load theory predicts and accounts for these frictional shearing forces. Flexible pipes include all thermoplastics, cast or ductile iron pipe, steel pipe, and reinforced fiberglass pipe.

6.2.3 Semirigid Pipe

Several pipe manufacturers describe that their products are *semirigid*, that is, the pipe displays properties that are characteristic of the behavior of both rigid and flexible pipe. One company in Australia has performed testing on their pipe that shows that although their product is considered a rigid pipe, under burial conditions the pipe exhibits flexible properties and in-ground deflections are significantly lower than those exhibited by the pipe in the unrestrained condition. The term *semirigid* has been argued by some engineers as nothing more than a marketing terminology. According to Dr. Moser of Utah State University, "Claims that a particular pipe is neither flexible nor rigid, but somewhere in between have little importance since current design standards are based either on the concept of a flexible conduit or on the concept of a rigid conduit."

6.3 PIPE SELECTION CONSIDERATIONS

Although economics is one of the engineering considerations during the design and selection of pipe material for a piping system, there are several other factors that are as important. These considerations include the following:

- Type of commodity to be transported (oil, gas, water, wastewater, and so on)
- Construction conditions and methods used
- Life expectancy
- Flow characteristics, such as corrosiveness and abrasion of wastewater
- Ease of handling and installation of pipe
- Pipe stiffness and strength, physical, and chemical properties
- Availability of diameter sizes, pipe section lengths, and joints for trenchless technology
- Construction and operational stresses in the pipe. In many cases for trenchless technology projects, the construction stresses in the pipe exceeds the operational stresses
- Location and pipe environment (inland, offshore, in-plant, corrosiveness of soil, and so on)
- Type of burial or support (underground, aboveground or elevated, underwater, and so on)

There are some unique pipe properties that may be required in a water pipeline and not in a sewer system, and vice versa. These topics are discussed in detail below.

6.3.1 Water System Pipelines

Water systems comprise a distribution and a transmission system as shown Fig. 6.3. In addition to the various sizes of pipes, a system also comprises various appurtenances including fittings, valves, joint restraints, pumps, and the like. The transmission portion, which is typically made of larger-diameter pipes, conveys water from the treatment plant to the distribution system. Although the transmission pipes contain very few interconnections, the

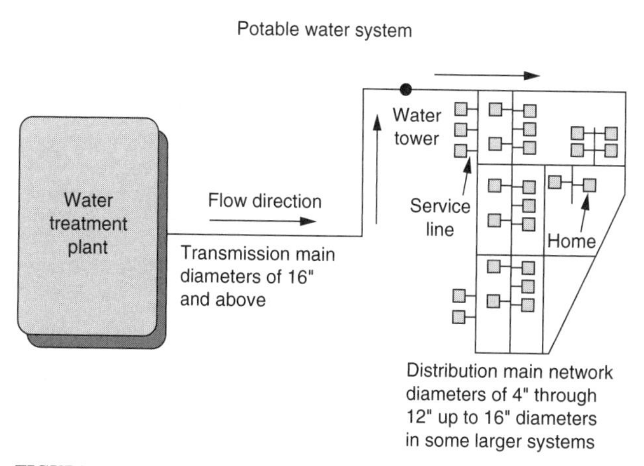

FIGURE 6.3 Schematic of a water system.

distribution system is a complex network of loops and interconnecting pipes of smaller diameters. Consequently, flow in distribution pipes is quite unpredictable and susceptible to frequent surging conditions.

Main design considerations in transmission pipes include internal pressure (pressure handling capacity of the pipes), longitudinal and radial stresses in the pipe walls, and thrust forces. For flexible pipes, ring deflections should also be taken into consideration. It is also important to ensure that air buildup within the system, particularly in hilly terrain, is accounted for in design by placing air valves particularly at the high points as well as other critical locations along the length of the transmission line. In a distribution system, all the above factors should be taken into account; additionally, more care should be taken with backfilling and appurtenances such as fittings and service connections. Other considerations should include proper methods and equipment for performing taps and other maintenance work. Preventing environmental effects such as corrosion or permeation of chemicals into pipe owing to gross contamination of surrounding soils is also of paramount importance. Both transmission and distribution lines are buried below frost line and usually at depths of no more than 6 to 7 ft.

Low frictional resistance within the pipe leads to a better flow, enabling the engineer to use smaller diameters and thereby reducing costs. Tight joints are also important to ensure that unaccounted-for-water is kept to a minimum. In recent years, an unusually high level of unaccounted-for-water has received particular attention in the waterworks industry. Significant financial losses to municipalities have resulted because of various types of leaking joints and cracked piping infrastructures. In some developing countries, unaccounted-for-water is as high as 45 percent.

6.3.2 Sewer System Pipelines

Sanitary sewers comprise a collection system and a treatment system as shown in Fig. 6.4. Sewer pipes called service laterals convey wastewater away from houses and transport it to a main sewer pipe, which transports the sewage to treatment plant before the final discharge into a stream or a larger body of water. The collection system in a large community comprises pipes that convey wastewater both by gravity and pressurized flow. Gravity pipes normally do not flow full at any time. Most municipalities and engineers design a

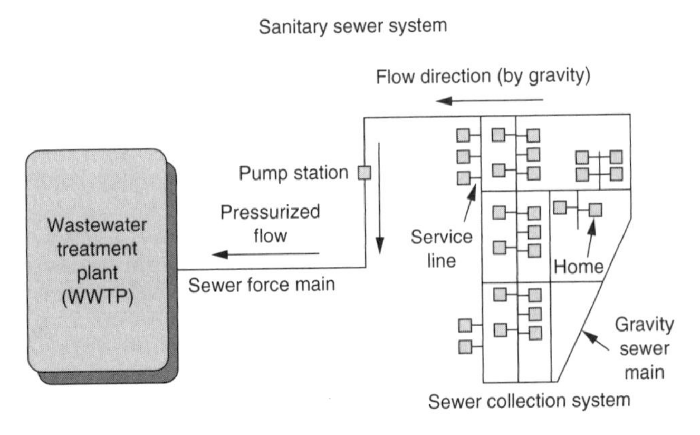

FIGURE 6.4 Schematic of a sanitary sewer.

gravity system to have an operating velocity of 2 ft/s, based on the conditions of the fluid and the specific site conditions. Once gravity flow has conveyed the wastewater to a pumping station, it is then conveyed by a pressurized system to the treatment plant. The section of pipe conveying the wastewater under pressure is called a sewer force main.

To allow for the pickup of wastewater flow from basements, the gravity fed portion of the system is usually buried at great depths. It is not unusual in some areas to have pipe depths in excess of 25 ft. Larger lines and systems are known to go as deep as 50 ft. Consequently, pipes buried at greater depths must be able to handle the added soil loading pressures. A firm consideration to the pipe-soil interaction mechanism must be given to ensure that excessive deflection in flexible pipe, or pipe cracking in rigid pipe is avoided.

The piping material used should have effective joints to minimize or eliminate any infiltration of groundwater into the piping system, or exfiltration of the wastewater into surrounding soils and groundwater sources. Flow within the piping system should be as efficient as possible. This is directly related to the friction coefficient of the piping material. A pipe with low Manning's n will enable the design engineer to use smaller diameters and still meet the design flows. The ability of the pipe material to handle high corrosive conditions that occur within a sanitary sewer system is another important characteristic. In systems where the wastewater is high in suspended solids, the piping material's resistance to scour is an important consideration. Life expectancy and experience with use of the pipe should also be an important decision factor; with a myriad of piping materials and long-term protection technologies available today, engineers should design for a useful life of at least 100 years.

All of the above considerations must also be given to sewer force-main design. In addition, the engineer must ensure that the force-main is capable of handling the maximum design pressures within the system; additional appurtenances such as air release valves should be used to prevent air buildup within the line, which may cause it to eventually fail. The effect of the cyclic surging on the pipe should also be included in design.

6.3.3 Importance of a 100-Year Design Life

In the past decades, it was reasonable to build a water or sewer system with a design life of 50 years. After all, the piping technology available in those days could usually sustain no more than a 50-year useful design life. Pipes were usually overcome by environmental effects such as corrosion; their joints were weakened because of root intrusion or joint separation resulting from ground movement or wall cracking causing leakage and eventual failure. Some materials such as cast iron that stayed in service for well over 100 years in many cases were able to do so on account of the sheer thickness of its walls. The thick walls played a sacrificial role as they gradually corroded. In the subsequent years however, to remain competitive with alternative materials, the wall thickness of ductile iron pipes saw significant reductions; a cast iron pipe made 100 years ago had walls that were remarkably thicker than the walls of a ductile iron pipe of equivalent pressure class today. Fortunately, various technologies that inhibit corrosion, such as polyethylene (PE) encasements, internal lining, and cathodic protection have become available as a means to prolong the useful lives of these metallic pipes. Furthermore, we now have at our disposal many other types of piping materials that can conservatively sustain a 100-year useful design life.

Rapid demographic changes throughout North America require municipal agencies, utilities, and other government entities to accommodate the steady rise in demand for potable water and wastewater services. Although capital improvement projects are increasing the size of existing water, sanitary sewer, and storm sewer piping systems, an equal effort is being made to renew and reconstruct the aging network of buried pipes. Some of the oldest underground piping infrastructures have already surpassed their 50th and even

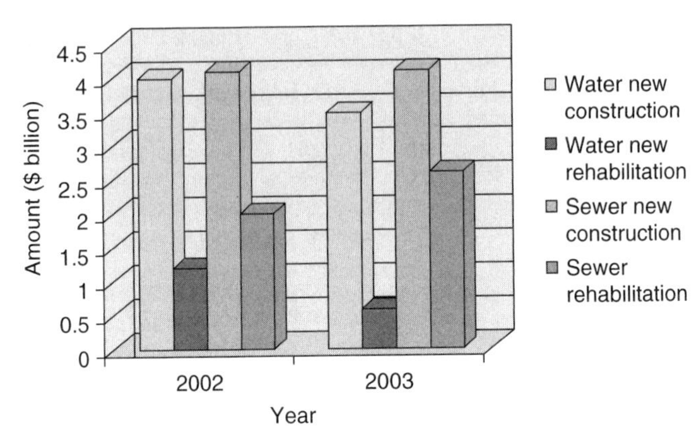

FIGURE 6.5 Water and sewer construction and renewal costs in the United States.

100th year of design life, while others are nearing a quarter or half century of their service lives, or have already surpassed their expected design lives and are in dire need of renewal. More alarming are those pipes installed in the past 20 to 30 years which have already failed owing to various environmental conditions. According to annual surveys of the municipal water and sewer infrastructure in the United States, conducted in the early 2003 and 2004 by *Underground Construction* magazine, US$1.2 billion and US$0.6 billion were spent in 2002 and 2003, respectively, for water system renewal, as shown in Fig. 6.5. For sewer renewal, US$2.0 billion and US$2.6 billion were spent in 2002 and 2003, respectively. For new construction of water systems, US$4 billion were spent in 2002 and US$3.5 billion were spent in 2003. New construction of sewers saw expenditures of US$4.1 billion in 2002 and US$4.14 billion in 2003. These costs are projected to continue an upward trend in the coming years.

As investments are made for building new pipeline systems and renewing older ones, the design engineer should actively strive for a *minimum* design life of 100 years. As we now realize, governments and private agencies cannot afford to replace and renew all buried pipe infrastructure on a 50-year basis. Engineers and owners of municipal piping systems should take into account that not only is pipe material itself important in the 100-year design goal, but pipe environmental effects along with loading conditions must be given equal consideration. It is simply not enough to have pipes capable of lasting 100 years; its joints must also serve their purpose of preventing infiltration/inflow (I/I) in sewers and unaccounted-for-water in potable water distribution systems. Rather than basing decisions on initial costs only, pipeline owners and municipalities should implement projects based on overall life cycle considerations for the system (see Chap. 2 for an analysis of the social costs).

There is no single pipe material that is best for all applications. However, on a relative merit basis for a particular project, there may be one material that clearly stands out as the best choice. Keeping an eye on the *long-term* (at least 100 years) performance of the chosen material, based on its current and past performance is a good selection strategy. The following sections on the different types of piping materials will focus on their general properties, but special emphasis will be placed on products designed for use in trenchless construction and renewal.

6.4 CEMENT-BASED PIPES

Pipes in this category include concrete pipe as well as asbestos-cement pipe. Both types incorporate Portland cement as the base material. Concrete pipes are designed with or without reinforcements and are used for both pressure and gravity applications. Asbestos-cement pipe is composed of a mixture of Portland cement, silica, and asbestos fiber and was used in North America mainly in potable water applications, though many sewer systems also have asbestos-cement pipes in service. Asbestos-cement pipes have not been manufactured or installed in the United States since the late 1980s, but large quantities of the pipe continue to be in service in water systems at various locations. It is therefore important for engineers to have a working knowledge of the properties of this material. Both concrete and asbestos-cement pipes are rigid conduits and are accordingly designed for installation.

6.4.1 Concrete Pipes

Trenchless construction with concrete pipe was first performed by the Northern Pacific Railroad between 1896 and 1900, when the pipe was installed by jacking. The jacking of concrete pipe is performed both for pressure and gravity applications. Concrete pressure pipes are also installed by tunneling methods. For small-diameter pipe in short tunnels, such as those under a highway or a railroad, it is a common practice to slide the pipe through a liner.

Manufacture. Concrete pipes can be made with any of the five grades of Portland cement. There are five manufacturing processes used in North America, four of which use mechanical means to place and compact a dry concrete mix into a form. A conventional wet mix and casting method is used in the fifth one. Each method is briefly discussed below.

1. *Centrifugal casting.* The form is rotated, acting as a centrifuge, while the concrete mix is fed into it. Vibration and compaction are also employed to consolidate the concrete mix. The centrifuge promotes extraction of water from the mix.
2. *Dry casting.* Low-frequency, high-amplitude vibrations are used to distribute and compact the dry mix in the form in this process. The form is then removed and the pipe cured.
3. *Packerhead process.* In this process, a high-speed rotating device, the packerhead, acts as the inner wall of the form and compacts the concrete against the walls of the form whereas the concrete is fed from the top. The packerhead gradually moves up.
4. *Tamping.* The concrete is mechanically compacted between the inner and outer walls of the form.
5. *West casting.* This method is typically employed for the manufacture of large-diameter pipes, using a concrete mix with higher water content than the other four methods. Again, the pipe is made between the inner and outer walls of a form.

Types of Concrete Pipes. There are five different types of concrete pipes, used for pressure and sewer applications.

1. *Nonreinforced concrete pipe (CP).* Used for nonpressure applications only, this pipe is manufactured in diameters of 4 in through 36 in, in lengths of 8 ft. Produced in three classes, its minimum strength requirement is the three-edge bearing per ASTM C-14.

2. *Reinforced concrete pipe (RCP)*. Three types of reinforcements are available for RCP-welded wire fabric, hot-rolled rod made of Grade 40 steel, and cold-drawn steel wire made from hot-rolled rod. It is used in low pressure applications of up to 55 psi. There are five load-bearing capacities, ranging from 1500 lb/ft to 22,500 lb/ft.
3. *Prestressed concrete cylinder pipe (PCCP)*. Often used in high-pressure applications, PCCP is a composite pipe of concrete and steel, capable of handling up to 500 psi. There are two types of PCCP—lined cylinder pipe and embedded cylinder pipe. A welded cylinder core with joint rings attached at both ends exists in both types. After curing, the pipes are wrapped with hard-drawn wire under high tensile stress, and then coated with cement slurry. Available diameters range from 24 in to 144 in. AWWA C301 is the governing standard.
4. *Reinforced concrete cylinder pipe.* This is similar to PCCP, but uses reinforcing cages in place of the hard-drawn wire. Standard AWWA C300 requires that the pipe be designed to withstand both internal and external pressure. Available diameters range from 24 in to 144 in. Pressure applications are its main use.
5. *Bar-wrapped steel-cylinder concrete pipe.* The steel cylinder is internally lined with a cement mortar lining. Once cured, a steel rod, under tension, is wrapped around the cylinder. Another cement mortar lining is then placed on the wrapped cylinder. Smaller-diameter pipes are considered rigid, whereas larger diameters behave as flexible conduits. AWWA C303 is the standard governing its manufacture and testing. Like the previous two types, bar-wrapped concrete pipe is also used in pressure applications. Figure 6.6 illustrates typical sections of polymer concrete pipes.
6. *Polymer concrete pipe (PCP)*. Polymer concrete originated more than 20 years ago in Germany. This type of pipe provides a corrosion-resistant concrete as needed for piping applications that require high concrete compressive strength and resistance to corrosive chemicals. These pipes are made by mixing a high-strength thermosetting resin with oven-dried aggregate to form a type of concrete. The resin within the mix provides for bonding the aggregate much like Portland cement does in traditional concrete pipes.

The pipe sections are cast vertically with an inner and outer form and are vibrated for compaction. After the forms are removed, the section is heated in a kiln to finish curing the resin. These pipes can typically be used to carry highly aggressive wastes, for pipe jacking as they have very high compressive strengths (up to 17,000 psi), or for microtunneling. They can also be used for gravity flow or pressure applications. Some manufacturers are also making polymer pipes in sizes appropriate for use as manholes. Also, semielliptical and circular liner pipes have been developed for sliplining sewers. There are several ASTM standards for the product.

FIGURE 6.6 Polymer concrete pipe.

Polymer concrete pipes have several benefits including high-strength corrosion resistance (they can be used in environments with pH ranges of 1 to 13), low wall roughness, and high abrasion resistance. The use of polymer concrete pipes is becoming more common in the following areas: direct bury, sliplining, jacking and microtunneling, tunneling, and above-the-ground applications. One of the main disadvantages in the past was the high cost associated with importing the pipe; its high cost relative to alternative materials limited its use to only niche markets where its superior qualities were needed. However, since 2002, polymer concrete pipe has been manufactured in the United States under the brand name of Meyer Pipe. This is now enabling competition with products such as GRP for jacking installations.

Fiberglass sleeve joints are made separately, with elastomer sealing and spacing rings laminated into the sleeve. Factory fitting of the couplings to the end of the pipe is performed and pressure testing is done up to 35 psi. As an alternative to the couplings, stainless steel collars are also available.

Applicable Standards. Table 6.1 lists RCP standards that are widely used in trenchless installations. Available diameter ranges are also given.

Joint Types. The concrete pipe industry has developed several different types of joints for the various types of pipes. Selection of the appropriate joint is based on the stringency of the application for water tightness. Although bell-and-spigot type joints are available for open-trench applications, trenchless methods such as microtunneling and pipe jacking dictate that the joint be flush with the outside wall of the pipe barrel. Therefore, the tongue and groove joint or the modified tongue and groove joint is ideal. Table 6.2 shows joint types that may be used in trenchless construction. Also available are mastic or mortar joints, which are not well-suited to prevent leakage.

A proprietary concrete and PVC composite pipe, PipeForm™, with an inner and outer PVC liner that gives it the strength of concrete and the corrosion resistance of PVC, has been used in microtunneling and pipe jacking projects for gravity applications. In addition to preventing corrosion, the PVC outer shell also reduces external skin friction during jacking, while the internal liner improves flow. Available diameters range from 18 in through 36 in. Distribution of jacking forces at the end surface is maximized by steel or plastic collars and rubber gaskets. Details of the joint are shown in Table 6.2.

Advantages and Limitations. Table 6.3 presents a summary of advantages and limitations of concrete gravity and pressure pipes. The widespread use of large-diameter concrete pipe in water and sewer applications is an indication of the material's acceptance at the municipal level. The vast selection of available diameters and pipe lengths is a convenience to the design engineer and contractor. On the pressure side, availability of various structural and pressure strengths also makes it convenient. The ability of manufacturers to make pipes

TABLE 6.1 Concrete Pipe Standards

Pressure standards	Nonpressure standards	Available diameters
ASTM C-361	–	12–106 in
AWWA C-300	–	–
AWWA C-302	–	–
–	ASTM C-76	12–144 in

TABLE 6.2 Concrete Pipe Joints

Joint type	Cross section
Compression type rubber gasket joints with tongue and groove. Intended for use with pipes meeting ASTM C-14, C-76, and C-655.	
Steel end ring joint. This is a high pressure joint, intended for use with ASTM C-361.	
PipeForm™ Joint. This proprietary product is a concrete-PVC composite, used in pipe jacking and microtunneling.	Exterior collar Plastic outer shell Steel reinforcement Compres-sion ring Concrete core Plastic inside liner & collar

capable of handling very high pressures in large diameters has resulted in its specification on some high-profile trenchless projects in recent years. Because a specialized work-crew is not needed; cost of labor for concrete pipe installation is not high. For trenchless construction, the high compressive strengths give it a definite advantage over some other pipe materials.

Many concrete pipes installed in gravity applications 20 to 30 years ago have shown signs of poor performance. Leakage through joints and cracks in the pipe has been a constant source of inconvenience to municipalities. For contractors, the heavy weight of concrete pipes makes it difficult to install when compared to alternative materials. Concrete pipe's sensitivity to bedding conditions in both shallow and deep installations has resulted in pipe failures by shear and beam breakage in many instances. The susceptibility of concrete pipe to both internal and external corrosion has also come to the forefront in recent years. Many engineers and city authorities now require concrete pipes to be internally lined because of their experience with its deterioration. Studies suggest that unlined RCP can show signs of deterioration in as little as 5 years after installation. Failures owing to hydrogen sulfide attacks and internal microbiological-induced corrosion at the crown of RCP have also been a topic of study in recent years. In pressure applications, the loss of prestress as a result of reinforcement wire breakage in PCCP can lead to catastrophic failures in high pressure lines.

TABLE 6.3 Concrete Pipe Advantages and Limitations

Advantages	Limitations
Specialized work crew not required for installation	Sensitive to bedding conditions—shear failure and beam breakage may occur
Large selection of available nominal diameters	Handling and installation difficulty because of heavy weight
Wide variety of pipe lengths available	Susceptible to external corrosion in acidic soil environments
Large selection of both structural and pressure strengths	Highly vulnerable to hydrogen sulfide attacks and internal microbiological-induced corrosion at crown
Relatively low cost of maintenance	Generally difficult to repair, particularly in cases of joint leakage or failure in pressure pipes
Capability to withstand very high pressures	Tendency to leak because of high pipe wall porosity and shrinkage cracking
Ideal for pipe jacking applications owing to high compressive strengths	Without internal lining, life span is significantly reduced
Internal corrosion can be significantly reduced by using thermoplastic lining	Low abrasion resistance—internal scouring can occur if solid content and flow velocities are high
External corrosion may be reduced by including sacrificial wall or by using Type V sulfate-resistant Portland cement	Reinforcements in PCCP can corrode or fail without little or no external evidence

A recent failure at the City of Austin, Texas, led engineers to reevaluate the use of not only concrete but also other piping materials. Research to assess the risks of such failures has been undertaken by various engineers and entities.

6.4.2 Asbestos-Cement (AC) Pipe

With the asbestos-related litigations of the 1980s, bans were placed on all construction materials that used asbestos. Pipe was no exception. In 1986, the U.S. EPA published a proposed regulation on the commercial uses of asbestos in which a ban on the manufacture and installation of AC pipe was proposed; this proposal was later carried out. Many asbestos-cement (AC) manufacturing plants relocated to other countries, and production in the United States came to a complete stop. In the early 1990s, the ban on the pipe was lifted, but today AC pipe is no longer manufactured in the United States. Many countries of the world, including Mexico, continue to specify and install asbestos-cement pipe for both water and sewer applications.

Worldwide usage of AC pipe increased from 200,000 mi in the 1950s to 2 million mi in 1988. In the United States, it was estimated that by 1988, over 300,000 mi of AC pipe was in service in water systems. According to the Association of Asbestos Cement Product Producers, AC pipe is still used in the Southwestern United States but it must be imported from Mexico. However, this usage is certainly an anomaly. The hazards of asbestos inhalation are increased during removal of the pipe, so utilities have chosen to keep them in the ground. These pipes are regularly tapped and maintained, so a working knowledge of the material is helpful. There are no known instances of trenchless municipal piping construction with AC pipe.

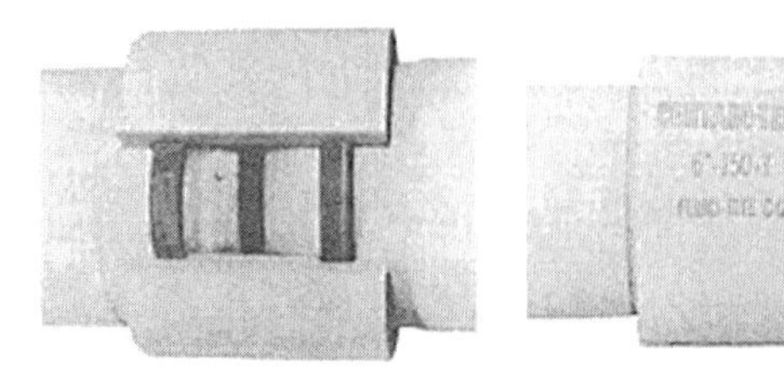

FIGURE 6.7 Joint cross section of coupling-joint in AC pipe.

Manufacture. AC water and sewer pipe is a fiber-reinforced, cementitious product, composed of an intimate mixture of Portland cement and silica. The controlled blending of these basic raw materials is built up on a rotating steel mandrel and then compacted with steel pressure rollers into a dense homogenous structure in which a strong bond is affected between the cement and the asbestos fibers. A smooth interior surface results from this process. Final curing of the product is done in an autoclave employing high pressure steam for dimensional and chemical stability.

Joint Types. AC pipes were joined together by means of a gasket-joint coupling (also referred to as a double-bell coupling joint). The ends of two pieces of pipe were slipped into the gasketed coupling, effectively creating two watertight seals. Some manufacturers permitted angular movements of up to 5° at joints. Figure 6.7 shows a typical AC pipe joint.

Advantages and Limitations. Among its advantages, its long operational life, immunity to corrosion, lightweight in smaller diameters, and watertight joints may be mentioned. Known limitations of the pipe include a low flexural resistance as a whole, easily damaged by construction equipment because of its brittle nature and a low chemical resistance. Significant research was done to study the effects of dissolved asbestos fibers in drinking water on human health. AC pipe manufacturers and the American Water Works Association (AWWA) did not recommend the use of this piping material where water was highly aggressive as corrosive water, such as acidic water with a low pH was more likely to attack piping products.

6.5 VITRIFIED CLAY PIPE (VCP)

The second group of rigid pipes among piping materials is VCP. The use of VCP in sanitary sewer systems throughout the United States is yet to be matched by any other piping material. For two centuries, VCP was the only commercially available material capable of withstanding the chemically aggressive environments of sanitary sewers. The earliest recorded use of clay pipe in the United States was in Washington, D.C., in 1815. Clay pipe is not used in pressure applications because of its inherently low tensile strength.

Only 50 years ago, the sanitary sewer engineering community followed the philosophy that leakage of wastewater through pipe joints was an acceptable methodology for effectively transporting suspended solids and reducing excessive flows within a sewer system. Clay pipes were therefore designed with a low emphasis on the effectiveness of their joints. This philosophy soon changed as engineers realized the hazards posed by wastewater leakage to soils and groundwater sources. The EPA's role in reducing I/I with the passing of several Congressional legislations such as the Water Pollution Control Act of 1972 and the Clean Water Act of 1977 were major factors in emphasis shifting to the requirement of watertight joints in sewer pipes. By the late 1960s, attempts were being made by the clay pipe industry to reduce leakage through its joints, while other materials such as thermoplastics were beginning to see a wider usage in sanitary sewers. In the arena of trenchless construction, clay pipe's ability to withstand high compressive loads and external abrasion has resulted in a significant rise in its acceptance and use in pipe jacking and microtunneling applications.

6.5.1 Manufacture

VCP is made of selected clay and shale that are aged to various degrees, and blended in specified combinations. Large crushing wheels grind the clay in a heavy perforated metal pan until the finely ground clay passes through the perforations. The ground raw materials are mixed with water in a pug mill. The mixture is then forced through a vacuum, deairing chamber until a smooth, dense mixture forms. The mixture is extruded under extremely high pressures to form the pipe. After drying, the newly formed pipe is placed in kilns and heated to temperatures of approximately 2000°F. The finished pipe then undergoes a QA/QC testing.

6.5.2 Applicable Standards

A relatively high minimum compressive strength of 7000 psi makes clay pipe a good contender for jacking and microtunneling installation. In 1994, ASTM C-1208 opened new doors for clay pipe in the trenchless construction arena (Table 6.4).

6.5.3 Joint Types

The evolution of joints in clay pipe is a shining example of the results of research, development and innovation within a piping industry. From its earliest days of having no joints to its present-day joint types, clay pipes have seen several iterations over the course of a century. To arrive at its present-day compression joints, clay pipes went from having no joints to field applied cement joints to field applied bitumastic joints to factory installed cement mortar or bitumastic joints, and finally to the compression seal joint. One manufacturer even makes a molded polyurethane joint attached to the spigot end and a PVC collar, which is shrunk and rapidly fitted and sealed to the socket-end of rigid vitrified clay pipe barrel as shown in Fig. 6.8.

TABLE 6.4 Product Standards

Pressure standards	Nonpressure standards	Available diameters
NA	ASTM C-1208: *Standard Specification for Vitrified Clay Pipe and Joints for Use in Microtunneling, Sliplining, Pipe Bursting, and Tunnels*	4–42 in

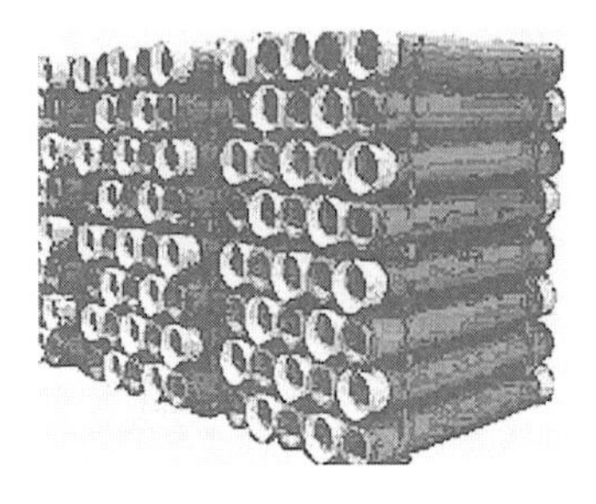

FIGURE 6.8 Innovative joints in clay pipe. (*Courtesy of Can Clay Corporation.*)

A recent study performed by the Central Contra Costa Sanitary District in California found that after 35 years of service of clay pipe in their sewer system, joint materials began to fail. The joints had fully failed at 50 years. According to the authors, this finding was contradictory to the literature values of 90 to 120 years for the life expectancy of clay pipe. Prior to 1955, clay sewer pipe used oakum and cement mortar, tar, or hot sulfur to seal joints.

The jointing system in ASTM C-1208 clay jacking pipe is a precision ground-recessed joint, ensuring dimensional accuracy and high end-bearing capacity. A stainless steel sleeve with elastomeric seals is used for jointing, as shown in Fig. 6.9 (a) and (b). The

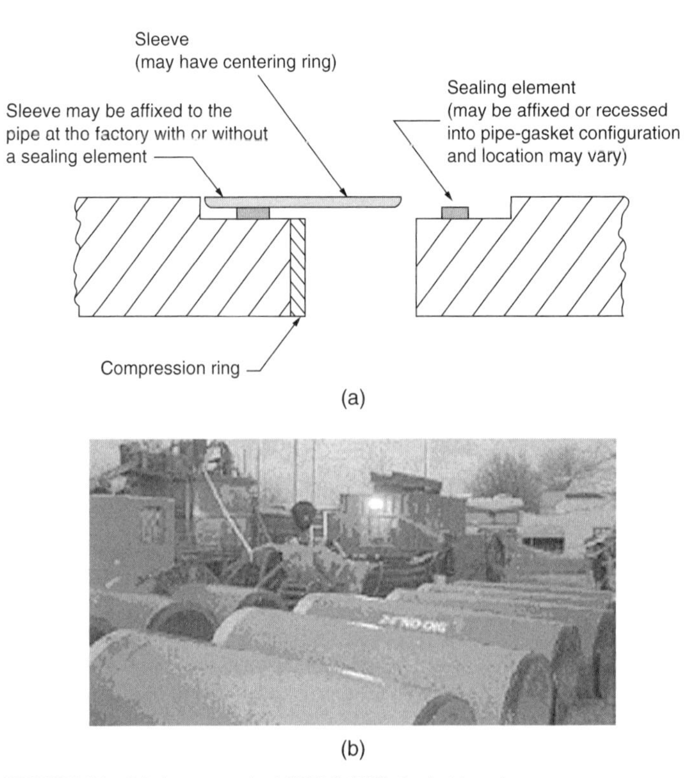

FIGURE 6.9 Jointing system in ASTM C-1208 clay jacking pipe.

TABLE 6.5 Clay Pipe's Advantages and Limitations

Advantages	Limitations
Resistant to both internal and external corrosion	Available for gravity applications only because of low inherent strength
Proven history of long life of the pipe itself	Sensitive to bedding conditions—may be subject to shear and beam failure
Improved joints have become available in recent years	Poor joints in pipe installed more than a decade ago, leading to leakage and root intrusion problems
Ability to handle high compressive forces, making it ideal for jacking installations	Short lengths, resulting in more pipe joints
Abrasion resistant	

sealing element is compressed between bearing surfaces to promote watertight integrity. The ends of the pipe are fitted with a compression ring to distribute the jacking forces of installation.

6.5.4 Advantages and Limitations

Table 6.5 summarizes some of the advantages and limitations of clay pipe. The inert nature of clay pipe was the reason behind its wide acceptance for use in sanitary sewer applications in years past. Resistance to both internal and external corrosion was its main advantage over other traditional piping materials such as concrete and cast iron. Clay pipes are also abrasion resistant making it a suitable material in sewers with high solid content. Its low sensitivity to temperature differentials prevents any significant expansion and contraction when buried in the ground. In recent years, the ability of clay pipes to withstand external abrasion and relatively high compressive forces has led to its wide usage in trenchless applications such as microtunneling and pipe jacking.

Although the inert characteristics of the material made it ideal for use in aggressive environments such as sanitary sewers, the inability of clay pipes until recently to manufacture effective joints has resulted in a drastic reduction of its use in sewer systems. Though innovation within the industry has led to better performing joints in recent years, the millions of feet of clay pipe installed in the ground decades ago are now responsible for I/I problems in sewer systems throughout the United States and Canada. Its susceptibility to shear and beam breakage owing to poor bedding conditions and ground movement has caused numerous leaks and cracks in a number of sewer systems where it was used. The short lengths in which clay pipes are manufactured (usually 8 ft or less) increases the number of joints in a sewer main line, thus raising the chances of leakage through joints. Root intrusion through clay pipe joints has led municipalities to institute annual root control measures in older systems.

6.6 PLASTIC PIPES

The introduction of plastic pipes in the late 1950s in North America was the beginning of an industry that would eventually revolutionize the field of municipal water and sewer piping. Plastics are formed by the polymerization of molecules containing hydrogen and carbon. The three main types of plastics pipes widely in use in North America include polyvinyl chloride (PVC), medium density or high density polyethylene (MDPE or HDPE), and glass-reinforced pipe (GRP, also called fiberglass pipe). PVC and PE fall into the group of *thermoplastics*, while GRP is a *thermoset* pipe. The unique quality of thermoplastics enable them to be heated, processed, formed, and reshaped many times, without any permanent changes taking place in the material's physical or mechanical properties. The properties of thermoplastics for construction material applications are better appreciated with an understanding of *viscoelastics*, discussed in the next section. Thermoset plastics are processed by a combination of chemicals and heat, and once formed, can not be reshaped. Thermoplastics such as PVC and PE used in pipe manufacture are referred to as *rigid plastics*. The term *rigid* indicates that these materials do not contain any plasticizers, which would make them more ductile, and hence unsuitable for buried municipal piping applications. Both thermoplastic and thermoset pipes are considered to be flexible conduits and are designed accordingly.

In the field of municipal trenchless construction, PE has been the dominant piping material in the past decade. By butt-fusion of successive lengths of PE pipe, a long *jointless* conduit is created, which can be installed by trenchless methods such as horizontal directional drilling (HDD). PE is also used in open-trench construction. Traditionally, PVC has been the most dominant material for open-cut installations in the North American water and sewer markets because of its bell-and-spigot gasket-joints, its light weight in smaller diameters, and ease to work with. In recent years, several manufacturers have also created proprietary PVC products with modified joints for trenchless installation. GRP pipes are used in the United States for both pressure and gravity applications, though the latter is the more prominent use.

6.6.1 Properties of Viscoelastic Construction Materials

PVC and PE are thermoplastics, and thus, viscoelastic materials. Viscoelastic materials exhibit elastic as well as viscous-like characteristics. A material that deforms under stress, but regains its original shape and size when the load is removed is classified as elastic. Viscous materials, on the other hand, after being subjected to a deforming load, do not recover their original shape and size once the load is removed. In reality, all materials deviate from the linear relationship between stress and strain (Hooke's Law) at some point in various ways.

Defining the direct relationship between stress and strain when a load is applied to a material is the most common way to evaluate the strength and stiffness of that material. Graph A in Fig. 6.10 illustrates the linear relationship between stress-strain in elastic materials. In an ideal elastic material, strain returns to zero as soon as the material is unloaded, and the linear relationship is not typically time-dependent. But it should be noted that in all materials, this behavior is valid only up to a certain stress point, called the yield point, after which the strain in the material will increase dramatically by creep, before finally failing.

In the set of curves, B, in Fig. 6.10, it can be seen that the stress-strain relationship is somewhat different for viscoelastic materials than it is for elastic materials. Clearly, we no longer see a directly linear relationship between stress-strain, and the gradients of the curves depend on the loading time. In other words, for a given stress level, the longer the loading time, the larger the strain reached. Creep is defined as continuing deformation (increasing strain) with time when the material is subjected to a constant stress. As a consequence of creep, failure of the material will occur after load is applied for a certain amount of time. So time dependency is a major factor to consider in viscoelastic material behavior. An important fact is that the time to failure is inversely proportional to the applied stress. In thermoplastic pressure pipe, it is therefore possible to find and apply a stress level that is low enough to ensure that the theoretical time to failure will surpass the design life of the pipeline.

In thermoplastic pipe applications, creep is prevented because the deflection of the pipe is kept constant, as is the case in buried PVC gravity (or pressure) pipe. Consequently, it

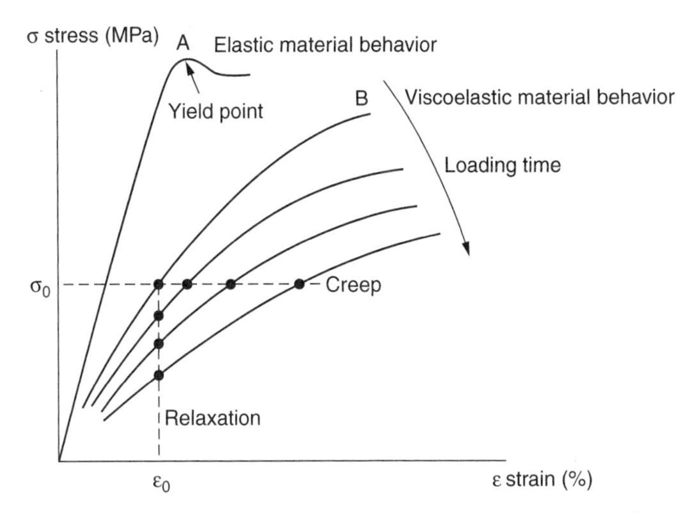

FIGURE 6.10 Stress-strain relationship in elastic and viscoelastic materials.[1]

can be seen from Fig. 6.10 that the initial stress decreases with time, and is referred to as the relaxation property of thermoplastic piping materials. These basic properties of viscoelastic materials such as PVC and PE enable engineers to design pipelines that ensure both structural integrity and the long-term design life of their municipal piping systems.

6.6.2 Polyvinyl Chloride (PVC) Pipe

PVC was discovered almost accidentally in the 19th century when German scientists, observing a newly created organic chemical gas, vinyl chloride (C_2H_3Cl), discovered that when it is exposed to sunlight, a chemical reaction took place, resulting in the creation of an off-white accumulation of solid material. Since then, scientists had observed the first polymerization and creation of a new plastic material, PVC. In 1839, a technical paper was published detailing the observations of the process. In 1912, several decades after its accidental discovery, Fritz Klatte, another German, laid the groundwork for the technical production of PVC. The oldest known PVC pipe was manufactured and installed in the 1930s in World War II in Germany and continues to be in service today. The technology was brought to the United States following World War II and by the mid-1950s ASTM groups were organized for plastic pipe standardization. In 1955, the total U.S. shipments of thermoplastic pipe (PVC, PE, and others) were estimated at about 40 million pounds. By 2000, the annual use of PVC alone for pipe production had reached 5 billion pounds.

Manufacture. A vinyl chloride molecule comprises carbon, hydrogen, and chlorine, configured as shown in Fig. 6.11 (a). PVC is obtained by polymerization of single units of the vinyl chloride molecule, which join to create long chains, and ultimately form PVC resin, Fig. 6.11 (b).

[1] This diagram was modified from Figure 3.2.23 on Page 61 in Janson, L.E., *Plastic Pipes for Water Supply and Sewage Disposal*, Borealis, Stockholm, Sweden (1999).

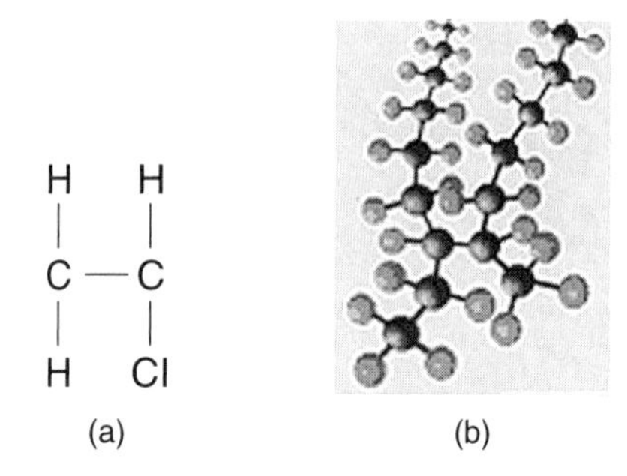

FIGURE 6.11 Vinyl chloride molecule and PVC chain.

PVC pipe is manufactured by first blending PVC resin with stabilizers, pigments, lubricants, processing aids, and functional additives: and heating this mixture to a temperature in the 400°F range. This causes the components to properly fuse and convert into a malleable state. In this molten form, the material is mechanically extruded into the pipe. Following the completion of the extrusion process, the pipe is allowed to cool, after which QA/QC testing is performed, before the final delivery to the end user.

Types of PVC Pipe. There are three distinct types of PVC pipes manufactured in the world, each differentiated by either the way in which it is manufactured (which dictates the directional orientation of the molecules), or by the content of modifiers in its chemical formulation (which affect the ability of the pipe to withstand large impacts by absorption and dissipation of the energy). The term *PVC* is a generic designation, which includes PVC-U (unplasticized PVC), PVCO (molecularly oriented PVC), and PVC-M (modified PVC).

1. *PVC-U.* Unplasticized PVC is the most widely used piping material in water and sewer systems in North America. The molecular structure of PVC pipe is a random arrangement of long chain molecules, where molecular entanglement is prevalent throughout the length of the pipe. In general, the PVC molecules do not exhibit any definite directional orientation, and therefore, a generally uniform strength prevails in both the radial (circumferential) and longitudinal directions. Testing has shown that the modulus of elasticity in 15-year-old PVC is only slightly higher in the longitudinal direction than in the radial directions. *For simplification, the term PVC as used in this chapter will denote PVC-U, the conventional type of PVC most used in pipe manufacture.* PVC pipes are manufactured for both pressure and gravity applications. All PVC pressure pipes (ASTM & AWWA standards) must meet the cell classification 12454, indicating a tensile strength of 7000 psi, and a modulus of elasticity of 400,000 psi. PVC pressure pipes have a hydrostatic design basis (HDB) of 4000, per ASTM D-2837, to which a factor of safety of either 2.0 or 2.5 is applied (depending on the standard). This effectively reduces the long-term design stress to either 2000 psi or 1600 psi. PVC gravity pipes are typically manufactured to cell class 12454 or 12364, where the latter has a tensile strength of 6000 psi and a modulus of elasticity of 440,000 psi.
2. *PVC-O.* Molecularly oriented PVC is made in the United States by the expansion of the conventional PVC pipe; during the expansion process, the molecules become oriented in a generally radial or circumferential direction. This molecular reorientation increases the strength of the pipe in the hoop direction. Also, the resulting HDB is increased from 4000 psi to 7100 psi. Consequently, this stronger material can have a thinner wall than a conventional PVC pipe of the same pressure capacity. The manufacture process of

PVC-O in the United States is called an offline process, whereas a second method, the online process, is more widely used in Europe. In the United States, PVC-O is considered a proprietary product because only one company makes it. PVC-O pipe is not used in gravity applications.

3. *PVC-M.* Modified PVC is produced by incorporation of additives or *impact modifiers* to enhance the toughness of the material. Resistance to fracture by absorption and dissipation of energy is an evidence of the toughness of the pipe material. PVC-M is made and used mainly in Europe and Australia, whereas only one manufacturer in the United States produces this type of pipe for nonburied applications.

Solid Wall and Profile Wall PVC Pipes. PVC pipes are available for pressure applications only as solid wall pipes. For gravity applications, both solid wall and profile wall pipes are manufactured. Solid wall pipe, as the name suggests, are made of a continuous wall of PVC of uniform thickness, as shown in Fig. 6.12(a). Profile wall pipe, on the other hand, is braced spirally or circumferentially with structural shapes, but provides a smooth-wall interior, as shown in Fig. 6.12(b). Profile wall pipes economize on the amount of material needed for fabrication; by altering the shape of the wall, the same stiffness as solid-wall pipe is achieved, using less material. Profile wall pipes generally fall into three categories—open profile (OP), closed profile (CP), and dual-wall corrugated profile (DWCP). OP pipes have their rib-enforcements exposed on the outside of the pipe. CP pipes make use of a closed profile that provides a continuous outer wall where the wall sections are hollow and are often described as an I-beam or honeycomb. DWCP pipes have a smooth-wall waterway, braced circumferentially with an external corrugated wall.

From a design perspective, both solid wall and profile wall PVC gravity pipes are limited to a vertical deflection of 7.5 percent, per ASTM and Uni-Bell recommendations. For pressure pipe, AWWA recommends a deflection limit of 5 percent.

Generally, profile wall pipes are used for open-cut installations as well as trenchless renewal processes such as sliplining. Pipes used for sliplining have modified joints that facilitate their installation by segmental sliplining.

PVC Applicable Standards and Products for Trenchless Construction. Although there are several widely used bell-and-spigot gasket-joint PVC pressure and gravity piping standards in North America written by organizations such as AWWA, ASTM, and Canadian Standard Association (CSA), products discussed in this chapter are specifically for trenchless applications. They are proprietary in nature in that they have uniquely designed jointing systems that enable the pipe to be pulled or pushed for various trenchless construction methods. Table 6.6 outlines details on each product along with pictures of joints. Also,

(a)

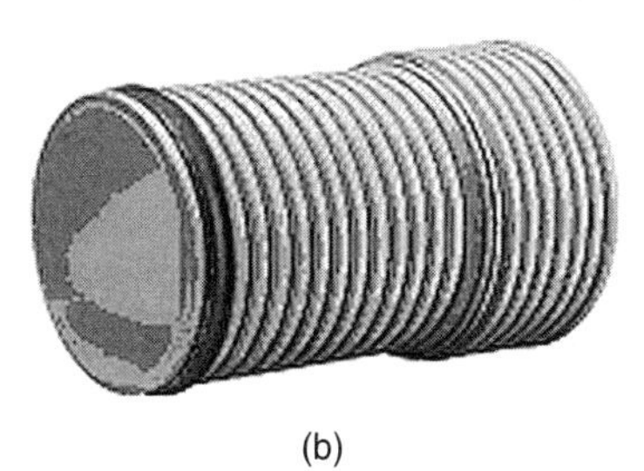

(b)

FIGURE 6.12 Solid wall and profile wall PVC pipe.

TABLE 6.6 Trenchless PVC Products and Standards

Pipe type	Standard compliance	Joint	Diameter
Fusible C-900™, Fusible C-905™, Fusible PVC™	AWWA C900/C905	 	4 in through 48 in
TerraBrute™	AWWA C900	 	4 in through 12 in, capability to go up to 48 in
CertaLok™	AWWA C900/C905	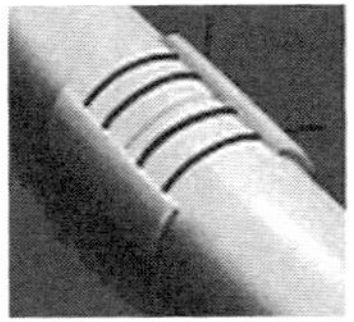 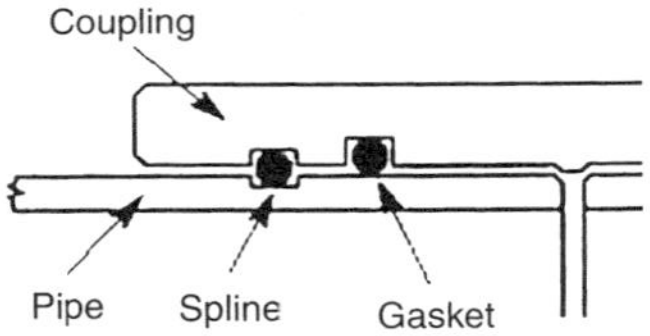	4 in through 16 in

there are several types of PVC piping products geared toward use in *trenchless renewal* market. Typically, these are profile wall pipes with proprietary joints. Information on these products is not included in this chapter but more details may be obtained from manufacturers or Uni-Bell PVC Pipe Association.

Fusible C-900™, Fusible C-905™, and Fusible PVC™. Until recently, HDPE was the only available thermoplastic pipe option that used butt-fused joints in the United States; this was the case for well over 30 years. In the late 2003, the water or wastewater industry saw the introduction of Fusible PVC™/C-900™/C-905™. This product combines a proprietary formulation with a fusion procedure that allows lengths of PVC pipe to be joined together in a continuous string for installation through a variety of methods, including HDD, sliplining, pipe bursting, as well as open-cut applications. Although Fusible PVC™ is primarily for nondrinking water applications like recycled water, force mains, gravity drains, and sewer applications, Fusible C-900™/C-905™ are specifically for use in the potable water systems. The pipes are manufactured to requirements of AWWA C900 and C905, and are NSF certified.

Fusible PVC pipe manufacture is possible because of two significant developments. The first development is a formulation that meets all guidelines for components as detailed in the Plastic Pipe Institute's (PPI) Technical Report No. 2, entitled *PPI PVC Range Composition of Qualified Ingredients*. This allows PVC pipe to be extruded in standard diameters and wall thicknesses. The pipe also has all of the basic characteristics of conventional PVC pipe for pressure capability, tensile strength, external load capability, and so on. The second development was the fusion procedure. The UGSI Fusible™ line uses standard butt fusion equipment for the joining of the pipe. A set of temperatures, pressures, and duration of fusion steps allow for the formulated pipe to fuse and create joints that are as strong as the original pipe material. The fusion procedure takes about the same length of time as other thermoplastics fusion joints.

With the advent of Fusible PVC, a new set of properties for PVC pipe needed to be defined. Most pipe properties developed for PVC have been for post installation conditions. This is because, until recently PVC was installed as bell and spigot open-cut construction. In open-cut application, there is no need for properties such as minimum *safe pulling force*, which are necessary for trenchless applications similar to horizontal directional drilling. These properties have now been tested and defined to deal with the installation parameters of a continuous string of PVC pipe in trenchless applications. Currently, additional testing is being done to define the realistic limits of Fusible PVC™.

Surface scratches of up to 10 percent of the pipes' wall thickness are accepted, per AWWA, during pull-in of the PVC pipe. A major advantage of the fusible pipe is that those municipalities already using conventional PVC can now use the same material in trenchless construction, enabling the design of a *complete PVC system*. It also allows municipal entities to have single PVC pipe and fittings for all applications. Tapping is also performed in the same manner as in gasket-joint PVC pipes. Furthermore, the internal flow area in Fusible PVC pipe is greater than that in an alternative thermoplastic pipe of an equivalent pressure capacity.

An available renewal system called Duraliner™ makes use of Fusible PVC to create a stand-alone structural expand-and-form system for renewing fully deteriorated and corroding metallic pipeline systems. See chapters on trenchless renewal methods for more information. Table 6.6 presents a summary of PVC pipe standards and jointing system.

Advantages and Limitations. Table 6.7 highlights the advantages and limitations of PVC pressure and sewer pipe. The greatest contributor to the rapid adoption of PVC pipe in water and sewer systems is its inherent ability to withstand both internal and external corrosion. Vast amounts of corrodible piping materials are being replaced each year by PVC. In sewer systems, the completely leak-free joints are a definite advantage over traditional

TABLE 6.7 Advantages and Limitations of PVC Pipe

Advantages	Limitations
Resistant to both internal and external corrosion	Sensitive to operating temperature, must be derated in case of long-term exposure to temperatures above room temperature
Gasket-joints have a better record of being leak-free than joints of substitute products	Sensitive to ultraviolet light if exposure is greater than 2 years (unless pipe is formulated with higher UV-inhibitor level)
Zero joint-leakage in fusible type PVC pipe	Less longitudinal flexibility than alternative thermoplastic piping material
All three restrained-joint PVC products have high tensile strengths for HDD and other trenchless processes	Gasket-joints can fail in the long term if spigot is over-inserted into the bell
Highly abrasion resistant for sewer applications	Thinner-wall sewer pipe is sensitive to bedding conditions.
Low internal frictional resistance for both pressure and nonpressure applications	Susceptible to chemical permeation in cases of gross contamination
At least 2.5 times stronger than other thermoplastic pipe (higher stiffness, higher HDB)	Susceptible to impact damage in cold temperatures
Expansion is significantly lower than in alternative thermoplastic piping material	

materials. The fusible-joint type PVC pipe is completely *jointless* and provides a definite advantage over jointed pipe. Pipes with proprietary joints for trenchless installation have high tensile strengths, allowing for long lengths to be pulled in at a time. Abrasion resistance has been an advantage over alternative cementitious piping materials. Low internal friction enables the use of smaller diameter pipe in both pressure and gravity applications. An HDB that is 2.5 times higher than alternative thermoplastic piping material allows for thinner walls and therefore higher flows. A much smaller expansion coefficient than alternative thermoplastics significantly reduces expansion and contraction concerns both open-trench and trenchless applications.

The sensitivity to temperatures on a long-term basis requires thermoplastics such as PVC to be derated in pressure applications. Pipe that is not formulated with a higher amount of ultraviolet inhibitor results in lowered impact strength after 2 years of continuous exposure to sunlight. PVC cannot be deflected longitudinally as much as alternative thermoplastics. Poor bedding can cause excessive deflection and failure in thinner wall pipe. Table 6.7 presents a summary of the advantages and limitations of PVC pipe.

6.6.3 Polyethylene (PE) Pipe

PE belongs to a group of thermoplastics known as polyolefins, materials made by polymerization of *olefin gases* including ethylene, propylene, and butylene. In Europe, polypropylene (PP) and polybutylene (PB) pipes are widely manufactured and used in gas and municipal applications, though this is not the case in North America. In the past decades, PP was used for service connections in water distribution systems in the United States, but various types of problems with the material such as cracking and chemical permeation has brought its use to an end.

PE was first invented in the United Kingdom in 1933 by the Imperial Chemical Company and later commercialized in 1939 to manufacture insulation for telephone and coaxial cables.[2] The use of PE pipe began in the United States in the 1950s, mainly in the gas industry. Today, more than 90 percent of new and replacement distribution pipes in the United States gas industry is PE pipe.[3] Other applications of PE pipe gradually spread from industrial to, in recent years, municipal applications. A steady rise in the use of PE in municipalities has resulted for two primary reasons: the acceptance of trenchless installation methods such as horizontal directional drilling, and the lack of alternative piping materials that can be fused together. However, development of restrained-joint gasketed PVC and ductile iron pipes in recent years has provided competition to PE in the trenchless construction arena. The most recent introduction of fusible PVC offers users of trenchless technology an alternative thermoplastic piping material to PE.

PE pipes in North America are classified into three groups, based on density and crystallinity, which is an indicator of the tensile strength. The higher the crystallinity, the greater the hardness, stiffness, tensile strength, and density. ASTM classifies Type I as a low-density PE (LDPE), Type II as a medium density PE (MDPE), and Type III as a high-density PE (HDPE). HDPE displays the highest stiffness whereas LDPE is the most flexible. With continued development of material manufacturing processes, the differences between HDPE and MDPE have become small. Most PE pipes for pressure applications in the United States are made of materials between the high end of medium density and the low end of high density, to achieve the best combination of required strength, flexibility, and toughness. It also allows pipe manufacturers to be competitive in the market place. Generally classified as HDPE, the material's ability to significantly reduce slow crack growth (environmental stress cracking) and crack propagation has been instrumental in its wide use for pressure and gravity applications.

Manufacture. The manufacture process of HDPE is similar to that of PVC; molten PE resin is mixed with additives such as heat stabilizers, antioxidants, pigments, carbon black and so on, and extruded through dies under pressure in specially designed extrusion machines. A second method for producing large-diameter solid wall, open profile, and closed profile PE pipes makes use of a spiral winding technique.

Resin is manufactured from the base ethylene molecule, Fig. 6.13 (a), a colorless gas composed of two double-bonded carbon atoms. Polymerization of the ethylene is performed with various catalysts, under heat and pressure, during which the double bond

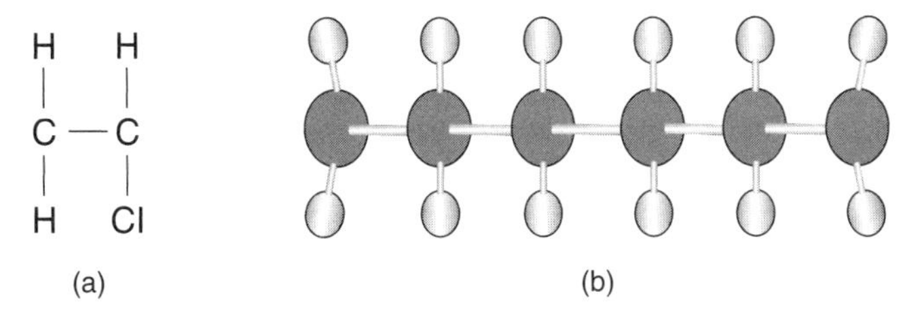

FIGURE 6.13 Ethylene molecule and polyethylene chain.[4]

[2] Hoechst Plastics – Pipes. Brochure No. HKR 111e-8122, Hoechst AG, Frankfort, Germany (1982).

[3] Water Environment Research Foundation, *New Pipes for Old: A Study of Recent Advances in Sewer Pipe Materials and Technology*, Project 97-CTS-3 Final Report, Alexandria, VA (2000).

[4] Figure 14b from Plastic Pipe Institute (PPI) *Handbook of Polyethylene Piping*, Chapter 3: Engineering Properties of Polyethylene, Washington, D.C. (2000).

between the carbons is broken, resulting in the formation of a bond with another carbon atom, Fig. 6.13 (b).

PE resins may be formed by copolymerization with other monomers such as butene, propylene, hexene, and octene; this results in small changes in chemical structure and properties such as density, ductility, hardness, and so on. Homoploymers are those resins manufactured without these comonomers.

Engineering Properties. Like PVC pressure pipe, HDPE also has a hydrostatic design basis (HDB) developed per ASTM D-2837. The HDB of HDPE is 1600 psi. In AWWA standards, a factor of safety of 2.0 is applied, reducing the long-term design stress to 800 psi. Based on the HDB, the strength of HDPE is 2.5 times less than that of PVC. Ordinarily, loads of small amplitude and short durations will cause the pipe to behave elastically, where initial deformations will be overcome as the pipe returns to its original shape. For installations where large tensile loads are applied, as is the case in HDD, it is important to recognize viscous deformations that will prevent the material from reverting to its original shape, particularly if the stresses are of large magnitudes and long durations. Therefore, safe design pull stresses for pull-in-place installations must be calculated accordingly. The higher stiffness of butt-fused PVC makes it significantly less sensitive to these considerations and must be evaluated against limitations of PVC and the benefits of PE pipe.

Types of HDPE Pipe. HDPE pipes are manufactured in both solid walls and profile walls, similar to PVC. Solid wall pipes are used for both pressure and gravity piping systems. Profile wall pipes on the other hand are used only for gravity systems. The low compressive strengths of profile wall HDPE pipe prevent their use in trenchless methods such as pipe jacking and microtunneling. Table 6.8 presents HDPE standards for trenchless construction methods.

Applicable Standards. Pipe manufactured to AWWA C906 must use grade 4 resin, which offers the highest resistance to environmental stress cracking. HDPE pipe manufactured to this AWWA standard is used for both pressure and gravity applications. Butt-fusion produces a continuous conduit ideal for pull-in installations such as HDD. Renewal methods such as pipe bursting also provide a good match for use of HDPE pipe. Testing has shown that surface scratches sustained during installation of up to 10 percent of the pipe's outside diameter does not cause any noticeable drop in the pipe's burst strength.

Joints. Butt-fusion of PE pipe is performed using available fusion machines made by several manufacturers (a McElroy machine used to fuse PVC pipe is shown in Table 6.6). The steps involve preparing ends of two pieces of pipe by planning the ends using special radial blades, heating the ends with a heat plate (after which the heat plate is removed), and finally

TABLE 6.8 Standards for Trenchless HDPE Products

Pressure standards	Nonpressure standards	Available diameters
AWWA C906: Standard for polyethylene (PE) pressure pipe and fittings, 4 in (100 mm) through 63 in (1600 mm), for water distribution	AWWA C906	4–63 in

fusing the ends together by application of pressure. Fusion parameters, that is, temperatures, pressure, and time interval, have traditionally differed slightly from pipe manufacturer to manufacturer. Similarly, PVC pipe also has its own set of recommended fusion parameters. A result of butt-fusion is the formation of a radial bead both internally and externally. See Table 6.6 for bead formed in butt-fused PVC pipe. Both internal and external beads can be removed prior to installation. For pressure applications, the internal bead has a negligible effect on flow. In gravity design, some engineers believe that the turbulent flow created by the bead is helpful in keeping suspended solids flowing; others feel it is detrimental to the flow.

Other types of joints in HDPE pipe include electrofusion welding and bell-and-spigot gasket joints. The latter is primarily used in HDPE gravity piping, installed via open-trenching. The low compressive strength of gasket-joint HDPE pipe makes it unsuitable for pipe jacking.

Advantages and Limitations. The continuous *jointless* conduit that results from the butt-fusion of HDPE pipe makes it an ideal piping material for pull-in installations such as HDPE and pipe bursting. HDPE pipe's ability to withstand both the internal and external corrosion is advantageous for both water and sewer systems. Its high flexibility is a favorable characteristic during trenchless installations; there is no need for very long entry pits. A very low internal resistance to flow makes it a good material for both pressure and gravity systems. In water systems, the expansive forces of freezing water do not cause the pipe to crack easily. In very cold temperatures, HDPE has a high resistance to failure by impact. The pipe also resists shatter-type or rapid crack-propagation failure.

Environmental stress cracking has traditionally been a concern for some thermoplastic pipes. Higher quality resins have successfully overcome this problem in HDPE pressure pipes. The butt-fusion of HDPE necessitates the use of skilled labor force. There is usually a 24-hour waiting time to allow the pipeline to stabilize after fusion of joints has been performed and before the pipe is buried. The high sensitivity of HDPE to temperature differentials may cause problems with appurtenances even after the pipe is buried. The lower hydrostatic design basis (HDB) of HDPE pipe results in the need for thicker walled pipe when compared to another thermoplastic pipe of the same pressure rating. The thicker wall may reduce the flow area within the pipe. Table 6.9 presents advantages and limitations of HDPE pipe.

6.6.4 Glass Reinforced Pipe (GRP or Fiberglass Pipe)

The third type of plastic pipe is GRP, also commonly referred to as fiberglass pipe. Unlike PVC and HDPE, GRP is made of a thermoset material. GRP was first manufactured in the United States in the 1950s, as an alternative to corrosion-prone concrete materials. Using a patented centrifugally cast manufacturing process; Perrault Fibercast Corporation of Oklahoma manufactured the first fiberglass reinforced polyester resin pipe. Traditionally used in industrial applications throughout the world, GRP is rapidly gaining market shares in North America for large-diameter municipal water and sewer applications. As a plastic material, GRP is designed as a flexible conduit, and is installed accordingly.

Manufacture. Fiberglass composites are made of glass fiber reinforcements, thermosetting resins, and other additives such as fillers, catalysts, hardeners, accelerators, and so on. Types of resin used include epoxy, polyester, and vinyl ester. The amount and orientation of the glass fibers provide the mechanical strength of the pipe. GRP is manufactured using one of the two methods:

TABLE 6.9 Advantages and Limitations of HDPE

Advantages	Limitations
Resistance to both internal and external corrosion	May be subject to environmental stress cracking
Butt-fused joints effectively create a continuous *jointless* conduit	Lower HDB than other thermoplastic material, requiring thicker walls, which results in smaller flow area
Abrasion resistant if used in sewer applications	Skilled labor and special equipment required for butt-fusion
High ductility and flexibility	More permeable to certain chemical contaminants than other thermoplastic pipe
Lightweight in smaller diameters	Not price-competitive with other thermoplastic pipe of same pressure capacity
Low internal friction	Cannot be located unless buried with metallic wire or tape
High resistance to failure by impact, even at very low temperatures	Very sensitive to temperature differentials, resulting in significant expansion and contraction even after burial
Resists shatter-type or rapid crack-propagation failure	High flexibility causes problems in retaining joint restraints, unless stiffener is inserted into pipe prior to attachment of restraint
Does not easily crack under expansive forces of freezing water	Degradation owing to ultraviolet light exposure has been seen in some HDPE pipes of low carbon black content

1. *Centrifugal casting.* This is the most widely used method of manufacture of GRP for municipal applications in North America. In this process, glass fiber reinforcements are placed in a steel mold, rotating slowly. As the rotation speed increases, catalyzed is introduced into the steel mold. The centrifugal action of the mold removes air from the resin and glass, resulting in a dense laminate, free of voids. Material properties can be altered by varying filler content, resin type, and cure and reinforcement type. The pipe is removed from the mold after heat curing.
2. *Filament winding.* There are two distinct processes for filament winding—continuous and discontinuous. Filament winding involves impregnating several glass reinforcing strands with a matrix resin and then the application of the wetted fibers to a mandrel under controlled tension in a predetermined pattern. Repeating this process results in the desired wall thickness. In the continuous process, an advancing mandrel causes the pipe to form. Fiber rovings, resin, and filler are added to make the pipe. The pipe is then cured and cut into desired lengths. In the discontinuous process, a standard length of mandrel is rotated, rovings, glass, and filler are added to produce pipe that is helically reinforced. The process allows for the formation of a continuous bell-end, monolithic with the pipe wall.

Applicable Standards. There are currently three widely used pressure and nonpressure piping standards, used in trenchless installations. Available pressure classes for AWWA C950 include 50, 100, 150, 200, 250, and over 250 psi. From the available pipe diameters, it can clearly be seen that GRP offers an alternative to other large-diameter traditional piping materials (Table 6.10).

TABLE 6.10 GRP Standards for Trenchless Construction

Pressure standards	Nonpressure standards	Available diameters
–	ASTM D-3262	8–144 in
ASTM D-3517	–	8–144 in
AWWA C-950	–	1–144 in

Joints. There are a variety of joints that have been developed over the years for GRP. The *Fiberglass Pipe Handbook* (1989) lists the following types: Coupling or bell-and-spigot joints, mechanically coupled joints, restrained coupling or bell-and-spigot joints, butt and wrap joints, bell-and-spigot with laminated overlay, bell-and-spigot adhesive joint, flanged joints, and mechanical joining systems.

Hobas, the leading manufacturer of municipal GRP, has variations of the above-described joints. For trenchless construction methods in general and pipe jacking in particular, they offer two separate types of joints: flush bell-spigot joint for gravity flow pipe jacking, and the flush fiber-wound collar (FWC) coupling for pressure pipe jacking. A third type of joint exists for use in the renewal of gravity pipe by sliplining. Refer to Chap. 14 for more information on sliplining.

In the flush bell-and-spigot joint type, as shown in Fig. 6.14, the sleeve is fitted to the pipe end, which has been machined down so that the joint outside the diameter is same as the pipe itself. An elastomeric gasket, contained in a groove on the spigot end of the pipe, effectively seals the joint. The flush FWC coupling is a modified version of the FWC coupling (a filament-wound sleeve with an EPDM elastomeric membrane, the coupling is bonded to one end of the pipe during manufacture).

Advantages and Limitations. Table 6.11 outlines some advantages and limitations of GRP. Excellent internal and external corrosion resistance in natural soils and corrosive wastewater and industrial applications has given GRP an advantage over traditional piping materials in large diameters. It also displays better abrasion resistance than cement-based pipes. It is significantly lighter than a concrete pipe of the same diameter. For pressure applications, a variety of available pressure classes makes it widely desirable to engineers and contractors.

Although GRP is noncorrodible, it is susceptible to strain corrosion in the presence of certain chemicals, such as those found in the sanitary sewers, where pH is less than 4. This can be overcome during design, by ensuring that the stresses are kept within a certain limit. Certain chemical contaminants can permeate the pipe. During installation, pipe may be damaged by an impact force.

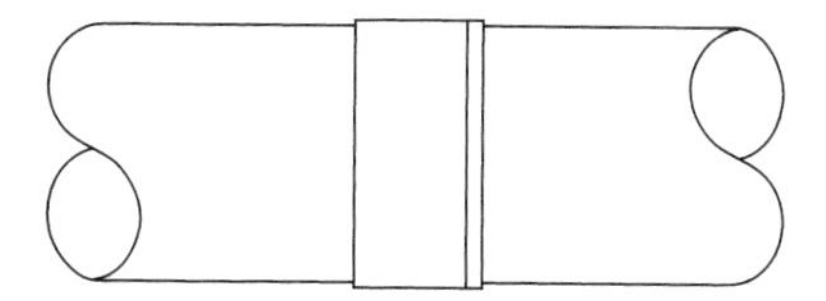

FIGURE 6.14 GRP flush bell-and-spigot joint for trenchless installation. (*Courtesy of Hobas Pipe USA.*)

TABLE 6.11 GRP Advantages and Limitations

Advantages	Limitations
Good internal and external corrosion resistance in ordinary soils	Strain corrosion can occur in acidic soils with pH less than 4
Better abrasion resistance than cement-based pipes	Permeable to certain contaminants
Lightweight compared to alternative materials	Susceptible to impact damage
Various pressure classes available for pressure pipe	

6.7 METALLIC PIPES

Metallic pipes in present-day use include ductile iron pipe and steel pipe. The precursor to ductile iron was gray cast iron pipe. Metallic pipes have traditionally been used in pressure applications, but some areas of North America use them for gravity sewer applications also. Most metallic pipes used in the past were installed by open-cut methods. Increasing acceptance of trenchless technologies has led to the utilization of metallic pipes with joints suited for pull-in or direct jacking installations. Steel pipe with welded joints is the product of choice for large-diameter HDD applications (usually more than 22-in diameters), horizontal auger boring (HAB), and pipe ramming (PR) projects.

6.7.1 Ductile Iron Pipe

Ductile iron pipe was developed in the 1940s from grey cast iron by distributing the graphite into a spherical form instead of a flake form. This was achieved by the addition of inoculants such as magnesium to molten iron. It resulted in the ductile nature of the new pipe, in addition to higher strength, impact resistance, and other improved properties. The commercial production of ductile iron pipe begun in 1955, and by the 1970s, it had almost completely replaced cast iron pipe in municipal applications. Since 1980, gray cast iron pipe for pressure pipe applications (not soil pipe) has not been produced in the United States.

The main use of ductile iron pipe is in water systems. However, the availability of thermoplastic pipes and newer substitute materials represents some significant competition to ductile iron since the late 1960s. The newer materials are lighter, in some cases less expensive, and perform many of the required functions within a distribution system. Above all, plastic pipes are inherently suited to withstand most forms of both internal and external corrosion that can be a concern to unprotected ductile iron pipe in some service environments.

However, corrosion problems for ductile iron pipe can overcome by the use of appropriate corrosion protection. The protection methods include cement mortar or polymer linings for internal corrosion protection, maintenance of flow properties, and polyethylene encasement per ANSI/AWWA C105/A21.5, or in uniquely aggressive environments cathodic protection, for external corrosion. A newer approach to iron pipe corrosion control, called the "Design Decision Model" (or "DDM"), has been developed by the Ductile Iron Pipe Research Association (DIPRA). This method tailors the chosen corrosion protection method of the pipeline based on several factors including substantial history, the corrosion potential of the specific environment involved, and the critical nature or priority of the specific pipeline involved.

In recent years, various manufacturers have also developed restrained joints for ductile iron pipe; therefore, the industry is able to offer its products to the trenchless construction industry. Today ductile iron pipe can be used for HDD, pipe bursting, casing and carrier pipe installation, and sliplining. The use of ductile iron pipe in sewer systems is less in comparison to its use in potable water distribution and transmission. Nevertheless, some municipal agencies use ductile iron pipe for their gravity flow applications. Flexible, restrained joint ductile iron pipes are particularly attractive for some horizontal directional drilling (HDD) installations. In HDD applications, joints can rapidly be assembled in field conditions with minimum equipment or highly skilled technicians. The pipe also can be installed with limited access, and need not be bent (which may result in high bending stress or strain) when installed in a normal curved drillpath.

Ductile iron pipes are flexible conduits and are designed and installed accordingly. However, when internal cement-mortar lining is specified, the vertical deflection of the pipe is limited to 3 percent instead of 5 percent. Ductile iron pipes with flexible linings, such as proprietary ceramic epoxy and fusion bonded epoxy primer or heat-fused polyethylene topcoat linings are limited by ASTM A-716 and A-746 standards to 5 percent deflection.

Cast Iron Pipe. It is important to have an understanding of the predecessor of ductile iron pipe, the cast iron pipe. Cast iron pipe has been used throughout the world for many years in portable water systems. In the United States, cast iron pipe was introduced in the early 1800s. Also referred to as gray cast iron, it accounts for a very large portion of buried water piping material throughout North America even today. Cast iron is a very strong, but brittle material. Some early unlined installations of cast iron pipe have lasted for more than a century. The thick walls played a sacrificial role as they slowly corroded over the years. Internal tuberculation in unlined cast iron pipe has caused severe hindrance to flow in some cases, where aggressive and tuberculation waters are encountered (Fig. 6.15).

Manufacture. The principal raw material used in producing ductile iron pipe is recycled ferrous material, including scrap steel, scrap iron, and other ferrous materials obtained from shredded automobiles, appliances, and so on. Although ductile iron is very similar in basic chemical make-up to gray cast iron, ductile iron is instead produced by treating molten low-sulfur base iron with magnesium under closely controlled conditions. The startling change in the metal is characterized by the free graphite in ductile iron being deposited in nodular form, instead of flake form as in gray iron. With the free graphite in nodular form, the continuity of the metal matrix is at a maximum, accounting for the formation of a far stronger, tougher

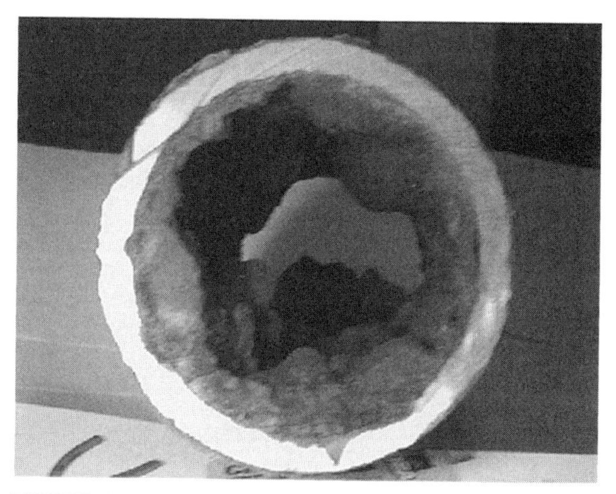

FIGURE 6.15 Internal tuberculation in unlined cast iron potable water pipe.

TABLE 6.12 Applicable Ductile Iron Standards for Trenchless Technology

Pressure standards	Nonpressure standards	Available diameters
AWWA C150/C151	ASTM A-716 ASTM A-746 AWWA C-150/C-151	4–64 in

ductile material. Ductile iron is roughly twice the strength of gray iron (ductile iron has greater yield strength than ASTM A-36 carbon steel) and further surpassing gray iron in ductility and impact characteristics. The pipe is cast using a centrifugal method, after which it is annealed in furnaces. An asphaltic coating is applied to the outside of the pipe, while the interior is coated with a cement-mortar lining. Unlike gray iron pipe, ductile iron pipe will bend significantly without breaking when subjected to even quite great loads, impacts, or deflections.

Applicable Standards. Table 6.12 lists a number of standards available for trenchless applications. It should be noted that these pipes can also be installed via open-cut methods. For trenchless pull-in installations, the joints must be restrained, as discussed in the following section. In recent years, ductile iron pipes with restrained joints have been used in trenchless construction methods (TCM) as well as for TRM.

Joints. Restrained joints in ductile iron pipe are available primarily to accommodate the thrust forces acting on a pipeline. However, pipes with these restrained joints have been used in recent years for various types of trenchless projects, both new constructions as well as renewals. HDD and pipe bursting have been the most common applications of the restrained joint ductile iron pipe. The joints are capable of withstanding tensile forces encountered during pull-in process.

Proprietary restrained joints have been designed by various manufacturers which incorporate a push-on gasket and special bell design in conjunction with their restraint mechanisms. Because of their proprietary nature, the push-on gaskets used in these joints may not be compatible with standard push-on gaskets. In the 350 psi allowable working pressure range, the joints are suitable for pipe diameters of 4 in through 24 in. In the 250 psi range, the joints are available for diameters of 30 in through 64 in. Figure 6.16 shows five such proprietary joints.

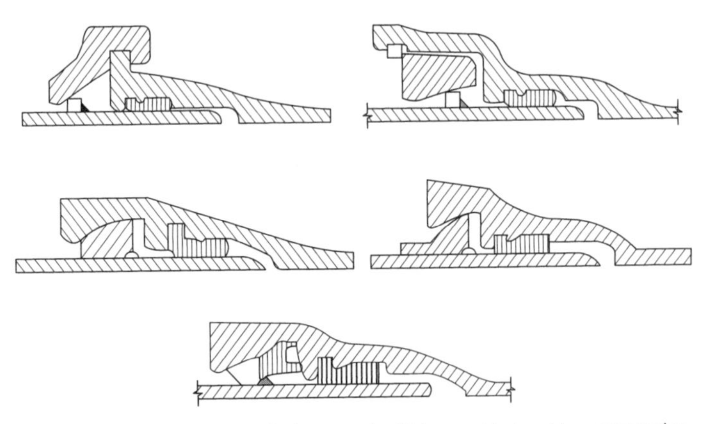

FIGURE 6.16 Proprietary ductile iron restrained joints used in trenchless construction.

Advantages and Limitations. Table 6.13 summarizes some of the advantages and limitations of ductile iron pipe. The high load bearing capacity, impact strength, and beam strength makes it most sturdy piping material for pressure applications. There are many different types of joints, including restrained joints, which are used in trenchless applications. The long lengths of ductile iron pipe (20 ft) minimize the number of joints within a water or sewer system. The wide availability of joining structures as well as various pressure classes and diameters along with its rugged features makes it easily desirable to engineers and specifiers. The pipe itself is impermeable. Also, there are various internal and external corrosion protection systems available, as well as various chemical, heat, and/or permeation resistance gasket materials. A substantial series of ANSI/AWWA and ASTM standards and manuals, including the comprehensive AWWA Manual M41 are good sources of information concerning ductile iron pipe.

Very large quantities of ductile iron pipe with standard internal cement lining and standard exterior asphaltic shop coating have been installed over the years with no other corrosion protection in most common soil environments. Unprotected ductile iron is however susceptible to both internal and external corrosion attacks when exposed in corrosive environments. Internal cement mortar lining for iron pipes, first installed in Charleston, SC in 1922, is proven effective in very long-term prevention of tuberculation and maintenance of flow properties in aggressive water pipelines. The design procedure for cement lined ductile iron pipe limits the deflection of the pipe to no more than 3 percent, although ductile iron is a flexible conduit and the pipe itself will tolerate far more deflection. ASTM standards A-716 and A-746, for ductile-iron culvert and gravity sewer pipe, respectively have the same limit for cement lined pipes and also allow 5 percent deflection for pipes with flexible linings. These pipes are frequently used for sewer service. The 5 percent deflection limit of ASTM standards is similar to many other flexible pipes.

A variety of exterior corrosion protection systems, with various attributes and performance, have been used for all metallic pipes over the years. The standard means of external corrosion protection for ductile iron pipes for nearly 40 years is PE encasement per ANSI/AWWA C105/A21.5. This standard also provides a soil evaluation procedure, developed over many decades of research and experience. Like other coatings, PE wrap can be damaged, for example during tapping, and must be installed carefully. However, a simple repair procedure and a special readily accomplished procedure for tapping wrapped pipes can take care of these issues. In very small pipe sizes, that per OSHA guidelines might allow a worker to safely lift and handle plastic pipes (where plastic pipes are allowed), the slightly heavier weight of ductile iron pipe may raise the cost of installation. Although ductile iron pipe prevents chemical (e.g., volatiles in hydrocarbons) permeation over the entire exposed area of the pipe itself, a standard gasket at the joint does not. However, special rubber gaskets are available for ductile iron pipe that are chemically penetration resistant. These special gaskets should be employed in areas of contaminated soil environments, or where the material conveyed inside the pipe contains chemicals. Table 6.13 presents a summary of advantages and limitations of ductile iron pipe.

6.7.2 Steel Pipe

Steel pipes, made from a versatile refinement of iron, have seen a wide range of usage for more than a century and a half. The development of high strength steel pipes has made it possible to transport fluids such as natural gas, crude oil, and petroleum products over long distances. Initially, all steel pipes had to be threaded together, which was difficult for large pipes, and they often leaked under high pressure. The development of electric arc welding machines in 1920s made it possible to construct leak-proof, high pressure, large-diameter pipelines. One of the earliest steel water pipe installations in the United States, still in service

TABLE 6.13 Advantages and Limitations of Ductile Iron Pipe

Advantages	Limitations
Excellent long-term performance history of standard pipes.	Unprotected pipe is susceptible to corrosion in corrosive environments, both internally and externally, unless suitably protected.
Wide variety of internal and external corrosion protection systems and tools available, including a *Design Decision Matrix* that takes into account the environment as well as the criticality of the pipeline.	Not all corrosion protection methods used over the decades have been effective in all environments.
Internal cement mortar lining prevents tuberculation and enhances hydraulic capabilities, as does an actual internal diameter that in most cases is greater than nominal diameters, compared to some other pipe materials. Any damage to lining can be readily repaired in the field with commonly available materials.	Polyethylene encasement can be damaged. Wrap must be installed properly, and damages, when they occur, must be repaired. Requires machine lifting in most cases, and may result in slightly higher installation cost in at least pipe sizes less than 8 in. It should be noted that most plastic pipes of approximate size 8-in and larger, as per OSHA guidelines, would also require machine lifting.
Strong material, with great ability to handle surge pressures, high load bearing strength, impact strength, beam strength, and ability to bend or deform when external loads are applied.	While a small size string of ductile iron pipe might be less flexible than plastic pipe, it is more flexible to bend/rotate than rigid steel drill rods which are used to drill the borehole in HDD applications. Also, if needed, flexible joint ductile iron pipe can be supplied in shorter lengths, at some increased cost, to accomplish tighter radii of curvature.
Wide varieties of joints, including rapidly assembling flexible restrained joints, enable various applications including trenchless methods. Commonly employed sealing designs have decades of successful applications and experience with minimum labor training and site conditions around the world.	Standard SBR gaskets are subject to chemical attack and permeation (over the small exposed area of the joints) in contaminated soils or when conveying contaminants (available special gaskets should be employed when contamination is present).
Available for pressure, gravity, and vacuum or higher external pressure applications	
Wide range of diameters and pressure classes are available. In some cases, with special considerations, ductile iron pipe has been successfully applied with actual working pressure conditions several times greater than the high pressure ratings.	
Long laying lengths reduce joints in the system.	
Pipe wall itself is impermeable to chemical permeation in contaminated soil areas. Also, special rubber gaskets that are chemical and permeation resistant are available (where this is the chosen option in contaminated areas).	

today, was in San Francisco in 1863. Developments in technology have given way to riveted steel pipes evolving to the automatically welded steel pipes of today. Various other developments have resulted in the creation of different types of joints as well as effective mechanisms for prevention of corrosion, making steel more versatile for trenchless and open-trench applications.

In municipalities, steel pressure pipes are used today in large-diameter potable water transmission applications. In municipal trenchless construction, steel pipes are used as casing pipe in processes such as microtunneling, pipe jacking, horizontal auger boring, HDD, and pipe-ramming because of their high stiffness and compressive strengths. There have even been several large-diameter spiral welded steel pressure pipe installations including projects in Texas, Washington, and Hawaii via HDD. Very detailed and comprehensive manuals and standards with regard to design, manufacture, and application of steel pipe are available from AWWA (the C2XX series), ASTM, ASCE and others. A comprehensive reference concerning many of these aspects, including installation of steel pipes, is AWWA manual M11.

Corrugated Steel Pipe. Corrugated steel pipes have been used for more than a century in gravity applications such as drainage and storm sewers. Though corrugated steel pipes have been used in some sanitary sewers, this is not the case today. Due to their relatively low compressive and tensile strengths, corrugated steel pipes are not used in trenchless or pressure applications. Therefore, a detailed discussion of corrugated steel piping products has been omitted in this chapter. The American Iron and Steel Institute's *Modern Sewer Design* is an excellent source for information on corrugated steel piping systems.

Manufacture. Steel pipes used in municipal applications are manufactured by an automatic welding process. There are generally three types of steel pipe, each identified by the way in which it is manufactured:

1. *Rolled and welded pipe.* This is one of the oldest methods of steel pipe production, where plates of steel are rolled into cylindrical pipes, usually 6 to 12 ft in length, then welded in the circumferential and longitudinal directions. The pipes used in casing applications for trenchless technology are of this type. They are also used in other types of applications.
2. *Electric resistance welded (ERW) pipe.* ERW pipes are generally manufactured and used in smaller diameters up to 24 in. ERW is a single straight seam welding process where continuous coils of treated, low carbon steel, called skelp, are shaped into cylindrical pipes by edge-forming, and then welded at the seam. These pipes can be manufactured in lengths of up to 100 ft. They are used in water systems, natural gas and hydrocarbon, as well as piling and other construction and industrial applications.
3. *Spiral welded pipe.* Starting with continuous rolls of steel similar to the type used for ERW pipe, the steel is fed into a machine and spirally wrapped against buttress rolls to form the pipe. The edges of the spiral pipe are then welded in and out by a double-submerged arc process. Spiral welded steel pipes are used in municipal water transmission applications in diameters of up to 156 in. Trenchless processes such as HDD have been used to install this type of pipe in the potable water systems.

Applicable Standards. ASTM A-139 Grade B is the type of casing pipe used in gravity trenchless construction. Pipe manufactured to the ASTM A-139 standard requires hydrostatic testing, because it is often used for medium internal pressure applications. As the use of the pipe for casing depends only on its structural capabilities, in most cases the hydrostatic test is not required. To exclude the hydrostatic test from the original standard, the casing pipe standard is referred to as ASTM A-139, Grade B (No Hydro). The principal standard for use in the water and

TABLE 6.14 Applicable Standards for Steel Pipe for Trenchless Construction

Pressure standards	Nonpressure standards	Available diameters
AWWA C-200 ASTM A-36, ASTM A-515, grade 60 or ASTM A-572, grade 42–with T7 type joint	ASTM A-139 Grade B ASTM A-36, ASTM A-515, grade 60 or ASTM A-572, grade 42–with Permalok™ T5 type joint	3–144 in and higher

waste industries is AWWA C200. This standard is referenced for use in large-diameter HDD projects for water transmission. Some challenging projects in the past in Texas, Virginia, Washington, and Hawaii have used this steel pipe standard. A special mechanical push-on joint, Permalok™, for steel pipe is manufactured as a proprietary product for casing pipe and trenchless applications. The ASTM standards for steel pipes are listed in Table 6.14. Further discussion of Permalok™ products is presented in the following section.

Joints. There are several types of welded joints available for steel pipe, each suited for a specific application:

1. Bell-and-spigot lap welded joints
2. Butt-welded joints (single-V butt-welded, and double-V butt-welded)
3. Butt strap welded joints
4. Mitered lap-welded joints

There are also a number of nonwelded joints:

1. Bell-and-spigot rubber gasket joints
2. Harness joints
3. Carnegie shape rubber gasket joints
4. Mechanical couplings
5. Split-sleeve mechanical coupling

For steel casing pipe used in boring and pipe jacking applications, it is important that there are no irregularities in alignment at the joints. If the casing is not straight, the ability of the contractor to keep the pipe in line and on grade is affected. Lap welded joint products are not recommended for jacking and boring installations. Though spiral welded pipes can be used for casing, it is a common practice to use a straight seam or seamless pipe. For jointing, both ends of the pipe can be beveled for welding, as in the single-V butt-welded joint, (Fig. 6.17). Another recommendation for achieving a good circumferential weld at the joint is to bevel one end of the casing pipe to a standard 37 degree bevel, and square cutting the other end.

For casing and pressure pipe trenchless applications, a patented product line, Permalok™, offers two mechanical push-on joints. It is particularly attractive to the trenchless excavation

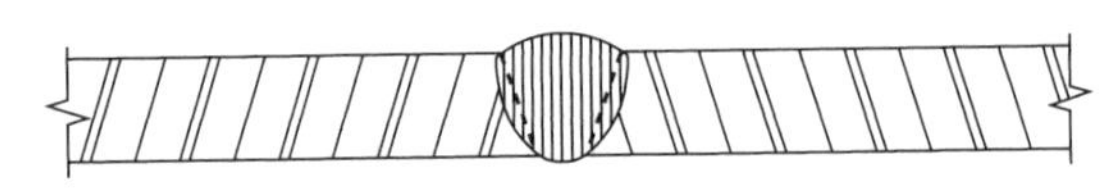

FIGURE 6.17 Single-V butt-welded joint.

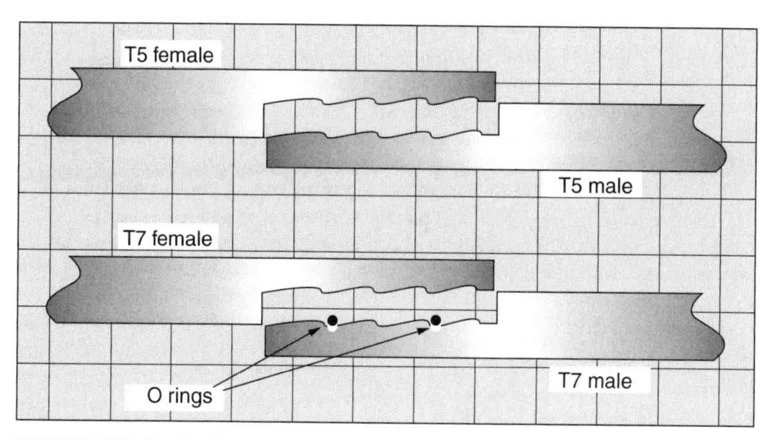

FIGURE 6.18 Interlocking jointing system. (*Courtesy of Permalok.*)

industry because the joints are designed to be flush with the interior and exterior surfaces of the pipe (Fig. 6.17). This joint type eliminates the time required for welding traditional-type steel pipes. Although using the Permalok™ pipe is approximately 2.5 times faster than welding, it costs more. Permalok joints are used for HDD, pipe ramming, and horizontal auger boring. Joint quality is further enhanced because the Permalok™ connector is consistently round, true, and perpendicular to the pipe axis. Its unique machined groove makes stabbing and aligning easy and quick. Also, the Permalok™ design, combined with the use of a sealant such as RTV Silicone, prevents leakage under considerable pressure. Standard specifications, listed in Table 6.14, ensure that all pipe manufactured by Permalok™ meet guidelines. The T5 profile was patented for steel pipe casing applications in 1993 and has been used in numerous trenchless excavation projects nationwide. The T7 profile is capable of withstanding pressures of up to 300 psi, and is used in trenchless pressure pipe applications. Fig. 6.18 illustrates the Permalok interlocking jointing system.

Advantages and Limitations. Table 6.15 lists some of the advantages and limitations of steel pipe. With its high tensile strength, steel pipes are capable of handling high pressures

TABLE 6.15 Advantages and Limitations of Steel Pipe

Advantages	Limitations
Various standards and methods are available for internal and external corrosion protection	Prone to internal tuberculation and external corrosion, subject to electrolysis
High tensile strength	Use of internal and external corrosion protection raises price of the product
High compressive strength	Low resistance to external pressures in large-diameter sizes
Easy to assemble, nonweld joints available	Air vacuum valves are necessary in large-diameter lines
Adopts well to locations where soil movements occur	Welding of joints require skilled labor and is time consuming
Good hydraulic properties when internally lined	Special care required to ensure proper alignment at joint in welded pipe

in water transmission applications. The high compressive strength of steel pipe makes it an excellent material for pipe jacking, microtunneling, pipe ramming, horizontal auger boring, and horizontal directional drilling. Assembly of nonweld joints is relatively easy owing to push-on bell-and-spigot gasket joints. Steel pipe can accommodate ground movements in case of seismic activities and difficult ground conditions during trenchless installations. Good hydraulic properties of internally lined steel pipe make it ideal for use in municipal water systems. The availability of various methods and standards for both internal and external corrosion significantly prolongs the useful design life of the steel pipe.

Though various corrosion protection mechanisms are available, the pipe is highly prone to corrosion attacks unless the protections are used. This can be costly. In large diameters, the pipe's ability to handle external pressure is low. Air vacuum valves must be used in large-diameter lines to eliminate the possibility of pipe collapse, but this is an added cost to the system. Welding of steel pipe joints requires skilled labor and is time consuming.

6.8 CORROSION PROTECTION

All common piping materials can be deteriorated or corroded and designers should consult with manufacturers to learn pipe product limitations. It should be noted that sulfates and chlorides as well as low resistivity soils in general can corrode various concrete pipes, and gasoline contamination has been known to result in swelling and bursting of some buried plastic pipes, whereas such contamination will not burst metallic pipes.

There have been many approaches to corrosion protection of all kinds of pipes through the centuries. There are decades of practical experience with many applications and corrosion protection systems of iron and steel pipes, in particular. There are field and laboratory studies of gray and ductile iron pipes in widely divergent soil types, and also in some notably very corrosive actual soil burial test sites, by the Ductile Iron Pipe Research Association (DIPRA) working in many cases in close conjunction with the utilities involved. Both unprotected iron and steel pipes will rapidly corrode in some soil environments, and in these environments suitable corrosion protection must be provided. It is also being discovered (in more recent applications of pipes that have not been around as long) that other piping materials, such as variously reinforced concrete, plastics, and composites, as well can also undergo forms of corrosion or environmental or stress-related deterioration that are perhaps not now quite as obvious or known to many pipeline practitioners.

American National Standard Institute (ANSI) and American Water Works Association (AWWA) have developed multiple standards for corrosion protection of iron pipe and fittings. These standards include ANSI/AWWA C104/A21.4 for internal cement mortar lining, ANSI/AWWA C105/A21.5 for polyethylene encasement (this standard also contains a soil evaluation procedure in the appendix that is helpful for practitioners to determine when standard pipes with thin asphaltic shop coatings can be directly buried in specific soils and when the supplementary polyethylene wrap should be applied as opposed to installing standard pipe without wrap), and AWWA C116 for fittings that in their normal production processes are coated inside and outside with fusion-bonded epoxy (FBE) instead of cement mortar.

More recently, building on these standards and other extensive experience, CORRPRO Company working with DIPRA as well as ductile iron pipe manufacturers and has conducted a 2-year study of corrosion and corrosion protection characteristics. This study has included field and laboratory evaluations related to short-term and long-term polarization rates under varying conditions; corrosion rate reduction, and corresponding cathodic current criterion. This information was then analyzed in conjunction with an extensive database from 1,379 physical inspections of buried iron water lines. The result of the study is a risk based corrosion protection design strategy for buried ductile iron pipelines referred to as the "Design Decision Model", or "DDM".[5,6]

The steel pipe industry has also been proactive in readily recommending and providing corrosion protection mechanisms to end users, despite the higher costs of some processes. According to the steel pipe industry, "Corrosion protection systems that include coatings, monitoring systems, and cathodic protection (installed incrementally as needed) are very cost-effective." There are a number of standards and processes for both internal and external corrosion protection of steel pipe, including cement mortar lining, paints and polyurethane linings, tape coatings, coal tar enamel coatings, cement mortar coatings, and epoxy and polyurethane coatings. Furthermore, the steel pipe industry acknowledges the use of cathodic protection as an effective and/or necessary method of protection to complete the corrosion protection process.

The coverage of the topic of corrosion in this chapter is intended to make the reader aware of one of the significant problems being faced by pipeline industry today. These problems include corroding, deteriorating, and/or failing water and sewer lines. There are currently huge amounts of pipeline system underground, some of which are without effective corrosion protection or other provisions for adverse environmental conditions. Pipe leakage and failure in these unprotected pipelines can significantly influence rate of unaccounted- for-water and I/I in future. It is essential for engineers to be aware of the problems and available engineering solutions as they design and renew the pipeline systems of the future. Proper corrosion protection will ensure that the full potential of all piping materials for use in specific applications is realized in the long run.

ACKNOWLEDGMENTS

This chapter was prepared by Mr. Shah Rahman, EIT, Regional Vice President, Underground Solutions, Inc., Dallas, Texas, and reviewed by Dr. Reynold K. Watkins, Professor Emeritus, Utah State University. The section on ductile iron pipe was provided by Mr. Ralph Carpenter of American Ductile Iron Pipe and American Spiralweld Pipe Company.

REVIEW QUESTIONS

1. Why is pipe material selection so important? Explain.
2. Describe rigid, semirigid, and flexible pipes.
3. How does the design process differ for rigid and flexible pipe?
4. Compare thermoset with thermoplastic materials.
5. Describe the advantages and the limitations of concrete pipe.
6. Describe the advantages and the limitations of clay pipe.

[5]D. H. Kroon, P.E., D. Lindemuth. P.E., S. Sampson, and T. Vincenzo. (2004). Corrosion protection of ductile iron pipe. *Proceedings of ASCE International Conference on Pipeline Engineering and Construction,* San Diego, Calif.

[6]R. W. Bonds, L. M. Barnard, A.M. Horton, and G. L. Oliver. (2004). Corrosion and corrosion control research of iron pipe. *Proceedings of ASCE International Conference on Pipeline Engineering and Construction,* San Diego, Calif.

7. Describe the advantages and the limitations of PE pipe.
8. Describe the advantages and the limitations of PVC pipe.
9. Describe the advantages and the limitations of GRP pipe.
10. Describe the advantages and the limitations of ductile iron pipe.
11. Describe the advantages and the limitations of steel pipe.
12. Describe the steel pipe's mechanical jointing system.

REFERENCES

Allouche, E. N., et al. (2003). Field evaluation of an innovative joint for pull-in-place trenchless installations, *Proceedings of the 5th Construction Specialty Conference of the Canadian Society for Civil Engineering*, New Brunswick, Canada.

American Concrete Pipe Association. (2004). *Concrete Pipe Design Manual*, Irving, Tex.

American Concrete Pipe Association. (2001). *Concrete Pipe Handbook*, Irving, Tex.

American Concrete Pressure Pipe Association. (2004). *Technical Information Paper: Concrete Pressure Pipe in Tunnel Installations.* Available at www.accpa.org.

American Iron and Steel Institute. (1989). *Steel Plate Engineering Data-Volume 3: Welded Steel Pipe*, Washington, DC.

American Water Works Association. (2003). *Manual of Water Supply Practices M41: Ductile-Iron Pipe and Fittings*, Denver, Colo.

ASTM C1208. (2001). *Standard Specification for Vitrified Clay Pipe and Joints for Use in Microtunneling, Sliplining, Pipe Bursting, and Tunnels, Volume 04.05*, Conshohocken, Pa.

AWWA C105 *Polyethylene Encasement for Ductile Iron Pipe Systems*, Denver, Colo.

AWWA Polyolefin Pressure Pipe and Fittings Committee. (1999). Committee report: Design and installation of PE pipe, *Journal of American Water Works Association*, 91(2):92 –100.

Buried Pipe Markets in North America 1999. (2000). Uni-Bell PVC Pipe Association, Dallas, Tex.

Carpenter, R. (2004). 7th annual municipal survey: Making sense of a complex market, *Underground Construction Magazine*, Houston, Tex., p. 22.

Certain-Teed. (1974). *Installation Guide: Certain-Teed Fluid-Title Pressure Pipe,* Document 40-21-13H.

Composites Institute of the Society of Plastics Industry. (1989). *Fiberglass Pipe Handbook*, New York, NY.

Dechant, D., and C. Perry. (2003). *External Corrosion Comparisons: Steel & Ductile-iron, Pipe*, Northwest Pipe, Portland, Oreg.

Dechant, D., and G. Smith. (January 2004). Present levels of corrosion protection on ferrous water piping in municipal infrastructure: A manufacturer's perspective, *Materials Performance*, 54–57.

Hitt, R., and D. Luckenbill. (2002). Microtunneling in Oklahoma, *Proceedings of Underground Construction Technology International Conference and Exhibition 2002*, Houston, Tex.

Holley, M., R. Diaz, and M. Giovanniello. (2001). Acoustic monitoring of prestressed concrete cylinder pipe: A case history, *Proceedings of ASCE Advances in Pipeline Engineering & Construction Conference,* San Diego, Calif.

Koch, G. H., M. Brongers, N. Thompson, Y. Virmani, and J. Payer. (2002). *FHWA-RD-01-156: Corrosion Cost and Preventive Strategies in the United States*, CC Technologies Laboratories, Dublin, Ohio.

Kwong, J., and W. Wanner. (2001). Construction of sewers by microtunneling in highly congested utility corridors, *Proceedings of ASCE Advances in Pipeline Engineering & Construction Conference*, San Diego, Calif.

Larson, J., and J. Pearl. (2003). Small diameter clay sewer pipe O&M strategy: Replace it now or run it to failure, *Proceedings of Collection Systems Conference 2003—Go cMOM*, Austin, Tex.

Liu, H. (2003). *Pipeline Engineering*, Lewis Publishers, New York, NY.

Luckenbill, M. (2004). Ask the engineer: Proper bedding for PVC pressure pipe, *PVC Pipe News*, Uni-Bell PVC Pipe Association, Dallas, Tex.

Marston, A., and A. O. Anderson. (1913). The theory of loads on pipes in ditches and tests of cement and clay drain tile and sewer pipe, *Bulletin 31*, Ames, Iowa.

Mielke, R. D. (2004). Use of dielectric coated spiral weld steel pipe in horizontal directional drilling, *Proceedings of NASTT No-Dig Conference 2004*, New Orleans, La.

Moser, A. P. (2001). *Buried Pipe Design*, U.S.A. Second Edition, McGraw-Hill, New York, NY.

Moser, A. P., and O. K. Shupe. (1989). *Testing Fifteen Year Old PVC Sewer Pipe*, Buried Structures Laboratory, Utah State University, Logan, Utah.

National Clay Pipe Institute. (1995). *Clay Pipe Engineering Manual*, U.S.A.

Iseley, D. T., M. Najafi, and R. Tanwani. (1999). *Trenchless Construction Methods and Soil Compatibility Manual*, National Utility Contractors Association (NUCA), Arlington, Va.

Nayyar, M. (Ed.) (1992). *Piping Handbook*, 6th ed., U.S.A., McGraw-Hill, New York.

Petroff, L., and M. Luckenbill. (1981). Flexibility of the design of fiberglass pipe, *Proceedings of ASCE International Conference on Underground Plastic Pipe*, New York, NY.

Pigg, B. J. (1998). Asbestos-cement pipe: The tried and true pipe, *ASTM Standardization News*.

Rahman, S. (2004). State-of-the-art review of municipal PVC pipe products, *Proceedings of ASCE Pipeline Engineering & Construction International Conference*, San Diego, Calif.

Rahman, S. (2004). Developments in North American PVC piping products for trenchless applications, *Proceedings of Plastics Pipes XII*, Milan, Italy.

Rahman, S. (2003). Municipal PVC piping products: A state-of-the-art review, *Proceedings of Texas Section-ASCE Fall 2003 Meeting*, Dallas, Tex.

Rahman, S. (2002). PVC pressure pipe: Past, present & future, *Proceedings of Center for Innovative Grouting Materials and Technology Conference*, Houston, Tex.

Shiells, D., and B. Zelenko. (2004). Microtunneling through the Capital Beltway, *Proceedings of North American Society for Trenchless Technology No-Dig 2004 Conference*, New Orleans, La.

Shrock, B. J., and J. Gumbel. (1997). Pipeline Renewal—1997, *Proceedings of North American No-Dig '97 Conference*, Seattle, Wash.

Sorensen, H. W., B. Tatum, M. Woolsey, and M. Young. (2003). Biogenic corrosion of full flowing concrete sewers: Case studies of failed pipes in Dallas, Tex; Phoenix, Ariz., and Laughlin, Nev., *Proceedings of Texas Water 2003 Annual Conference*, Corpus Christi, Tex.

Spangler, M. G. (1941). The structural design of flexible pipe culverts, *Bulletin 153*, Ames, Iowa.

Staples, L. B. (1995). New tools for the condition evaluation of waterlines, *1995 Conference on Corrosion and Infrastructure, Practical Applications and Case Histories*, NACE International and U.S. Department of Transportation, Federal Highway Administration.

Sutherland, P. (2002). Semirigid pipes—Research and design, *NSW IPWEA State Conference*, New South Wales, Australia.

Uni-Bell PVC Pipe Association. (2001). *Handbook of PVC Pipe: Design & Construction*, Dallas, Tex.

Uni-Bell PVC Pipe Association. (2003). *Uni-TR-5-03: The Effects of Ultraviolet Radiation on PVC Pipe*, Dallas, Tex.

Water Environment Research Foundation. (2000). New pipes for old: A study of recent advances in sewer pipe materials and technology, *Project 97-CTS-3 Final Report*, Alexandria, Va.

Water Industry Database: Utility Profiles. (1992). *Report by the American Water Works Association*, AWWA, Denver, Colo.

Watkins, R. K., and M. G. Spangler. (1955). *Some Characteristics of the Modulus of Passive Resistance of Soil—A Study in Similitude*, Logan, Utah.

Zarghamee, M., and R. Ojdrovic. (2001). Risk assessment and repair priority of PCCP with broken wires, *Proceedings of ASCE Advances in Pipeline Engineering & Construction Conference*, San Diego, Calif.

CHAPTER 7
HORIZONTAL AUGER BORING

7.1 INTRODUCTION

Horizontal auger boring is a well-established trenchless method that is widely used for the installation of steel pipes and casings under railway and road embankments. It is an economical pipe installation method that can be used in a variety of ground conditions to prevent open-cuts on pavements, and reduce disruptions to traffic.

Although there are numerous benefits of horizontal auger boring method, there are also some limitations. This chapter provides an overview of the method and includes information essential for the planning and execution of a successful horizontal auger boring project.

7.2 BRIEF HISTORY

Vin Carthy, Salem Tool Company, and Charlie Kandal formally developed the horizontal auger boring method, simultaneously and independently. The original machines were used for drilling horizontal blast holes for coalmines in Somerset, Pennsylvania (ASCE, 2004).

In the 1940s, Charlie Kandal founded the Ka-Mo Company that manufactured the Ka-Mo auger boring machines. The early machines were electrically operated, but the second-generation machines were gasoline powered. The large shape and size of the machine required a deep pit for making a bore. Ka-Mo machines were designed to run on a track. The early bores were typically not cased and the auger had a tendency to screw (dig) itself in the ground. By 1951, the auger machines were capable of boring 5-in diameter, uncased holes up to 230 ft in length.

Both Salem and Ka-Mo dominated the horizontal earth boring (HEB) business until 1961 when Al Richmond started manufacturing smaller HEB machines. Richmond's company, Wert & Starn Pipeline, began building boring machines under the trade name of Tornado. The tornado was built with the engine on top of the gearbox and was driven with a chain. Richmond's smaller units were convenient and quickly became popular. The 24-in machine, the "Power Midget," was the most popular model, however 30- and 36-in machines were also available for larger installations.

As the technology advanced into cased bores, the thrust requirements increased and the design was modified so that the machines were split for ease of handling. Ernie Coppica, of Wixom, Michigan, invented the steering systems for horizontal auger boring in the 1960s. In 1970, Leo Barbera founded American Augers. American Augers started by building hydrostatic drive machines with a built-in slip clutch. If the machine hit a boulder or some other obstruction, it would switch to the maximum torque but at the same time block out the thrust. The machine would advance only when the condition was relieved.

The current manufacturers of HEB machines include American Augers, Barbco, Bor-It Manufacturing, Horizontal Equipment Manufacturing Inc., McLaughlin Boring Systems, Michael Byrnes Manufacturing, and several other smaller manufacturing firms. In addition, several contractors manufacture their own boring equipment.

7.3 METHOD DESCRIPTION

Horizontal auger boring is a trenchless technique employed to drill horizontal bore holes by rotating the cutting head. The cutting head is attached to the augers that stay inside the casings. Auger boring machine generates torque that is transmitted to the cutting head through the flighted tube. The auger boring operation requires a driving shaft and reception shaft. The boring equipment including auger boring machine, augers, and cutting head is located in the driving shaft and drills horizontal bore holes in the ground. Spoil is removed from the borehole to the backside of the casing by the movements of helically-wound auger flights. The vertical alignment of the auger boring operation can be controlled by using a water level. However, it is difficult to control the horizontal alignment in the auger boring operation without special instrumentation (See Sec. 7.6).

It is possible to use horizontal auger boring to construct an uncased borehole by using a cutting head and auger. However, this practice results in an unsupported hole, and the unprotected augers rotating in the drive shaft create hazards for the workers. Therefore, common practice is to simultaneously jack the steel casing with the boring operation. If uncased auger boring is permitted, it should be limited to soil conditions with sufficient stand-up time and short, small-diameter bores.

7.3.1 Track-Type Horizontal Auger Boring Method

There are two main types of horizontal auger boring methods (Fig. 7.1). One is the track-type horizontal auger boring method and the other is the cradle-type horizontal auger boring method. The track-type horizontal auger boring operation comprises other equipment such as boring machine, casings, cutting head, and augers. The track type also may employ casing lubrication system, steering system, locating system, and casing leading-edge band for its operation. The auger boring machine is located on the track and moves back and forth along the track while providing jacking and rotating force to the augers and casings during the boring operation. The layout of track-type auger boring operation is shown in Fig. 7.2.

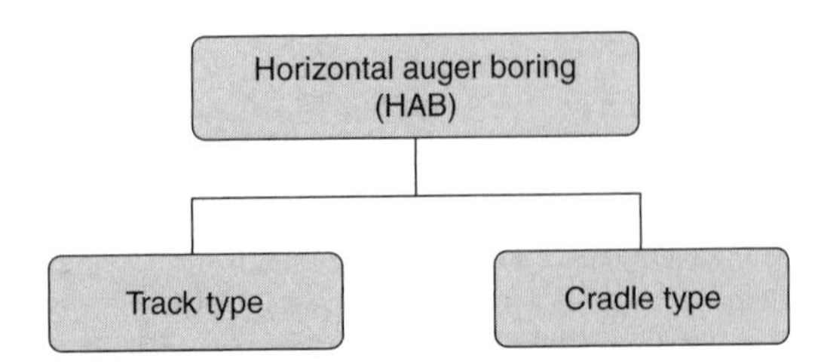

FIGURE 7.1 Horizontal auger boring categories.

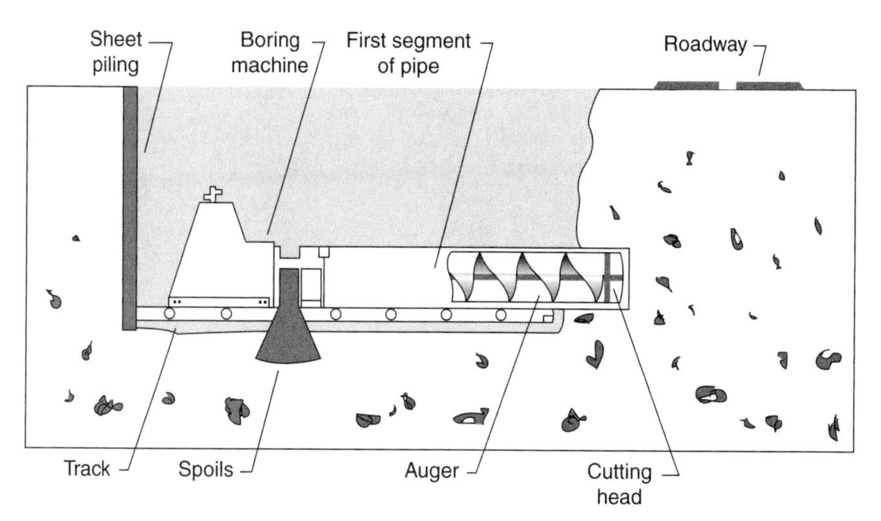

FIGURE 7.2 Track-type horizontal auger boring (Iseley and Gokhale, 1997).

The auger string is composed of connected augers end to end. One end of the auger string is connected to the boring machine and the other end is linked to the cutting head. The torque and thrust force generated by the boring machine are transported through the auger string to the cutting head. The rotation of the cutting head and augers can cut out the ground and remove the spoils from the front of the casing to the back. At the same time, boring machine can proceed forward using the hydraulic jacking force supported by the thrust block. By repeating this operation, casing can be installed in the ground. Figure 7.3 shows the augers before connection.

FIGURE 7.3 Augers (Abraham and Gokhale, 2002).

FIGURE 7.4 Track system for horizontal auger boring (Abraham and Gokhale, 2002).

Two main factors that affect horizontal auger boring are torque and thrust. The power source that can be pneumatic, hydraulic, or an internal combustion engine through a mechanical gearbox creates the torque. The torque rotates the auger that in turn rotates the cutting head. One end of the ram is attached to the boring machine while the other end is attached to lugs that lock into the track system.

As horizontal auger boring operation has a limited line and grade control, the initial setup of the track system in the driving shaft is critical to the accuracy of the auger boring operation. Therefore, a properly constructed drive shaft is important for the success of a track-type horizontal auger boring project. The shaft requires a stable foundation and adequate thrust block. The foundation must support the tracks, permitting the machine to move forward and backward without vertical movement. The track system must be placed on the same line and grade as the desired bore hole. If the track foundation settles, accuracy will be affected and binding forces could result within the borehole. Often this foundation will require crushed stone or concrete as shown in Fig. 7.4.

The thrust block transmits the horizontal jacking forces from the tracks to the ground at the rear of the drive shaft. The thrust block must be designed to distribute the jacking force over a sufficient area so that the allowable compressive strength of the soil is not exceeded. If the thrust block fails or moves, borehole accuracy will be compromised and binding forces could result within the borehole. The track-type horizontal auger boring operation involves procedures discussed in the following sections.

7.3.2 Jobsite Preparation

This step involves the investigation of underground utilities and designing layout of job site. A detailed survey of the existing site conditions, both surface and subsurface, should be performed in the initial stages of planning. This information plays a vital role in determining the feasibility, alignment, gradient, and physical constraints of the proposed horizontal auger boring project.

The jobsite should be thoroughly surveyed for surface features such as overhead power lines and other obstructions, highway or railway crossings, existing grading contours, water drainage problems, availability of easements and rights of way, job access, and work space limitations. A survey should be done to determine if there is adequate space for boring pits, setting up equipment and stockpiling of material. Natural drainage gradient should be examined and considered to prevent flooding of the jobsite.

The subsurface investigation can be initiated once the general site conditions and the surface survey data are obtained. The most critical aspect of subsurface survey is the location of all the existing utilities in and around the design path. All the utilities on the proposed path should be contacted, located, and plotted to avoid potential underground obstruction to the bore. This is especially true beneath the roadway to be crossed as the potential of hitting a utility is very high and damaging the utility lines will cause serious disruptions to traffic and great inconvenience to the public.

Geotechnical investigation of the site should be performed to identify the general and any special subsurface conditions. The extent of the investigation may vary depending on the knowledge of known local geological conditions.

In most areas where there is uniform soil formation, it may be possible to use the existing data to identify the type of soil conditions to be encountered. In areas where the soil conditions are not well-known, a preliminary investigation comprising surface borings, soil sampling, and laboratory sieve analysis should be undertaken.

One of the problematic ground conditions for horizontal auger boring is boulders. Geotechnical reports should comment on the potential for encountering boulders and their size and frequency. The horizontal auger boring method can handle boulder sizes up to $^1/_3$ of the diameter of the casing, and therefore, casing sizes should be designed to accommodate the largest size boulder that is expected to be encountered. Obviously, it is difficult to predetermine the size of boulders with a certainty, hence it is better to err on the conservative side by upsizing your casing. It is not unusual for a horizontal auger boring operation to come to a temporary standstill upon encountering unanticipated obstacles. This usually requires, removing the augers from the casing, sending a person to the front end of the bore to remove the obstacle using a jackhammer, and in some instances even blasting.

A geotechnical baseline report (GBR) prepared by the engineer should be made available prior to bidding for the project. The report should contain the following information:

- Project description
- Available existing information or reports
- Description of the geology
- Generalized geologic profile
- Groundwater conditions
- Presence of contaminants
- Geological features that may impact the project
- Man-made features that may impact the project
- Anticipated ground behavior at the shaft locations
- Anticipated ground behavior along the pipeline

Subsurface parameters that are important to a successful horizontal auger boring project are listed in Table 7.1.

"Geotechnical Baseline Reports for Underground Construction," a joint publication of the Underground Technology Research Council (UTRC), ASCE and American Institute of Mining, Metallurgical and Petroleum Engineers (AIME), describes these parameters in

TABLE 7.1 Important Subsurface Parameters for a Successful Horizontal Auger Boring Project

Soil parameters	Rock parameters
Gradation	Color
Permeability	Grain size
Density	Composition
Standard penetration value (blow count)	Intact rock strength
Cohesion	Hardness
Moisture content	Quartz content
Atterberg limits	Fracture frequency
Unconfined compressive strength	Rock quality designation (RQD)
—	Total core recovery

great detail. The ASCE Manual of Practice (MOP) No. 106 for Horizontal Auger Boring Projects provides detailed information on HAB planning, design, and construction.

7.3.3 Bore Pit Excavation and Preparation

Boring pits should be located at a safe distance from the existing roadway. The distance of the pit from the roadway should be adequate to allow sloping of the pit if necessary. If sufficient sloping cannot be accomplished because of constraints, earth support system of pit walls should be considered. Adequate room should be provided for safe loading and unloading of equipment, and spoil removal. Construction of the pit is the responsibility of the contractor and should conform to the rules set forth in the Occupational Safety & Health Administration (OSHA) Code of Federal Regulations, Construction Standards for Excavations, 29 CFR part 1926, subpart P. There are specific requirements for pit construction, protection, barricades, traffic control, installation, and type of ladders used in the pit and personal safety equipment.

Excavation can begin once the utilities are located and marked. The jacking pit should be offset more on the side of the bore line where spoil exits the auger. This facilitates the access for spoil removal (Fig. 7.5). The utilities that are encountered in the pit must be properly supported. Adequate dewatering system must be installed if wet ground conditions are to be expected.

The bottom of the boring pit should be filled with crushed stone or gravel to make it firm enough to support the boring machine tracks, boring machine, casing, and the auger. Usually 2 in × 8 in × 16 ft wooden planks are placed parallel to the track rail under the track for support. A concrete floor may be placed if the bore is of considerable length and size, and/or soil conditions warrant it. A concrete floor is recommended when boring in rock. In all cases, the track support must be set to the proposed grade of the bore.

The boring machine exerts thrust to the backside of the boring pit. A backing plate should be installed against the wall opposite to the bore in the pit to withstand the thrust exerted by the boring machine. The backing plate can be steel sheeting, steel plate, or timber for low to medium thrust pressures or a concrete backstop in addition to steel plate if the thrust pressure is high. Adequate care should be taken to ensure that the thrust pressure developed by the operation does not affect the existing utilities near the bore pit.

In most cases, an exit pit is required at the end of the bore. The safety requirements for an exit pit are similar to those for the entrance pit. Unless absolutely necessary, no personnel should be allowed in the exit pit during the boring operation. As the casing pipe

FIGURE 7.5 Spoil removal using backhoe (Abraham and Gokhale, 2002).

approaches the exit pit, care should be taken to prevent collapse of the pit wall above the pipe.

The possibility of flooding always exists during the boring operation. The location of a pit sump for pumping should be considered during the design of the pit. The location of the sump is dependent on the slope of the pit floor, but generally to the rear of the pit and away from the ejected spoil.

7.3.4 Setting the Boring Machine

The most critical part of the bore is the setting of the machine track on line and grade. If the alignment is not right when the bore is started, it is not likely to improve during the boring process. Figure 7.6 shows the installed horizontal auger boring machine and track system.

Different types of equipment may be required on or around the boring site. Excavators or cranes are needed to dig the boring pit and set the equipment. Boring machine and tracks appropriate for the job are required. Augers must be placed in the casing sections. A cutting head is selected depending on the ground conditions and is installed in front of the first auger section. Other optional systems may be employed for the horizontal auger boring operation. These include:

Lubrication system: To reduce the friction between the casing and soil, a lubricant may be applied to the outer skin of the casing. This can also reduce the requirement for the thrust capacity of boring machine. Two basic types of lubricants are bentonite and polymers.

Water level: Water level is a device to measure the grade of pipe casing as it is being installed. It permits the monitoring of grade by using a water level sensing head attached

FIGURE 7.6 Horizontal auger boring machine on track (Abraham and Gokhale, 2002).

to the top of the leading edge of the casing. A hose connects the bottom of the indicator tube to a water pipe running along the top of the casing as shown in Fig. 7.7.

Grade control head: The grade control head is used for making minor corrections in the grade. It can be used to make vertical corrections only. During the boring process, the actual grade can be monitored with the water level and the necessary adjustments can be made with the grade control head.

FIGURE 7.7 Water level (Abraham and Gokhale, 2002).

FIGURE 7.8 Cutting head and partial banding (Abraham and Gokhale, 2002).

7.3.5 Preparation of Casing

In most cases, the lead casing is prepared in the yard prior to its transport to the jobsite and arrives at the jobsite with the auger inside and the cutting head attached to the leading end of the auger. A partial band at or near the head end of the casing is recommended when boring in most soil conditions. The band compacts the soil and relieves pressure on the casing by decreasing the skin friction. The cutting head and auger inside the casing as well as partial band are shown in Fig. 7.8.

7.3.6 Installation of Casing

When casings are prepared and the horizontal auger boring is setup, the leading casing is moved onto the track and connected to the boring machine by welding as shown in Fig. 7.9. *Collaring*, which is the first operation, pushes the cutting head into the ground without lifting the casing out of the saddle. When about 4 ft of casing has entered the ground, the engine is shut down, the saddle is removed, and the line and grade of the casing is checked.

After the first section of the casing has been installed in the ground, rotating the auger until all the spoil is removed cleans the casing. The machine is then shut down and the auger pin in the spoil chamber is removed. The machine is then moved to the rear of the track and is again shut down. Then the next section of the casing and auger are lowered into position. The augers at the face are aligned flight to flight, the hexagonal joint is coupled and the auger pin is installed. Once the casing to be installed is aligned with the installed casing, the two are tacked together then fully welded. The process is then repeated until the bore is completed. Figure 7.10 shows the soil removal during horizontal auger boring.

Wing cutters are devices attached to the cutting head that open and close as necessary. When the cutting head is rotated clockwise, the wing cutters open up to provide over-excavation of the borehole. The over-excavation of the borehole allows the casing to enter more easily because it reduces the casing skin friction. Wing cutters are used only in stable soil conditions and are never used with the cutter head inside the casing. The wing cutters are adjustable to control the amount of overexcavation. The standard overcut is $^3/_4$ in when not using a steering head and 1 in when using steering head. When the cutting head is rotated

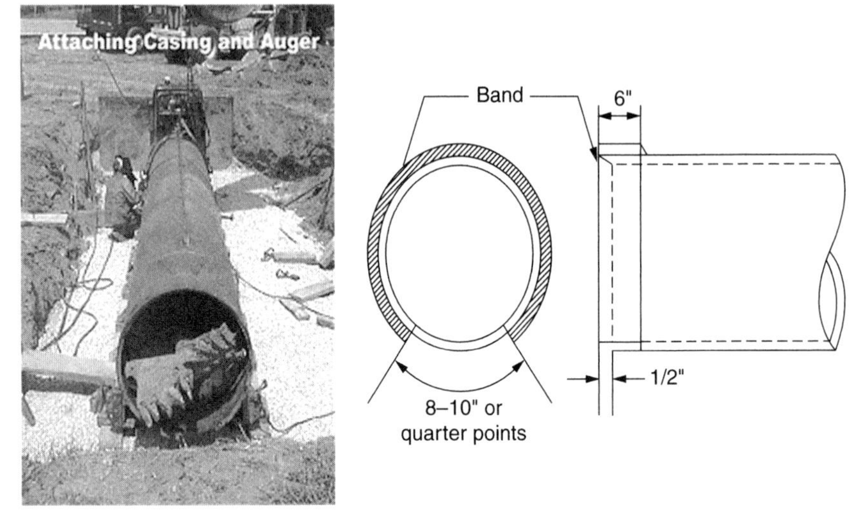

FIGURE 7.9 Connection of casing and auger (Abraham and Gokhale, 2002).

FIGURE 7.10 Soil removal (Abraham and Gokhale 2002).

counterclockwise, the wing cutters close up so that the cutting head can slide back inside the casing for auger removal purpose. The wing cutters must be set so as not to overexcavate the bottom of the casing, as this would cause the bore to drift downward. Keeping the boring head centered can prevent overexcavation of the bottom. This is accomplished by using new or built-up augers in the lead section of the casing. A worn auger in the lead section will allow the head too much freedom and the wing cutter pattern will be erratic.

Once the bore is completed, the machine is shut down and the cutting head is removed. Rotating the augers then cleans the casing. The torque plates are then removed to detach the machine from the casing and the augers are retracted till the coupling is well outside the casing. The auger section is uncoupled from the machine and the other auger sections and is then removed. The machine is coupled to the next auger and the process is repeated until all the auger sections are removed.

7.3.7 Carrier Pipe Installation Using Blocking

For gravity sewer installations, where installation of the carrier pipe on line and precise grade is required, it is extremely important that the carrier pipe be blocked into position to prevent floatation. This can be accomplished by using a wood blocking banded to the carrier pipe to allow for adjustment of the grade inside the casing. Usually redwood (in the western states of the United States) or other hardwood skids banded to the carrier pipe are used. It should be noted that a line and grade bore would allow spacers or regular blocking. Both methods protect the bells of the carrier pipe during installation and support the pipe off the bottom of the casing. If floatation or *hammering* is a concern, *centering* spacers can be used to hold the pipe down. Skids can also be banded along the top of the carrier for the same effect. It is recommended that three or four skids, 2-ft long, be evenly spaced around the carrier at each joint and in the center (for plastic pipe) of the carrier pipe. Ideally, the casing should be designed to allow approximately 1 in of clearance between the top of skids and the inside diameter of the casing. Casing spacers that center the carrier pipe inside the casing do not allow for a differential blocking.

Another method of carrier pipe installation is the use of premanufactured spacers or casing insulators. These spacers come in plastic, fiberglass, stainless steel, and carbon steel. They can also be coated in epoxy, rubber, and various other materials. Manufacturers provide recommendations for design and spacing of the spacers. If the carrier pipe is properly supported with spacers, there is no need for any fill inside the annular space between the carrier pipe and the casing. A further advantage is the ease with which the carrier pipe can be removed if future maintenance becomes necessary.

7.3.8 Grouting of Casing

After the carrier pipe is blocked inside the casing, grout or sand backfill may be used in the annulus between the carrier and casing pipes. This is especially true in situations where high groundwater exists. When using sand, blocking the pipe down is important, as sand and pea gravel do not have sufficient strength to hold the carrier pipe down even if the annular space between the carrier and casing is entirely filled. Sand and pea gravel are abrasive and may damage the pipe. The installation process involves using air to jet the material inside of the casing under high pressure. Smaller casings that do not allow personnel-entry are very difficult, if not impossible, to fill completely.

If sand-cement grout is used, care must be taken when grouting small pipes in a large annulus, as the heat of hydration of the sand-cement grouts can damage some plastic pipes. Another disadvantage of sand-cement grouts is that it is very difficult if not impossible, to remove the carrier pipe for future maintenance.

Lightweight cellular grouts are a popular alternative for this application. As compared to regular sand-cement grouts, cellular grouts have a lower density that reduce the possibility of floatation of the carrier pipe, and have superior fluidity allowing for low installation pressure. Usually a compressive strength of 150 psi is recommended in most applications.

7.3.9 Site Restoration

Once all the augers are removed, the boring machine and the tracks are removed from the pit, and the desired utilities are installed through the casing and the required connections are made. The entrance and exit shafts are then backfilled and the site restored to condition required by project specifications.

7.4 CRADLE-TYPE HORIZONTAL AUGER BORING

Cradle-type horizontal auger boring method is suitable for projects that have adequate room. The bore pit size is a function of the bore diameter and the length of the bore. This method is commonly used in petroleum pipeline projects where large rights-of-way are essential.

This method offers the advantage that all work is performed at the ground level rather than in the pit. The bore pit is excavated several feet deeper than the invert of the casing pipe to allow space for the collection of spoil and water as the borehole is excavated. The method does not require any thrust structures; however, a jacking lug must be securely installed at the bore entrance embankment. Figure 7.11 shows the operation of cradle-type horizontal auger boring method.

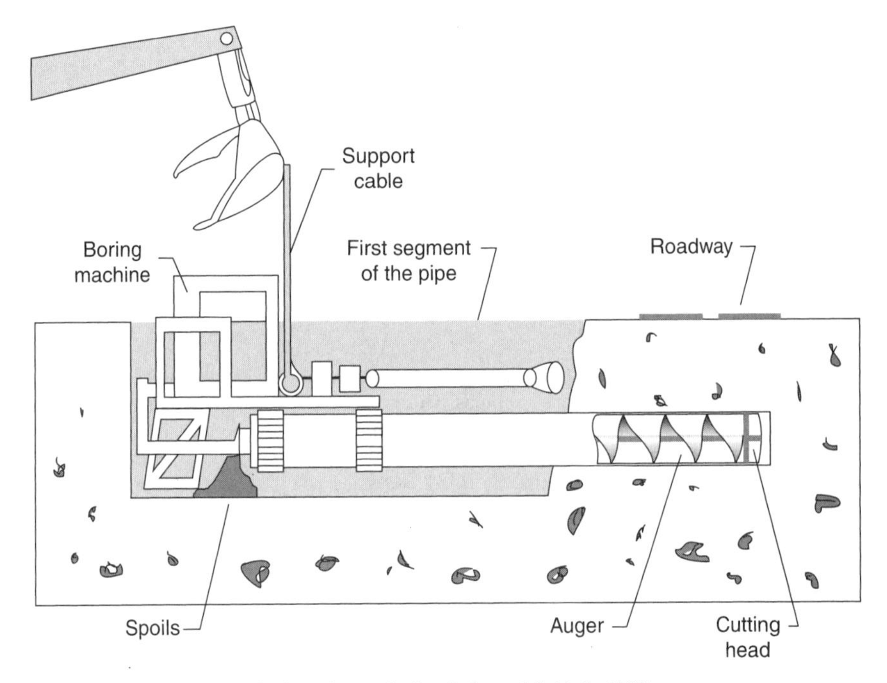

FIGURE 7.11 Cradle-type horizontal auger boring (Iseley and Gokhale, 1997).

7.5 MAIN FEATURES AND APPLICATION RANGE

7.5.1 Diameter Range

Horizontal auger boring can be used to install casing pipe ranging from 4 to over 60 in. in diameter, with the most common diameters ranging from 8 to 36 in. When the diameter of the pipe to be installed is less than 8 in, other trenchless technologies become more appropriate and economical, especially, where the line and grade are not very critical. For larger diameters where the line and grade are more critical and considering all project conditions, pipe jacking and microtunneling can be the better alternatives as they provide greater accuracy.

7.5.2 Drive Length

Horizontal auger boring was initially developed to cross under a two-lane roadway with an average length of 40 ft and a maximum length of 70 ft. However, typical project lengths range from 100 to 600 ft, with the demand for longer installations increasing. The longest continuous track-type horizontal auger boring project to date is 900 ft.

7.5.3 Type of Casing

Because the augers rotate inside the pipe, the pipe and the coating material must resist potential damage caused by rotating augers. Therefore, a typical casing pipe is made of steel. The product or carrier pipe installed inside the casing can be made of any material suitable for the product being carried. The pipe wall thickness commonly ranges from $^1/_4$ to $1\,^1/_2$ in based on the pipe sizes, project conditions, and the required strength. Generally, the casing thickness that is used for horizontal auger boring is sufficient, so that additional corrosion protection coatings are not required. As any coating on the casing pipe will be damaged during installation from the rotation of the augers, or the casing being jacked through the soil, or both, it is recommended that a pipe with greater wall thickness be used to increase the longevity of the pipe.

7.5.4 Workspace Requirements

Shafts are required at both ends of the bore. The drive shaft is the primary working shaft. The size of the shaft is determined by the diameter of the borehole and the length of the casing segments to be used. Typically, casing segments are 10, 20, or 40 ft in length; the most common length is 20 ft. If casing segments of length 20 ft are used, the shaft size will be 30 to 35 ft in length by 8 to 12 ft in width. The surface area should be approximately, 75 by 150 ft. The minimum surface area should be 30 by 82 ft. Sufficient space should be available for loading, unloading, and storage of materials and equipment.

7.5.5 Soil Conditions

Horizontal auger boring can be used in a wide range of soil conditions, from dry sand to firm dry clay to solid rock. Firm sandy clay is the most compatible soil condition for using this method. Boulders or cobbles as large as one-third of the casing diameter can be accomplished.

TABLE 7.2 Applicability of Different Soil Conditions for Horizontal Auger Boring

Type of soil	Applicability
Soft to very soft clays, silt, and organic deposits	Yes
Medium to very stiff clays and silts	Yes
Hard clays and highly weathered shales	Yes
Very loose to loose sands (above water table)	Marginal
Medium to dense sands (below the water table)	No
Medium to dense sands (above the water table)	Yes
Gravels and cobbles less than 50 to 100 mm diameter	Yes
Soils with significant cobbles, boulders, and obstructions larger than 100 to 150 mm diameter	Marginal
Weathered rocks, marls, chalks, and firmly cemented soils	Yes
Significantly weathered to unweathered rocks	Yes

In case of unstable soils, care should be taken regarding the cutting edge leading the casing edge as this may result in spoil being removed without any advancement in the casing which means that excessive spoil is being removed. This situation can create a void between the casing and the borehole, leading to surface subsidence. Table 7.2 presents applicability of different soil conditions.

7.5.6 Productivity

The success of the project depends to a large extent on the quality of the drive shaft. Shaft construction may take 1 day for shafts less than 10 ft when the excavation embankments can be sloped. Shaft construction could take several weeks if the shaft is greater than 30 ft and excavation support systems, such as steel sheet piling, has to be used.

The auger boring operation takes a four-person crew 3 to 4 h to setup the horizontal auger boring equipment for a steel casing project 24 in. in diameter using 20 ft segments. A typical production rate for such a project is 40 ft to 60 ft for each 8-hr shift. Depending on soil conditions and casing diameter and length, horizontal auger boring typically takes place at a rate of 3 to 12 ft/h.

7.5.7 Accuracy

If a steering head is not used in the horizontal auger boring system, accuracy depends on groundwater conditions, length of drive, initial setup, and operator skills. An accuracy of ±1 percent of the length of the bore can typically be achieved. For projects that require a higher accuracy, an oversized casing is installed to provide maneuvering room for the carrier pipe inside the casing to obtain the specified tolerance. Recent innovations in the guidance systems, makes it possible to achieve greater accuracy, but at an increased cost.

7.5.8 Major Advantages

The major advantage of horizontal auger boring is that the casing is installed as the borehole excavation takes place. Hence, there is no uncased borehole that substantially reduces the probability of a cave-in, which in turn could cause surface subsidence. Also this method can be used in a wide variety of soil types—making it a very versatile method.

7.5.9 Major Limitations

The horizontal auger boring method requires different size cutting heads and augers for each casing, which entails substantial investment in equipment. This method requires construction of a bore pit and an accurate initial setup. The horizontal auger boring operation may not be successful in running sands and unstable soils and may require dewatering under water table. The accuracy in line and grade is limited unless modified methods are used.

7.6 GUIDANCE SYSTEMS

7.6.1 Waterline System (Grade Only)

The water level is a device to measure the grade of pipe casing as it is being installed. It permits the monitoring of grade by using a water level sensing head attached to the top of the leading edge of the casing. The level operates in the same way as the sight tube on a boiler. Both ends of the system are vented to ambient pressure. A pit mounted control and indicator board is located at some convenient point in the pit near the operator. A hose connects the bottom of the indicator tube to a water pipe running along the top of the casing. Water is used to fill the system. The level of water in the pit indicator will then show the level of the valve at the end of the casing as it is pushed into the ground. One should be careful when using this system to ensure that the system is full so that an incorrect reading is not obtained.

7.6.2 Mechanical Line and Grade Control

Controlling both line and grade is achieved by using biaxial hinges. In the mechanical line and grade control head, a water level is used to monitor the grade while the alignment is monitored using a sonde or transmitter. There are two sets of encased steering rods that are used for making the corrections for line and grade. Miniature gearboxes are used to reduce the effort to turn the rods for line and grade control on long bores. The leading edge of the casing should be properly prepared and care should be taken to minimize torque and thrust and to keep the casing along the design alignment. To control the line, a proposed alignment is marked on the surface above the proposed bore. As the bore progresses, the offset is measured from the proposed line and the corrections made to compensate for it. The limitation of this method is that the surface above the bore must be accessible to take readings for controlling the line.

7.6.3 Electrical Line and Grade Control

Computer aided technology permits the steering head to be controlled hydraulically whereas using sensitive gyroscopes monitors the location of the steering head. The gyroscopes measure the deviation of the pipe installation from planned line and grade and display the information on the computer screen. Based on this information, the line and grade can be corrected manually by using a joystick or the corrections can be automatically made by the computer. The advantage of this method is that as this is an inertial system that uses gyroscopes to measure the deviation from planned line and grade, the readings are not affected by any other magnetic fields. This permits installations to a very precise line and grade even under buildup areas or areas that are affected by magnetic influences. Currently, the size of the tracking devices limits the minimum diameter of installation to 36 in, whereas theoretically there is no upper limit to it.

7.6.4 Walkover Systems

The system comprises a walkover receiver, a cable-ready remote display with power supply, and a sensitive pitch cable transmitter. This particular transmitter is used exclusively for installing gravity flow pipes, where precise grade control is required. The transmitter will read both positive and negative grades with equal accuracy.

The transmitter is connected with a wire the length of the bore to a display located at the operator's station. The wire powers the transmitter and carries the pitch or roll data back to the display. The steel of the casing acts as the ground connection. Because the transmitter is mounted on the lead casing, the display provides the operator with its real-time pitch and roll position. This instantaneous information allows the operator to react to any grade deviations as well as identify if the forward shield has begun to roll in either direction.

The auger's heading can be determined by using the locate points. The front locate point is found approximately 70 percent of the depth out in front of the transmitter above the ground surface. The transmitter can be located using the walkover receiver at depths up to 150 ft.

7.7 *RECENT INNOVATIONS*

The more important innovations in horizontal auger boring technology in recent years have been in the following areas:

7.7.1 Guided Boring Method (GBM)

The guided boring method (GBM), also known as pilot tube method is used in conjunction with horizontal auger boring machine to install small-diameter pipes with greater grade and alignment precision. The GBM comprises a specialty designed theodolite guidance system to guide the installation of pipes. Accurate pipeline installation is achieved through video monitor surveillance of an illuminated target via theodolite. Pilot head steering is accomplished by aligning an angled pilot head to the desired course and thrusting forward. Pilot tubes are installed behind the steering head and rotated while thrusting forward. After the steering head has reached the reception shaft, a reaming head and auger tubes with flighting are installed behind the pilot tubes. With the addition of each section of the auger tube in the launch shaft, a section of pilot tube is removed in the reception shaft. The process is repeated until all pilot sections have been removed. A pipe adapter is then installed on the last section of auger casing and subsequent pipes thrust into place while the auger tubes are removed from the reception shaft. The success of this method depends on the ground condition. This method is not recommended for soils with boulders as they could deflect the pilot tubes.

7.7.2 Controlled Boring System (CBS): Steerable Line and Grade System

The controlled boring system (CBS) is a steering system to control line and grade for steel casing bores up to 48 in. in diameter. CBS uses a steering system that uses a walkover transmitter locating system that monitors line and grade throughout the length of the bore. The boring machine operator can view any deflections throughout the bore path and make proper steering adjustments with an electric over hydraulic push button controlled steering system with digital readout. The CBS can be used in conjunction with most boring machines.

7.7.3 Steel Pipe Interlocking Joining System

The interlocking steel pipe joining system developed by Permalok Corporation, St. Louis, Missouri, provides a preinstalled precision joint connection that provides quick, easy, and permanent joints. This joining system practically eliminates the need for in-field welding and the down time associated with it, thereby increasing the productivity.

7.7.4 Laser Guided Tunnel Attachment

A laser guidance system is used with the tunnel attachment to enhance the line and grade capability of auger boring machines and to accommodate larger-diameter pipes and greater lengths.

7.7.5 Mechanical Line and Grade Control Head

The mechanical line and grade control head is used for making minor corrections both on line and grade. See Sec. 7.6.2.

7.7.6 Electronic Line and Grade Control Head

The electronic line and grade control head is a technology that permits installation of pipes with increased accuracy by horizontal auger boring method under appropriate site conditions. See Sec. 7.6.3.

REVIEW QUESTIONS

1. What are the main applications of HEB method? What type of pipe is used in this method?
2. Compare capabilities, limitations, and applications of track-type and cradle-type horizontal auger boring.
3. Describe the main features of HAB.
4. Name the geotechnical investigation requirements for HAB.
5. Name and describe the main features of HAB.
6. What is guided boring method? What is its advantage over conventional horizontal auger boring method?

7. Describe the mechanical line and grade control head
8. Describe the electronic line and grade control head.
9. How a carrier pipe is installed using blocking?
10. What types of soil conditions are more appropriate for the HAB method?

REFERENCES

Abraham, D., and S. Gokhale (2002). *Development of a Decision Support System for Selection of Trenchless Technologies to Minimize Impact of Utility Construction on Roadways*, FHWA/IND/JTRP-2002/7, SPR-2453, National Technical Information Service, Springfield, Va.

ASCE (2004). Manual of Practice for Horizontal Auger Boring Projects, *ASCE Manuals and Reports on Engineering Practice No. 106*, American Society of Civil Engineers, Reston, Va.

Iseley, D. T., and R. Tanwani (1993). *Trenchless Excavation of Construction Equipment and Methods Manual*, 2nd ed., National Utility Contractors Association (NUCA), Arlington, Va.

Iseley, D. T., M. Najafi, and R. Tanwani (1999). *Trenchless Construction Methods and Soil Compatibility Manual*, National Utility Contractors Association (NUCA), Arlington, Va.

Iseley, T., and S. Gokhale (1997). Trenchless installation of conduits beneath roadways, *Synthesis of Highway Practice 242*, Transportation Research Board, National Academy Press, Washington, DC.

CHAPTER 8
PIPE RAMMING

8.1 INTRODUCTION

Pipe ramming involves the use of the dynamic force and energy transmitted by a percussion hammer attached to the end of a pipe. The basic procedure generally comprises ramming a steel pipe through the soil by using an air compressor. Pipe ramming permits the installation of large steel casings in a wide range of soil conditions. It provides continuous casing support during the drive with no overexcavation, and it does not require the jetting action of water or drilling fluids.

Pipe ramming is most valuable for installing larger pipes over shorter distances and for installations at shallower depths. It is suitable for all ground conditions except solid rock, and is often safe where some other trenchless methods can lead to unacceptable surface settling (e.g., open-face augering in loose soils). A further application of pipe ramming is the installation of steel pipes to form a roof support for tunnel construction beneath an existing infrastructure such as railroad tracks. Pipe ramming is typically used for horizontal installations, but can also be applied for vertical projects, such as piling driving or micropiling. An example of vertical application is an installation of vertical supporting piles from a bridge through a body of water, when the bridge cannot support the weight of a crane necessary in a traditional method of installation of such piles (Atalah et al., 1998).

8.2 METHOD DESCRIPTION

The two major categories of pipe ramming are closed-face and open-face (Fig. 8.1). With the closed-face pipe ramming technique, a cone-shaped head is welded to the leading end of the first segment of the pipe to be rammed. This head penetrates and compresses the surrounding soil as the casing is rammed forward. The soil-pipe installation interaction that results when this method is used is similar to the interaction that takes place when soil compaction methods are used. The wedge or cone-shaped end can be used for pipe diameters up to 8 in.

With the open-face pipe ramming technique, the front of the leading end of the steel casing or conduit remains open so that a borehole of the same size as the casing (i.e., a cookie-cutter effect) can be cut. This allows most of the in-line soil particles to remain in place, with only a small amount of soil compaction occurring during the ramming process. This technique is employed for pipes larger than 8 in. Figure 8.2 shows the open-face pipe ramming process, and Fig. 8.3 shows a closed end on an 8-in casing. Figure 8.4 shows a rammer used for pipe ramming projects.

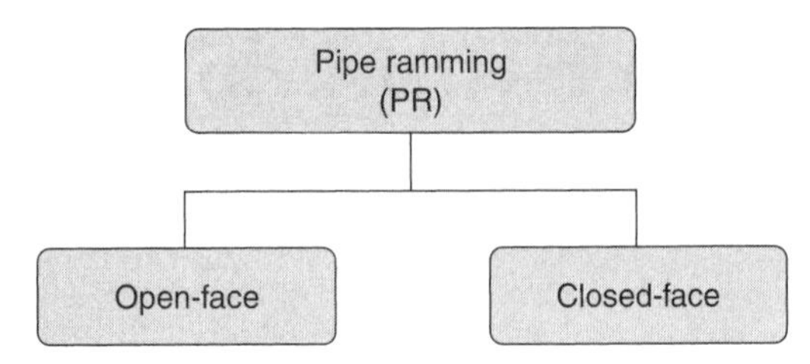

FIGURE 8.1 Pipe ramming categories.

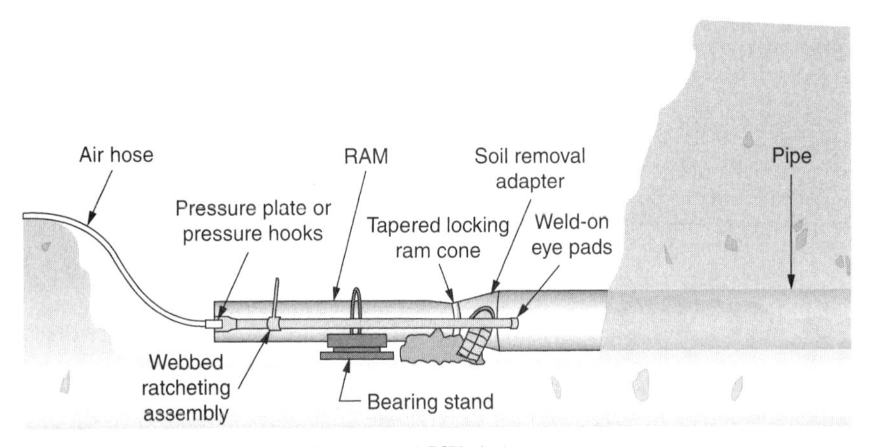

FIGURE 8.2 Open-face pipe ramming process (ACCU-pipe).

FIGURE 8.3 Closed-end pipe with multiple reducers (Abraham and Gokhale, 2002).

FIGURE 8.4 Rammer for pipe ramming operation (Abraham and Gokhale, 2002).

To facilitate the pipe ramming process, the leading edge of the first casing is usually reinforced by welding a steel band 12 to 24 in wide around the exterior surface of the pipe. The banding provides two advantages: (1) it reinforces the leading edge and (2) it decreases the friction around the casing. A band can also be installed on the inside edge of the leading section of the pipe. This band also reinforces the leading edge of the pipe and creates a clearance for the soil to move inside the casing. This clearance will help during the cleanout process as well as reduce the friction that exists inside the casing.

After the casing installation process is complete, the soil that has entered the casing is removed by applying compressed air or water from either end for small-diameter casings. For large casings, augers can be used to mechanically remove the soil from the inside of the pipe. Steel casings and augers used for pipe ramming projects are shown in Fig. 8.5.

For large-diameter pipes with long lengths or in certain soil conditions (such as stiff clays or sands), a steel pipe is installed on the top of the pipe being installed at a point approximately 24 in from the front of the casing. This line is used to supply water, bentonite, or other drilling lubricants inside, or outside, or both of the casings to facilitate spoil removal, reduce friction, and maintain the integrity of the hole being cut.

The pipe ramming procedure is as follows:

1. Construct an adequate shaft.
2. Install a cone or band on the leading edge of the casing.
3. Place casing in the drive shaft and adjust for the desired line and grade. When the line and grade are not critical, the pipe can be supported by construction equipment such as backhoes, cranes, side-boom tractors, by wood or block supports, or it can be supported directly on the pit floor. When the casing pipe is unguided, the line and grade accuracy is determined by the initial setup as well as by the ground conditions

FIGURE 8.5 Steel casings and augers for pipe ramming projects (Abraham and Gokhale, 2002).

FIGURE 8.6 Rammer and casing supported by a backhoe (Abraham and Gokhale, 2002).

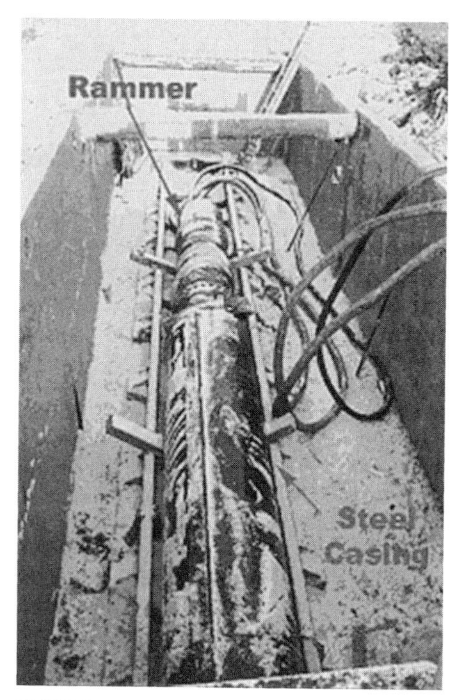

FIGURE 8.7 Rammer connected to the casing (Abraham and Gokhale, 2002).

encountered. Thus, a properly constructed driving shaft is a major component of a successful pipe ramming project. In cases where the line and grade are critical, the pipe is supported by adjustable bearing stands, launch cradles or platforms, I-beams, and auger boring machine tracks. Figure 8.6 shows that backhoe is supporting the rammer and casing.

4. Attach hammer device and connect to pneumatic or hydraulic power source. This is accomplished by using special adapters for each size of pipe. After the adapters are in place, the tool is connected to the pipe with lugs welded to the pipe. These lugs are used to hold the straps, chains, or hoists linked to the rammer tool. A rammer connected to leading casing is shown in Fig. 8.7.
5. Initiate the drive and continue until installation is complete. (If multiple pipe segments are being used, after each segment is installed, remove the hammer, weld another pipe segment to the end of the previous casing, and repeat the cycle until the installation is complete.)
6. Remove the cone, if used, or clean out the casing as required.
7. Remove the equipment.
8. Restore the area as required.

8.3 MAIN FEATURES AND APPLICATION RANGE

8.3.1 Diameter Range

Typical diameters of pipe installed by pipe ramming are 4 to 60 in for open-face pipe ramming and 4 to 8 in for closed-face pipe ramming. However, casings of 120 in and larger have been installed in recent years.

8.3.2 Drive Length

Typical drive lengths are more than 200 ft, although crossings have been up to 400 ft. However, site and project conditions such as diameter of pipe and most importantly soil conditions must be considered to determine the drive length and to provide an accurate bore.

8.3.3 Type of Casing

The type of casing and conduit is limited to steel pipe. The pipe must be able to endure the repeated impact loads of the percussive hammer. Therefore, the pipe's wall thickness is a very important design consideration.

8.3.4 Required Working Space

Adequate site access and working space are essential for a successful installation. The location of insertion and receiving pit is determined by the job requirements, right-of-way access, and regulations of the authorities with regulatory authority over the site. The length of insertion pit can be calculated from the length of the pipe sections to be installed, with adding approximately 10 ft to pipe section length. (The ramming tools are usually between 6 and 10 ft long). For example, a 100 ft drive could be accomplished in a single drive as long as the casing pipe is welded together before the drive begins, or it could be accomplished using five 20-ft pipe segments. If the surface area is not confined, the insertion pit may be considerably longer thus allowing the pipe or casing to be installed in fewer sections. The typical width of the insertion pit is up to 10 ft, with about 3 ft of clearance on each side of the tool. The pit depth should be at least 4 ft, and the minimum clearance above the ramming tool should be between 5 and 7 ft. The required working space at the drive shaft is typically 8 to 12 ft in width and 30 to 60 ft in length.

8.3.5 Soil Conditions

A significant feature of the pipe ramming technique is its versatility. It is suitable for a wide range of soil conditions, from stable to unstable, with or without the presence of high groundwater. Although pipe ramming can be applied in a wide variety of soils, some soils are better suited for this method than others. The most suitable soil conditions for pipe ramming are soft to very soft clays, silts, and organic deposits, all sands (very loose to dense) above the water table, and soils with cobbles, boulders, and other obstacles of significant size but smaller than the pipe diameter (soils with cobbles can be in extremely wet conditions, even with running water). Pipe ramming is more difficult in medium to dense sands below the water table, medium to very stiff clays, hard clays, highly weathered shale, soft

TABLE 8.1 Applicability of Different Soil Conditions for Pipe Ramming

Type of soil	Applicability
Soft to very soft clays, silt, and organic deposits	Yes
Medium to very stiff clays and silts	Yes
Hard clays and highly weathered shales	Marginal
Very loose to loose sands (above water table)	Yes
Medium to dense sands (below the water table)	No
Medium to dense sands (above the water table)	Yes
Gravels and cobbles less than 2 to 4 in diameter	Yes
Soils with significant cobbles, boulders, and obstructions larger than 2 to 4 in diameter	Yes
Weathered rocks, marls, chalks, and firmly cemented soils	Marginal
Significantly weathered to unweathered rocks	Marginal

or highly fractured rocks, marls, chalks, and firmly cemented soils. The only soil condition that pipe ramming is completely unsuitable for is solid rock. However, in rocky ground conditions, a pneumatic tool can be used to break up the smaller rocks and boulders or force them out of the path either to the outside or inside of the casing. Table 8.1 presents applicability of pipe ramming method for different soil conditions.

8.3.6 Productivity

Usually a 2- to 3-person crew is all that is needed for small applications. Under suitable soil conditions, the typical rate of penetration ranges from 2 to 10 in/min.

8.3.7 Accuracy

The accuracy of pipe ramming method depends on the initial setup. Once the ramming has begun, there is a limited amount of control in changing the direction of the bore. Occasionally, a wedge or shoe can be placed in larger-diameter pipes at the leading edge and at the required location to help redirect the bore. This shoe is generally made of metal or wood. Also, grade control of the pipe can be aided by removing a portion of the soil during the ramming process. The removal of this soil reduces the weight of the casing, reduces friction, and can help redirect the bore.

8.3.8 Major Advantages

The pipe ramming method is an effective method for installing medium-to-large diameter pipes. The versatile pit sizes, maximum drive length, and ability to handle different soil conditions makes this method a practical and economical technique for installing steel pipes. This method does not require any thrust reaction structure as the ramming action is because of impulses induced in the pipe by the percussion tool. The pipe ramming method is also multifunctional. A single size of pipe ramming tool and the air compressor can be used to install a wide variety of pipe lengths and sizes. Ramming can also be used for vertical pile driving, angular ramming, or pipe replacement.

8.3.9 Major Limitations

The major disadvantage of the pipe ramming method is the minimal amount of control over line and grade. Therefore, the initial setup is of major importance. Also, in the case of obstructions, like boulders or cobbles, especially for pipes with small diameter, the pipe may be deflected. Therefore, sufficient information on the existing soil conditions must be available to determine the proper size of casing to be used. Other drawbacks include high noise levels that are typical for pipe ramming (if no noise suppression is used) and sometimes a significant soil disturbance that can happen if a blockage is created at the end of the installed pipe.

8.4 EFFECTS OF PIPE RAMMING ON SURROUNDING ENVIRONMENT

Under each dynamic application of the force by the pipe ramming equipment, the pipe vibrates and the generated vibrations are transferred from the pipe to the soil particles. Ground vibrations associated with pipe ramming have not been studied so far but are expected to be similar to those in the pipe bursting operation, because both methods use similar equipment, operating generally at 200 to 500 blows/min. In pipe bursting, ground vibrations are rapidly attenuated with the distance from the source and are not likely to be damaging to the nearby underground objects, except at very close distances from the origin of vibrations, that is, two to three pipe diameters of a pipe being burst for buried pipes, and approximately, eight pipe diameters for surface structures. Thus, ground vibrations from pipe ramming are not expected to damage nearby objects at similar distances. It should be noted; however, that increase in diameter and power, or the presence of rock may cause high levels of vibration at much greater distances.

Surface disruption associated with pipe ramming happens rarely because a solid, steel pipe is in the ground all the time and the soil within the pipe is not removed until later in the process. As a result, the danger of creation of voids during construction and postproject settlements is greatly diminished. However, pavement sags or humps occur occasionally on the surface above an installation. The type and extent of surface disruption depend on the soil conditions, type of pipe ramming (open-end pipe vs. closed-end pipe), and the depth of installation.

REVIEW QUESTIONS

1. Describe the pipe ramming process.
2. Describe and compare the two main categories of the pipe ramming method.
3. Outline the basic installation procedure for pipe ramming.
4. What soil conditions are most applicable to pipe ramming method?
5. Discuss the main benefits and limitations of pipe ramming technique.
6. Describe the range of applications of the pipe ramming technique with respect to pipe diameter, drive length, and workspace requirements.
7. Discuss the effects of pipe ramming on the surrounding environment.
8. In what conditions should an engineer select the pipe ramming method? Explain.

REFERENCES

Abraham, D., and S. Gokhale, (2002). *Development of a Decision Support System for Selection of Trenchless Technologies to Minimize Impact of Utility Construction on Roadways*. FHWA/IND/JTRP-2002/7, SPR-2453, National Technical Information Service, Springfield, Va.

ACCU Pipe Ramming Systems, Alberta, Canada. Available at http://www.accupipe.com.

Atalah, A., R. S. Sterling, P. Hadala, and F. Akl, (1998). *The Effect of Pipe Bursting on Nearby Utilities, Pavement and Structures*. TTC Report No. 98-01, Trenchless Technology Center, Louisiana Tech University, Ruston, La.

Iseley, D. T., M. Najafi, and R. Tanwani, (1999). *Trenchless Construction Methods and Soil Compatibility Manual*, 3rd ed., National Utility Contractors Association, Arlington, Va.

Iseley, T., and S. Gokhale, (1997). *Trenchless Installation of Conduits beneath Roadways*. Synthesis of Highway Practice 242, Transportation Research Board, National Academy Press, Washington, DC.

CHAPTER 9

PIPE JACKING AND UTILITY TUNNELING

9.1 INTRODUCTION TO PIPE JACKING

The term *pipe jacking* can be used to describe a specific installation technique as well as a process applicable to other trenchless technology methods. When referred to as a process, it implies a tunneling operation with use of thrust boring and pushing pipes with hydraulic jacking force. This concept of *jacking system* is adopted by many trenchless technologies including auger boring and microtunneling. However, in this chapter the term pipe jacking is regarded as an installation technique.

Pipe jacking is a trenchless technology method for installing a prefabricated pipe through the ground from a drive shaft to a reception shaft. The first use of pipe jacking was at the end of the 19th century. In the 1950s and 1960s, new capabilities were added to pipe jacking by the European and the Japanese companies, including extended drive length, upgraded line and grade accuracy, enhanced joint mechanism, new pipe materials, and improved excavation and face-stabilizing shields. These developments as well as the improved operator skills and experience have enabled pipe jacking to be a popular trenchless technology.

In the pipe jacking operation, jacks located in the drive shaft propel the pipe. The jacking force is transmitted through the pipe-to-pipe interaction, to the excavating face. When the excavation is accomplished, the spoil is transported through the jacking pipe to the drive shaft by manual or mechanical means. Both excavation and spoil removal processes require workers to be inside the pipe during the jacking operation. This is essentially what separates pipe jacking from microtunneling. Although it is theoretically possible for a person to enter a 36-in diameter pipe, from a practical standpoint it is very difficult for the person to work in it. Therefore, the minimum recommended diameter for pipe installed by pipe jacking is 42 in. However, it is feasible to install reinforced concrete (RC), centrifugally cast glass-fiber-reinforced polymer mortar (CCFRPM), and polymer concrete pipe (PCP) with 36 in inside diameter (ID) and 42 in outside diameter (OD).

9.2 METHOD DESCRIPTION

Figure 9.1 illustrates the typical components of a pipe jacking operation. The cyclic procedure uses the thrust power of the hydraulic jacks to force the pipe forward through the ground as the pipe jacking face is excavated. The spoil is transported through the inside of the pipe to the drive shaft, where it is removed and disposed off. After each pipe segment

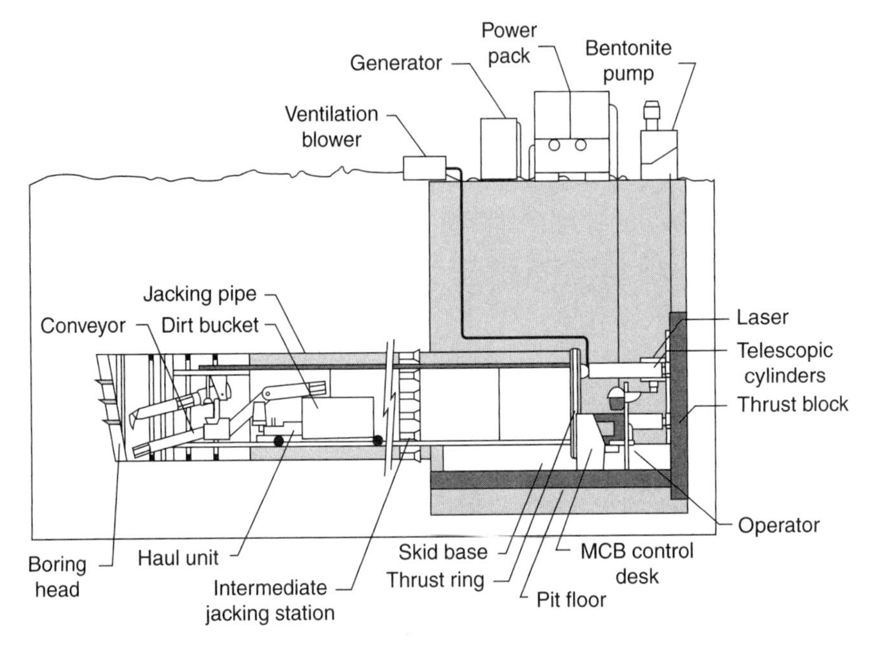

FIGURE 9.1 Typical components of a pipe jacking operation (Iseley and Gokhale, 1997).

has been installed, the rams of the jacks are retracted so that another pipe segment can be placed in position for the jacking cycle to begin again.

Excavation is accomplished by hand mining or mechanical excavation within a shield or by a tunnel boring machine (TBM) as shown in Fig. 9.2. The excavation method selection is based on an assessment of the subsurface for instability. If there is any possibility of excavation face collapse, soil stabilization techniques must be considered. The common soil stabilization techniques are dewatering and grouting. Alternatively, closed-face earth pressure balance or slurry microtunneling methods may be appropriate (Fig. 9.2).

Because of the large jacking forces required to push large-diameter pipe through the ground, the design and construction of the jacking shaft are critical to the success of the project. The shaft floor and thrust reaction structure must be designed to withstand the weight of the heavy pipe segments being placed on them repeatedly.

Important optional equipment available for the pipe jacking method includes a pipe lubrication system and intermediate jacking stations (IJSs). The pipe lubrication system comprises mixing and pumping equipment necessary for applying bentonite or polymer slurry to the external surface of the pipe. An adequate lubrication system can decrease jacking forces by 20 to 50 percent; however, the most common reduction factor range would probably be 20 to 30 percent (Terzaghi, 1950). IJSs are used for pipes, 36 in. in diameter or larger, between the drive shaft jacking plate and the jacking shield or TBM to redistribute the total required jacking force on the pipe. IJSs comprise a steel cylinder installed between two pipe segments in the pipeline being jacked. Hydraulic jacks are then placed around the internal periphery of the steel cylinder. The IJS is pushed forward through the ground with the pipeline until its operation is necessary. When the main jacks reach

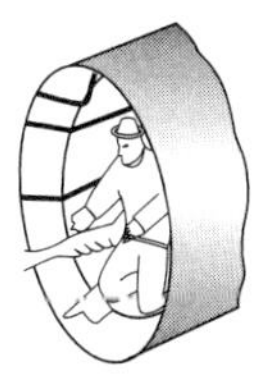

Hand shield: An open-face shield for manual excavation

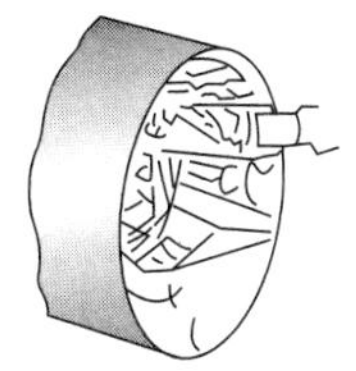

Backacter: An open-face shield with a mechanical backacter

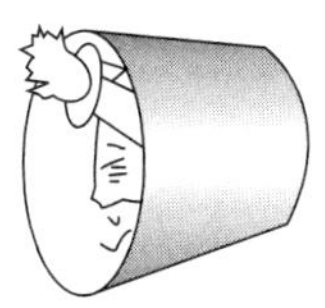

Cutter boom: An open-face shield with a cutter boom or road header

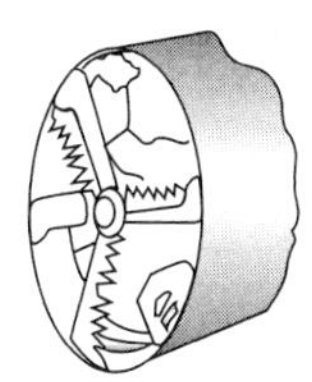

Tunnel boring machine (TBM): A shield with a rotating cutting head

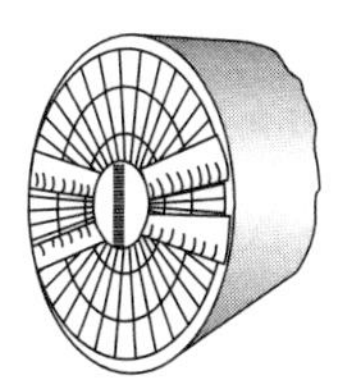

Earth pressure balance machine (EPBM): A full-face tunnel boring machine with a balanced screw auger to control the face pressure

(a)

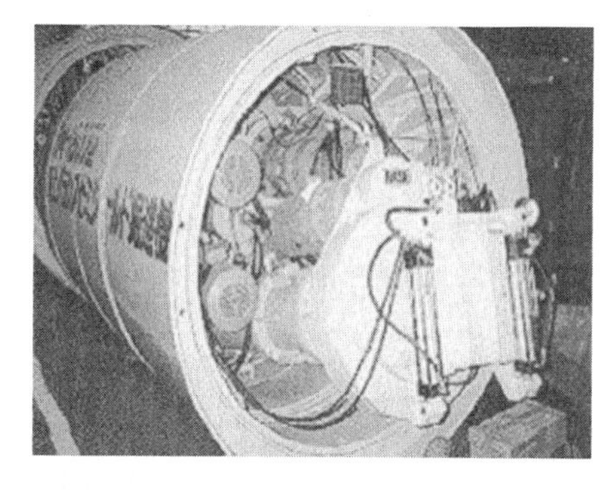

(b)

FIGURE 9.2 Tunnel boring machines and excavation techniques (Iseley and Gokhale, 1997); (a) Excavation techniques; (b) EPBM (earth pressure balance machine).

FIGURE 9.3 Intermediate jacking station (Abraham and Gokhale, 2002).

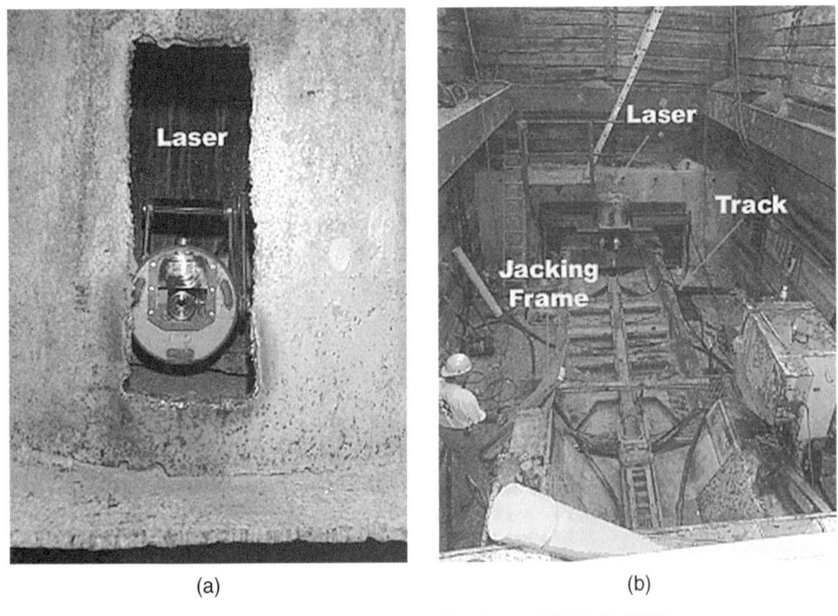

FIGURE 9.4 Laser guidance system for pipe jacking (Abraham and Gokhale, 2002).

FIGURE 9.5 Laser point for alignment control (Abraham and Gokhale, 2002).

approximately 80 percent of the design load, the jacking force on the pipe behind the IJS is held constant, and the jacks in the IJS are activated to propel the forward section of the pipeline. Figure 9.3 shows the IJS.

The basic pipe jacking procedure can be described as follows (Iseley and Gokhale, 1997):

1. Excavate and prepare the driving shaft.
2. Setup the jacking frame and the hydraulic jacks to adjust to the proposed design line and grade.
3. Install laser guidance system in the driving shaft as shown in Fig. 9.4. During the drilling operation, the operator who stays inside the boring machine continuously checks the mark on the steering head and the laser point. If the operator detects a deviation, the operator will articulate the steering head back to the correct alignment. The laser point for the alignment is shown in Fig. 9.5.
4. Lower the boring machine into the driving shaft and set it up. Figure 9.6 shows the different views inside and outside the pipe jacking boring machine. Figure 9.6 (f) shows pipe jacking in progress inside the launch shaft, while Fig. 9.6 (g) shows the arrival of the tunneling machine at the reception shaft.
5. Mate jacking push plate (thrust ring) to shield or TBM. The thrust ring is the frame that the main cylinders push against to advance the boring head and pipe. The ring provides a 360-degree surface against the pipe to minimize the point pressure and reduce the chance of breakage.
6. Advance shield or TBM through the prepared opening in the forward shaft support structure. Begin the excavation and spoil removal process. Continue excavation, spoil removal, and forward advancement until the shield or the TBM is installed. The

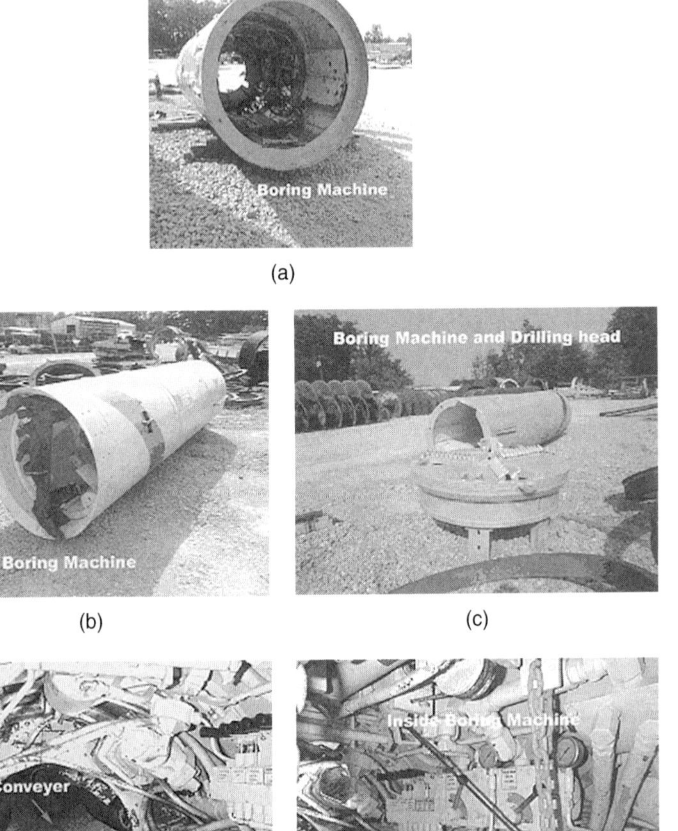

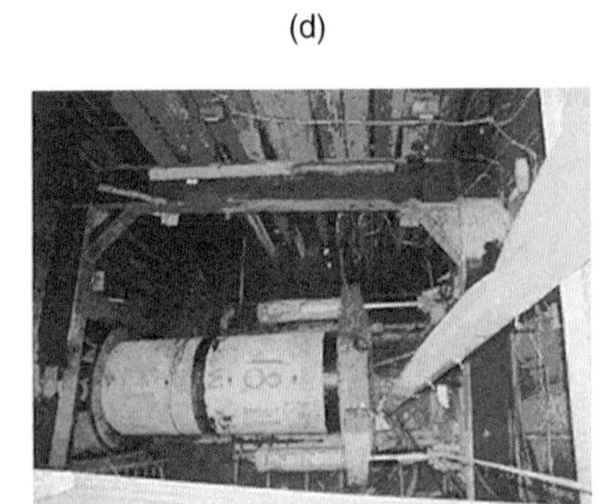

(f) (g)

FIGURE 9.6 Pipe jacking tunnel boring machine (Abraham and Gokhale, 2002).

FIGURE 9.7 Control panel for the jacking machine (Abraham and Gokhale, 2002).

control panel outside the boring machine controls the movement of the jacking machine, whereas the control levers inside the boring machine control the tunneling operation. The control panel for the jacking machine is shown in Fig. 9.7.

7. Retract jacks and push plate to provide a space for the pipe segment.
8. Place the first pipe segment on the jacking tracks.
9. Mate the push plate to the pipe and the pipe to the shield or TBM. Figure 9.8 shows the TBM, IJS, casing, and jacking machine.
10. Initiate forward advancement, excavation, and spoil removal. There are five main approaches for removal of the excavated soil from the excavation face to the drive shaft for further disposal. These soil conveyance systems include: (i) wheeled carts or skips, (ii) belt or chained conveyors, (iii) slurry system, (iv) auger system, and (v) vacuum extraction system (Fig. 9.6).
11. Repeat pipe jacking cycles until the complete line is installed.
12. Remove the shield or the TBM from the reception shaft.
13. Remove the jacking equipment, IJS, and the tracks from the drive shaft.
14. Restore the site as required.

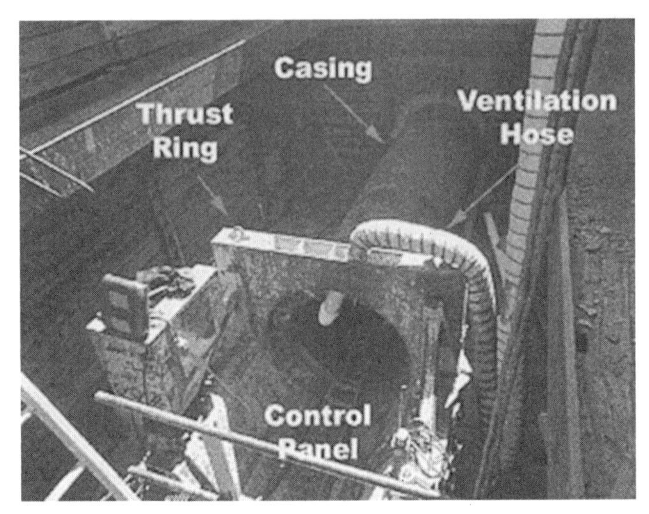

FIGURE 9.8 Setup for pipe jacking operation (Abraham and Gokhale, 2002).

9.3 MAIN FEATURES AND APPLICATION RANGE

9.3.1 Diameter Range

As the method requires people working inside the jacking pipe, it is limited to person-entry pipe sizes. This has been accomplished in pipes as small as 30 in. However, the minimum recommended diameter for a pipe installed by the pipe jacking is 42 in OD or 36 in ID. If frictional forces can be overcome, theoretically, there is no limit to the size of the pipe that can be jacked; however, usually the largest is approximately 12 ft in diameter, with the most common sizes ranging from 48 to 72 in.

9.3.2 Drive Length

The length of a pipe jacking drive is determined by the amount of available jacking thrust and the compressive strength of the pipe. The jacking thrust can be minimized or managed by providing an adequate overcut, applying adequate lubrication between the outside surface of the pipe and the bore hole, maintaining accurate line and grade control, using high-quality pipe products, and using IJSs. The longest pipe jacking project in the United States had a continuous jacking length from drive shaft to reception shaft of approximately 3500 ft. The most common drive lengths range from 500 ft to more than1000 ft.

9.3.3 Type of Pipe

The type of pipe used for the pipe jacking method must be capable of transmitting the required jacking forces from the thrust plate in the jacking shaft to the jacking shield or TBM. Steel pipe, reinforced concrete (RCP), CCFRPM, and PCP with 36 in ID and 42 in OD are the most common types of pipe used in pipe jacking. A cushioning material should be used between the pipe segments to assist in distributing the jacking loads evenly over the cross section of the pipe. The most common type of material used as a cushion material is particleboard.

9.3.4 Required Working Space

The site must provide space for storage and handling of pipe and spoil and adequate space for the shaft. The size of the jacking shaft is determined by the pipe diameter, pipe segment length, jacking shield dimensions, jacking system dimensions, thrust wall design, pressure rings, and guide rail system. For example, the drive shaft size for a pipe jacking project using pipe 60 in. in diameter with segments 10 ft in length would require a 12 ft wide and 25 to 32 ft long shaft, depending on selection of the jacking and the excavation equipment.

9.3.5 Soil Condition

Cohesive soils provide the most favorable soil conditions for pipe jacking. It is possible to use pipe jacking in unstable soil conditions as long as special precautions are taken, such as dewatering and using closed-face machines and earth pressure balance machines to counterbalance the ground pressure (Fig. 9.2). Table 9.1 presents applicability of pipe jacking methods for different soil conditions.

9.3.6 Productivity

A reasonable productivity range for pipe jacking projects is 33 to 60 ft per 8-h shift with a four- or five-person crew. Factors that can affect productivity include the presence of groundwater, unanticipated obstructions such as boulders or other utilities, and changed conditions such as encountering wet silty sand after selecting equipment for stable sandy clay.

TABLE 9.1 Applicability of Pipe Jacking Method for Different Soil Conditions

Type of soil	Applicability
Soft to very soft clays, silt, and organic deposits	Marginal
Medium to very stiff clays and silts	Yes
Hard clays and highly weathered shales	Yes
Very loose to loose sands (above water table)	Marginal
Medium to dense sands (below the water table)	No
Medium to dense sands (above the water table)	Yes
Gravels and cobbles less than 2 to 4 in diameter	Yes
Soils with significant cobbles, boulders, and obstructions larger than 4 to 6 in diameter	Marginal
Weathered rocks, marls, chalks, and firmly cemented soils	Marginal
Significantly weathered to unweathered rocks	No

9.3.7 Accuracy

Pipe jacking is capable of installations to a high degree of accuracy. As laser is used for controlling the line and grade, installation to an accuracy of within an inch are common. A reasonable anticipated tolerance is ± 3 in for alignment and ± 2 in for grade.

9.3.8 Major Advantages

Pipe jacking can be accomplished through almost all types of soils. A high degree of accuracy can be obtained. Being located at the excavation face, the operator can see what is taking place and take immediate corrective action in case of changing subsurface conditions. The face can be readily inspected manually or by using a video camera. When unforeseen obstacles are encountered, they can be identified and removed by pulling back the cutter head.

9.3.9 Major Limitations

Pipe jacking is a specialized operation and like many other trenchless technologies requires a lot of planning and coordination. Although these operations can be conducted on a radius, it is recommended that all directional changes be made at the shafts. The pipe and liners used for the operation should be strong enough to resist the jacking forces. Therefore, not all types of pipes can be used for this operation.

9.4 INTRODUCTION TO UTILITY TUNNELING

UT is differentiated from general tunneling by virtue of the tunnel's size and application. Utility tunnels are primarily conduits for pipelines and utilities rather than passage for pedestrians or vehicular traffic. Although methods of soil excavation for utility tunneling and pipe jacking remain similar, the difference is in the tunnel lining. In the pipe jacking technique, the pipe is the lining, whereas in utility tunneling special liner plates or rib and lagging systems are used to provide temporary ground support.

UT can therefore be defined as the general approach of constructing underground utility lines by removing the excavated soil from the front of cutting face and installing liner segments to form a continuous ground support structure. Currently utility tunneling methods require person-entry inside the lining during the tunneling process.

The big breakthrough for soft ground tunneling came in the 1970s and 1980s with the innovation of pressure balance shields and improved permanent linings. These and more recent developments have made tunneling faster, safer, and more economical.

9.5 METHOD DESCRIPTION

Figure 9.9 illustrates the typical components of a utility tunneling operation. A utility tunnel is normally constructed and lined under ground surface between two access shafts. Excluding the preparation work such as the construction of pits and field setup, a typical cycle of utility tunneling procedure comprises four major steps, namely, (i) soil excavation, (ii) soil removal, (iii) segmental liner installation, and (iv) direction steering and tunnel advancing.

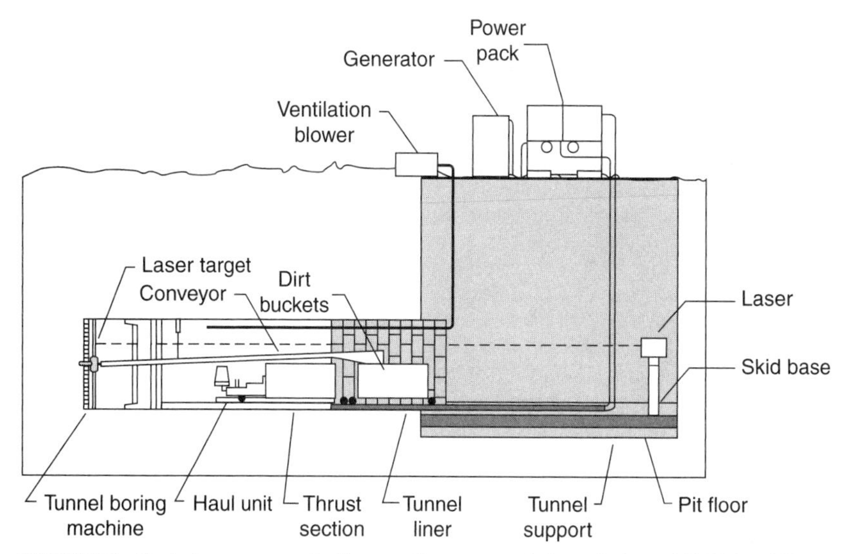

FIGURE 9.9 Typical components of utility tunneling system techniques (Iseley and Gokhale, 1997).

9.5.1 Soil Excavation

Depending on the soil characteristics, soil excavation can be accomplished by hand mining, open-faced mechanical excavation, closed-face TBMs, or by the new Australian tunneling method (NATM).

Hand mining is the simplest form of soil excavation, which is carried out by tunneling worker(s) at the front face using picks, shovels, or pneumatic hand-held tools. Under unstable soil conditions, a protective shield is normally required, which may have a forward projection to provide additional face stability during soil excavation. In an articulated shield, line and grade corrections can be accomplished by activating the hydraulic cylinders. In a fixed shield, minor line and grade changes are accomplished by differential excavation in the desired direction. When a high underground water table is present, compressed air can be applied to prevent or minimize ground water inflow. Generally, hand mining operation is slow and therefore limited to short drives, but it provides the advantage of simple operation and the capability to handle difficult and varying soil conditions. In addition, hand mining requires minimum work space and can be used to install linings as small as 30 in diameter.

Open-faced mechanical excavation is accomplished by using special shields equipped with powered excavation devices. Such soil cutting devices can be rotary cutter beams mounted on the front of the shield, a modified hydraulic backhoe, a rotary boom cutter, or any combination. In case of unstable soil conditions or high underground water table, compressed air is normally used, and with proper system design, the tunneling operator is not required to work inside the pressurized zone. Most open-faced shields still provide personnel access to the front face if the need arises. This enables the on-line adjustment of cutter head configurations to accommodate varying ground conditions as well as the manual handling of unexpected obstructions. The soil excavation rate is much improved compared to the case of hand mining.

Closed-face tunneling shields are normally referred to as TBMs, which are equipped with hydraulically or electrically driven rotary cutter heads or disc cutters. The cut soil is forced inside the shield through slits or other kinds of openings in the cutter head as the shield is advanced. Generally, closed-face TBMs provide much improved face stability during soil excavation, and are thus most suitable for noncohesive soils below water table. However, in practice, they have much wider applications. Some systems have crushing devices to cope with gravel and boulders. The major drawbacks of TBMs include relatively high equipment cost, limited face access, and the fact that it can only be used to install circular tunnels.

More sophisticated TBM systems incorporate a pressure chamber that provides a balance between the soil face pressure with the external water head and the mixed soil pressure inside the chamber. Such systems have been more commonly adopted in non-person-entry microtunneling methods. The working mechanism of such chamber is addressed in Chap. 13, "Microtunneling."

The NATM is a popular method used for tunneling in rocks. The basic principle of NATM is to allow the ground surrounding the tunnel to deform just enough to mobilize its shear strength. Frequently, this limited deformation is achieved by using a flexible tunnel liner constructed out of steel ribs and wire reinforced shotcrete. NATM technique essentially requires that the soil be strong enough to be self-supporting once the required deformation occurs. This method is not recommended for shallow covers because soil is unable to arch properly.

9.5.2 Spoil Removal

As discussed in Sec. 9.2, there are currently six main approaches for conveying excavated soil from the front cutting face back to the drive shaft for disposal. These spoil conveyance systems include: (i) wheeled carts or skips, (ii) belt and chain conveyors, (iii) positive displacement pumping device, (iv) slurry system, (v) auger system, and (vi) vacuum extraction system. The selection of appropriate soil removal system is dependent on the available space inside the tunnel, the manner of soil excavation, the mechanism of the face pressure balance, and the total tunnel length.

9.5.3 Steering Control and Tunneling (Shield) Advancement

Using theodolite and laser systems for direction measurement are the two most common approaches. Theodolite is the classic survey instrument that can monitor the current tunnel front face positions directly and also other 3-D position data of the tunneling shield, if used. The equipment cost is low; however, the surveying process requires a skilled operator, a light source and can not be used for continuous monitoring as it causes temporary interruption to the tunneling work.

A laser monitoring system allows any direction variation to be adjusted immediately. However, the laser beam is sensitive to temperature variation along the line and might become dispersed over a long distance of dusty air. Sometimes the combination of laser and theodolite methods will yield more satisfactory results. Other more sophisticated inertia surveying systems like gyroscope have been adopted for curved tunneling.

The steering control is accomplished during the soil excavation and the shield advancing process. A minor direction change is much easier when no protection shield is used under stable soil conditions. A tunneling shield is normally equipped with a few jacking cylinders at its rear portion, which propelled the shield forward by jacking against the already erected liner sections as the face excavation is proceeding. These jacking cylinders can apply different forces and extend at different speeds during one forward tunneling cycle to correct the direction of the shield. After the shield has been advanced a certain distance, these jacks are retracted to leave room at the rear of the shield for the in situ installation of new segmental liners.

9.5.4 Liner Installation

As long as the front face soil is excavated and removed, or the shield is advanced with the jacking cylinders retracted, new segmental liners are transported through the erected lining to the front face. These liners are erected in situ and connected to the existing lining; and so the tunnel length increases. If a tunneling shield is used, the retracted jacking cylinders are extended and contact against the front profile of the new lining. This completes a single forward tunneling cycle. Such procedure repeats until the tunnel has reached the target location. It is to be noted that in case of tunneling through the rock formation, there might be no lining needed.

The currently used liner materials are of two major types, namely prefabricated liner plate and rib and lagging systems with the latter more extensively used today as shown in Fig. 9.10. Liner plates are typically made of steel or precast reinforced concrete that can virtually be of any shape. The steel liner plates have flanged edges that allow the overlapping and bolting together of successive liner plates to form an integrated lining. Concrete plates are bolted together through precast holes. Steel plates are more widely used than concrete plates, for utility tunnel applications because of their higher strength-weight ratio. There are sometimes predrilled holes in the liner plates for grouting purposes.

(a) Wood lagging

(b) Tunnel liner plates

FIGURE 9.10 Utility tunnel linings. (*Courtesy of Midwest Mole, Inc.*)

9.6 MAIN FEATURES AND APPLICATION RANGE

In utility tunneling, the lining is formed by in situ installation of segmental liners, and the tunneling shield, if used, is the only portion that is jacked forward along the whole tunnel length. The required jacking force is relatively small and there is theoretically no upper limit for the length of a single-pass tunneling process. As personnel front face access is normally available during the tunneling process, these methods are suitable for various soil conditions and ground stability can be well-controlled. Besides, high accuracy in steering is achievable and curved tunnel alignment can also be accommodated. However, its main disadvantage is that the original tunnel lining can normally serve as temporary supporting structure only to allow final carrier pipelines to be transported and installed inside. The annular space between the liner and the pipe must be grouted.

9.6.1 Diameter Range

The tunnel liner materials are normally steel ribs with wooden lagging or steel plates; however, there is no restriction on the final carrier pipeline material. Although theoretically, it is possible for a person to enter a 36-in diameter pipe, from a practical stand point it is very difficult for the person to work in this environment for any extended period of time. Hence, the minimum tunnel diameter recommended is 42 in. For longer drives, the recommended diameter is 48 in. Currently, the upper limit for utility tunneling is considered to be 132 in.

9.6.2 Productivity and Special Concerns

Although remote control of many elements of the tunneling process has been realized in the recent years, the installation of segmental liners always requires manual operation at the front face of the tunnel. Generally speaking, utility tunneling is a relatively slow and labor-intensive process. The actual tunnel advance rate is a function of soil conditions encountered, method of soil excavation and removal, liner materials, as well as the field coordination and skill level of the tunneling personnel. The productivity will be further impacted if personnel operation is required in the uncomfortable compressed air zone most of the time. Also, as the tunneling operation needs high level of skill and coordination, a long learning curve is expected when difficult ground conditions are present.

9.6.3 Soil Conditions

Table 9.2 presents applicability of pipe jacking methods for different soil conditions.

9.6.4 Emerging Technologies

Utility tunneling techniques are well-established as the history of contemporary tunneling practice can date back to the 1800s, when the first tunneling shield was invented. New technologies available have enabled more efficient TBMs and powerful devices to facilitate the soil removal over a long distance. Gyroscope steering systems or similar technologies are under development to provide more dependable direction control for long distance and curved-alignment drives.

TABLE 9.2 Applicability of Utility Tunneling Method for Different Soil Conditions

Type of soil	Applicability
Soft to very soft clays, silt, and organic deposits	Yes
Medium to very stiff clays and silts	Yes
Hard clays and highly weathered shales	Yes
Very loose to loose sands (above water table)	Yes
Medium to dense sands (below the water table)	No
Medium to dense sands (above the water table)	Yes
Gravels and cobbles less than 2 to 4 in diameter	Yes
Soils with significant cobbles, boulders, and obstructions larger than 4 to 6 in diameter	Marginal
Weathered rocks, marls, chalks, and firmly cemented soils	Yes
Significantly weathered to unweathered rocks	Yes

REVIEW QUESTIONS

1. What is pipe jacking? What is the distinction between pipe jacking process and pipe jacking installation?
2. Describe the different lining methods used in utility tunneling.
3. Describe the utility tunneling process.
4. Describe the main features of pipe jacking and its possible range of applications.
5. List the major advantages and limitations of the pipe jacking installation technique.
6. How is it utility tunneling is differentiated from pipe jacking? Explain.
7. Describe the different techniques used for soil excavation and spoil removal for pipe jacking.
8. Describe the two common methods for steering control and tunneling (shield) advancement.
9. Describe how you would recommend pipe jacking or utility tunneling for different site and project conditions?
10. Outline the main features and the range of application for utility tunneling.

REFERENCES

Abraham, D., and S. Gokhale (2002). *Development of a Decision Support System for Selection of Trenchless Technologies to Minimize Impact of Utility Construction on Roadways*. FHWA/IND/JTRP-2002/7, SPR-2453, National Technical Information Service, Springfield, Va.

ASTM, *American Society for Testing and Materials* (ASTM International), West Conshohocken, Pa. Available at http://www.astm.org.

Bennett, R. D., L. K. Guice, S. Khan, and K. Staheli (1995). *Guidelines for Trenchless Technology: CIPP, FFP, Mini-HDD, and Microtunneling*. Construction Productivity Advancement Research (CPAR) Program Technical Report, U.S. Army Corps of Engineers. Washington, DC.

Iseley, T., and S. Gokhale (1997). *Trenchless Installation of Conduits beneath Roadways*. Synthesis of Highway Practice 242, Transportation Research Board, National Academy Press, Washington, DC.

Iseley, T., M. Najafi, and R. Tanwani (1999). *Trenchless Construction Methods and Soil Compatibility Manual*. Trenchless Technology Committee, National Utility Contractors Association (NUCA), Arlington, Va.

Kramer, S., W. McDonald, and H. Thomson (1992). *An Introduction to Trenchless Technology*, Van Nostrand Renhold, New York, NY.

Terzaghi, K. (1950). *Geologic Aspect of Soft Ground Tunneling, Applied Sedimentation*. P. Trask (Ed.). John Wiley and Sons, New York, NY.

CHAPTER 10
HORIZONTAL DIRECTIONAL DRILLING

10.1 INTRODUCTION AND BACKGROUND

Horizontal Directional Drilling (HDD) technology originated from the oil fields in the 1970s and was evolved by merging technologies used in utilities and water well industries. Since then, HDD has been widely used in the pipeline installation industries.

The first known river crossing using the HDD method took place in 1971. Approximately, 615 ft of 4-in diameter steel pipe was installed across the Pajaro River near Watsonville, California, for the Pacific Gas and Electric Co. (DCCA, 1994). By integrating the existing technology from the oil well drilling industry and modern surveying and steering techniques, today's directional drilling methods have become the preferred approach for installing utility lines, ranging from large-size pipeline river crossings to small-diameter cable conduits.

Martin Cherrington introduced HDD in 1971. This new technology provided a practical alternative to conventional trenching methods. It would ultimately revolutionize the way trenchless construction industry would come to install underground utilities and pipelines in cities and under large natural obstacles like the Mississippi River. One of the early forms of horizontal boring operation is shown in Fig. 10.1.

Modern HDD technology evolved from the boring techniques employed in the 1960s to install cables and conduits underground in urban areas. Trenching or open-cut methods were commonly used at this time to install a variety of utilities underground.

In 1964, Cherrington built his first utility boring drill rig and formed Titan Contractors that same year to specialize in utility road boring in Sacramento, California (Fig. 10.2). A unique combination of events contributed to Titan Contractors' initial success at that time. A building boom in Sacramento coupled with a national movement to clean up America sponsored by President Lyndon Johnson's wife, Lady Bird Johnson, motivated unsightly utility lines to be placed underground. The County of Sacramento mandated that all utilities in its service area be placed underground to comply with this new program. With streets, sidewalks, curbs, and gutters already in place in many new subdivisions, Titan Contractors was presented a unique business opportunity. The local power company, Sacramento Municipal Utility District (SMUD) contracted Titan Contractors to drill in new underground cables throughout its service area. The workload was so great that Titan reinvested available profits from its operations to develop new equipment and technology to expand its business. New rigs and downhole drilling tools were designed and developed to create operating techniques that would ultimately contribute to horizontal drilling technology becoming an accepted construction technique in the underground construction industry.

FIGURE 10.1 Early horizontal boring operation. (*Courtesy of Cherrington Corp.*)

Titan Contractors gained considerable experience and completed several significant projects over the next few years. One such project required drilling and pulling back a power cable along a curved street 1530 ft in length. Titan used a 2000 lb push-pull rig with 500 ft-lb of torque and $1^{1}/_{8}$ in outer diameter (OD) drill pipe to drill the directional hole. Considering that no directional control technology existed for accomplishing this type of borehole, the successful completion of the project was quite remarkable. Despite achieving such a feat, no recognition of significance to horizontal drilling was ever reported.

In 1971, Titan Contractors was invited to look at and bid on several road crossings for PG&E (Pacific Gas and Electric, a major California gas and electric utility) near Watsonville, California, just south of San Francisco. While physically investigating the potential job sites,

FIGURE 10.2 Utility boring in Sacramento. (*Courtesy of Cherrington Corp.*)

a PG&E engineer asked Martin Cherrington to look at a site where the utility needed to cross the Pajaro River, a dry river bed. The company was interested in a solution to take a gas line across the river without trenching it. Upon investigating the site it became obvious there would be problem with using conventional trenching methods. The river had a steep high bank, approximately 20 to 25 ft high. The bottom half of the bank was composed of sand and the top half consisted of rich topsoil. The adjacent field supported rows of the artichokes that are a common crop for the area. On the opposite side of the river channel, a small bank, 5 to 6 ft high, flattened out and extended eastward 30 to 40 ft to a levee, approximately 10 ft high. Beyond the levee was a field of potatoes. The bottom of the channel consisted of loose unconsolidated sand.

To trench the river, parallel double sheet pile would need to be driven deep enough so that the river bottom could be excavated sufficiently to lay the 4 in gas pipeline between the piles. Once installed across the river, the trench would have to be backfilled and the piles extracted. Based on the cost of a similar project a mile down river, PG&E reasoned that a drilling operation might be a more cost-effective solution. Figure 10.3 illustrates the first HDD rig on PG&E Pajaro River crossing.

Cherrington and PG&E initially considered driving two vertical caissons on either side of the river. One would be placed on the top of the high West bank and the other just outside the East levee. The product pipe could then be drilled, bored, or jacked from near the bottom of one caisson to the bottom of the other.

After careful consideration, Cherrington realized that there might be another alternative. Experience to date had revealed a curious phenomenon that had plagued attempts to drill a straight hole from pothole to pothole on many utility boring projects. With no directional control technology available, some types of drill stem tool configurations had a tendency to drill upward into the existing substructures, often coming out unexpectedly in the middle of a busy street. These drill stem tools were discarded as design failures. It was not realized that these discarded downhole-drilling tools held the key to solving a significant problem that confronted the pipeline contractors—a way to conveniently cross the major rivers without disrupting them.

FIGURE 10.3 First HDD rig on PG&E Pajaro River crossing, 1971. (*Courtesy of Cherrington Corp.*)

Thinking that these discarded drill stem tools would work, Cherrington took his horizontal drilling equipment to the Feather River, a few miles north of Sacramento. The sand and soil characteristics at the site on the Feather River were similar to the Pajaro River near Watsonville. Rather than drilling across the river, Cherrington decided to test-drill parallel to it on one bank. The entry angle for the first hole was approximately 10 degrees from horizontal. After drilling about 60 ft, the drill bit surfaced. Increasing the entry angle of the second hole to approximately 15 degrees, the drill bit surfaced well over 100 ft from the entrance. On the third and final test the entry angle was increased to 30 degrees. Joint after joint was fed into the hole. Tension was high when finally the bit surfaced nearly 300 ft away and 40 ft offline from the planned bore path. Cherrington was both relieved and elated that the discarded drilling tools he had rescued from the scrap pile actually performed as expected. The tests confirmed that given the optimum entry angle, proper drilling techniques and the right downhole drilling assembly, a barrier such as a river could be crossed using horizontal drilling techniques. Horizontal drilling would be a revolutionary step to eliminate all the problems typically associated with conventional trenching methods across rivers. With confidence that the new technique would work, Cherrington made plans to return to the Pajaro River and attempt to drill under it.

Convinced now that it was possible to traverse the Pajaro River using horizontal drilling techniques, the two-vertical caisson scheme was abandoned. Before starting the job, however, Cherrington decided to investigate the available oil field directional drilling technology and the methods that might be adapted to horizontal drilling. After learning what was available, he decided to change his original drilling plan to incorporate oil field tools and drilling practices that might increase his chance of success. This would be the first attempt to directionally drill, surface to surface, under a river using directional drilling tools. Cherrington used a 5 in OD mud motor with a bent sub above and a single shot survey system. The single shot device, although crude compared to today's downhole electronic survey instrumentation, would provide a reasonably accurate measurement of azimuth, inclination, and tool-face readings of the bottom-hole assembly while drilling. A single shot survey tool comprises a gimbaled, free-floating compass incorporating a concave glass hemisphere scribed with azimuth and inclination lines.

Housed in a $1^3/_8$ ft OD, nonmagnetic cylinder and liquid filled, the force of gravity on a small ball rolling in the glass hemisphere indicated the direction of the bottom-hole assembly relative to the magnetic north and its inclination relative to the vertical. The tool was pumped down the drill pipe and docked in a locating receptacle just behind the bent sub and mud motor. Once the tool was in position, a miniature camera with light and a timer attached took a picture of the compass and ball at a preset time. The single shot was then retrieved to the surface via an attached wire line and the film was developed. The reading or survey provided azimuth, inclination, and tool face (a measurement of the direction that the bent-sub is pointing about the axis of the borehole relative to the high side of the hole).

However, using the oilfield directional drilling tools proved disappointing. Hole angle could not be sufficiently built nor even maintained, so the approach was quickly abandoned in favor of more familiar utility boring techniques. Reverting to the lessons learned while drilling adjacent to the Feather River, Cherrington successfully crossed the Pajaro River using the discarded drilling tools that tended to drill back to the surface. Figure 10.4 shows the pipe section successfully drilled using HDD.

News of the PG&E project soon spread to the pipeline industry. Dow Chemical, faced with placing a series of 4 to 12-in chemical lines under a variety of waterways in Louisiana, invited Cherrington's company, Titan Contractors, to apply the newly discovered HDD technique to their projects. HDD had never been attempted on such a scale and more importantly no equipment or proven technology was available to execute these projects. An early version of Titan Contractors' HDD rig is shown in Fig. 10.5. In spite of this overwhelming challenge, Cherrington succeeded with the support and encouragement of Dow Chemical.

FIGURE 10.4 Celebrating the first successful HDD crossing (note the pipe in the foreground). (*Courtesy of Cherrington Corp.*)

As each Dow Chemical project was successfully completed, the benefits of using HDD spread rapidly throughout the pipeline industry. Other companies soon adopted Cherrington's new method thus spawning a new industry.

The first generation HDD process used a technique called the *washover* method (Fig. 10.6). Initially, a pilot hole was drilled with a small-diameter bit with $1^3/_4$ in drill pipe. A $4^1/_2$ in drill pipe was then washed over the pilot string followed by the product pipe. Each product pipe-joint was welded, x-rayed, and coated before running into the hole. A special drive sub, with a drilling mud hose attached, was welded to the top of the product pipe. Mud was pumped through the pipe while it was rotated and thrust over the $4^1/_2$ in drill pipe until it was completely installed.

FIGURE 10.5 Early Titan Contractors HDD rig. (*Courtesy of Cherrington Corp.*)

FIGURE 10.6 Washing product pipe over $4^1/_2$ in drill pipe. (*Courtesy of Cherrington Corp.*)

In 1978, Cherrington developed a method to pull multiple pipes or conduits through the same hole, simultaneously. The success of this technique led to the conclusion that it could be applied to large-diameter pipes as well. The new HDD technique developed included drilling a pilot hole, opening the hole with a reamer and then pulling multiple strings or one large-diameter pipe in place. This rapidly replaced the old *washover* technique and became the accepted practice in the HDD industry. Until this time, the washover technique was only capable of installing product pipes up to $12^1/_2$ in. in diameter. Today, the *pre-ream* or *pullback* method (Fig. 10.7) is the accepted practice to install pipelines and conduits on crossings worldwide.

The HDD method evolved from relatively simple utility boring to installing large-diameter (more than 50 in) pipes for as much as 4000 to 5000 ft. Rivers and beaches can now be crossed with pipelines and conduits without the risk of environmental damage, so commonly associated with trenching operations.

10.2 HDD CLASSIFICATIONS

HDD is defined as a steerable system for the installation of pipes, conduits, and cables using a surface launched drilling rig. Traditionally, HDD was used in large-scale crossings such as rivers in which a fluid-filled pilot bore is drilled without rotating the drill string, and this is then enlarged by a washover pipe and back reamer to the size required for the product pipe. However, today HDD has become a very versatile technique used for installing 2-in conduits under streets and driveways, as well as 60-in casings under rivers, dams, and levies.

The HDD industry is divided into three major sectors—large-diameter HDD (maxi-HDD), medium-diameter HDD (midi-HDD), and small-diameter HDD (mini-HDD, also called guided boring)—according to their typical application areas. Although there is no significant difference in the operation mechanisms among these systems, the different

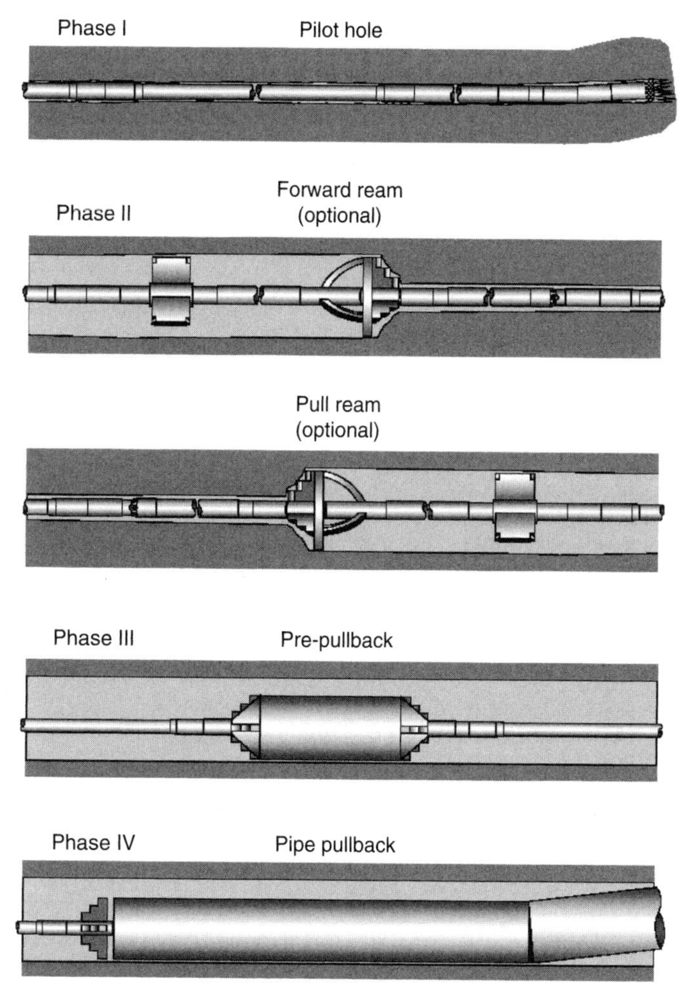

FIGURE 10.7 Different phases of HDD: Pream and pullback. (*Courtesy of Cherrington Corp.*)

application ranges often require corresponding modification to the system configuration and capacities, mode of spoil removal, and directional control methods to achieve optimal cost-efficiency. Table 10.1 compares typical maxi-, midi-, and mini-HDD systems.

10.3 METHOD DESCRIPTION

Directional drilling methods utilize steerable soil drilling systems to install both small- and large-diameter lines. In most cases, HDD is a two-stage process. Stage 1 involves drilling a pilot hole approximately 1 to 5 in. in diameter along the proposed design centerline. In stage 2, the pilot hole is enlarged to the desired diameter to accommodate the pipeline. At the same time, the product pipe is connected to the end of the drilling rod by swivel and

TABLE 10.1 Comparison of the Main Features for Typical HDD Methods

Type	Diameter	Depth	Drive length	Torque	Thrust/ pullback	Machine weight	Typical application
Maxi	24 to 48 in	≤200 ft	≤6000 ft	≤ 80,000 ft-lb	≤ 100,000 lb	≤ 30 ton	River, highway crossings
Midi	12 to 24 in	≤75 ft	≤ 1000 ft	900 to 7000 ft-lb	20,000 to 100,000 lb	≤ 18 ton	Under rivers and roadways
Mini	2 to 12 in	≤ 15 ft	≤600 ft	≤950 ft-lb	≤20,000 lb	≤9 ton	Telecom and power cables, and gas lines

pulled through the pilot hole. However, for large-diameter pipes, several passes of the back reaming may be necessary to enlarge the hole to the desired size, and additionally the back-reaming and pullback operations are performed separately.

The pilot hole is drilled with a surface-launched rig with an inclined carriage, typically adjusted at an angle of 8 to 18 degrees with the ground for entrance and 8 to 12 degrees for the exit angle. The preferred minimum radius in feet for steel pipe is 100 times of the diameter of the pipe in inches. For plastic pipe, the multiplication factor is 40, that is, 40 times of the diameter of pipe in inches.

10.3.1 Fluid-Assisted Mechanical Drilling

Rotating the drill bit, assisted by the thrust force transferred from the drill string, performs soil cutting in the mechanical drilling process. The mechanical drill bits may vary from a slim cutting head with a slanted face for small and short bore applications to a diamond-mounted roller cutter used with mud motors for large and long crossings. For small systems used for mini-HDD, directional steering control is accomplished mainly by the bias caused by the slanted cutter head face. For large systems used for maxi-HDD, a bent housing (a slightly bent section between 0.5 and 1.5 degrees of the drill rod) is used to deflect the cutter head axis from the following drill string. In both small and large systems, a curved path can be followed by pushing the drill head without rotating, and a straight path can be drilled by applying simultaneous thrust and torque to the drill head.

High-pressure jetting is used in utility boring applications (another trenchless technology technique that is seldom used today). It is an effective method to cut through soft formations and lends itself to directionally controlling the course of the borehole due to the unique shape of the jet bit. The utility boring method utilizes high-pressure, low-volume drilling fluid, usually bentonite mud, to jet through shallow, soft-soil formations to ensure the surrounding formation will not become saturated and unstable. HDD, on the other hand, uses mechanical cutting and high-volume and high-pressure fluid through soft formations with a bentonite and/or polymer mud as the bit advances. High volume and velocity in HDD are necessary to carry cuttings back to the surface via the annulus.

In mini-HDD, the torque transferred from the drill string usually rotates drill bits. For larger systems, the required drilling torque can be derived from a down-hole mud motor located just behind the drill bit. A medium-pressure, low-volume (1 to 2 GPM) drilling fluid is used to assist in the mechanical drilling process. There are two variants of drilling fluid use: fluid recirculation and fluid suspension. Fluid recirculation involves (1) moving the soil cutting from the bore hole in the form of slurry with a larger volume of drilling fluid, (2) cleaning the hole, and (3) refilling the hole with the slurry. The fluid suspension method that uses only a small amount of fluid keeps the soil cuttings in the slurry, with few or none removed from the hole. Theoretically, the choice between these two approaches depends on the soil conditions; however, in practice, the fluid recirculation method typically used in maxi-HDD systems and the fluid suspension method is used extensively in mini-HDD systems.

Midi-HDD systems employ a combination of recirculation and suspension methods. For long crossings requiring the use of a down-hole mud motor, high flow rates and large amount of drilling fluid are necessary for providing the soil cutting torque. Such large volumes of fluid can act as the conveyance medium for spoil removal. Recirculation reduces the extra stress in the drill string caused by suspended soil cutting that might be very high for a long drive. For small, short bores at a shallow depth, a down-hole mud motor is not used and the spoil removal usually is not required because the soil cuttings can be kept in the fluid suspension.

A unique technique for maxi-HDD involves the use of a washover pipe or casing with a large internal diameter, to be slid over the drill string during the pilot bore drilling process. When in place, the washover pipe can significantly reduce the friction around the drill string and provide stiffness to the drilling system. It also can be used to perform the prereaming, final reaming, and pullback operation (Fig. 10.7).

Incorporating offset jets and direction sensing and steering devices into the system achieves directional steering capacity. The deflection force created by the offset and angled fluid jets are used to form a curved drill path. An alternative to the offset jets is a special steerable head that will bend slightly under increased fluid pressure. Rotation of the jetting head can be accomplished by using a hydraulically or electrically driven down-hole motor, rotating a string of steel drilling rods, or attaching a special auger-type fin device behind the jetting head.

10.3.2 Drilling Process

Preconstruction Preparation. A design plan and profile drawings have to be prepared for each crossing. Owners typically provide these design, drawings, and relevant data such as soil conditions. After the design work is complete, site preparation is performed. A drilling rig is setup at the proper location. Slurry is prepared to stabilize the borehole and to lubricate the surface of the borehole. A transmitter is inserted into the housing provided on the pilot drilling string near the cutting head. Other equipment and facilities such as generators, pumps, storages, and offices are prepared at this stage.

A drilling rig is shown in Fig. 10.8. The drill strings are connected one after another by pushing and rotating them clockwise. To remove the strings, they are pulled and rotated counterclockwise.

On the other side of the proposed alignment, pipelines, reamer, and storage spaces that are required for prereaming and pullback are prepared.

Pilot Hole. Drilling of the pilot hole is the most important phase of an HDD project, because it determines the ultimate position of the installed pipe. A small-diameter (1 to 5 in) drilling string penetrates the ground at the prescribed entry point at a predetermined angle routinely between 8 to 18 degrees. The drilling continues under and across the obstacles along a design profile (Fig. 10.8).

Concurrent to drilling pilot hole, a larger-diameter pipe, called *wash pipe*, can be installed for maxi-HDD. The wash pipe follows and encases pilot drill string. The wash pipe protects the small-diameter pilot drill string from the surrounding ground, and reduces

FIGURE 10.8 Drilling rig (Abraham and Gokhale, 2002).

the friction around the drilling string. It also preserves the drilled hole in case the drill string is retracted for bit change.

Fluid-assisted mechanical cutting is the most typical method in drilling process. Using fluid-assisted mechanical cutting method, the drilling process is performed by rotating the drill bit and thrusting force from the drill string. The fluid action penetrates the ground by injecting a small amount of fluid with high pressure and high velocity. This fluid causes the void to create a space for the drill string to proceed. The typical jetting fluid is bentonite or polymer-based slurry whereas water may be used for short bores with stable soil conditions.

The drill path is monitored by a special electronic tracking system housed in the pilot drill string near the cutting head. The electronic tracking system detects the relation of the drill string to the earth's magnetic field and its inclination. The location data are transmitted to the receiver that calculates the location of the cutting head. It is recommended that the measurements be made at least every 30 ft. If the underground condition is complex, more frequent measurements may be required. By comparing the detected location and designed location, the direction of next drill is determined. See Chap. 4 for more information on HDD tacking and locating.

Once the drill head surfaces at the exit point, the location of the drill head is compared with planned location to determine that the actual location is within the allowable tolerance. A reasonable drill target at the pilot hole exit location is 10 ft left or right, and –10 to +30 ft in length. This accuracy is improving with the enhancement in equipment and operation skills. If the exit point is out of tolerance, some part of the bore should be redrilled. When the exit location is acceptable, the drill head is removed to prepare the next phase, prereaming and pullback.

Prereaming. In general, the final size of the bore should be at least 50 percent larger than the OD of the product pipe. This overcut is necessary to allow for an annular void for the return of drilling fluids and spoils and to allow for the bend radius of the pipeline. To create a hole that accommodates the required size of pipe, prereaming is necessary (Fig. 10.1).

Typically, the reamer is attached to the drill string at the pipe side and pulled back into the pilot hole. Large quantities of slurry are pumped into the hole to maintain the borehole and to

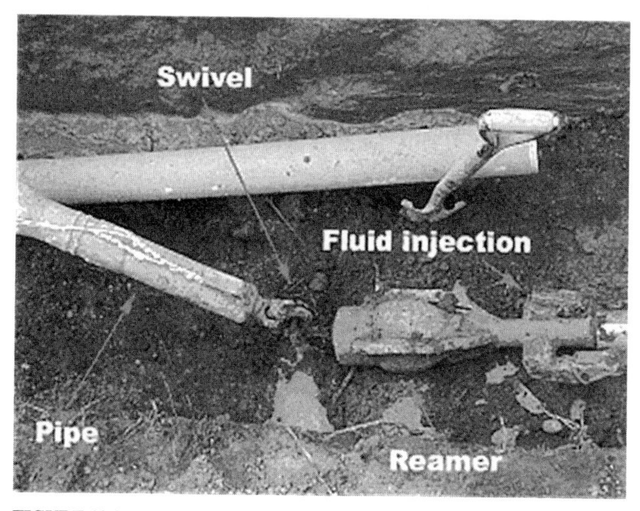

FIGURE 10.9 Components of pullback operation. (Abraham and Gokhale, 2002).

flush out the soil cuttings. The type of reamer varies based on the soil type. A blade reamer is used for soft soils; a barrel reamer for mixed soils; and a rock reamer with tungsten carbide inserts is used for rock formations. The soil condition, type of reamer, and the correct amount of drilling fluid are critical to the successful and economical completion of the project.

Pullback. Once the prereaming is completed, the pipe or conduit can be pulled back into the reamed hole filled with drilling fluid (Fig. 10.1). The pipe is prefabricated and tested at the pipe side. If the pipe is made of steel, it is recommended that the pipe be placed on rollers to reduce the friction and to protect pipe coating. However, this operation is usually not required for HDPE pipe installation.

The drill pipe is connected to the product pipe using a pull head or pulling eye and a swivel. The swivel is a device used to prevent the rotation of the pipeline during pullback. A reamer is also located between the pull head and the drill string to ensure that the hole remains open and to allow lubricating fluid to be pumped into the hole during the pullback. The pullback operation will continue until the pipe or conduit surface at the drill rig. The pull head is disconnected; the drill rig removed, and cleaned up and tie-ins are started. The components used for pullback operation are shown in Fig. 10.9.

10.3.3 Tracking System

The greatest technological potential and development for directional drilling lies in the area of tracking systems. Wireless steering tool systems are an example of the development. The walkover system and the wireline steering system are the most common tracking systems currently in use. However, other tracking systems such as the electromagnetic telemetry (EMT) system, and the mud-pulse-telemetry system are also available for tracking the drilling path.

Basic features of walkover and wireline tracking systems are briefly described in the following sections. Refer to Chap. 4 for detailed information on HDD tracking and locating.

FIGURE 10.10 Receiver for walkover tracking system (Abraham and Gokhale, 2002).

Walkover System. The walkover system is the most widely used system in drilling operation. A transmitter or sonde equipped in housing behind the drill bit is a major component of this system. The sonde transmits a signal to the surface. On the surface, a hand-held receiver picks up the signal and analyzes the data. Remote receiver also can be used for this data collection and analysis. As the walkover system is generally regarded as the most economical tracking method, it is commonly employed in jobs using small to mid-size drill bits. This system has been adopted from the cable locating technology, whereas the other tracking systems have been adopted from the oil and gas exploration industry. Fig. 10.10 shows the tracking receiver.

The advantage of the walkover tracking system is low cost. After the initial investment, the only major expense is the replacement of batteries and maintenance of sonde. This system has a higher productivity than the other systems. However, the tracking is restricted by geological conditions. For instance, if the drilling work crosses the freeway or river, it is not an easy task to walk over. The signal transmitted from the sonde often interferes with signals from other media such as overhead power lines, traffic signals, rebar in foundations, and so on.

Wireline System. The wireline runs with a steering tool located in a nonmagnetic bottom hole assembly. Thus, the location can be positioned with the signal from the transmitter to the receiver through the wire. The remote device displays the position information.

This system overcomes the depth limitation, because the power and signals are transmitted through the wire. It also provides better accuracy than the walkover system, because other materials do not interfere with the signal. The record keeping is easy when the system is hooked to a computer. It is more efficient than the walkover system, considering the time required for replacement of batteries, which frequently occurs during hard rock drilling. Also, productivity is impacted because the wire interferes with threading pieces of drill rods. The relatively high initial cost for purchasing or rental fee of manpower and equipment is the primary obstacle when using this system.

10.4 MAIN FEATURES AND APPLICATION RANGE

10.4.1 Diameter Range

In maxi- and midi-HDD, the size of pipes installed can range from 3 in to 48 in. in diameter. Multiple lines can be installed in a single pull, but only in the case of small-diameter pipes. The installation procedure for multiple lines is the same as for single lines, with the bundle being pulled back as a single unit along the prereamed profile. A significant multiple line crossing is more than 2000 ft in bore length and comprises five separate lines, pulled as one, ranging in size from 6 in to 16 in. The maximum size pipe that can be installed by the mini-HDD system is 12 in. in diameter.

10.4.2 Depth of Installation

Mini-HDD can install pipes up to 15 ft in depth. This depth limitation comes from the restriction in the capacity of walkover tracking system. However, for larger machines, such as midi- and maxi-HDD, the maximum installation depth for HDD is 200 ft.

10.4.3 Drive Length

The type of soil and site conditions determines the length of bore in HDD. Bore spans can range from 400 ft to 6000 ft for maxi- and midi-HDD. However, small lengths are not economically feasible because of the high operational costs of these systems. Mini-HDD is capable of installing pipelines and utilities 600 ft in one continuous pass to a specified tolerance.

10.4.4 Type of Casing

In general, the pipe to be installed is limited to one that can be joined together continuously, while maintaining sufficient strength to resist the high tensile stresses imposed during the pullback operation. In maxi- and midi-HDD, steel pipe is the most common type of casing used. However, butt-fused HDPE pipe also can be used. HDPE pipe, small-diameter steel pipe, copper service lines, and flexible cables are some of the common types of pipe materials being used today in mini-HDD.

10.4.5 Required Working Space

The directional drilling process is a surface-launched method; therefore, it usually does not require access pits or exit pits. If utility installation is being undertaken, pits may be required to make connections with the existing utility. The rig working area should be reasonably level, firm, and suitable for movement of the rig. For maxi- and midi-HDD, an area of 400 ft by 200 ft is considered adequate. The equipment used in mini-HDD is portable, self-contained, and designed to work in congested areas. Adequate space is required on the pullback side to have long string of welded steel pipe or butt-fused PE pipe ready for pullback operation.

TABLE 10.2 Applicability of Mini-HDD Method for Different Soil Conditions

Type of soil	Applicability
Soft to very soft clays, silt, and organic deposits	Yes
Medium to very stiff clays and silts	Yes
Hard clays and highly weathered shales	Yes
Very loose to loose sands (above water table)	Yes
Medium to dense sands (below the water table)	Yes
Medium to dense sands (above the water table)	Yes
Gravels and cobbles less than 2 to 4 in diameter	Marginal
Soils with significant cobbles, boulders, and obstructions larger than 4 to 6 in diameter	No
Weathered rocks, marls, chalks, and firmly cemented soils	Yes
Significantly weathered to unweathered rocks	Marginal

10.4.6 Soil Conditions

Clay is considered ideal for HDD methods. Cohesionless fine sand and silt generally behave in a fluid manner and stay suspended in the drill fluid for a sufficient amount of time; therefore, they are also suitable for HDD.

Generally, mechanical drilling systems can be applied in a wide range of soil conditions. A pilot hole can be drilled through soil particles ranging from sand or clay to gravel, and even in continuous rock formations, by using suitable drill bits; however, problems might occur in spoil removal, pilot hole stabilization, and backreaming operations. Today's technology enables large drilling operations to be conducted in soil formations comprising up to 50 percent gravel. Tables 10.2 and 10.3 present applicability of HDD in different soil conditions.

10.4.7 Productivity

HDD systems have the highest pilot hole–boring rate of advancement among all trenchless construction methods. For mini-HDD rigs, a three-person crew is sufficient. In suitable ground conditions, a regular work crew can install a 600 ft conduit in one day.

TABLE 10.3 Applicability of Midi- and Maxi-HDD Methods for Different Soil Conditions

Type of soil	Applicability
Soft to very soft clays, silt, and organic deposits	Yes
Medium to very stiff clays and silts	Yes
Hard clays and highly weathered shales	Yes
Very loose to loose sands (above water table)	Yes
Medium to dense sands (below the water table)	Yes
Medium to dense sands (above the water table)	Yes
Gravels and cobbles less than 2 to 4 in diameter	Marginal
Soils with significant cobbles, boulders, and obstructions larger than 4 to 6 in diameter	Marginal
Weathered rocks, marls, chalks, and firmly cemented soils	Yes
Significantly weathered to unweathered rocks	Marginal

10.4.8 Accuracy

The accuracy of installation for maxi- and midi-HDD depends on the tracking system being used and the skill of the operator. However, the reported accuracy is within 1 percent of the length. For mini-HDD, the accuracy depends on the methods employed. When using fluid-assisted mechanical cutting, the drill head can be located within 6 in range. The steering accuracy for this case is up to 12 in range. If a higher accuracy is desired, it can be achieved by reducing the interval at which the location readings are taken. However, this process will take more time.

10.4.9 Drilling Fluids

Drilling fluids are an important component of HDD process. They are used to stabilize the borehole, assist in the cutting process, cool down the transmitter, carry the spoil out of the borehole, and lubricate the borehole for reducing the frictional resistance of the product pipe. See Chap. 4 for a complete discussion of drilling fluids.

10.4.10 Major Advantages

The major advantage of HDD is its steering capability. In case of hitting obstacles, the drill head can be pulled back and guided around the obstacle. As HDD system can launch from the ground surface, no drive and reception pits are required. Therefore, the setup time is relatively shorter than other trenchless construction methods. As no shafts are required, the project costs are reduced. The single HDD drive length that can be achieved is longer than any other nonworker-entry trenchless method.

10.4.11 Major Limitations

The disposal of the slurry mixed soil cuttings needs to be considered in advance, especially when no fluid circulation method is to be used. Although the U.S. Environmental Protection Agency (EPA) does not consider bentonite a toxic material, the acceptability of such spoil material varies among local agencies as well as landfill owners.

Thorough site investigation is of extreme importance, as corrective measures applied midway in the drilling or backreaming operation can be very time consuming and costly. When boring under roadways or other environmentally sensitive areas, the use of pressured fluid may cause serious concerns regarding the possible deleterious effect of bentonite through slurry migration laterally and vertically. Care should also be taken to prevent possible ground movement and loss of slurry to the pavement for installations with shallow soil cover (Fig. 10.11).

10.4.12 Potential Problems

As is true with any type of construction, some potential problems may be encountered during the horizontal drilling projects. Table 10.4 gives a summary of the potential problems with the possible causes and the necessary actions required to remedy these problems.

10.4.13 Pipe Materials

Pipe in HDD installations should be smooth, flexible, and have sufficient strength to resist tension, bending, and external pressure installation loads. The pipe must also have a joining method that will allow individual joints to be fabricated into a long pull section without

FIGURE 10.11 Pavement distress owing to uncontrolled slurry pressures in HDD operation (Abraham and Gokhale, 2002).

creating a significant external upset or reducing strength. Two materials are typically used for HDD installation, steel and HDPE. Usually for diameters up to approximately 23 in, HDPE is method of choice. Recently other pipe materials, such as fuseable PVC or restrained joint PVC and ductile iron pipe have become available for HDD use.

The majority of maxi-HDD installations have been completed using welded steel pipe. This probably results from the fact that HDD grew out of the petroleum pipeline industry where the use of steel was dictated by high-pressure service. Although installation loads need to be checked by the design engineer, the strength of steel eliminates problems with installation loads in most cases. The high strength of steel also provides contractors with a margin for error during installation.

For mini- and midi-HDD applications, HDPE pipe can provide several constructability benefits over steel pipe on an HDD installation. While steel pipe often necessitates a substantial "breakover" radius during pull back requiring the pull section to be lifted into an arc, HDPE pipe can typically be pulled into borehole directly off of pipe rollers. If space is not available to fabricate the pull section in one continuous segment, this reduction in breakover length can reduce the number of tie-ins required. The flexibility of HDPE pipe also provides more options for laying out the pull section as it can be bent around obstacles. Radius of curvature is generally not a concern when installing HDPE pipe because HDPE can normally withstand a tighter radius than can be achieved with the steel drill pipe used to drill the pilot hole. Therefore, the steel drill pipe limits borehole curvature. Also, the use of HDPE pipe eliminates the need for field joint coating and fabrication of HDPE is typically faster and less expensive than fabrication of steel. However, the tensile and pressure capacities of HDPE pipe are significantly less than those of steel. As a result, analysis of installation and operating stresses is critical especially in midi-HDD operations in order to determine if HDPE pipe is suitable.

Midi- and maxi-HDD applications typically require detailed analysis of the pipe or conduit in relation to its intended application. Because of the large anticipated pulling loads, high-strength pipe, including joints essentially as strong as the original pipe, must be used. Thus, PE or steel are appropriate materials. A careful analysis of the PE or steel pipe must be performed, subject to the route geometry, to verify or determine an appropriate pipe or conduit configuration.

TABLE 10.4 Summary of HDD Potential Problems

Problem	Probable cause	Solution
Lost position of drill head	Locator showing inaccurate readings	Check locator performance. Try push and pullback of the drill head to track it.
Difficulties in product pipe pullback	Product pipe pushed into the sidewalls of the curved bore hole	Alternatively push and pull to free pipe.
Drill head exits off target	Steering difficulties and/or inaccurate locator	Pull back head reasonable distance and redrill.
Back reaming difficulties	Possible blockage owing to cobbles or gravel	Push reamer back out. Detach pipe and reamer. Pullback with drill bit to clear obstruction.
Steering difficulties	Hit bedrock or a hard layer at steep angle	Drill very slowly to pass through hard ground.
Fluid migrates to surface	Fissured rock or hydraulic fracture	Lower the fluid pressure.
Alignment too tight for product pipe	Difficulty steering section	Enlarge the section of the bore hole.
Loss of bore hole stability	Fluid pressure fluctuation between rig and drill face	Increase applied fluid pressure to just below maximum permissible value.
Groundwater seepage washes out drilling fluid	High groundwater pressure or low drilling fluid rate	Adjust drilling fluid weight and flow rate.
Plugged fluid jets	Debris in drill string	Remove and clean.
Separation of drill string immediately behind reamer	Damaged swivel assembly	Blind push backward and dig up.
High drill torque requirements	Worn bit or cutting head	Replace.
Increased torque overnight	Collapsed hole or cohesive soil	Drill continuously or rotate periodically overnight.
High pullback forces	Radius too small	Flatten drilling path curves.
Warning siren and/or flashing lights	Advance electric strike system activated because drill head is too close to or struck a live electric underground line	Do not move while on the protective mat. Call electric company to disconnect power line. Always wear safety shoes and gloves.

PE Pipe. In particular, (HDPE, Specification D-1248) is recommended because of its superior mechanical strength and low-frictional characteristics. If such pipe is provided in short segments, the individual units should be joined using a butt-fusion technique in accordance with Practice D-2657. This will allow the inherent strength of the PE pipe to be maintained during the placement process and when subjected to other operational stresses.

Steel Pipe. Welding must join steel pipe or conduit sections. The steel pipe should be provided with a protective coating to resist corrosion. Typically, fusion bonded epoxy coating may be used, even for the field joints. Shrink sleeves are also acceptable if properly applied and suitably bonded. For crossing rocky or partially rocky areas, additional coating thickness is required. A coating to provide additional weight for ballast is generally not required. Any necessary pipe fabrication process should be performed and checked, including pressure tightness and coating integrity, prior to completion of the boring operation. This is necessary to avoid delays or interruptions during the pipe pullback operation. Nonpressurized (e.g., ultrasonic) techniques are available for verifying air or hydrotest integrity of joints.

Cable Conduit Applications. For cable conduit applications, including electric power and telecommunications, small-diameter HDPE pipe may be supplied on a continuous reel including internal pull line or the cable itself, as preinstalled by the manufacturer. In addition, the pipe may be provided with the interior surface prelubricated. In accordance to that, the owner or engineer will specify such features. TR-NWT-000356 specifies requirements for telecommunications applications, including HDPE pipe with various internal surface profiles, including smooth wall or ribbed.

Operational and Installation Loads. Load and stress analysis for an HDD pipeline installation is different from similar analyses of conventionally buried pipelines because of the relatively high-tension loads, bending, and external fluid pressures acting on the pipeline during the installation process. In some cases, these loads may be higher than the design service loads. Pipe properties such as strength and wall thickness must be selected such that the pipeline can be both installed and operated within customary risks of failure. Analysis of the loads and stresses that govern pipe specification can most easily be accomplished by breaking the problem into two distinct events: installation and operation.

The pipe will be subject to loads during its long-term operation and short-term installation process. It is the responsibility of the designer to determine the design and selection of the pipe, to serve the function it is intended for and that it withstands the operational stresses at the directionally drilled section as well as at other sections along the pipeline. This practice deals primarily with the loads imposed during the directional drilling process and other issues related to this aspect of the installation.

The installation loads during HDD installation include tension, bending, and external pressure as it is pulled through a prereamed hole. The stresses imposed on the pipe are result of combined effects of these loads. A successful installation depends on good estimation of these loads. The tensile forces on the pipe include frictional drag, fluidic drag, and effective weight of the pipe. The bending stress depends on the bending radius of the pipe and the bore path design. The external pressure includes (1) hydrostatic pressure from the weight of the drilling fluid surrounding the pipe in the drilled annulus, (2) hydrokinetic pressure required to produce drilling mud flow from the reaming assembly through the reamed annulus to the surface, (3) hydrokinetic pressure produced by surge or plunger action involved with pulling the pipe into the reamed hole, and (4) bearing pressure of the pipe against the hole wall produced to force the pipe to conform to the drilled path.

The operating loads on pipes installed with HDD operations are basically same as pipelines installed by open-cut installations. However, a pipeline installed by HDD will contain elastic bends. Flexural stresses imposed by elastic bending should be checked in combination with other longitudinal and hoop stresses to evaluate if acceptable limits are exceeded. The operating loads include internal pressure, external pressure, and thermal expansion. ASTM F 1962-99 should be used to include arching effects when calculating soil loads on the pipe installed by HDD. It should be noted that a pipeline installed by HDD will contain elastic bends. Flexural stresses imposed by elastic bending should be checked, in combination with other longitudinal and hoop stresses to evaluate if acceptable limits are exceeded.

10.5 EMERGING TECHNOLOGIES

Traditionally, gas and electric distribution pipeline construction has been the major market sector for HDD. In the 1990s the telecommunication boom, in particular the fiber optic cable installation, provided an impetus to the HDD market, resulting in double-digit growth in terms of the number of directional drilling units sold, and also the number of companies and contractors engaged in HDD construction. By the end of the decade much of the fiber optic work had been completed in North America, causing many contractors to declare bankruptcy owing to lack of business. The number of drilling rigs sold in 2001 was 65 percent less than those sold in 2000 (UCT, 2002). Today, the market continues to remain soft, with companies and contractors looking for alternate markets for HDD.

10.5.1 HDD for Gravity Installations

One of the most promising market segments, long thought to be unsuitable for HDD, may prove to be the next boom; thanks to the recent innovation in the HDD technology. Water and sanitary sewer construction appears to be the next frontier for HDD application. Engineers who design sanitary sewer systems have been slow to embrace HDD technology, as many do not consider directional drilling appropriate for their projects because of their concerns that the equipment cannot install pipe to grade—an essential requirement for gravity-flow systems, and many hold to the mistaken belief that directional machines cannot accommodate pipe of the sizes used in sewer systems. Although it is true that the early directional machines were limited in the sizes of product they could install, today relatively compact models can easily place pipe to diameters of 30 in, making them capable of handling a large portion of sewer pipe being used today. Grade control is the primary roadblock for making sewer installations with directional drilling equipment. Until recently, most available HDD tracking electronics could measure grade only in 1 percent increment, not precise enough for grade work.

In 2001, researchers at the UTILX Corporation were looking to extend the application of HDD to gravity sewers, through the development of a new generation of guided boring equipment featuring improved tracking, boring, and installation technology is under development. A trial project to install a gravity sewer line with a gradient of 1.5 percent and manhole-to-manhole distances of 330 ft was undertaken in 2002. Soon thereafter, the method was proposed commercially.

The ArrowBore™ method complements the existing electronics and incorporates physical verification of the location of the directional boring machine stem as it is advanced through the ground. Excavations, termed sight-relief holes are used at certain points throughout the bore path to provide both the contractor and the utility owner's inspector with a way to visually inspect that the drill stem is installed in the correct place prior to final installation of the main. Once the pilot bore is completed a back reamer sized just slightly larger than the main line, typically $^1/_4$ to $^1/_2$ in larger than the OD of the pipe, is connected and a hole is reamed ahead of the pipe to be installed. The sight-relief holes allow for relief of pressure while a portable vacuum machine removes the excess spoils created during the placement of the main. The tight-fit hole eliminates the possible deflection and flotation from line and grade that can occur when a hole is reamed to a size 1.25 to 1.5 times the OD of the pipe as is the general practice with conventional directional boring methods. The process also allows a project to be installed at depths ranging from 15 to 25 ft with little change in cost.

Several installations in the United States have attested to the feasibility of this method. In early 2004, Precision Directional Boring LLC, Brunswick, Ohio, used the technique in conjunction with electronic locating equipment to install a 400 ft stretch of gravity sewer with accuracy. The project involved installing the 12 in PVC pipe to serving as a sanitary sewer for a private subdivision. The process used laser technology and vertical boreholes to establish line and grade coordinates every 30 ft. The 30 18 in vertical relief holes on the Berea project, were drilled to the depth of the installed pipe, and lined with 16-in HDPE liner for stability.

10.5.2 HDD for Pipe Removal

Another recent development involves the use of HDD in conjunction with pipe renewal industry. The patented Rotary Impactor™ developed by Vermeer Corporation uses an 8-in pipe removal tool, designed for use with vermeer horizontal directional drills. Unlike traditional pneumatic pipe bursting systems (see Chap. 18), there are no air hoses involved in this method. The result is a significant reduction in setup time. Additionally, the need for large entry and exit pits is also reduced. The HDD operator drills from an entry point through the existing pipe to the exit pit. The pneumatic tool is attached to the drill rod, and the drill operator begins the pullback process. The rotation of the drill stem activates an internal cam, which produces repetitive impacts to the front of the tool thus bursting the existing pipe, in much the same way as a pneumatic pipe bursting operation but at a much higher frequency. The impact rate, which is dependent on the rotational rate of the drill, averages 750 to 1000 impacts/min. With the HDD providing pullback tension and rotation, the pneumatic tool can burst through the existing pipe while pulling the new pipe into the compacted bore hole. Lubricating fluid is delivered down the drill stem and through internal passages in the unit, providing lubrication and easing the stress on the new pipe.

The inniream system developed by Nowak Pipe Reaming, Inc., Goddard, Kansas, is a new pipe removal development using directional drilling equipment to upsize an existing pipe. The system consists of removal of the existing clay, PVC, asbestos-cement, or non-reinforced concrete pipe and simultaneous replacement with a new pipe of equal or greater diameter. The removal process is accomplished by back reaming using a regular directional drilling machine. The directional drilling machine is equipped with a cutterhead having spirally placed carbide tipped teeth that grind and pulverize the existing pipe into pieces. The pipe particles and excess materials, resulting from upsizing, are carried with drilling fluid to manholes or receiving pits and are retrieved with a vacuum truck or a slurry pump for disposal. The new pipe (PVC or HDPE) is attached behind the mandrel and follows the cutterhead as it progresses. Thus the destruction of the existing pipe, cutting the soil to the required size, and the installation of the new pipe is a simultaneous process. Another advantage of this system is that the primary equipment is a directional drilling system that can also be used in other trenchless technology projects.

10.5.3 New HDD Capabilities

Other developments in directional drilling are on two fronts. On one hand, for the large drilling systems, more powerful rigs are under development to enable a continuous installation of up to 25,000 ft in length; on the other hand, more compact small drilling rigs are more available to be used within confined spaces: a frequently encountered situation in congested urban areas. Also, a middle-class or so called *midi-HDD* systems has emerged to fill the application range gap between those of HDD and mini-HDD, such as highway, short river crossings for relatively large-diameter pipelines.

Developments are also focusing on the computer-aided automatic steering control and extending the application of HDD and mini-HDD to more challenging soil conditions. With such systems, the detailed as-built record can be easily obtained for future use.

A recent innovation in the drilling fluid has been the use of polymer gels to replace bentonite slurries. Typically 1 gal of gel is required for 800 gal of water. This eliminates the need for carrying a large towing truck to mix bentonite slurry.

Gas Research Institute (GRI) has recently developed guidelines and computer-aided design tools for the installation of PE gas pipe using directional horizontal drilling. The interactive software developed in this program will facilitate the design and planning tasks and contribute to the improvement of engineering and final project quality.

An area, currently under investigation, that has the potential of revolutionizing the directional drilling industry is *obstacle detection*. The current directional systems are not capable of detecting obstacles such as boulders, cobbles, pipelines, cables, and so on. Future developments could include the capability of detecting the type and the location of obstacles thus permitting the operator to choose a route that avoids or maneuvers around obstructions. Such capability would help eliminate accidents.

REVIEW QUESTIONS

1. Describe horizontal directional drilling (HDD) and its main classifications.
2. Explain the fluid-assisted mechanical cutting process.
3. Describe major HDD steps.
4. Discuss why the greatest technological potential and development for directional drilling lies in the area of tracking systems.
5. Describe the main features and the application range of HDD.
6. What are the main advantages and limitations of the HDD process?
7. List some of the potential problems with the HDD process and some possible solutions.
8. What are some of the new technologies emerging in the HDD industry today?
9. Describe in what type of ground and project conditions you would recommend HDD method and why.
10. Describe emerging technologies and new HDD capabilities.
11. Describe pipe removal process using HDD.
12. Discuss pipe load considerations for HDD operations.

REFERENCES

Abraham, D., and S. Gokhale. (2002). *Development of a Decision Support System for Selection of Trenchless Technologies to Minimize Impact of Utility Construction on Roadways.* FHWA/IND/JTRP-2002/7, SPR-2453, National Technical Information Service, Springfield, Va.

American Society of Civil Engineers. (2005). *Pipeline Design For Installation by Horizontal Directional Drilling*, Reston, Va.

Digital Control Incorporated (DCI), Kent, Wash. (2002). *Product Manual.* Available at http://www.digital-control.com./homepage4.html.

Directional Crossing Contractors Association (DCCA), Dallas, Tex. (1994). *Guidelines for a Successful Directional Crossing Bid Package.* Available at http://www.dcca.org/.

Directional Crossing Contractors Association. (DCCA), Dallas, Tex. (2000). *Guidelines for Successful Mid-sized Directional Drilling Projects.* Available at http://www.dcca.org/.

Gas Research Institute (GRI). (1991). *Guidelines for Pipelines Crossing Highways.* Gas Research Institute, Chicago, Ill.

Iseley, T., and S. Gokhale. (1997). *Trenchless Installation of Conduits beneath Roadways.* Synthesis of Highway Practice 242, Transportation Research Board, National Academy Press, Washington, DC.

Iseley, T., R. Tanwani, and M. Najafi. (1999). *Trenchless Construction Methods and Soil Compatibility Manual.* Trenchless Technology Committee, National Utility Contractors Association (NUCA), Arlington, Va.

Subsite Tracking Equipment, Ditch Witch, Charles Machine Works, Perry, Okla. (2002). Available at http://www.ditchwitch.com/dwcom/Product/ProductView/138.

CHAPTER 11
MICROTUNNELING METHODS

11.1 INTRODUCTION

Microtunneling (MT) is a method of installing pipes below the ground, by jacking the pipe behind a remotely-controlled, steerable, guided, articulated microtunnel boring machine (MTBM). The MTBM, which is connected to and followed by the pipe being installed, ensures that the soils being excavated are fully controlled with the rate of advancement of the boring machine at all times. The minimum depth of cover to the pipe being installed using the MT process is normally 6 ft or 1.5 times the outer diameter of the pipe being installed, whichever is greater. MT method minimizes ground surface settlement or heave. The overcut of the MTBM will be determined by the need to satisfy different ground conditions to allow steering and to introduce lubricant. Overcut, unless excessive, is not connected to settlement or heave. Overcut usually should not exceed 24 mm (1 in) on the outside radius of the pipe. The annular space created by the overcut normally is filled with lubrication material that is suitable for a particular soil condition.

According to the American Society of Civil Engineers' (ASCE's) *Standard Construction Guidelines for Microtunneling*, MT can be defined as "a remotely-controlled, guided pipe jacking technique that provides continuous support to the excavation face and does not require personnel entry into the tunnel" (ASCE 2001). MTBM is operated from a control panel, normally located on the surface. The system simultaneously installs pipe as spoil is excavated and removed. Personnel entry is required only for working inside the shaft. The guidance system usually references a laser beam projected onto a target in the MTBM, capable of installing gravity sewers or other types of pipelines to the required tolerance, for line and grade. Although in Europe, the term *microtunneling* is used for small diameters of less than 36 in, there is no size limit used in the United States.

11.1.1 History

MT technology originated in Japan. In the mid-1960s, there was an increasing public demand in Japan to control water pollution and to improve the quality of the environment. Iseki Inc. started building worker-entry-sized-tunneling shields in the 1970s. However, the market was soon demanding nonworker-entry-sized systems, which could deal with a wide range of soil conditions. This demand led to the development of remotely controlled small-diameter slurry tunneling systems, that is, MT. Komatsu in Japan developed MT in 1975. Iseki Inc. introduced the first MT equipment in 1976. The development of the MT technique allowed tunneling in soft, unstable soil conditions. Iseki introduced the crunching mole in 1981, which could crush boulders as large as 20 percent of the OD of the pipe.

United Kingdom was the first foreign country where Iseki used an MTBM in the late 1970s. In Germany, Dr. Soltau was the first to develop MTBM systems in 1982. Since then, Germany has been a major user and manufacturer of MTBM in the world. In 1984, MT was first introduced into the North America. The first project was the installation of 615 ft of 72-in diameter pipe under I-95, Fort Lauderdale, Florida, for the Miami-Dada Water and Sewer Authority. Since 1984, there has been a growing demand for MT in North America. In 1995, the first U.S. made MT machine was manufactured by Akkerman, Inc., Brownsdale, Minnesota.

11.2 METHOD DESCRIPTION

MT is a trenchless construction method for installing conduits beneath the roadways in a wide range of soil conditions, while maintaining close tolerances to line and grade from the drive shaft to the reception shaft. The MT process is a cyclic pipe jacking process.

Based on the mode of operation, the MT method can be subdivided into two major groups: (1) slurry method and (2) auger method. In the slurry-type method, slurry is pumped to the face of the MTBM. Excavated materials mixed with slurry are transported to the driving shaft, and discharged at the soil separation unit above the ground. In an auger-type method, excavated materials are transported to the drive shaft by the auger in a casing pipe, and then hoisted to the ground surface by a crane. However, as slurry MT is more versatile because it protects the tunnel face by slurry pressure (especially under water table and unstable ground), the auger-type MTBM is not common in the United States. Therefore, the slurry-type MTBM will be discussed in this book in detail. Both MT systems comprise the following five major components:

1. MTBM
2. Jacking system (pipe jacking equipment suitable for the direct installation of the product pipe)
3. Automated spoil transportation and rate of excavation controls
4. Guidance and remote-control system
5. Active direction control
6. Jacking pipe

11.2.1 Slurry MTBM

In this method, a rotating cutting head excavates soil mechanically. The rotation of the cutting head can be eccentric or centric, and the speed of rotation (RPM) can be constant or variable. Cutter heads are bi-rotational. The head normally rotates in a clockwise direction when looking from the rear of the machine. Reverse rotation can provide more flexibility in overcoming obstructions and difficult ground conditions. The spoil excavated at the face is extruded through small parts located at the rear of the MTBM face into the mixing chamber. The main functions of this chamber are to mix the spoil with clean water from the separation system and control hydrostatic head imposed on the MTBM face by a body of water or groundwater. When the spoil and water are mixed to form slurry with suitable pumping consistency—typically less than 60 percent solids—the slurry is transported to the solids separation system hydraulically. Figure 11.1 illustrates the inside structure of slurry-type

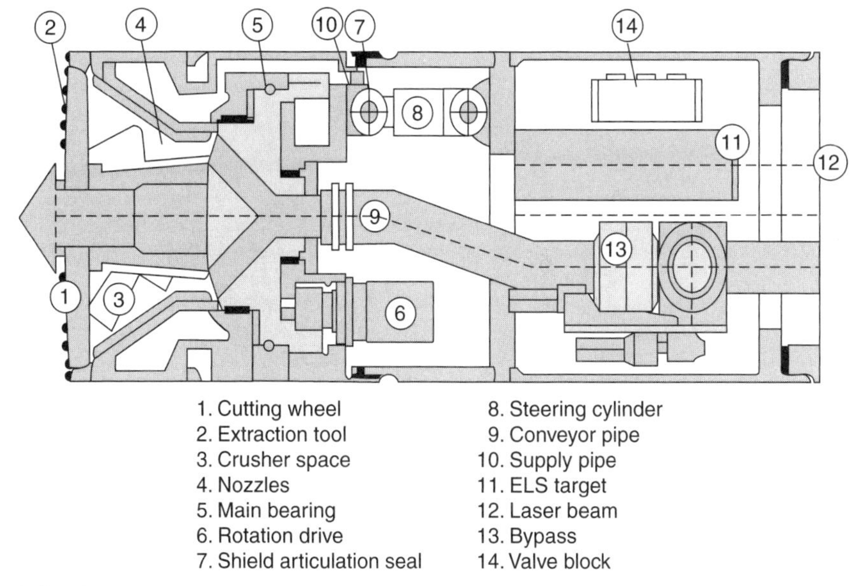

FIGURE 11.1 Typical slurry type MTBM. (*Courtesy of Herrenknecht Inc.*)

MTBM. Drives of up to 1200 ft have been completed in full-face solid granite by MTBM with rock strengths exceeding 20,000 psi. Virtually, all ground conditions can be completed with large slurry MTBM.

Some pictures of slurry-type MTBMs are shown in Figs. 11.2 to 11.4 (Abraham and Gokhale, 2002).

FIGURE 11.2 MTBM.

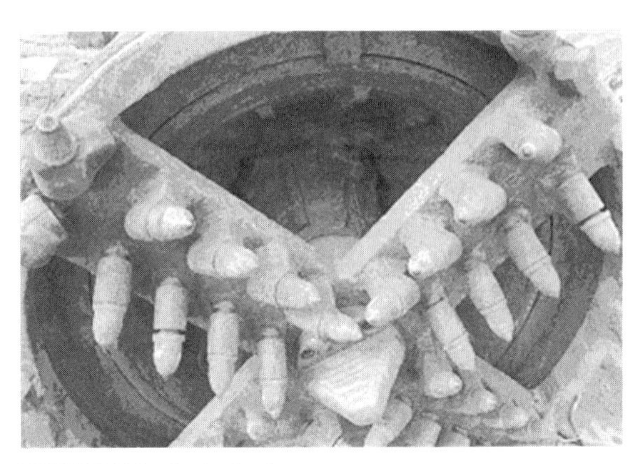

FIGURE 11.3 Cutting head.

11.2.2 Jacking System

The jacking system comprises the jacking frame and jacks. A jacking frame is also shown in Fig. 11.5. Figure 11.6 shows a 42-in steel casing with 20-ft long section that is being jacked.

The jacking capacity ranges from approximately 100 tons to over 1000 tons. The jacking capacity is mainly determined by the length and diameter of the bore and the soil. The soil resistances are generated from face pressure, friction, and adhesion along the length of the steering head and pipe string. The jacking system determines two major factors of MT operation: the total force or hydraulic pressure and the penetration rate of pipe. The total jacking force and the penetration rate are critical to control the counterbalancing forces of the MTBM.

11.2.3 Spoil Removal System

The spoil is mixed into the slurry in a chamber located behind the cutting head of the MTBM. This mixed material is transported through the slurry discharge pipes and discharged into a

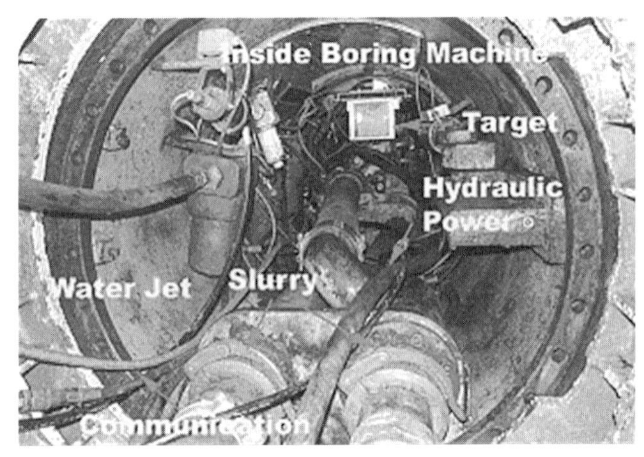

FIGURE 11.4 Inside of MTBM.

FIGURE 11.5 Jacking frame for microtunneling (Abraham and Gokhale, 2002).

separation system. This system is a closed-loop system because the slurry is recycled. The velocity of the flow and the pressure should be carefully regulated because the slurry chamber pressure is used to counterbalance the groundwater pressure. The machine can be sealed off from external water pressure, allowing underwater retrieval. Slurry is a mixture of bentonite (a clay material) in a powder form and water. The bentonite is used to increase the density of water so that it can transport heavy spoil particles. These heavy particles are filtered from the slurry at the separation units. The filtered slurry is sent to the storage tanks, which is recirculated through the system. Figure 11.7 (a) shows the soil separation system. One of the three screens for the separation system is shown in Fig. 11.7 (b).

11.2.4 Guidance and Remote-Control System

Laser is the most commonly used guidance system for MT. It gives line and grade information for the pipe installation. Laser is installed in the driving shaft and gives a fixed

FIGURE 11.6 Steel casing being jacked. (*Courtesy of Kerr Construction Inc.*)

(a) Soil separation system

(b) Screen for soil separation system

FIGURE 11.7 Soil separation system (Abraham and Gokhale, 2002).

reference point. Laser target and a closed circuit television (CCTV) camera are installed in the MTBM. There should not be any obstruction along the laser beam pathway from the driving shaft to the laser target. There are two types of laser targets: the passive system and the active system. In the passive system, a target grid is mounted in the steering head. CCTV monitors this target and the information obtained by this CCTV is transferred back to the operator's control panel. The operator can make any steering correction based on the information. In the active system, photosensitive cells are installed on the target and these cells convert information into digital data. Those data are electronically transmitted to the control panel and give the operator digital information of the location. Both active and passive systems are commonly used. Figure 11.8 shows the laser used for the Soltau MT system. The target mounted in the MTBM is shown in Fig. 11.9.

FIGURE 11.8 Laser for guidance of MTBM (Abraham and Gokhale, 2002).

FIGURE 11.9 Target mounted in the MTBM (Abraham and Gokhale, 2002).

Operation boards are usually located in a standard container with 8 by 20 ft dimensions. Operation board comprises control panel, computer, monitor, and a printer. Through the operation board, all the MT operations such as tunneling machine, main jacks, interjack stations, direction or speed of the cutting wheel, bentonite lubrication equipment, and so on can be controlled. An example of operation board of Soltau microtunneling is shown in Fig. 11.10. The screen of the computer in operation board is presented in Fig. 11.11.

In addition to the computer monitor, two other monitors are used in the MT operation. One is for communication purpose, and the other one is for monitoring the inside of MTBM. A small camera with a microphone is installed at the top of sheet pile at the driving shaft, which provides the overview of the operation. The operator in the cabin can see and hear the tunneling site and so can control the equipment based on the input from the crews on the site. Another small camera is installed inside the MTBM. This camera provides a view inside the MTBM. These two monitors are shown in Figs. 11.12 and 11.13.

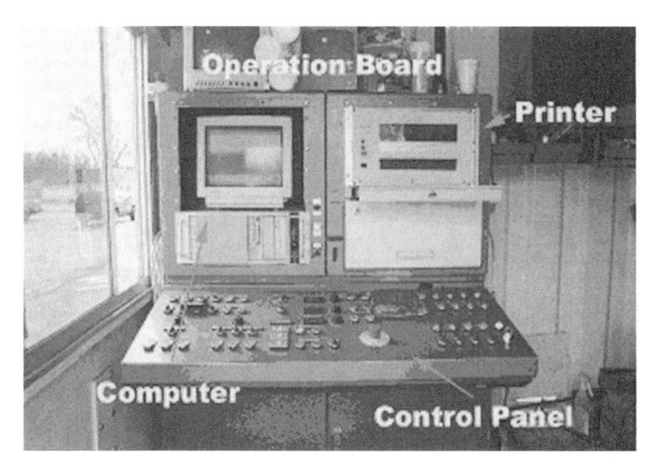

FIGURE 11.10 Operation board of an MTBM (Abraham and Gokhale, 2002).

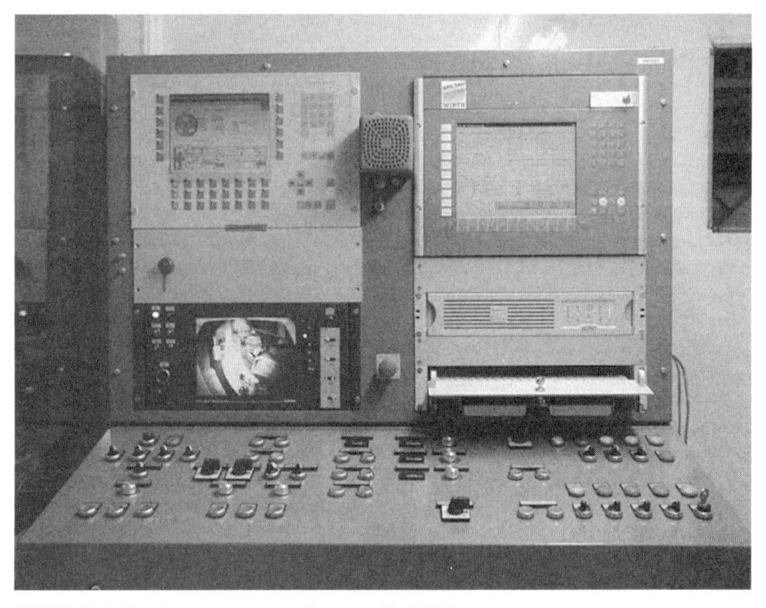

FIGURE 11.11 Computer screen. (*Courtesy Wirth Soltan.*)

11.2.5 Active Direction Control

Line and grade is controlled by a guidance system that relates the actual position of the MTBM to a design reference, by a laser beam transmitted from the jacking shaft along the centerline of the pipe to a target mounted in the shield. The MTBM is capable of maintaining the grade to within 1 in and line to within 1.5 in. It should be noted that as with any other trenchless project, the line and grade tolerances are subject to project and ground conditions.

FIGURE 11.12 Monitor for communication (Abraham and Gokhale, 2002).

FIGURE 11.13 Monitor showing a view inside the MTBM (Abraham and Gokhale, 2002).

The active steering information is monitored and transmitted to the operation console. The minimum steering information available to the operator on the control console usually includes the position relative to the reference, role, inclination, attitude, rate of advance, installed length, thrust force, and cutter head torque.

11.2.6 Jacking Pipe

In general, a pipe used for jacking must be round and have a smooth, uniform outer surface, with watertight joints that also allow for easy connections between pipes. Pipe lengths must be within specified tolerances and the pipe ends must be square and smooth so that jacking loads are evenly distributed around the entire pipe joint such that point loads will not occur when the pipe is jacked in a reasonably straight alignment. Pipe used for pipe jacking is capable of withstanding all the forces that will be imposed by the process of installation, as well as the final in-place loading conditions. The driving ends of the pipe and intermediate joints are protected against damage as specified by the manufacturer. The detailed method proposed to cushion and distribute the jacking forces is specified for each particular pipe material.

Any pipe showing signs of failure may be required to be jacked through to the reception shaft and removed. The pipe manufacturer's design jacking loads should not be exceeded during the installation process. The ultimate axial compressive strength of the pipe must be a minimum of 2.5 times the design jacking loads of the pipe. Presently, the following pipe materials specially manufactured for MT operations are available: (1) reinforced concrete pipe (RCP), (2) centrifugally cast glass-fiber-reinforced polymer mortar (CCFRPM), (3) polymer concrete pipe (PCP), (4) vitrified clay pipe (VCP), (5) steel pipe, and (6) ductile iron pipe (DIP). PVC pipe is installed with MT on experimental basis.

11.3 MICROTUNNELING PROCESS

The typical layout of construction site for slurry-type MT is shown in Fig. 11.14. Two shafts are required for the MT operation: a driving shaft and a reception shaft. An MTBM is setup on the guide rail of the jacking frame in the driving shaft. The main jack pushes the

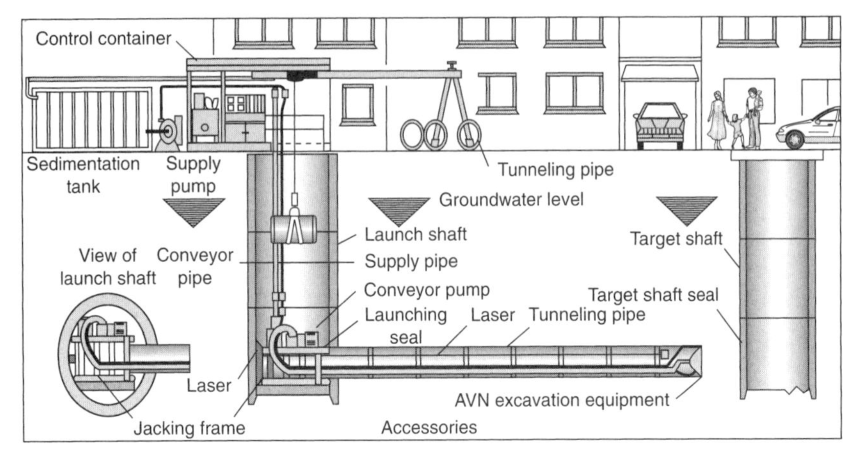

FIGURE 11.14 Overview of the construction site for slurry-type method. (*Courtesy of Herrenknecht Inc.*)

machine, and excavation starts. After the machine is pushed into the ground, the first segment of the pipe is lowered. As main jack pushes the pipe, the MTBM simultaneously excavates soil (Ueki et al., 1999).

The tunneling process for slurry type is as follows:

1. Excavate and prepare the driving shaft.
2. Setup the control container and any other auxiliary equipment beside the jacking shaft.
3. Setup the jacking frame and the hydraulic jacks.
4. Lower the MTBM into the driving shaft and set it up.
5. Setup laser guidance system and the MTBM in the driving shaft.
6. Setup the slurry lines and hydraulic hoses on the MTBM as shown in Fig. 11.15.
7. The main jack pushes the MTBM.

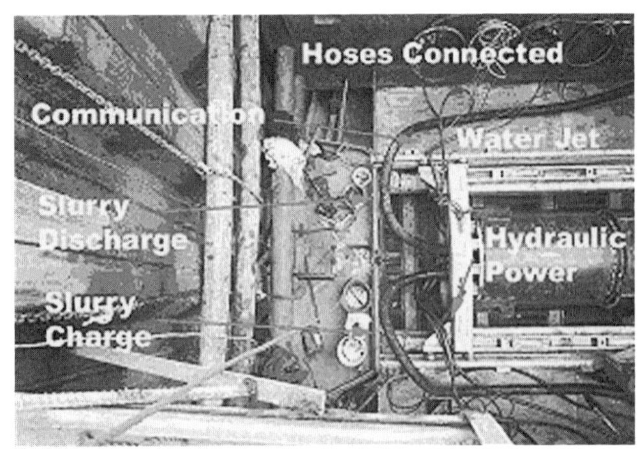

FIGURE 11.15 Slurry lines and hydraulic hoses (Abraham and Gokhale 2002).

FIGURE 11.16 MTBM at the receiving shaft. (*Courtesy of Kerr Construction Inc.*)

8. After the MTBM is pushed into the ground, the slurry lines and hydraulic hoses are disconnected from the jacked section (or MTBM).
9. The hydraulic jacks are retracted.
10. A new pipe segment is lowered in the driving shaft.
11. Connect the slurry lines and hydraulic hoses in the new pipe segment to the ones in the previously jacked segment (or MTBM).
12. Jack the new pipe segment and excavate, while removing the spoil.
13. Excavate and prepare the receiving shaft.
14. Repeat steps 8 to 12 as required until the pipeline is installed.
15. Remove the MTBM through the receiving shaft. Figure 11.16 shows the MTBM entering the receiving shaft.
16. Remove jacking frame and other equipment from the driving shaft.
17. Grout the annular space between the exterior pipe surface and the tunnel.
18. In case of sewer applications, install manholes at the shaft locations.
19. Remove shoring, lining, or casing from the shaft and backfill them.

11.4 MAIN FEATURES AND APPLICATION RANGE

11.4.1 Diameter Range

Based on the experiences in the United States, the range in diameter for MT is from 10 to 136 in. The most common range is between 24 and 48 in. Slurry MT systems can be applied for larger-size pipes than auger MT systems.

11.4.2 Depth of Installation

As the MT operation is performed remotely, there is no theoretical limitation for the maximum depth of installation for MT. However, a minimum of 5 ft of cover or a depth-of-cover to diameter ratio of 3 is usually recommended for MT to avoid possibility of any heave or settlement of the surface and preventing slurry to escape.

11.4.3 Drive Length

The most common drive lengths for slurry MT systems is from 500 to 1500 ft and from 200 to 500 ft for auger MT systems. Curved drives are possible but to date one has not been completed in the United States.

11.4.4 Required Working Space

Adequate working space needs to be provided at the drive shaft to accommodate the required equipment and materials for the MT operation. The space requirement is determined by the drive shaft size, which can range from 16 by 33 ft to 50 by 100 ft, depending on pipe diameter and length and equipment dimensions. Adequate working space typically would range from 20 to 40 ft in width and from 75 to 150 ft in length.

11.4.5 Soil Conditions

The most favorable ground condition for slurry MT is wet sand, and the most favorable ground condition for auger MT is the stable sandy clay. However, a wide selection of MTBM cutter heads are available that provide the capability to handle a range of soil conditions, including boulders and solid rock. Typically, boulders of 20 to 30 percent of the machine diameter can be removed by MT by crushing the boulders into particle sizes of 1 in and smaller. Table 11.1 presents applicability of slurry MT for different soil conditions.

TABLE 11.1 Applicability of slurry microtunneling for different soil conditions

Type of soil	Applicability
Soft to very soft clays, silt, and organic deposits	Yes
Medium to very stiff clays and silts	Yes
Hard clays and highly weathered shales	Yes
Very loose to loose sands (above water table)	Yes
Medium to dense sands (below the water table)	Yes
Medium to dense sands (above the water table)	Yes
Gravels and cobbles less than 2 to 4 in. diameter	Yes
Soils with significant cobbles, boulders, and obstructions larger than 4 to 6 in. diameter	Marginal
Weathered rocks, marls, chalks, and firmly cemented soils	Yes
Significantly weathered to unweathered rocks	No

11.4.6 Productivity

Crews of four to eight can obtain a production rate of 30 to 60 ft per 8-h shift in MT operations (dependent on soil and project conditions).

11.4.7 Accuracy

The method is capable of installing gravity flow pipes to a high precision. Hence, these are mainly used for the installation of gravity flow lines where a high degree of precision is required. The laser system for controlling the alignment permits systems to be installed to an accuracy of ±1 in.

11.4.8 Major Advantages

MT methods are capable of installing pipes to accurate line and grade tolerances. They have the capability of performing in difficult ground conditions without expensive dewatering systems or compressed air. Pipelines can be installed at a great depth without a drastic effect on the cost. The depth factor becomes increasingly important as underground congestion is increased or a high water table is encountered. Safety is enhanced as workers are not required to enter trenches or tunnels. The product pipe, with sufficient axial load capacity, can be jacked directly without the need of a separate casing pipe.

11.4.9 Major Limitations

The capital cost in equipment is high. However, on projects where these methods have been competitively bid against other tunneling methods, the unit price costs have been in line. Some MTBM systems have difficulty in soils with boulders which size more than 20 or 30 percent of the machine diameter. One of the major disadvantages of MT methods has been their inability to use flexible or low-strength pipes such as PVC. Other limitation includes problems caused by obstructions, such as large boulders, roots, or manmade structures.

REVIEW QUESTIONS

1. What is MT? Describe the major components of the MT process.
2. Compare slurry- and auger-type MTBM.
3. Describe the MT guidance system.
4. Describe the spoil removal system in slurry MT method.
5. In what project conditions a MT method is recommended? Explain.
6. Describe the main characteristics of MTBM.
7. What are the main applications of the MT method?
8. What are the principle advantages and limitations of the MT method?

REFERENCES

Abraham, D., and S. Gokhale (2002). *Development of a Decision Support System for Selection of Trenchless Technologies to Minimize Impact of Utility Construction on Roadways.* FHWA/IND/JTRP-2002/7, SPR-2453, National Technical Information Service, Springfield, Va.

American Society of Civil Engineers (ASCE) (2001). *Standard Construction Guidelines for Microtunneling.* CI/ASCE Standard 36-01, Reston, Va.

Atalah, A., and P. Hadala (June 1996). Microtunneling database for the USA and Canada from 1984 to 1995. *Proceedings of the Pipeline Crossings*, Burlington, Vt., ASCE, Reston, Va.

Bennett, R. D., L. K. Guice, S. Khan, and K. Staheli (1995). *Guidelines for Trenchless Technology: CIPP, FFP, Mini-HDD, and Microtunneling.* Construction Productivity Advancement Research (CPAR) Program Technical Report. U.S. Army Corps of Engineers. Vicksburg, Miss.

Essex, R. J. (1993). *Subsurface Exploration Considerations for Microtunneling/Pipe Jacking Projects. Proc. Of Trenchless Technology: An Advanced Technical Seminar for Trenchless Pipeline Rehabilitation, Horizontal Directional Drilling, and Microtunneling*, Vicksburg, Miss., January 26–30.

Iseley, T., and Gokhale, S. (1997). *Trenchless Installation of Conduits Beneath Roadways. Synthesis of Highway Practice 242, Transportation Research Board*, National Academy Press, Washington, DC.

Iseley, T., Najafi, M., and Tanwani, R. (1999). *Trenchless Construction Methods and Soil Compatibility Manual.* Trenchless Technology Committee, National Utility Contractors Association (NUCA), Arlington, Va.

Najafi, M. (1993). *Evaluation of a New Microtunneling Propulsion System,* Ph.D. Dissertation, Louisiana Tech University, Ruston, La.

Nido, A. (1999). *Productivity Projection Model for Microtunneling Operations Based on a Quantitative Analysis of Expert Evaluation.* MS Independent Research Study, School of Civil Engineering, Purdue University, West Lafayette, Ind.

Staheli, K., and G. E. Hermanson, (1996). Microtunneling, when, where, and how to use it. *Water Environment and Technology*, 8(3):31–36. Alexandria, Va.

Ueki, M. (1999). A Decision Tool for Microtunneling Method Selection. MS Thesis, University of Texas at Austin, Austin, Tex.

Ueki, M., C. T. Haas, and J. Seo (1999). Decision tool for microtunneling method selection. *ASCE Journal of Construction Engineering and Management,* March/April 1252:123–131.

CHAPTER 12
PILOT TUBE MICROTUNNELING

12.1 INTRODUCTION

Pilot tube microtunneling (PTMT) was introduced to the market in the 1990s. PTMT should *not* be confused with methods such as horizontal directional drilling and horizontal auger boring. The ASCE 36-01 Standard Construction Guidelines for Microtunneling defines PTMT as a "multistage method of accurately installing a product pipe to line and grade by use of a guided pilot tube followed by upsizing to install the product pipe."

PTMT is a *hybrid* version of conventional microtunneling. PTMT combines the accuracy of microtunneling, the steering mechanism of a directional drill, and the spoil removal system of an auger-boring machine. PTMT employs augers to transport spoil and a guidance system that includes a camera mounted theodolite. The target uses electric light emitting diodes (LEDs) to secure high accuracy in line and grade. When project conditions are suitable, PTMT can be a cost-effective tool for the installation of small-diameter pipes of sewer lines or water lines. This technique can also be used for house connections direct from the main line sewers. Typically, pilot tube machines can be used in soft soils and at relatively shallow depths. Jacking distances of 400 ft have been accomplished with newer guidance systems.

12.2 METHOD DESCRIPTION

PTMT can be applied for the installation of small-diameter pipes, which require high accuracy in line and grade. PTMT employs augers for excavation and soil removal and the jacking system for pushing the pipes, as does horizontal auger boring. It uses a theodolite with camera for the accurate guidance system. The target with LEDs is mounted in the steering head and is monitored through the TV monitor, which is similar to the guidance system of microtunneling.

This method cuts a borehole with a steering head connected to pilot tubes, the size of which is smaller than the required size. Then a reamer and auger casing with augers inside enlarge the borehole. The product pipes are then to follow the auger casing to be installed in the ground. As the application of this technique is for smaller-size pipes, the equipment and the required space for the operation is smaller than that of other jacking methods such as horizontal auger boring (HAB), pipe jacking (PJ), and microtunneling (MT).

PTMT employs a steering head for boring and adjustment of alignment and grade. The steering head has a slant on one side and several types of steering heads are available based on the degree of the slant. For instance, Akkerman Inc. provides three types of steering

FIGURE 12.1 Steering heads for PTMT. (*Courtesy of Akkerman Inc.*)

heads with different slants of 30, 45, and 60 degrees. Two different types of steering heads are shown in Fig. 12.1.

For accurate boring using the PTMT, the target is one of the critical components. Figure 12.2(a) shows the target used for PTMT operation, and Fig. 12.2(b) shows the target mounted in the steering head. LEDs are arrayed to compose two circles and one line from the center. It is operated by a battery that lasts for about 10 days when charged. When the target is turned on, the LEDs are illuminated, as shown in Fig. 12.2(b), and they can be seen through the theodolite even though the target and the steering head are in the borehole.

The guidance system, comprising target, theodolite, camera, and monitor, can detect the deviation of the drilling profile, and gives the operator continuous information about the location of the steering head. The structure of the guidance system is illustrated in Fig. 12.3. If deviations are detected though the monitor, the operator can modify the direction of the steering head by thrusting it using the characteristic of slanted surface of steering head. As

(a) Target (b) Target in the steering head

FIGURE 12.2 Theodolite and target for PTMT. (*Courtesy of WIRTH GmbH, division of SOLTAU Microtunneling.*)

FIGURE 12.3 Guidance system for PTMT. (*Courtesy of Herrenknecht Inc.*)

the accuracy of the theodolite setup and target determines the accuracy of the entire project, prudent setup is required.

The boring process in PTMT can be described as follows:

1. Excavate and prepare the driving and receiving shafts.
2. Lower the thrust frame into the driving shaft and set it up. The thrust frame for PTMT is shown in Fig. 12.4.

FIGURE 12.4 Installation of thrust frame. (*Abraham and Gokhale, 2002.*)

FIGURE 12.5 Installation of a theodolite. (*Abraham and Gokhale, 2002.*)

3. Setup the guidance system in the driving shaft. Figure 12.5 shows the setup of a camera-mounted theodolite. The camera will be connected to the video monitor installed at the driving shaft, as shown in Fig. 12.6, and will be used for monitoring the line and grade of the drilling profile.
4. Setup the steering head and target. Different steering head can be deployed based on the soil conditions.
5. Install the pilot tube behind the steering head. The boring process proceeds with the rotation and the thrust of the pilot tube. The deviations are continuously adjusted

FIGURE 12.6 Video monitor. (*Abraham and Gokhale, 2002.*)

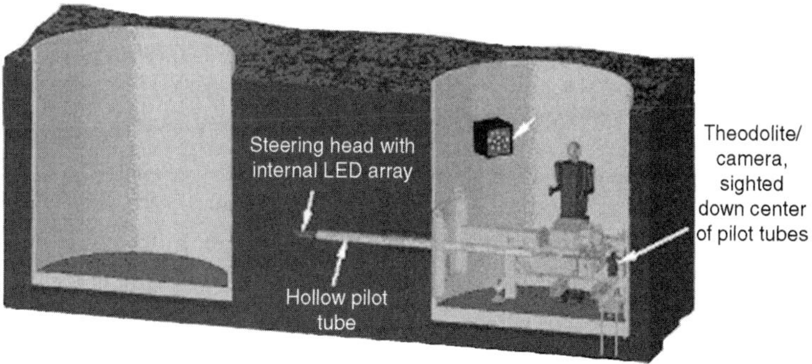

FIGURE 12.7 Pilot tube boring. (*Courtesy of Akkerman Inc.*)

through the video monitor surveillance of the illuminated target via the theodolite. The process of pilot tube boring is illustrated in Fig. 12.7. Pilot tubes used for this process are shown in Fig. 12.8.

6. When the steering head reaches the receiving shaft, the reamer and the casing with auger inside are connected to the last segment of the pilot tube. Then the reamer and the auger enlarge the pilot borehole by rotating and forward thrusting the reamer and casing. The steering head and the pilot tubes can be retrieved at the receiving shaft. This process is repeated until all pilot tubes are removed. The illustration for this process is presented in Fig. 12.9. Reamer, augers, and casings used for this process are shown in Figs. 12.10, 12.11, and 12.12, respectively. Figure 12.13 presents the PTMT control board.
7. After the reamer has reached the receiving shaft, an adapter is installed at the end of the last casing. This adapter connects the casing and the product pipe. The jacking frame pushes the product pipe while the casings are retrieved at the receiving shaft. In this operation, the augers remove the soil in the casings, and are then retrieved at the driving shaft.

(a)

(b)

FIGURE 12.8 Pilot tubes. (*Abraham and Gokhale, 2002.*)

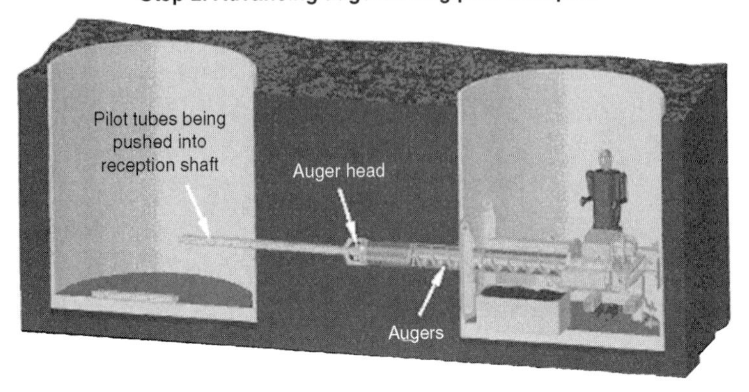

FIGURE 12.9 Reaming process. (*Courtesy of Akkerman Inc.*)

FIGURE 12.10 Reamer for PTMT. (*Courtesy of Akkerman Inc.*)

FIGURE 12.11 Augers for PTMT. (*Courtesy of Akkerman Inc.*)

FIGURE 12.12 Casings for PTMT. (*Courtesy of Akkerman Inc.*)

This process is repeated until all the casings and augers are removed. Figure 12.14 illustrates the installation of product pipes. The pipe adapter is shown in Fig. 12.15.

12.2.1 Different Variations of PTMT

There are two variations of PTMT as described below:

1. *Three-phase PTMT*: This method is most common as 80 percent of the PTMT pipes are installed by this method. It is used for small diameters between 6 in and 10 in. Figure 12.16

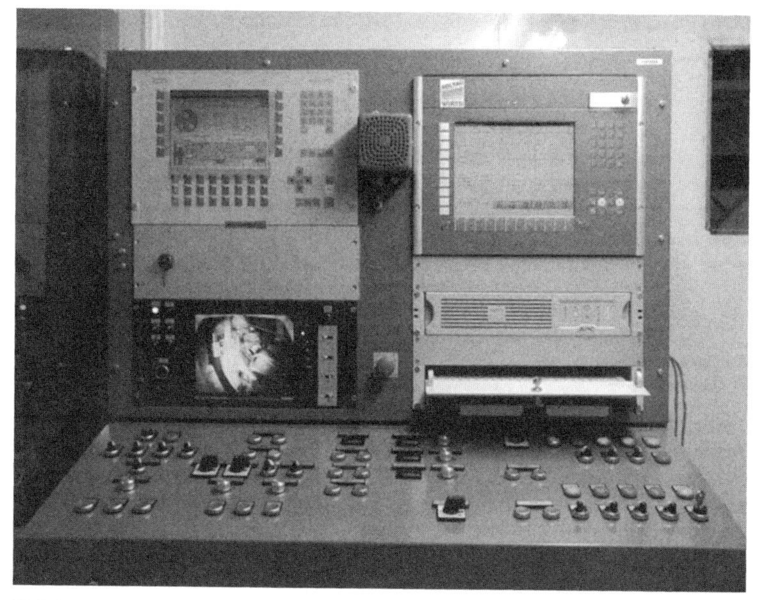

FIGURE 12.13 PTMT control board. (*Courtesy of WIRTH GmbH, division of SOLTAU Microtunneling*).

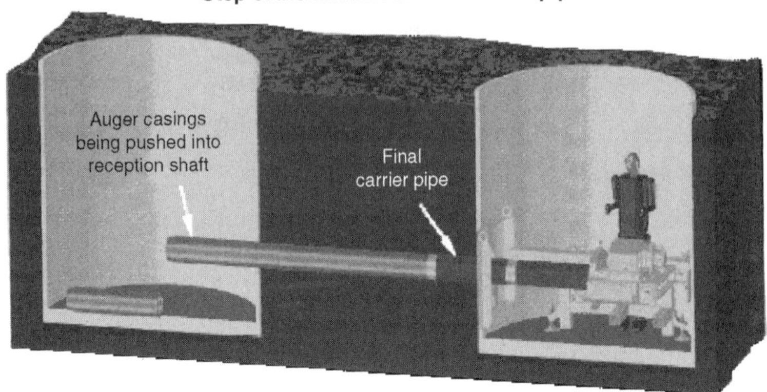

FIGURE 12.14 Installation of pipes using PTMT. (*Courtesy of Akkerman Inc.*)

presents the three-phase process. In the first phase, a steerable pilot tube is jacked from an entry pit into the ground displacing the soil. In the installation of the pilot tube, there is no spoil removal, as the pilot tube displaces the soil. The pilot is installed by thrust alone and spoil is displaced. This system is generally suitable without cutting soil in a ground condition with a standard penetration test (SPT) of less than 50. Only during the steering process, rotation is used. The second phase installs a temporary steel casing. In the third phase, the product pipe is installed and at the same time the casing is pushed out and collected at the exit shaft.

FIGURE 12.15 Pipe adapter for PTMT. (*Abraham and Gokhale 2002.*)

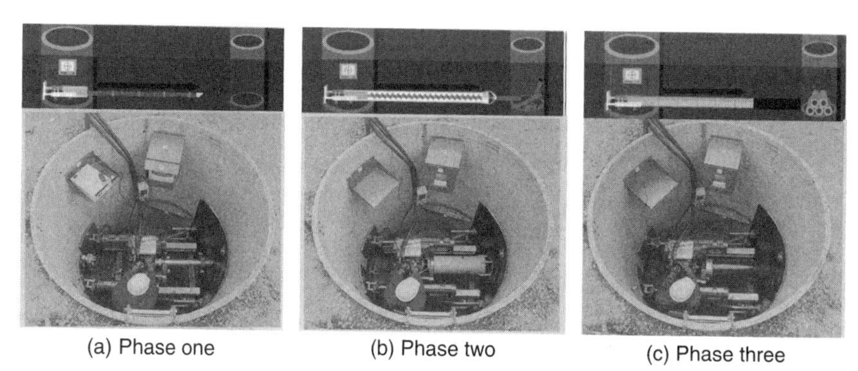

(a) Phase one (b) Phase two (c) Phase three

FIGURE 12.16 The three-phase PTMT process. (*Courtesy of WIRTH GmbH, division of SOLTAU Microtunneling.*)

a. Phase one
b. Phase two
c. Phase three

2. *Two-phase PTMT*: For diameters 10 in and above, the three-phase is used. This method uses augers and casing inside the product pipe being installed at the same time during reaming. More than 160,000 ft of pipe has been installed with this method in the U.S.A. Figure 12.17 presents the new PTMT accessories.

12.3 MAIN FEATURES AND APPLICATION RANGE

PTMT can install pipes from 4 in to 30 in inside diameter (ID). The maximum drive length of PTMT is 300 ft, but drive lengths of 400 ft have been installed in suitable ground conditions and with latest technology available. PTMT operations have an accuracy of 0.25 in for 300-ft pipe installations. However, the actual accuracy achieved at the jobsite depends on the capability of theodolite and the operator's skills. The productivity of a 250-ft PTMT

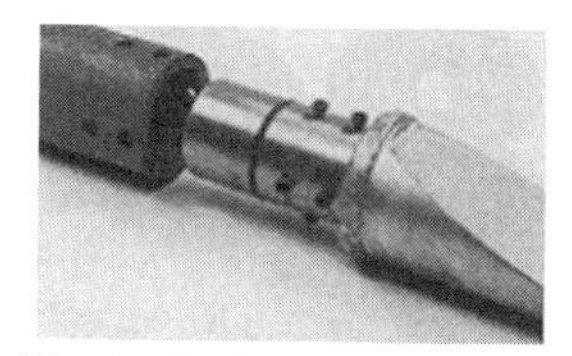

The steering head with integrated target and tapered tip for steering control.

The pilot tube assembly with new couplings. (*Patent pending*)

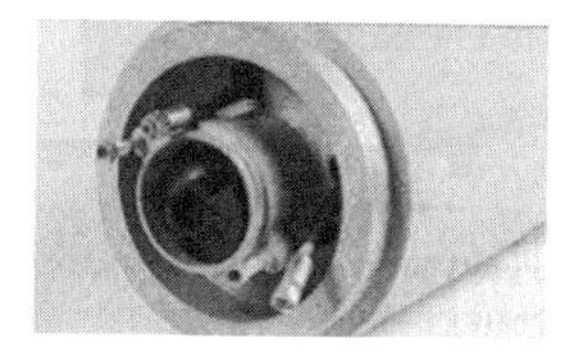

Reamer with auger supply lines for operation below the water table.

FIGURE 12.17 The PTMT accessories. (*Courtesy of WIRTH GmbH, division of SOLTAU Microtunneling.*)

TABLE 12.1 Applicability of Pilot Tube Microtunneling for Different Soil Conditions

Type of soil	Applicability
Soft to very soft clays, silt, and organic deposits	Marginal
Medium to very stiff clays and silts	Yes
Hard clays and highly weathered shales	Yes
Very loose to loose sands (above water table)	Marginal
Medium to dense sands (below the water table)	Marginal
Medium to dense sands (above the water table)	Yes
Gravels and cobbles less than 2 to 4 in. diameter	Yes
Soils with significant cobbles, boulders, and obstructions larger than 4 to 6 in. diameter	Marginal
Weathered rocks, marls, chalks, and firmly cemented soils	No
Significantly weathered to unweathered rocks	No

installation from setup to completion (including pilot bore, reaming, and product pipe installation) is possible in three days. PTMT can be applied in a variety of soft soil conditions. However, hard soil with relatively large boulders and rocks can cause some challenges to the performance, as can running sands and unstable soils. Different types of reaming heads are available, such as reaming heads suitable for clay soil and also the ones with flow control; to work in running sand up to 15 ft below the water table. There are also swivels and reaming heads available with full-face cutters capable of excavating harder ground. Table 12.1 presents applicability of PTMT for different soil conditions.

12.4 ADVANTAGES

The PTMT is very accurate and can be used above or below the water table. There is a small workspace requirement for setup container and the overhead hoist. Choice of hydraulic power packs, diesel, or electric is possible.

12.5 LIMITATIONS

The method is applicable in soft grounds where displacement of soil for pilot tube is possible. Entry and exit pits are required.

REVIEW QUESTIONS

1. What is PTMT?
2. Compare the three-phase and the two-phase PTMT methods.
3. For what types of ground and project conditions would you recommend PTMT? Explain.

4. What guidance system is used for PTMT? Briefly describe its primary components and setup procedure.
5. How does PTMT method differ from MTBM method? Explain.
6. How does PTMT method differ from horizontal auger boring method? Explain.
7. How does PTMT method differ from horizontal directional drilling method? Explain.
8. Check with your local utility contractors' association or your city's public works department to see if PTMT has been used in your area. If PTMT has been used, obtain all the project information, such as, product pipe size and type, drive distance, location, type of ground conditions encountered, project duration, type of application, any challenges encountered, and accuracy obtained.

REFERENCES

Abraham, D., and S. Gokhale. (2002). *Development of a Decision Support System for Selection of Trenchless Technologies to Minimize Impact of Utility Construction on Roadways.* FHWA/IND/JTRP-2002/7, SPR-2453, National Technical Information Service, Springfield, Va.

Akkerman Inc., Brownsdale, Minn. (2004). Available at www.akkerman.com.

American Society of Civil Engineers (ASCE). (2001). *Standard Guidelines for Microtunneling*, Arlington, Va.

Borhtec BmbH, Alsdorf, Germany. (2004). Available at http://www.bohrtec.de/.

Herrenknecht, A. G., Schwanau, Germany. Available at http://www.herrenknecht.com.

North American Society for Trenchless Technology (NASTT) (2004). Arlington, Va. Available at www.nastt.org.

WIRTH GmbH, Division SOLTAU Microtunneling. Available at http://www.microtunneling.com/soltau/-rvs80.htm.

CHAPTER 13
CURED-IN-PLACE-PIPE

13.1 INTRODUCTION

Cured-in-place-pipe (CIPP) installation is one of the most widely used methods of trenchless pipeline renewal and for both structural and nonstructural purposes. The CIPP process involves a liquid thermoset resin-saturated material that is inserted into the existing pipeline by hydrostatic, air inversion, or mechanically pulling with a winch and a cable and inflating. The material is heat-cured in place resulting in a CIPP product. Insituform introduced CIPP in the United Kingdom in 1971 and entered the U.S. market in 1977. In 1989, the InLiner process was introduced in Houston, Texas. In the 1980s, Ashimori Industries developed a method called hose lining basically for pressure pipe applications in the Japan gas industries. This increased number of available methods, provides many technical and economical benefits for utility owners through increased competition. Chapter 5 presented an overview of the CIPP design process. This chapter presents an overview of the CIPP method.

13.2 SITE COMPATIBILITY AND APPLICATIONS

CIPP can be used effectively for a wide range of applications that include sewers, gas pipelines, potable water pipelines, chemical and industrial pipelines, and pressure pipes. The physical properties of CIPP make it especially suitable for different types of pipe geometries including straight pipes, pipes with bends, pipes with different cross-sectional geometries, pipes with varying cross sections, pipes with lateral connections, and deformed and misaligned pipelines. However, several factors must be evaluated before choosing CIPP as the method of renewal for an individual project. Factors such as space availability, chemical composition of the fluid carried by the pipeline, the number of laterals, the number of manholes, installation distance, renewal objectives, structural capabilities of the old pipe, and the like must be assessed before making a choice on the renewal system. CIPP is also used for localized repairs in a wide range of systems. This is discussed in more detail in Chap. 20 where several localized repair techniques are described.

The possibility of negotiating bends depends on the installation and curing process of the various systems. Liners that are inverted during insertion can usually negotiate bends up to 90 degrees. There are, however, limitations to the degree of bending that UV-cured liners can manage. Bends can also present problems for liners that are pulled into place because the liner must remain on the underlying protective foil and must under no circumstance become twisted. It is important to be aware that vertical creases may occur in the liner when negotiating sharp bends. Table 13.1 presents an overview of CIPP applications and possible limitations.

TABLE 13.1 Overview of CIPP Applications and Possible Limitations

Pipeline type	Suitable	Notes
Sewers	Yes	
Gas pipelines	No	
Potable water pipelines	Yes	
Chemical or industrial pipelines	Yes	A
Straight pipelines	Yes	
Pipelines with bends	Yes	B
Circular pipes	Yes	
Noncircular pipes	Yes	C
Pipelines with varying cross sections	Yes	D
Pipelines with lateral connections	Yes	E
Pipelines with deformations	Yes	F
Pressure pipelines	Yes	

Notes: A. As each manufacturer uses different types of resin, reinforcement, and the like, the chemical resistance of liners varies with manufacturer and service provider. However, all systems using felt impregnation with polyester or other resins are suitable for aggressive wastewater. If special characteristics are required, individual manufacturers should be contacted to help finding the most suitable product for specific requirements with regard to chemical content, concentration, temperature, and so on. B. The possibility of negotiating bends depends on the installation and curing processes of the various systems. Liners that are inverted during insertion can usually negotiate bends up to 90 degrees. However, the number of bends that can be negotiated is limited by the pipeline geometry. Bends can also present problems for liners that are pulled into place because the liner pipe must remain on the underlying protective foil and must under no circumstance become twisted. It is important to be aware that vertical creases may occur in the liner when negotiating sharp bends. C. All the described methods are suitable for lining main pipelines with circular, egg-shaped, or V-shaped cross sections. D. Some proprietary liner pipe systems allow CIPP techniques to be used in pipelines with varying cross-sectional areas. E. Laterals can be reinstated using a robotic cutter. It is presently possible to reopen laterals on pipelines with a diameter of 3 in or more. In person-entry systems, laterals are reinstated manually. With CIPP systems, watertight connectors can be fitted between laterals and main lines. See Chap. 18 for localized repair methods. F. CIPP can be used where deformities are minor. However, the general advice when ovality is more than 10 percent is to dig the pipe up. Although a CIPP liner can get through the existing pipe deformities, all the deformities will be noticeable after renewal, and there will be wrinkling in the CIPP liner.

13.3 MAIN CHARACTERISTICS

The primary components of CIPP are a flexible fabric tube and a thermosetting resin system. For typical CIPP applications, the resin is the primary structural component of the system. These resins generally fall into one of the following generic groups, each of which has distinct chemical resistance and structural properties. The most common resin types used for CIPP applications are unsaturated polyester, vinyl ester, and epoxy. All of these resins have distinct chemical resistance to domestic sewage.

Unsaturated polyester resins were originally selected for the first CIPP installations because of their chemical resistance to municipal sewage, good physical properties in CIPP composite, excellent working characteristics for CIPP installation procedures, and economic feasibility. Unsaturated polyester resins have remained the most widely used systems for the CIPP processes for over two decades.

TABLE 13.2 Typical CIPP Design Properties

Property	ASTM test method	Polyester (psi)	Vinyl ester (psi)	Epoxy (psi)
Tensile strength	D638	2,000–3,000	2,500–3,500	4,000
Flexural strength	D790	4,000–5,000	4,000–5,000	5,000
Flexural modulus	D790	250,000–500,000	250,000–500,000	300,000

Vinyl ester and epoxy resin systems are used in industrial and pressure pipeline applications, where their special corrosion, solvent resistance, and higher temperature performance are needed. These systems can be used in residential areas. However, it is typically not necessary and the costs will increase.

The primary function of the fabric tube is to carry and support the resins until it is installed and cured. This requires that the fabric tube withstand installation stresses with a controlled amount of stretch but with enough flexibility to dimple at side connections and expand to fit the existing pipeline irregularities. The fabric tube material can be woven or nonwoven, with the most common material being a nonwoven, needled felt. Impermeable polyethylene and polyurethane coatings are commonly used on the exterior, or interior, or both of the fabric tube to protect the resin during installation. The layers of the fabric tube can be seamless, as with some woven material, or longitudinally joined with stitching or heat bonding.

The primary differences between various CIPP systems are in the composition and structure of the tube, method of resin impregnation (by hand or by vacuum), installation procedure, and curing process. Typical CIPP design properties are shown in Table 13.2. It should be pointed out that these values might not represent properties of each resin after strained corrosion test.

13.4 METHOD DESCRIPTION

CIPPs are installed either by inverting them into the pipeline using water or air pressure, or by pulling them in with the help of a winch. The system is then cured using water, steam, or UV light. The process involves several stages each of which is briefly described below.

13.4.1 Initial Setup: Cleaning, Televising, and Resin Impregnation (Wet-Out Process)

First, the existing pipe to be lined is televised and cleaned. The distance between inspection manholes is measured and all the damaged sections and service inlets are carefully noted. If an analysis of the old pipe justifies use of CIPP method, an order is placed with the CIPP manufacturer based on the actual length and diameter measurement and the product is manufactured for insertion into the particular pipe. After the pipeline has been cleaned and televised, a flexible fabric tube, manufactured and cut according to specific project conditions is impregnated with the designed hardening material (resin and catalysts), to prepare it for insertion through a manhole or another convenient entry point. It is critical that the fabric tube is totally saturated to provide consistent finished physical properties required by design. Figure 13.1 presents the resin impregnation (wet-out) process.

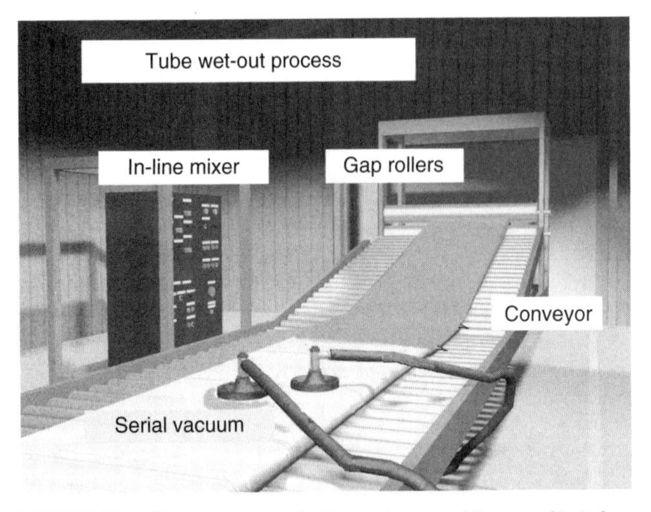

FIGURE 13.1 The resin impregnation (wet-out) process. (*Courtesy of Insituform Technologies.*)

13.4.2 Installation

There are two basic methods of CIPP installation: the inversion method and the winch method. Specific variations of these techniques exist depending on the manufacturer.

Inversion Method. In the inversion method (Fig. 13.2), the liner is clamped around an inversion ring and turned inside out (inverted) down the length of the pipe using compressed air or water to pressurize it. The flexible tube is then forced by the compressed air or water against the wall of the pipe to be rehabilitated.

Water Inversion. Water inversion is the oldest method and can be used for both small and large diameters. The tube is inverted by a water head located in the manhole. The inversion head is fitted to an inversion standpipe, which provides the necessary head to invert the liner into the existing pipeline. For large-diameter pipes, the tube is inverted from top of the inversion standpipe. A cable is attached to the foremost end of the tube to guide it and ensure correct inversion into the pipeline. Strengthening strips may be sandwiched between the layers of the laminate to prevent longitudinal stretching during installation.

Compressed Air Inversion. Compressed air inversion techniques differ depending on which curing method is to be used. If the CIPP liner is to be cured with steam, it is inverted into the existing pipe via an inversion using compressed air with an optional guide cable attached. When the CIPP liner is in place, pressure is released and the compressed air is replaced by steam or water for curing.

If the CIPP liner is to be cured with UV light, installation is accomplished from a truck containing the liner positioned above the launch pit. The CIPP liner is fitted to a launching device that is lowered into the pit. Mobile installation equipment is used if the truck cannot be positioned above the pit. The liner is inverted into the pipe using compressed air and guide cable, the air pressure and speed of advance being controlled during the whole period of time. Once the CIPP liner is in place, the compressed air pressure is maintained while UV light source is fitted into place. Water inversion, and to a certain extent compressed air

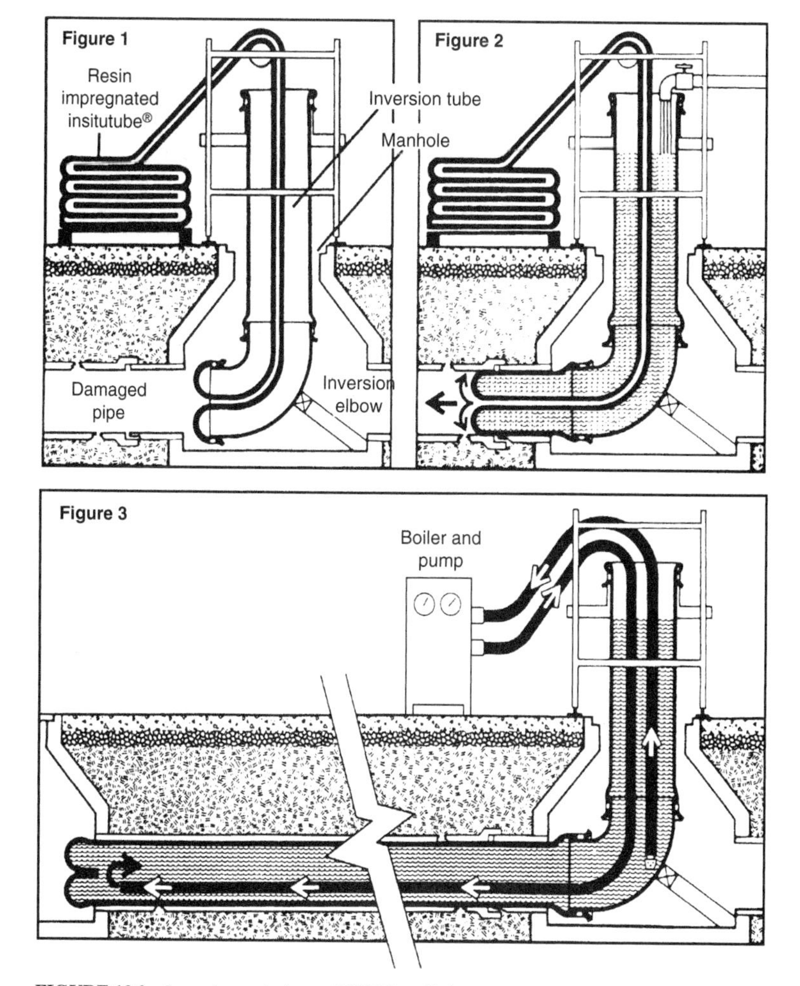

FIGURE 13.2 Inversion technique of CIPP installation.

inversion, ensure that any water collected in pipelines with structural irregularities is forced out and that the pipeline is not damaged in any way during installation.

Winched Insertion. In the winch installation technique, liners are pulled into place on protective membranes, and pulled through the length of the pipe. The purpose of the membranes is to prevent the resin from being damaged when it is dragged over cracks, fractures, corroded sections of the pipe, and the like, to reduce friction. There is also a risk that loose pipe fragments will become dislodged when the membrane and liner are winched into pipe. It should be noted that tubes with outer coating can be pulled into place without the membrane.

Then the pipe is inflated using compressed air or pressurized water and allowed to cure. Winched insertion has limitations with regard to pipe diameter and length. The laminate design prevents the liner from being torn apart during installation (Fig. 13.3).

FIGURE 13.3 The winched installation process. (*Courtesy of Insituform Technologies.*)

13.4.3 Curing

Curing of the resin is done by hot water, steam, or UV light. The specific choice between these techniques should be determined before installation, because this may affect the choice of the installation process. Each of these installation methods is described below.

Water Curing. Hot water curing is the oldest method of curing CIPP. The curing process can be documented and continuous registration of water temperature ensures the correct laminate curing. It is also possible to control laminate cooling so that tensile stresses in the material can be minimized. Curing with water allows long runs of large-diameter piping to be lined.

The main limitation of this process is that it can be slow and that water used for curing is typically wasted. It is also necessary to consider the recipient of the large quantities of water used during lining.

Steam Curing. Curing with steam has been in use since the beginning of the 1990s. The main benefits of the method are that curing occurs rapidly and it can be used for steep sewers having a fall of up to 200 ft. However, this method has the following disadvantages:

- It can be difficult to control the steam supply and the liner may therefore *boil*.
- Cooling occurs suddenly, increasing the tensile stress in the laminate.
- It is difficult to check whether the laminate is thoroughly cured, and it can also be difficult to ensure complete curing in pipe sections containing structural irregularities or where groundwater infiltrates the pipe.
- When lining pipes of limited gradient or pipes with structural irregularities, it is important to be aware of the fact that *puddles* of condensation can cause insufficient curing.

UV Light Curing. Curing with UV light has been used since the early 1980s. The method provides rapid curing and is relatively simple to control. The curing process is controlled and documented by continuous registration of air pressure, laminate temperature, UV lamp output or light intensity, and curing rate. With UV curing, cooling occurs continuously as the UV light source gradually passes through the pipe, thus reducing tensile stresses in the liner. However, water and steam curing are widely used because of their ease of operation, and long history of use in the CIPP industry.

13.4.4 Final Steps

After curing, the new pipe is cooled down and drained. The ends of the cured pipe are then trimmed to form a smooth, seamless CIPP. A special cutting device or robot is used in conjunction with a closed-circuit television (CCTV) camera to reopen service connections, locating the dimples visually with the camera. The manhole openings and lateral cut outs are then sealed and the final CCTV is provided to the customer if required.

Reopening the service connections can be done by person-entry for larger-diameter pipes or robotically for smaller diameters. Normally, a small dimple is left in the CIPP liner directly over each service connection, allowing them to be easily located and reopened. However, the number and locations of the service connections should be noted during the prelining televising process to ensure that all connections are reopened, and to aid in locating those that are difficult to identify.

The new pipe or CIPP is of slightly smaller inside diameter fitted in configuration with the shape of the existing pipe. The CIPP has a very smooth interior surface, which generally improves the flow capacity despite the slight decrease in diameter.

13.5 MAJOR ADVANTAGES

- Grouting is not normally required because of the tight fit of the CIPP against the existing pipe except at manhole openings and at lateral openings if required.
- The CIPP has no joints and has a very smooth interior surface, which typically improves the flow capacity despite the slight decrease in diameter.
- Noncircular shapes can be accommodated often without a decrease in flow capacity.
- Lining is capable of accommodating bends and the existing pipe deformations. Most shapes of pipes can be lined.
- Entry is possible via existing manholes or through minor excavations. Remote-control internal lateral connection is employed.
- Although there is a negligible decrease in the existing pipe cross section, the hydraulic capacity is improved because of a smooth CIPP surface.

13.6 MAJOR LIMITATIONS

- The tube or the hose must be specially constructed for each project (some manufacturers provide stock materials to their licensees for small diameters, such as 8, 10, and 12 in).
- The existing flow must be bypassed during the installation process.
- Sealing may be required at liner pipe ends and at lateral connections to prevent infiltration.

- Amount and the type of resin used is a contractor's function and the owners should have technical specifications and inspection procedures to ensure proper resin quality and handling.
- The curing process must be carefully monitored, inspected, and tested to ensure obtaining the specified physical and chemical properties.
- The curing process may take 1 to 5 h and styrene might be present with polyester resin during the curing. The circulated water used for curing may contain styrene and must be removed from the job site. On sanitary sewer jobs, water is routinely discharged to the sanitary sewer. If the project is not a sanitary sewer (such as storm sewer, culvert, and the like), curing water must be removed from the job site. However, it should be noted that majority of CIPP work is in sanitary sewers, where discharge of curing water is not a problem.
- Obstructions can inhibit the lining process.
- The cost of CIPP installations compared with other trenchless renewal methods can be a concern. Depending on the situation, the cost may be more than such methods as sliplining, thermoformed, and close-fit methods. Therefore, it is generally recommended to evaluate the existing pipe problems and select the most cost-effective method for the conditions of the existing pipe and the long-term pipeline applications.

13.7 CASE STUDY

13.7.1 Introduction and Overview

The City of Beloit, Wisconsin, had a chronic infiltration problem during periods of high river levels and the associated high groundwater in the valleys along the Rock River and the two tributary creeks. There was a direct correlation between the water pollution control facility (WPCF) flow and river level readings. A concentrated sewer renewal program was undertaken in 2001 to address the greatest concerns of infiltration in the sewer system by focusing renewal efforts on the sewers that were at or below the annual flood stage of the river.

Project Objectives. The project had the following goals:

- Achieve electrical cost reductions
- Extend the useful life of the water pollution control facility
- Extend the life of all related machinery
- Minimize disinfectant chemical consumption
- Enhance influent strength to optimize biological phosphorus removal

The intent was to achieve these goals with minimum social costs such as traffic disruption, pavement restoration, and disruption to adjacent underground utilities in the congested downtown business district. Consultants were hired to assist with the flow monitoring, report writing, inspection, material testing, and tracking of quantities used.

Project Overview and Results. The majority of the renewal work was performed over the summer and early fall of 2001. Both contracts were completed on time and under budget. Initial infiltration reduction estimates were expected to be between approximately 0.5 and 1 million gal/day. The project results exceeded preliminary estimations by removing approximately 2.5 million gal/day of infiltration, which amounted to nearly 30 percent of the WPCF's prerenewal daily flow as witnessed in the flow monitoring results. With a total cost of US$714,559.34 and a conservative 20-year capitalized present worth figure of US$1.50 per thousand gallons to transport and treat water to calculate the cost savings, the payback on this project (in 2001 dollars) was about 6 months. The power consumption cost reduction alone was approximately US$150,000 annually.

Project Planning. Proper project planning, design, management, execution, and quality assurance and quality control (QA/QC) are crucial to the success of a sewer renewal project, especially when using trenchless technologies. These key factors, which made the project such a success, are highlighted below. Details of the project are elaborated on later.

From the very early planning stages and investigation of infiltration and inflow (I/I) problems through the design of the project, specification writing, and construction, and attention to details is necessary. Without all the pieces of the puzzle studied, planned, and executed properly, it could lead to a less than satisfactory project and a significant waste of money. However, when the plan, the specifications, and the work are good, a major success can be realized.

There are many things to consider when undertaking a large trenchless sewer renewal project. Projects vary greatly from one to another, but many of the key concepts are applicable to all. The intent here is to describe what was done on a particular project that had a very successful outcome in reducing a significant amount of infiltration. During the planning stage the procedures and specifications are laid out in detail, and major concepts and key points that were critical to the project's success are discussed.

When dealing with a sanitary sewer collection system, the mindset of "out of sight, out of mind" often comes up. This unfortunate mindset is a major factor that has led our nation into the state it is in with a significant amount of the country's underground infrastructure in a seriously deteriorated condition. The sanitary sewers, as well as other underground utility piping, have been left out of sight and out of mind for far too long in most communities around the country. Much of the piping is in need of serious repair or renewal.

13.7.2 Predesign Field Investigations

Like many projects, the research, planning, and investigation stages for a sewer renewal project are often time-consuming and must be initiated well ahead of the actual construction phase of the project. Starting by cleaning and televising all sanitary sewers, and all storm sewers if possible, in suspected problem areas of the collection system, might be a good idea. This provides first-hand visual information on the conditions of the collection system piping. Clean pipes should also provide for easier installation of flow monitoring equipment and more reliable results. Flow monitoring can help determine which basins or areas of the collection system have significant I/I problems during wet weather events or during groundwater or river level fluctuations. If it is economically feasible, the flow-monitoring equipment can also be used to track flows for a time period during and after the sewer renewal project for the basin(s) determined to have I/I problems. This allows for tracking the effectiveness of the sewer renewal work and provides quantitative data for the results of the project. Clean sewers should also provide for better CIPP sewer lining and sewer grouting contract pricing. Televising the storm sewers in the same area often allows for viewing and confirming the location of cross connections between the sanitary and storm sewer systems that cannot be seen by televising just the sanitary sewers.

Dye water flushing of services should be done in conjunction with sewer televising to determine which service lateral connections are in use. Use of city personnel and equipment (if available) can be beneficial because this is a very time-consuming process.

City personnel are also familiar with the residents, traffic, and politics of the community and that can be helpful. Checking the city records for locations of building service connections may be marginally helpful depending on how complete and accurate the record set is.

The *big picture* approach to looking at the collection system as a whole, or at least by drainage basins, works best for initially trying to identify potential problem areas. Identify such areas by using what is visible and known about the system and any events such as surcharges or basement backups that have occurred in the past. The topography and age of the piping should also be considered. The combination of field work investigations using tools such as televising, flow monitoring, smoke and dye testing, and visual inspection, along with the research of the history of the system and any known correlation of events that have caused problems within the system are important. This investigation work can help to determine the most critically deteriorated parts of the system, or at least the most highly suspect parts of the system. Repairing individually identified active leaks, unless major, is usually an exercise in futility and monetarily wasteful in correcting significant I/I problems for a community. An entire area, or basin, should be viewed collectively as the potential problem, and if there is a confirmation through further investigation, then a comprehensive renewal plan should be initiated. Such an approach has proven itself to be effective in many cases of sewer renewal.

13.7.3 Evaluation of Optional Technologies

A comprehensive renewal plan involves multiple strategies and methodologies in repairing the problems that are the root cause of I/I in sewer systems. In addition to the traditional open-cut methods of sewer repair and replacement, there is a wide variety of trenchless technologies available for underground infrastructure renewal. Selecting the right renewal method for the right application and having it performed properly can achieve success. Certainly open-cut methods still have their place and sometimes are appropriate in certain situations, but in many cases trenchless technologies are available and are more cost-effective than open-cut methods.

There are many questions that must be answered and many issues to be addressed when considering and evaluating the various renewal methods. First, the appropriate methods for repairing a particular defect or type of defect must be identified. Then they must be evaluated against one another to determine the most cost-effective and acceptable method. When considering different options, there are a number of factors to take into account:

- Determine whether or not a section of the pipe or a manhole is too deteriorated to save with the help of trenchless technology.
- Assess whether the hydraulic capacity of the sewer needs to be increased in the near future.
- When the pipe capacity is an issue, increased hydraulic capacity of CIPP or other renewal methods must be considered with a possibility of installing a larger-diameter pipe using an in-line replacement and the cost benefit of each method should be evaluated.
- Although it is *not* usually the case, evaluate whether CIPP lining of the existing pipe with a future parallel relief sewer line is economically feasible.
- If the sewers being evaluated are a part of a total street reconstruction program, then it might be better to relay all the sewers and laterals out behind the curb line prior to placement of a new pavement.
- Compare the cost of one or two CIPP localized or point repairs in conjunction with chemical grouting versus total line (manhole-to-manhole) CIPP.

When comparing trenchless technology of any kind versus open-cut methods, the social costs associated with open-cut construction should be factored in. Things such as traffic disruption, dirt and noise pollution, damage to pavement, possible damage to adjacent underground utilities, completion timeline, and related issues such as public relations must be considered as well. These factors are important in making the decision on whether or not to go trenchless, depending on how critical the social cost aspect is for a given location.

In weighing the various options of trenchless pipeline renewal technology against one another, there are many factors to consider. Not all of these factors will be discussed here (see Chap. 5 for a complete discussion of method selection criteria) and only the methods considered for the case study project are focused on.

For example, comparing CIPP lining versus chemical sewer grouting involves different comparisons. Although grouting can be cheaper by about 50 percent, it offers little structural renewal. The only structurally related benefit it provides is that the grout conglomeration with the soil outside the pipe helps protect the pipe possibly from damage that might occur if soil continues to be washed away from outside the pipe resulting in a void. This void could eventually cave in and damage the pipe. In a clay pipe that has one or two isolated structural problems that can be short-lined or relayed, depending on current costs, CIPP point repair might be more feasible and economical for a long-term than grouting. Even if a concrete pipe has no structural defects, but is leaking at every joint, CIPP lining may still be the best choice over grouting if corrosion is a concern. Long-term cost analysis often suggests that CIPP is preferable over grouting. Grouting may be preferable for short-term cost-effective renewal projects looking for a quick payback.

Grouting, or other appropriate renewal methods, of the service lateral connections is highly recommended to make the sewer renewal project a comprehensive one. The U.S. Environmental Protection Agency (EPA) studies have shown that oftentimes the infiltration at the service lateral connections to the sewer main can be a significant portion of the overall I/I problem. Another factor that may make lateral renewal an important part of a comprehensive renewal project is that once the leaks in the manholes and main lines have been effectively sealed, the groundwater level may actually rise. This can occur because groundwater no longer drains into the sewer system. This same scenario can cause leaks in higher elevation areas of the collection system, which had showed no visible leaks before, to start leaking because what was once in dry ground is now in a saturated groundwater zone. The water will always seek openings into the system, and what may start as a minor leak can become more significant over time. Meanwhile, the soil surrounding the pipe may get washed away and can eventually cause sink holes or the ground above it to collapse. This could potentially collapse the pipe and cause a *domino effect* of further problems depending on the location of the leak.

13.7.4 Project Design, Specifications, and QA/QC

Once all the parts of the collection system to be rehabilitated have been identified and the repair and replacement methods have been decided upon, the next step is writing a comprehensive set of engineering specifications for the project. The specifications should be written to include all the items and details of the project deemed to be critical in making it a success. The specifications should be based on the preliminary investigation and research work performed and the proper materials and procedures for the construction and the installation for the renewal methods selected. In order to make it a comprehensive project, the specifications need to be comprehensive as well. Detailed specifications help to eliminate gray areas and opportunities for contractor misinterpretation can be avoided. In doing so, it can be made clear exactly what is to take place during every phase of the project: who is to do what, exactly what work is to be performed and how, what equipment and products are to be used, what each party's responsibility is, what the work schedule is, how the payment will be made, and all related issues. If something is thought to be a questionable or debatable item, then it should be addressed in the specifications to eliminate any possible confusion.

Some important design and specification factors are discussed here. A section on televising work should be included in the specifications for CIPP and sewer grouting work to ensure control over the final pre- and postrenewal video quality.

Field observation by a resident project representative can be very important to a project's success. If the project is large enough and the budget allows for a full-time inspection, it is highly recommended. When any two of the following are done concurrently, manhole lining, sewer grouting, or sewer lining, at least two field observers are recommended. When bidding the project by the vertical foot for manhole lining, by the gallon for manhole grouting, by the lineal foot for test and sealing, and by the gallon for sewer sealing, having a full-time inspector can be important. However, if full-time field observation cannot be afforded, then bidding this way is placing a lot of trust on the integrity of the contractor.

Field observation adds costs to the project, but helps ensure that the project work is performed properly and that the contractor follows the design specifications. This should provide for a quality finished product that will meet the intended goals and provide a renewed system that should be successful. Field observation of all aspects of sewer renewal is often worth the expense.

The quality assurance portion of the specification should include, or address, the following concerns at a minimum:

Defects. The CIPP liner should be free of foreign inclusions, dry spots, pinholes, and delaminations.

Wrinkles. Wrinkles, or *fins*, in the bottom half of the CIPP should be limited to a percentage of the pipe diameter or its removal should be required.

Televising. Pre- and postrenewal televising needs to be complete and of high quality.

Material testing. Third-party material testing laboratories need to be used to ensure the CIPP and the cementitious manhole materials meet the specifications.

CIPP material and design parameters. The following design parameters in accordance with ASTM F1216 should be mentioned and *project specific* values assigned. The values listed in Table 13.3 are those that were used for the City of Beloit, Wisconsin, Phase 2 project in 2001.

CIPP Liner thickness. The engineer should design the minimum and the maximum thickness range for each CIPP project and include them either in the bidding forms or in the specifications. A minimum thickness of 6 mm on all 8-in or larger pipes is recommended. As a good rule of thumb, it is always best to maintain a dimension ratio (DR) of 50 or less.

Penalties. Penalty clauses for CIPP thickness and flexural modulus of elasticity should be included. A 15 percent margin on thickness and a 10 percent margin on the modulus of elasticity were used for the City of Beloit project and are recommended. A penalty clause for over-cutting services is also a very good idea.

Test samples. It is also recommended to specify the minimum length of CIPP pipe required for preparing a sample for testing. This was not specified for the City of Beloit project, but has since been deemed necessary and the City of Beloit will require a minimum of a 1-ft sample for all future projects.

CIPP liner inversion. The liner inversion rate should be restricted to a recommended 2 ft/s.

Annular seal at manholes. A specification calling for a hydrophilic elastomer watertight seal installation where groundwater is a concern for infiltration is recommended.

TABLE 13.3 *Sample* CIPP Design Parameters

Project parameter	Design value
Design factor of safety	2.0
Long-term *creep* effect factor	0.50
Ovality	4 percent
Structural enhancement factor, K	7
Groundwater depth (next to the river)	+2 ft
Groundwater depth (in the structural rehabilitation areas only)	4 ft
Soil depth (above crown)	To the surface
Soil modulus	1000 psi
Soil unit weight	120 pcf
Live load	0.0
Design condition	Fully deteriorated
Flexural modulus (D790)	400,000 psi
Design results	
CIPP thickness for resin system of Es = 400,000 psi	0.17 in
CIPP thickness for resin system of Es = 300,000 psi	0.18 in
CIPP thickness for resin system of Es = 250,000 psi	0.19 in

Exfiltration testing. Specifying a hydrostatic exfiltration test that essentially meets the Ten State Standard of 200 gal of permissible leakage per inch mile of CIPP sewer pipe per day is recommended. Specify that the original pipe diameter will be used in the calculations.

Lateral connection sealing. Where sewers are above the groundwater table and most services are factory wyes or tees, the cost-effectiveness of grouting the lateral connection is doubtful. However, where the sewer is below the water table occasionally, or all the time, and especially where old hammer tap connections were made, then those services should be sealed. There is often a much debated political problem here in regard to the question, "Whose problem is it?" In some cities, the lateral including the connection to the main line sewer is the property owner's responsibility. If the city goes ahead and corrects the problem, then they are setting precedent for the rest of the city's private services. Different communities have used different approaches to addressing and financing this issue. One example would be having a section in the city's Sewer Use Ordinance stating that cost of any repairs determined to be necessary at a lateral connection would be an expense shared by the city and the user. The user's share of the cost would be paid for over a period of time through a special assessment added to the property taxes.

Pressure testing. The minimum test pressure to be used should be specified on all the lines that are to be grouted.

Maximum grout use. It may be wise to specify the maximum amount of grout to be used in sealing joints and bulkheading dead service laterals. A minimum amount specified for lateral bulkheading may be necessary, as was the case in Beloit.

Follow-up QA on grouting. This includes a requirement for televising each line and retesting a percentage of all lines previously grouted, and include penalty clauses, in the specifications.

Manhole cleaning or preparation. Manhole preparation for cementitious liner application should include removing loose debris, pressure washing (with caustic if slime coat present), prepatching large holes, and grouting any active leaks prior to lining. It may be helpful to address excessive moisture control that could cause adhesion problems for the cementitious liner. The manufacturer's application procedures should be understood and followed.

Cementitious liner re-tempering. If cementitious manhole lining product has started to set before being applied, it should not be used. In this case, the remainder of the batch should be discarded. As high air temperature and humidity may cause problems with liner application, it may be beneficial to specify scheduling of manhole lining based on weather conditions. The manufacturer's application procedures should be understood and followed.

Cleaning excess cementitious liner material. It should be specified that the contractor performing the manhole lining is responsible for preventing liner material from being sprayed or washed into the sewer and that any significant debris from manhole cleaning, repair, or lining that falls into the sewer line should be removed immediately.

Warranties. In this case study, a minimum of 5-year warranty for all sewer grouting and manhole lining was required.

13.7.5 Project Results

The majority of the project renewal work was done in phase 2 in the summer of 2001 and focused primarily on the following trenchless renewal technologies:

- CIPP lining from manhole to manhole
- CIPP spot repairs (4 ft to 30 ft)
- Sewer pipe chemical grouting
- Manhole chemical grouting
- Cementitious manhole lining

There was a high degree of QA/QC on all aspects of the renewal project as deemed necessary by the owner. Some of the more critical parts of QA/QC for the project are noted here: cleaning the lines with video inspection prior to lining for the CIPP work, the use of gaskets to seal the annular space between the liner and the existing pipe at the manhole connection, conducting the exfiltration test (less than 200 gal/in-mi diameter/day), and testing a cured CIPP sample for minimum strength (400,000 psi modulus E) and minimum thickness. Reinstating the live laterals was also closely monitored so that over-cutting of services did not occur. All lining operations were closely observed.

For the grouting operations, QA/QC involved a full-time observation of the pregrouting cleaning of the line to be grouted, and an observation of all joints, cracks, and dead service laterals being tested, sealed, and bulk headed. The sealed joints and cracks were immediately retested after being sealed. Follow-up video inspection a year later was done as specified, and during high groundwater levels when possible, to see if any visible leaks also appeared. A specified percentage (10 percent) of the total grouting project footage was also retested a year later and passed the minimum failure percentage (10 percent). Any joint that failed was then also regrouted and retested. During observation of the grouting work, all payments were closely tracked and documented.

The previous televising records were extremely helpful in the determination of live and dead service lateral connections. This made reinstating laterals after lining relatively straightforward. For the grouting operation, the televising records identified which laterals could be grouted shut.

The manhole grouting and lining portion of the project also required a full-time observation. There were many critical factors that had to be watched to ensure a successful liner application. Some of the important issues of concern for the owner were

- Step removal
- Proper cleaning and manhole preparation
- Elimination of active leaks by grouting
- Excess moisture control after cleaning for proper liner adhesion
- Proper liner application according to specified guidelines
- Proper liner thickness and finish
- No sloughing of liner after application
- No attachment of liner to casting
- Proper bench formation
- Representative liner sample collection and storage

In addition to the items noted above, QA/QC for the manhole liner application included collecting and storing representative samples of the cementitious liner material for the third-party strength testing to meet the minimum compressive strength as tested according to the ASTM C-109 strength test. After several days of cure time, the manholes were then also entered and checked for signs of leaks, cracking, and hollow spots. The hollow spots were located simply by tapping around the manhole walls with a steel crowbar. Any fine cracks or hollow spots were marked and noted and will be monitored over the next several years of freeze-thaw cycles to determine if any signs of damage or failure appear.

Results from the flow monitoring showed that significant flow reduction was achieved because of the renewal efforts. Since the project work in 2001, additional, smaller renewal projects have been carried out (in 2002 and 2003) to continue to further reduce the infiltration problems. Obviously, the cost-effectiveness of the renewal work diminishes as work progresses because of less infiltration reduction per dollar spent and longer payback period. Altogether, only about 1.3 percent of Beloit's sanitary sewer collection system was renewed. The trenchless technologies used amounted to almost 12,500 ft of CIPP-lined sewers, approximately 8250 ft of sewer lines chemically grouted, and 75 manholes had a structural cementitious liner applied. Additionally, approximately 850 ft of sewer was replaced by means of the traditional open-cut trench relay methods, three manholes were completely removed, another 18 abandoned, and two new ones constructed.

The cost of sewer renewal work from 2001 through 2003 was approximately US$965,000. This does not include the costs for flow monitoring, observation services, third-party material testing, and some other incidental services and expenses.

The overall renewal project efforts, including the work from 1998 through 2003, have resulted in a net influent flow reduction to the WPCF approaching 60 percent, or approximately 4.5 million gal/day of I/I reduction. The dramatic drop in collection system I/I is reflected in the graph shown in Fig. 13.4 where the WPCF influent flow is now significantly less than what is shown prior to the summer of 2001 renewal project work. Similarly, evidence can be seen of dramatic infiltration reduction by the installation of one CIPP liner that was installed on July 10, 2001, in a significantly deteriorated line that was leaking heavily. Also note the much more consistent baseline influent flow that is no longer significantly affected by even significant increases in river elevation.

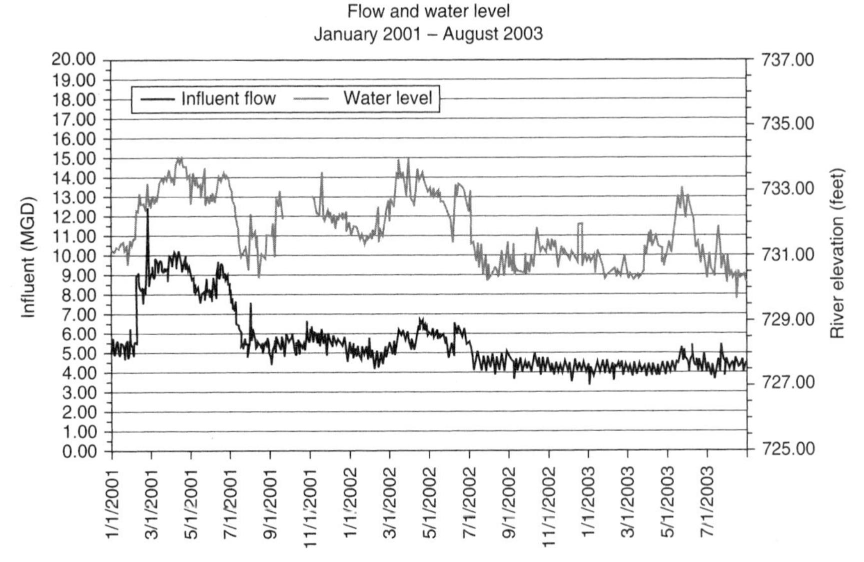

FIGURE 13.4 Evidence of the I/I reduction by installation of one CIPP segment.

REVIEW QUESTIONS

1. What is a cured-in-place pipe (CIPP)? Explain.
2. List CIPP applications.
3. Describe the "wet-out" process.
4. What are the primary components of CIPP? Briefly discuss their use and the main characteristics such as strength, flexural modulus, and so on.
5. What are the primary methods for saturated fabric installation? Describe each method in detail.
6. Discuss the importance of initial setup in CIPP installation.
7. Explain the different methods of curing.
8. What final steps are necessary after curing the CIPP?
9. List the main advantages and limitations of CIPP method for trenchless renewal.
10. A 24-in sanitary sewer, 8000 ft long, carries raw sewage to the city's wastewater treatment plant. The pipe is 45 years old and is made of concrete. There are nine manholes on the way and no laterals. In recent years, there have been approximately five failures per year costing the city approximately US$25,000 for each emergency repair and bypassing the raw sewage to a nearby river. The top of the pipe on the average is 10 ft deep from the ground surface and the water table is assumed to be 4 ft deep. Although the pipe location is out of the city limits, it is adjacent to the railroad tracks that make the open-cut replacement option impractical. The city has employed you to select the best and the most cost-effective method to renew this sewer (include installing a new line within your options). Assume 1000 ft of the pipe will be renewed at one time. Consider all cost factors

such as cleaning and inspection, bypass pumping during the renewal process, and testing and inspection after the work is complete. Obtain your cost information from local contractors and make assumptions as necessary. Present your results in a technical report and include justifications for eliminating or selecting each option. Consider for each option such factors as capital cost, applicability, constructability, life expectancy, hydraulic capacity, and structural integrity. How much would be the cost to the city if the city decides to postpone the renewal work for 20 years?

REFERENCES

Iseley, D. T., and M. Najafi (1995). *Trenchless Pipeline Renewal,* The National Utility Contractors Association (NUCA), Arlington, Va.

Lanzo Lining Services (2004). *Engineering Design Manual for Renewal of Cured-in-Place Pipe*, Company Manual.

Najafi, M. (1994). *Trenchless Pipeline Renewal: State-of-the-Art Review*, Trenchless Technology Center (TTC), Louisiana Tech University, Ruston, La.

Scandinavian Society for Trenchless Technology (SSTT) (2002). *No-Dig Handbook*, Copenhagen, Denmark.

Statewide Urban Design and Specifications (SUDAS) (2004). *Design Manual: Chapter 14*, Ames, Iowa.

CHAPTER 14
SLIPLINING

14.1 INTRODUCTION

Sliplining (SL) is one of the earliest and simplest forms of trenchless pipeline renewal methods and can be used for structural or nonstructural purposes. This technique has been used since the 1940s for renewal of deteriorated pipes. More than 60 years of experience has shown that SL is a proven cost-effective renewal technique with all benefits of trenchless technology such as minimum disruption of service, surface traffic, and minimum property damage that would otherwise be caused by open-cut method. SL involves insertion of a new pipe of smaller diameter into the existing pipe and the grouting of the annular space between the existing pipe and the new pipe. The pipe materials most commonly used in SL are polyethylene (PE), glassfiber reinforced polyester (GRP), and PVC, although use of any other pipe material is possible.

14.2 SITE COMPATIBILITY AND APPLICATIONS

SL can be used to renew gravity pipelines; however, the decrease in existing pipe's cross-sectional area needs to be checked against any gain in flow capacity due to better smoothness of the new pipe. Also, it is usually necessary to excavate connections from surface and disconnect them prior to liner pipe installation. Service lateral reconnection from inside the pipe is possible; although the process is more complicated when compared to CIPP (see Chap. 13) and close-fit pipe renewal methods (see Chap. 15). For potable water pipes, approval of the relevant regulatory agency is needed for all materials in contact with the potable water. In chemical or industrial pipelines, the pipe must be subjected to material compatibility with relevant chemicals. In industrial applications, the pipe material's resistance to chemicals, high temperature conditions, and other adverse environmental conditions must be evaluated.

Severe bends cannot usually be negotiated, especially for larger diameters. All bends add to the friction between the existing and new pipes during installation. As a result, this reduces the length of the new pipe that can be pulled in without overstressing the pipe. For pipes with varying cross sections, the liner pipe must be sized to minimum dimensions of the existing pipe, unless tapers are incorporated. It is unusual to pull a new pipe into a worker-entry pipe as a continuous string because of the high weight of the new pipe. Table 14.1 presents an overview of SL capabilities and possible limitations.

TABLE 14.1 Overview of Sliplining Capabilities and Possible Limitations

Pipeline type	Suitable	Notes
Sewers	No	A
Gas pipelines	Yes	
Potable water pipelines	Yes	B
Chemical or industrial pipelines	Yes	C
Straight pipelines	Yes	
Pipelines with bends	Yes	D
Circular pipes	Yes	
Noncircular pipes	No	
Pipelines with varying cross sections	No	E
Pipelines with service lateral connections	No	F
Pipelines with deformations	Yes	G

Notes: A. SL can be used to renew sewers, but is not usually the first-choice renewal method for gravity pipelines because of reduction in cross-sectional area of the existing pipe. B. Approval of the relevant regulatory agency is needed for all materials in contact with potable water. C. The pipe material must be resistant to chemicals, extreme temperatures, and other requirements of industrial applications. D. Normally, SL method cannot accommodate bends. E. The new pipe must fit the smallest diameter in the existing pipeline unless adapters are built into the line. F. For sewers, it is necessary to excavate lateral connections and disconnect them before injecting grout. Connection from inside the pipe is a possibility, although the process is more complicated than for CIPP and close fit renewal methods. For potable waterlines, it is always necessary to excavate service connections. G. SL is not suitable for pipelines with considerable deformations.

14.3 MAIN CHARACTERISTICS

SL or pipe insertion can be used to install a flexible new pipe of up to 10 percent smaller diameter compared to the host pipe. This is done by pulling or pushing the new pipe into the deteriorated existing pipeline. After installation and grouting, the new pipe can form a continuous, watertight pipe inside the existing one. Where the new pipe has to be laid to an even grade, use of plastic, metal locators or spacers is essential. Spacers also help maintain the pipe location during annular grouting with a cementitious material.

14.4 SLIPLINING METHODS

SL can be categorized into two main categories: continuous and segmental. Each of these methods is explained in the following sections.

14.4.1 Continuous Sliplining

The continuous SL method (Fig. 14.1) involves accessing the deteriorated pipe at strategic points and inserting high-density polyethylene (HDPE) or PVC pipe, joined into a continuous line, through the existing pipe. This technique has been used to renew gravity sewers, sanitary force mains, water mains, outfall lines, gas mains, highway and drainage

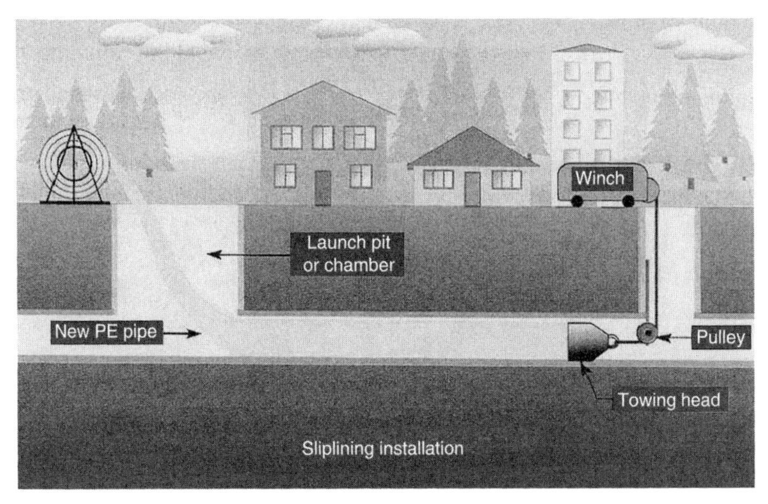

FIGURE 14.1 Typical continuous sliplining process (Gokhale and Hastak, 2003).

culverts, and other pipelines. It has been used to restore pipes as small as 1 in with the maximum pipe diameter limited by the availability of pipe material.

This method can be used both for thin-walled and thick-walled liners. HDPE (or fuseable PVC) solid-wall pipe is slipped into an existing pipeline after the joints are butt-fused. The method requires a guiding trench for an insertion pit. There is typically a slight-to-moderate loss of hydraulic capacity. It leaves an annular space, which for gravity sewers, water mains, and other applications depending on the specific project conditions and type of loading, may require grouting. Excavation is required to reestablish each lateral connection.

Polyethylene Pipe. One of the most common materials used for continuous SL is PE pipe. Its ability to be fusion welded into long jointless sections that can be quickly pulled into place makes it a popular choice. The reason for the popularity of PE may be attributed to its ability to butt-fuse small sections or join them by collarless joints into very long continuous lengths that can be quickly pulled into place. Moreover, it is abrasion resistant and is sufficiently flexible to negotiate minor bends during installation. It should be noted that a fuseable PVC pipe now is available that can be used in continuous SL method.

Other Pipe Materials. Additional pipe products with restrained joints such as PVC or ductile iron can also be pulled into the line in a manner similar to that for PE pipes. These materials are usually installed a section at a time, rather than being completely assembled prior to installation like polyethylene. Again, flow bypassing is not normally required. Because of ring stiffness, for ductile iron pipe, grouting of the annular space is not required, and it is optional for PVC liner pipe depending on the condition of the original pipe.

14.4.2 Segmental Sliplining

This method involves the use of short sections of pipe that incorporate a flush sleeve joint commonly used in microtunneling and pipe jacking processes. A number of plastic pipe

FIGURE 14.2 Segmental sliplining process. (*Courtesy of Hobas Pipe, USA.*)

products, such as GRP, PVC, PP, and PE that include short-length sections with a variety of propriety smooth joints (both inside and outside) have been specially developed for SL sewers. This method is applicable for diameters greater than 24 in.

Segments of the new pipe are assembled at entry points and forced into the host pipe. After the segmental pipe is installed, the annular space is grouted. The laterals are usually reconnected by excavation from outside. Figure 14.2 illustrates a segmental SL process.

Pipe Materials. It is possible to use any common sewer pipe material that can be inserted into the existing pipe. To minimize the reduction in cross-sectional area of the existing pipe, the pipe must have smooth inside and outside joint. There is a wide variety of pipe products that meet these criteria. Most are intended for pipe jacking, microtunneling, or directional boring; and their bell-less or low-profile bell configurations also make them well-suited for SL.

Pipe products such as vitrified clay, concrete, ductile iron, and centrifugally cast glass-fiber reinforced polymer mortar (CCFRPM) pipe can all be pushed into the pipe from a relatively small access pit. These products are pushed into the pipe using jacking equipment. Flow bypassing is not normally required because the line remains open during the insertion process. In addition, because of the inherent structural integrity of these pipe products, grouting the annular space is not as critical as for the flexible plastic pipe; however, it is still normally done to lock the pipe in place.

14.5 INSTALLATION

During the SL process the pipe is pulled or pushed into the host pipe from a launch pit to a reception pit. Installation procedure is slightly different for each of the two methods of SL (i.e., continuous or segmental). However, certain typical tasks common to all of these methods are listed below.

- Inspecting the existing pipe
- Cleaning and clearing the host pipe from any obstruction

- Joining (by butt fusion or gasketed bell and spigot) lengths of PE pipe
- Providing access to the host pipe
- Inserting the liner pipe and positioning it inside the existing pipe
- Stabilizing (grouting) annular space (if required)
- Construction service and lateral connections
- Construction of terminal connections

14.5.1 Installation: Continuous Sliplining

As discussed previously, PE pipes can be fused together to form long segments. This fusing may be done either aboveground or in the launch pit. Fusion aboveground may require long launch pits because of the limitations on bending set by the minimum permissible radius of curvature for PE pipes. This is especially true for deeply buried pipes and large-diameter pipes. Fusion in the launch pit allows the pit to be shorter, but the rate of installation is limited by the time taken to fuse pipe segments and cool welds. The cooling phase in particular is important for service life of the finished pipeline. This is because too short cooling times weaken the strength of the pipeline during installation and in the long term.

Beads are formed internally and externally on PE pipes during fusion welding. In the case of sewer pipelines, both internal and external beads are often removed before pipe insertion. In the case of water pipelines, internal beads are not removed because of the risk of contamination from the use of debeading tools.

When pipes are pulled into position, the towing head is a very important component. It grips the new pipe and transfers the pulling force from the winch cable. The towing head should provide a secure connection without producing high levels of localized tensile stress. In some systems, the end of the pipe is closed off to prevent soil or other debris from entering the pipe. This is especially important in the case of potable water pipes.

A breakaway connector can be fitted between the winch cable and towing head to avoid subjecting the PE pipe to stress. Such connectors can be adjusted to separate at loads lower than the permissible load on the pipe. Although undesirable, a broken connector is usually preferable to the pipe being damaged with subsequent service failure. The use of breakaway connectors also induces operators to avoid using excessive pulling force during pipe insertion.

Installation by Pulling. Small-diameter pipes are often pulled into position with the aid of a sleeve made of diamond-shaped mesh. The sleeve is fastened to the winch cable and pulled over the end of the PE pipe. During insertion, the sleeve tightens and grips the PE pipe, which can then be pulled into position in the host pipe. Short sections of small PE piping can be pulled into place manually, but most pipes must be pulled into place by the winch cable. The winch must provide a steady, progressive pull without snatching or uncontrollably varying the force. The winch must be carefully positioned and the cable carefully guided, and it is often necessary to place additional pulleys in the manhole or reception pit to ensure that the cable has an unobstructed path and does not rub against any part of the launch pit.

Installation by Pushing. Numerous types of manually or hydraulically powered machines are available for pushing new pipes into existing ones. Some are designed to operate from launch pit, whereas others are positioned on the ground immediately behind the pit. The machine grips the new PE pipe and pushes it forward into the host pipe. The grip mechanism is then released and returned to the starting position, and the process is repeated.

14.5.2 Installation: Segmental Sliplining

The exact installation procedure is dependent on the pipe size and the materials used, for example whether reinforced pipe materials or plastics are used. Pipe segments are prefabricated sections specifically designed for each project. They are constructed from two pieces, joined longitudinally and circumferentially, and can be made from glass reinforced cement (GRC), glassfiber reinforced polyester (GRP), polymer concrete (PC), and concrete. Units are assembled manually, prior to grouting of the annular gap. Individual lining units are passed into the pipeline via manholes or special access shafts. For worker-entry pipelines the lining begins at the furthermost point, from where the units are positioned and joined. Each existing pipe segment is renewed, and then grouted before proceeding to the next existing pipe section. For nonworker-entry pipes, the segments are pushed or pulled into pipeline until entire length has been lined. Excavation is required to restore lateral connections.

14.5.3 Connecting Service Laterals

Excavation is normally necessary to connect laterals and branch lines while SL gravity sewers. To connect internally, openings can be cut in the PE pipes before injecting grout, and laterals and branch lines can be sealed off using inflatable bags. Sealing the laterals and branch lines off would prevent any grout from being forced into them. However, the complexity of such operations only justifies their use in situations where outside access via excavation is difficult or impossible. Moreover, it is only possible to use the procedure in large-diameter pipes.

Connections must be excavated and branch lines disconnected before grout is injected. Connecting laterals to PE liners is carried out in the same way as for new installations. Special connectors must be used to connect the new junction to the existing branch. Where waterlines are concerned, excavation is always necessary for connecting service lines or for making any other type of connection to the existing pipe.

14.5.4 Grouting

SL systems in which the liner pipe bonds to the existing pipe to form a composite pipeline, and systems in which the new pipe only functions as a permanent mold for the annular grout require the use of cementitious grout with a compressive strength of 1.5 to 3 psi. Liner pipes that are held in place by host pipe but do not need to bond to it, only require a grouting material capable of transferring loads from one pipe to the other. The strength of some of the injection grouts used for this purpose corresponds to that of stiff clay.

Injection grout made of ordinary Portland cement and pulverized fly ash is commonly used although numerous special grouting materials are available. One of these is a grout with extremely low viscosity that is capable of flowing through the annular space with the aid of gravity or minimum pressure and which sets within 20 min. One benefit of these grouts is that they allow stage grouting to proceed more rapidly than with conventional materials.

The forces applied to liner pipes during grout injection are often greater than those experienced during normal service. Failures owing to grout pressure and floatation forces are related to the weight of the grout displaced by the liner pipe (i.e., the volume of the liner pipe multiplied by the unit weight of the grout) rather than the weight of the grout in the annular space.

It is a common practice to fill up the liner pipe with water during grouting as this helps counteract the floatation forces and resist external pressure. It may still be necessary to inject grout in stages even with liner pipe filled with water. This is especially true for large gravity sewers, where the gradient of the sewer is critical and floatation is unacceptable.

14.6 DESIGN CONSIDERATIONS

Pipes used for SL are generally, but not always, similar type and specification to those used for new construction. PE pipes are usually aimed at applications where internal pressure is the main criterion, and the design of PE slipliners in pressure pipes should follow the same principles as for new pipes. Thin-walled (nonstructural) slipliners may be used, provided the existing pipe has sufficient load bearing capacity, and that complete grout fill of the annulus can be achieved so that the new pipe is fully supported. This is often difficult to guarantee, and thin-walled liner pipes are therefore not favored for basic SL, although they are frequently used in modified (close-fit) SL described in Chap. 15, Close-Fit Pipe (CFP).

Annulus grouting may not be required when renewing pressure pipes, but is usually necessary for gravity pipelines to increase the ring stiffness of the new liner pipe. Slipliners in sewers are usually designed to be restrained by the host pipe and the annulus grout, but do not form a bond with the existing pipe wall. In such cases, the grout acts only as filler, and does not require high structural strength. Systems that rely on the host pipe for some measure of structural support are sometimes known as *interactive lining* techniques. Because of the relatively low flexural modulus of PE, thick-walled pipes may be needed to withstand high external loading. This may be a significant factor with gravity pipes, which are laid at considerable depths or are subjected to high vehicle loading. In such cases, it may be more economical to design the PE liner as a permanent support for high-strength grout, rather than to increase the wall thickness of the liner pipe itself. In this type of lining system, the grout is the main structural element. In all cases, the liner pipe must be designed to withstand not only the internal and external forces in service, but also the loads during installation—particularly winching forces and grout pressure.

14.7 SLIPLINING GAS LINES

Several techniques are developed to allow the insertion of a new PE pipeline into the existing gas mains or lateral service lines without interrupting the gas supply. These methods generally rely on gas flowing through the annular space between the existing and the new pipelines during installation, and so take advantage of a reduction in pipe diameter. This may be acceptable in the case of the existing mains originally designed for gas pressures lower than gas pressures currently available. It is outside the scope of this book to describe the many proprietary systems for live insertion. Strict and detailed safety procedures must be laid down for installation, and the following section is intended *only* as a general guide to the basic principles.

Renewal systems are available for low and medium pressure gas mains. The first stage is to isolate the section of main to be renewed by bypassing at one or both ends of the isolated section. The new PE pipe is then fed through seals attached to the existing main at the entry pit, and is pushed using pneumatic or hydraulic machines through the entire length of main to be renewed. Typical insertion lengths are between 300 to more than 1500 ft.

There are many installation techniques. In the simplest method, the new PE pipe is passed through seals in the exit pit, and can then be connected either to the existing pipe or to a new, generally higher pressure, network. In all methods, during the installation, the annular space between the existing pipe and the new pipe is utilized to maintain gas supply to customers. In order to facilitate the transfer gas flow to the new PE pipe, polyurethane foam is injected into the annular space to stop the flow of gas. This process would allow the existing main to be closed and gas flow diverted to new pipe.

Gas mains from 3 to 18 in diameter can be renewed using the above method. For the renewal of gas service pipes, a technique is available that allows the existing gas meter position to be maintained by enabling the insertion of PE pipe through a 90 degrees elbow, around a tee, or through a number of long-radius bends. After removal of the meter and the main stopcock, the line-blowing assembly is fitted to the service connection at the meter position. Air is blown through the existing service pipe to remove any loose rust. The pipe receiver, bend, and standpipe are fitted to the service, and air is allowed quickly into the pipe to blow a line through to the far end. This is then used to pull back the winch cable, and the winch is fitted to the top of the pipe receiver. A short length of PE pipe is winched through to remove any further rust or encrustation. Full-length pipe is installed by using the winch in combination with a pushing force applied manually from the other end. A test is conducted after a brief period to allow the pipe to recover from any stretching. The technique can also be modified for the renewal of water services.

A method of live insertion for gas service pipes has been developed in which a new PE pipe is pushed into an existing cast iron or other service line through a sealing system attached to the existing pipe, either inside the customer's premise or by means of a small excavation outside the building. No excavation is necessary at the service connection with the gas main in the road or street. The annular spacc between the existing and the new pipes is filled with a permanent sealant that is prevented from entering the mains system by a type of nose cone fitted to the leading end of the PE pipe. The system is available for services from 1 to 2 in diameter. Adaptations for use of this technique at higher diameters and in water networks are under development.

14.8 ADVANTAGES

- As a renewal method, SL does not require investment in costly specialized equipment.
- Jacking pipes and fittings, as used for trenchless construction methods can also be used for SL.
- SL is a conceptually simple technique that can be applied to either pressure or gravity pipelines.
- SL can be used for structural or nonstructural purposes.
- Live insertion (with the existing flow) is possible.

14.9 LIMITATIONS

- The most important limitation of SL is the reduction of pipe diameter inherent in the system. It is therefore necessary to establish whether the use of SL is compatible with the capacity requirements of the renewed pipeline. However, a loss of cross-sectional area does not proportionally represent loss of capacity because of improved hydraulic characteristics of the liner pipe.
- Where manhole access is not possible, a pit excavation is required for access during installation process.
- For lateral connections, open-cut excavation is required.
- Grouting is generally required.

14.10 CASE STUDY: LARGE DIAMETER SEGMENTAL SLIPLINING

Name of project: Big Creek Interceptor Renewal

Project location: Cleveland, Ohio

Owner: Northeast Ohio Regional Sewer District (NEORSD)

Engineer: *Metcalf & Eddy, Inc. of Ohio*

Contract amount: $16,000,000

Project Description. The Big Creek Interceptor is an approximately 45,000 ft long combined sewer interceptor conveying sanitary sewage and some storm water flows from the western inner suburbs of Greater Cleveland Ohio to the NEORSD's Southerly Wastewater Treatment Plant located south of downtown Cleveland on the Cuyahoga River. The Big Creek Interceptor was constructed in the 1930s mostly by hand tunneling through soft shale rock at a depth of 35 to 100 ft below the surface. The tunnel was lined with brick and clay tiles.

During a 1-h storm event that occurs once every 5 years, hydraulic modeling of the Big Creek Interceptor predicts that flows through the interceptor range from 18 to 21 million gal with approximately 3 million gal exiting the interceptor through combined sewer overflows into the local streams and rivers. In addition, during these large storm events, the interceptor is surcharged from 9 to 12 ft above the crown of the pipe. This translates into an internal pressure of approximately 750 lb/ft2 over internal pipe wall. This pressure has dislodged brick and clay tiles from the tunnel liner system, exposing the soft shale rock to cycles of exposure to sewage and causing the shale rock to flake off and fall into the invert of the interceptor. This in turn has caused blockages in the normal sanitary flows through the interceptor and allowed for the buildup of hydrogen sulfides.

To alleviate this problem, the NEORSD determined that the Big Creek Interceptor should be renewed to improve the structural integrity and hydraulic characteristics of the interceptor. The renewal project would also significantly reduce combined sewer overflows into the local streams. Another objective was reducing odor control problems caused by the buildup of hydrogen sulfides.

This project was planned over a number of phases. This case study presents the evaluation of the existing condition of approximately 17,200 ft of 66-in and 75-in brick sewer interceptor. The evaluation work included providing recommendations for renewal method, construction materials, schedule, and cost, and technical and permitting issues. Gunite was used on approximately 2150 ft of the sewer pipe to protect the brick sewer during installation of a new Southwest Interceptor tunneled under the Big Creek Interceptor. The pipe trestle carrying the Big Creek Interceptor across the Cuyahoga River was replaced in 1992 and was in good condition.

Technical Challenges

- Limited methods for flow diversion of the existing flows in pipe
- No practical method for bypass pumping of flows
- Deep excavations in urban areas to access pipe
- Steep embankments through private land to access pipe trestles
- High levels of sulfides and chlorides in flow stream in suspension from land use (manufacturing) in upstream reaches of interceptor
- Collapsed section of brick and clay tile tunnel liner

Type of Trenchless Method. A number of trenchless renewal methods and materials were considered for renewing the Big Creek Interceptor. CIPP, GRP panels, gunite liner system, epoxy coating systems, and a number of slipline pipe materials were considered. Based on the technical challenges identified above, the renewal method and material had to be resistant to sulfide attack, be able to be installed with limited worker-entry into the pipe, can be installed with live flows in the pipe, and limit the number of deep access shafts (in the urban area) required to install the method.

The renewal method recommended was to renew approximately, 11,100 ft of the Big Creek Interceptor sewer with 54-, 57-, and 66-in CCFRPM pipe. The recommended renewal also included 11 existing manholes by SL with CCFRPM pipe, installing two new manholes at creek crossing locations, renewing three manhole structures by applying a cementitious product and an epoxy overcoat, abandoning two manholes, and replacing one drop structure. Plans and specifications were prepared for the recommended improvements.

In addition, repairs of the previously gunite section would be accomplished by adding epoxy overcoat over the existing gunite to provide protection to the cementitous gunite material from sulfide attacks. This could be accomplished because it was downstream of one location where flows from within the Big Creek Interceptor could be diverted to a nearby large-diameter interceptor.

Preliminary Design Activities. The preliminary design activities included a review of the record construction drawings for original construction of the Big Creek Interceptor and any drawings of the construction work that made physical changes to the interceptor. In addition, a review of the local sewer collection system contributing sanitary and combined sewer flows to the Big Creek Interceptor was also completed. This was done to determine how best to divert flows away from the Big Creek Interceptor. With this information, a hydraulic model of the Big Creek Interceptor was completed to determine the flow conditions during dry weather and specific storm events. A review of available geotechnical information was also completed to determine subsurface conditions. Available closed circuit television (CCTV) inspection tapes and inspection logs were evaluated to determine the internal pipe conditions.

Field Investigations. During the preliminary design phase, field investigation work such as subsurface geotechnical borings, to verify assumptions and information obtained on the subsurface conditions, was completed. In one area where the Big Creek Interceptor passes near a steep railroad embankment, and in between a small pipe trestle carrying the pipe over a drainage creek and the large pipe trestle carrying the pipe over the Cuyahoga River, the CCTV inspection indicated possible movement of the interceptor. A detailed investigation of inclinometer readings and groundwater elevations was conducted to determine the possible movement of steep railroad embankment. Each of the manholes connected directly to the Big Creek Interceptor were inspected to determine the physical conditions of the manholes. In addition, specific reaches of the interceptor were inspected by worker-entry to verify the physical conditions of the interceptor; and to collect samples of the pipe sediment, wastewater, and concrete mortar between the bricks. Additional samples were taken at the gunite section and the bricks. Each of these samples was tested.

The wastewater in the Big Creek Interceptor was tested to determine the chemical makeup of wastewater to determine the corrosive characteristics of the wastewater. The concrete mortar and bricks were also tested for chemical attack. These tests were done to help the engineers understand the reasons for the failure of the pipe, and to determine what new materials would best be suited to resist the corrosive conditions. In addition, the testing of the pipe sediments was completed to determine not only the corrosive conditions, but also to determine the requirements for removal, handling, and disposal of the pipe settlements, should any hazardous material be found.

Finally, a field walk along the surface above the Big Creek Interceptor was completed to determine the best locations for access shaft excavations. Land uses such as whether or not the area was residential, institutional, commercial, or industrial were reviewed. Street traffic patterns were also reviewed to determine the impact of shaft construction on traffic and emergency vehicle routes.

Final Design Considerations. During the final design phase, the locations of access shafts were reviewed. Because of the average depth of 60 ft, and the existing soft soil conditions, the cost of each shaft was extensive. Therefore, it was decided to reduce the number of shafts and their depths. Specifying a segmental slipline pipe that could be jacked a long distance from a single working shaft yielded this purpose. In addition, suitability of constructing access shafts in areas where the Big Creek Interceptor passed through or under the existing drainage streams and were relatively shallow in depth, were investigated. Most of these areas were on private lands and temporary and permanent easements for truck access to the sites needed to be obtained.

Construction Contract Documents. Construction contract documents including drawings and specifications have been prepared for this project. The project will be ready for bid after NEORSD obtains all the necessary permits and easements. In addition, excavation support systems were planned in soft soil for residential areas to limit impact on local residents. To reduce impact on local residents, contractors will have specific work hours. Easement acquisition was planned to limit heavy construction traffic in the residential neighborhoods.

ACKNOWLEDGMENTS

Sections of this chapter have been excerpted from the references listed. The reader is referred to these references for more information.

REVIEW QUESTIONS

1. What is SL? In what conditions is it suitable?
2. What are the primary materials used in SL? Explain.
3. Describe the slipline grouting process and its objectives.
4. List the main advantages and disadvantages of SL method.
5. Explain the installation procedure for two main SL methods.
6. How are lateral connections reinstated in the SL method?
7. What kinds of SL systems require grouting? Explain.
8. Discuss some of the design considerations of the SL method.
9. Discuss gasline SL methods.
10. For what project and site conditions would you recommend SL method?

REFERENCES

Australian Society for Trenchless Technology (2004), *ASTT–ISTT Trenchless Technology Guidelines: Sliplining*. Available at http://www.astt.com.au/Sliplining.pdf.

Gokhale, S. and M. Hastak (2003). Automated assessment technologies for sanitary sewer evaluation, *Proceedings of ASCE New Pipeline Technologies, Security, and Safety Conference*, Baltimore, Md.

Scandinavian Society for Trenchless Technology (2002). *No-Dig Handbook*, SSTT, Scandinavia.

Statewide Urban Design and Specifications (SUDAS) (2004). *Design Manual: Chapter 14*, Ames, Iowa.

CHAPTER 15
CLOSE-FIT PIPE

15.1 INTRODUCTION

Close-fit pipe (CFP) can be used for structural or nonstructural purposes in sewer systems, potable water, gas supply lines, and industrial applications. In structural applications, using the reduced diameter pipe (RDP) method, the applicable diameters are from 4 to 30 in. For nonstructural applications using mechanically folded pipe (MFP) method, diameters up to 64 in are possible with lengths up to 1000 ft.

The RDP, under patented name of rolldown, was originally invented in United Kingdom by Stewarts & Lloyds Plastics (SLP), and subsequently by the British Hydromechanics Research Association. Rolldown was first introduced for lining gas mains in 1986, and has since been developed and used for water supply and other pressure pipeline applications. To date, more than 140 mi of rolldown polyethylene (PE) liner pipes have been successfully installed.

15.1.1 Primary Characteristics

CFP trenchless pipeline renewal uses a new PE pipe that is modified in cross section before it is installed. After placement or insertion into the existing pipe, it is reformed to its original size and shape to provide a close-fit with the existing pipe. There are two methods of CFP as shown in Fig. 15.1.

In the first method, called MFP, the thin-wall polyethylene pipes are butt-fused and mechanically folded at the job site prior to insertion (Fig. 15.2). The MFP operation requires an insertion pit and an access chamber or pit at the far end for pulling the pipe through with a winch. Once in place, the new pipe is reverted back to its circular form by pressurization with water at an ambient temperature that breaks the temporary restraining bands to form a close-fit within the host pipe, sealing leakage and preventing corrosion. After installation is completed, service connections are generally reestablished by excavation. Figure 15.3 illustrates the diameter reduction, banding, and insertion of MFP process.

The second CFP method is RDP, a technique that involves the use of long, butt-fused sections of PE pipe. This method is primarily used for pressure pipelines and provides a standalone and fully pressure-rated pipe. This technology also can be used to provide a thin wall lining to eliminate leakage in an otherwise sound pipe. After preparation of butt-fused pipe and construction of entry and receiving pits, the diameter of the PE pipe is substantially reduced from the original PE pipe diameter. The diameter reduction is performed either by a mechanical method (rolldown) or by a thermal method (swage). The reduction operation is performed just before the new pipe is inserted into the existing pipe. After the new pipe is inserted into the existing pipe, it is reverted to its extruded diameter and thus fits snugly into the existing pipe,

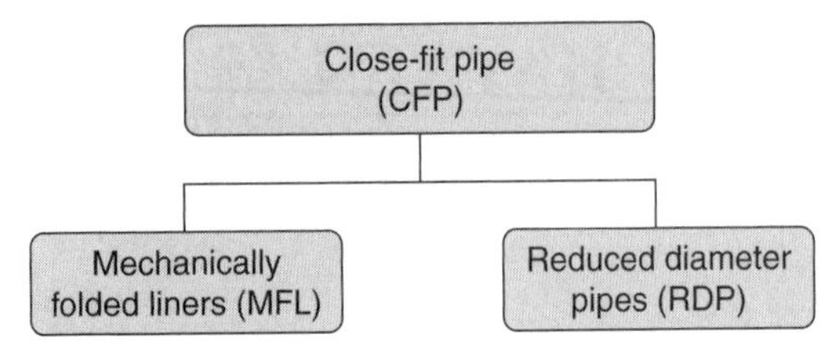

FIGURE 15.1 Two main variations of CFP process.

FIGURE 15.2 Mechanically folded pipe. (*Courtesy of Subterra.*)

FIGURE 15.3 The diameter reduction and banding process for the mechanically folded pipe. (*Courtesy of Subterra.*)

FIGURE 15.4 The reduced diameter pipe (RDP) process. (*Courtesy of Subterra.*)

making a close-fit with no gaps or annular space. As with the MFPs, connections are generally reestablished by excavation to ensure a complete seal with pressure pipe. In nonpressure applications, lateral connections may be reopened using a remote-controlled robotic cutter. Methods are in development for remote-controlled reinstatement and sealing of connections for pressure applications. The RDP process is shown in Figs. 15.4 and 15.5.

15.2 CLOSE-FIT PIPE APPLICATIONS

As for any other renewal method, it is important to choose the best renewal method for each individual project. The type of application, working space available, chemical composition of the fluid carried by the existing pipeline, the pressure and temperature in the pipeline, number of service laterals, depth, diameter, number of bends, any misalignments or joint settlements, number of manholes, and so on, must be assessed for each specific project. Although CFP can be used for structural or nonstructural purposes to renew both gravity and pressure pipelines,

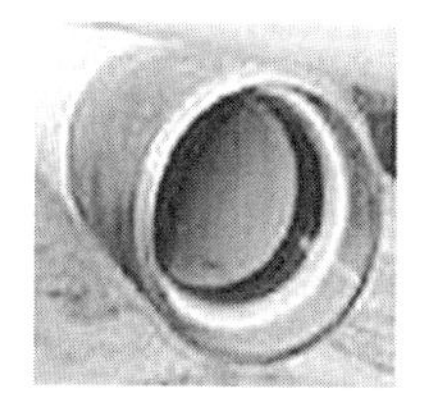

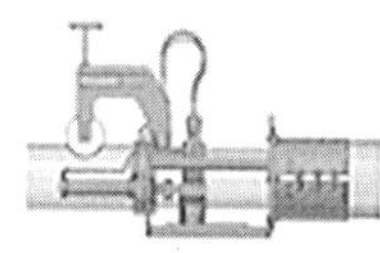

FIGURE 15.5 The reduced diameter pipe (RDP) process. (*Courtesy of Advantica.*)

TABLE 15.1 Overview of Close-Fit Capabilities and Possible Limitations

Pipeline type	Suitable	Notes
Sewers	Yes	A
Gas pipelines	Yes	
Potable water pipelines	Yes	B
Chemical or industrial pipelines	Yes	
Straight pipelines	Yes	
Pipelines with bends	Yes	C
Circular pipes	Yes	
Noncircular pipes	No	
Pipelines with varying cross sections	No	
Pipelines with lateral connections	Yes	D
Pipelines with deformations and misalignments	No	E
Pressure pipelines	Yes	

Notes: A The MFPs are generally not applicable to sewer applications. B. All local and national codes for all materials that come into contact with potable water needs to be checked. Both MTL and RDP with butt-fused joint PE pipes are suitable for renewing potable water pipelines. Both methods usually require the excavation of entry and reception pits. In addition, valves, service connections, and sharp bends must be dug free before renewal work is begun. C. The new pipe must fit the smallest diameter in the existing pipeline unless adapters are built into the line. The MFP process can be used for bends up to 45 degrees. D. For sewers, it is not necessary to excavate lateral connections, as they can be reopened from inside the new pipe using robotic cutters. It is always necessary to excavate service line connections on potable water pipelines. E. The method is not directly suitable for pipelines with considerable deformations and misalignments.

it is mostly used for pressure pipes such as water and gas applications. Range of applications includes water supply and distribution (potable and raw), gas transmission and distribution, sewer force-mains, oilfield water injection lines, and industrial water pipelines. Table 15.1 provides an overview of the close-fit applications and possible limitations.

15.3 METHOD DESCRIPTION

The installation of a CFP renewal method includes several steps. The first step is site selection and excavation of inspection and entry/exit insertion pits. The second step is inspection, cleaning, and preparation of existing pipeline (including the removal of protruding fittings). The third step is preparation and butt fusion of PE liner pipe string, and removal of external beads. It should be noted that initial PE liner outside diameter (OD) must be greater than internal host pipe diameter to ensure a close-fit after reversion.

The fourth step is feeding the PE liner through the MFP machine or RDP machine. The MFP shape requires securing with temporary restraining bands. The RDP is performed either by a mechanical method (rolldown) or by a thermal method (swage).

The fifth step is inserting liner into the existing pipe. To decrease pulling forces, lubrication before insertion is required. Step six includes sealing ends of liner pipe and filling with water at an ambient temperature. The next step includes pressurizing the liner pipe to bring it to a close-fit with the existing pipe and hold it for a specified reversion period (usually 12 h). The last step includes depressurization and complete liner end terminations and connections.

15.3.1 Preliminary Evaluation

To conduct a productive and successful close-fit project, it is important to carry out a preliminary evaluation. First, using an appropriate inspection method, a detailed preliminary surveying is conducted to determine condition of the existing pipeline and the need for cleaning (see Chap. 3). Service lateral data, including locations, dimensions, and manhole conditions and any unused service lines and intruding service line connections are reported in this evaluation. The existing pipeline material, length, and depth, dimensions, deformations, and any displaced joints are evaluated. Also, surface and subsurface conditions including traffic loads and water table depth and its fluctuations are investigated. Finally, locations and sizes of access pits, based on specific site conditions and project needs are determined.

15.3.2 Preparatory Work

After necessary preliminary evaluation has been completed, there are several preparatory procedures before start of renewal project. Usually, the preparatory work would depend on the flow characteristics, and type of application such as whether it is a sewer or pressure line. For a sewer, for example, preparatory work would include:

- Inspection and possible grinding of intruding lateral service line connections, pipe section lengths, and displaced joints on the existing pipeline
- Inspection and possible grinding or cutting away of scattered deposits
- Closing or blocking off unused service lines
- Using localized repair method (see Chap. 18) for restoring badly deformed existing pipe sections
- Preventing groundwater infiltration (temporary sealing to allow CFP renewal)

15.3.3 Installation Phase

As for any renewal project, installation is an important phase of CFP process. It is important for both the pipeline owner and the contractor to ensure that both the preparation of the new pipe and its installation are carried out according to specifications.

As CFP methods are usually used for potable water and gas supply lines, it is necessary to excavate valves, service connections, and sharp bends before installing the line.

Diameter Reduction Process. With the MFP, the butt-fused PE pipe is fed through a shaper to mechanically fold it into a *heart* shape that is temporarily held by restraining bands. The reduced cross section creates a clearance to facilitate the installation of the PE pipe into the existing pipe.

There are two methods of RDP: (1) the mechanical, rolldown method and (2) the thermal swage method. In the rolldown process the PE pipe is fed through concentric rollers to reduce the diameter of the renewal pipe. In the swage process, heat is used to assist the diameter reduction.

Rerounding Process. Once inserted, the MFP is reverted back to its circular form by pressurization with ambient temperature water. This pressure breaks the temporary restraining bands, thereby unfolding the pipe and creating a close-fit within the existing pipe.

After insertion the RDP are pressurized with water to revert the liner pipe to its original extruded diameter, by pressing against the inside of the existing pipe.

15.3.4 Project Inspection and Delivery

After installation, it is often necessary to carry out the finish work, such as the following:

- For the case of water mains, end terminations, and installation of tees or branches
- Inspection and leak testing
- Service line reinstatement
- Customer notification
- Customer survey

The main purpose of the above tasks is to document that the renewal process has been completed in accordance with the contractor's own quality control and according to the technical specifications and contract documents.

15.4 QUALITY DOCUMENTATION

A final quality review compares the completed renewal work with the requirements of technical specifications and contract documents. A final review would expose any variations or changes between the contract documents and the work of the contractor.

When a renewal project is completed, the contractor provides the pipeline owner with documentation of the quality of the work performed. In this respect, it is important that requirements for such documentation be included in the bid documents. The documentation should include the following:

- Information on the products and the installation method used
- Pre- and postinspection results
- Test results
- Any changes or deviations from contract documents and possible corrective work performed
- Information on service lateral reinstatements and pipe ends or manhole connections and possible grouting process performed
- Recommendations for future inspections, cleaning, maintenance, and other renewal works to be completed as part of a comprehensive renewal work
- Survey of residents, businesses, and customers about the renewal project

15.5 ADVANTAGES

The CFP method has the following advantages over comparable renewal technologies:

- The product pipe is manufactured at the factory; therefore, the installation process in the field has fewer variables impacting a successful installation.
- For the same reason above, quality assurance process is more efficient. A PE pipe with recognized physical and chemical properties is installed.

- Being a CFP, the reduction in the existing pipe cross-sectional area is minimal.
- The CFP method can solve pipeline internal corrosion problems, water quality problems associated with pipeline internal corrosion, leakage from corrosion holes, cracks, and failed pipeline joints, and flow capacity problems arising from pipe tuberculation and deposits.
- The MFP is capable of accommodating bends up to 45 degrees.
- The new pipe can be installed up to 1000 ft; therefore, it will have a smooth and jointless pipe with better flow characteristics than the existing pipe.
- Internal lateral connections are possible.

15.6 LIMITATIONS

CFP has the following limitations:

- Limited diameter range and installation length are available.
- A large working space is required to lay the butt-fused string of PE pipe before feeding into the diameter reduction machine and insertion into the existing pipe.
- The method may not be suitable if there are variations in cross-sectional area of the existing pipe and for existing pipes with significant bends.
- Usually an insertion pit is required.
- In most cases bypassing the flow during the installation is required.
- For water mains, locations of valves and connections may have to be excavated.

ACKNOWLEDGMENTS

Sections of this chapter have been excerpted from the references listed. The reader is referred to these references for more information.

REVIEW QUESTIONS

1. What are the primary characteristics of a CFP?
2. In what conditions would you recommend a CFP?
3. How does CFP differ from other renewal methods?
4. Describe and compare the two main variations of CFP method.
5. What type of preparatory work is usually performed for CFP?
6. What are the CFP construction considerations?
7. What are the advantages and limitations of CFP?
8. Describe in detail the CFP installation process.
9. Check your local city phone directory to see which companies or contractors offer CFP method. Interview one of these companies to see how many feet of pipe they have installed and which variation of CFP they have used. Check the prices they offer and compare it with other renewal methods.

10. Check your local city public works department to see if they have used CFP method. Obtain specific project conditions (location, diameter, existing pipe type, length, application, and the like) and their decision criteria to select this method over other methods.

REFERENCES

Advantica Web Site (2004). Available at http://www.advantica.biz/hardware_licensing/swagelining/why_swagelining.htm.

Iseley D. T., and M. Najafi (1995). *Trenchless Pipeline Renewal*, The National Utility Contractors Association (NUCA), Arlington, Va.

Najafi, M. (1994). *Trenchless Pipeline Renewal: State-of-the-Art Review*, Trenchless Technology Center (TTC), Louisiana Tech University, Ruston, La.

Scandinavian Society for Trenchless Technology (SSTT) (2002). *No-Dig Handbook*, Copenhagen, Denmark.

Subterra (2004). *Rolldown Design Guide.* Available at http://www.subterra.co.uk/pipe_rehabilitation_frameset.html.

Subterra (2004). *Subline Design Guide.* Available at http://www.subterra.co.uk/pipe_rehabilitation_frameset.html.

CHAPTER 16
IN-LINE REPLACEMENT

16.1 INTRODUCTION

In-line replacement (ILR) technology originated in the 1970s and can be used to *replace* water, wastewater, and gas pipelines. This method is most cost-effective if a new pipe with a larger diameter is required. The ILR methods can be used to replace virtually all types of pipe materials such as concrete, clay, steel, ductile iron, cast iron, asbestos, polyvinyl chloride (PVC), and polyethylene (PE) pipe. Diameters of up to 48 in and drive lengths of 1500 ft have been completed with pipe bursting. The new pipe can be PE (most common); PVC, ductile iron, glassfiber reinforced polyester mortar, and vitrified clay. The ILR can be divided into two main categories of pipe bursting and pipe removal. Each of these categories is further divided into different methods (see Fig. 16.1).

16.1.1 Pipe Bursting (PB)

PB can be used to renew sewer, water, gas, and other pipelines, and is used particularly to upsize the capacity of the existing pipelines. This method is also applicable where existing pipelines are in such poor condition that renewal process with other trenchless technologies is not possible and replacement is therefore required. PB is used in situations where it is not possible to use open-cut technique on the entire stretch of the pipeline or where excavation is very expensive and social costs are high.

British Gas, currently Advantica, developed PB technology to replace deteriorated gas pipelines and currently holds patent rights to this method. In a typical PB operation, a cone-shaped tool (*bursting head*) is inserted into the existing pipe and forced through it, fracturing the pipe and pushing its fragments into the surrounding soil. At the same time, a new pipe is either pulled or pushed (depending on the type of new pipe) in the annulus left by the expanding operation. In the vast majority of PB operations, the new pipe is pulled into place. The new pipe can be of same size or larger than the replaced pipe. The rear of the bursting head is connected to the new pipe, and the front end of the bursting head is attached to either a winching cable or a pulling rod assembly. The bursting head and the new pipe are launched from the insertion pit. The cable or rod assembly is pulled from the pulling or reception pit. Figure 16.2 presents a PB process.

Dependent on the type of bursting head used, PB can be further divided into pneumatic, hydraulic, static, pipe splitting, and pipe insertion. Different methods of PB are described in Sec. 16.2.

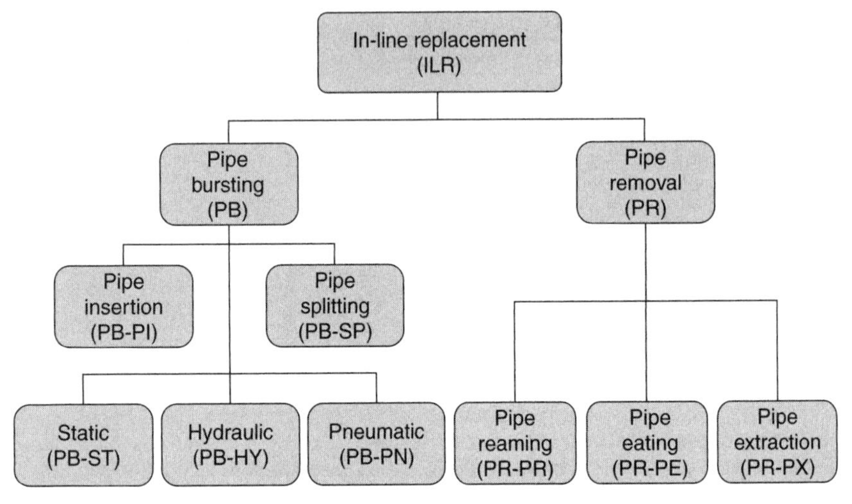

FIGURE 16.1 Categories of in-line replacement methods.

16.1.2 Pipe Removal (PR)

PR (also known as pipe eating) is a replacement technique, which is based on horizontal directional drilling (HDD) technology or microtunneling (MT) technology. This method excavates the existing pipe in fragments and removes them rather than displacing them into the surrounding ground. The pipe removal process is similar to new trenchless construction method installation. Pipe removal is further divided into several methods as follows:

- *Pipe reaming.* A technique that uses a specially designed variation of the reaming process from HDD, to excavate the existing pipe in fragments and remove them rather than displace them, and pulls a new pipe in to replace the existing pipe.

FIGURE 16.2 Pipe bursting process. (*Courtesy of Vermeer Manufacturing Company.*)

- *Pipe eating.* This technique comprises an MT machine, which has been specially adapted to allow the existing pipeline to be broken up, and removed. The microtunnel boring machine (MTBM) has a larger diameter than the existing pipeline.
- *Pipe ejection or pipe extraction.* A technique that removes the existing pipe as a whole from the ground, by pushing or pulling it toward a reception pit where it is broken and taken out.

16.2 PIPE BURSTING METHODS

16.2.1 Pneumatic Pipe Bursting (PB-PN)

Pneumatic PB is the most frequently used type of PB. It is used in majority of PB projects worldwide. In the pneumatic PB, the bursting head is a cone-shaped soil displacement hammer (see Fig. 16.3). The bursting head is driven by compressed air, and operated at a rate of 180 to 580 blows/min. The percussive action of the bursting head is similar to hammering a nail into a wall, where each impact pushes the nail a small distance farther into the wall. In a like manner, the bursting head creates a small fracture with every stroke, and thus continuously cracks and breaks the existing pipe. The percussive action of the bursting head is combined with the tension from the winch cable, which is inserted through the existing pipe and attached to the front of the bursting head. The winch cable keeps the bursting head pressed against the existing pipe wall, and pulls the new pipe behind the head. The air pressure required for the percussion is supplied from the air compressor through a hose, which is inserted through the new pipe and connected to the rear of the bursting tool. The air compressor and the winch are kept at constant pressure and tension values, respectively. The bursting process continues with little operator intervention, until the bursting head comes to the reception pit.

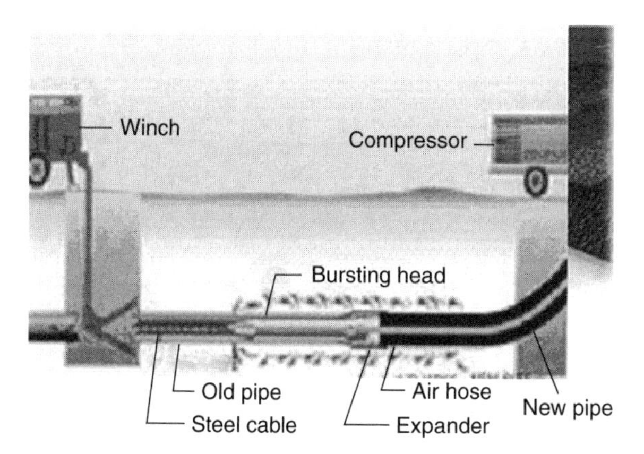

FIGURE 16.3 The pneumatic pipe bursting method. (*Courtesy of TT Technologies.*)

FIGURE 16.4 Hydraulic pipe bursting methods. (*Xpandit, Courtesy of Miller Pipeline Corporation.*)

16.2.2 Hydraulic Pipe Bursting (PB-HY)

In the hydraulic bursting system, the process advances from the insertion pit to the reception (pulling) pit in sequences, which are repeated until the full length of the existing pipe is replaced (see Fig. 16.4). In each sequence, one segment of the pipe (which matches the length of the bursting head) is burst in two steps: first the bursting head is pulled into the existing pipe for the length of the segment, and then the head is expanded laterally to break the pipe. The bursting head is pulled forward with a winch cable, which is inserted through the existing pipe from the reception pit, and attached to the front of the bursting head. The rear of the bursting head is connected to the new pipe and also the hydraulic supply lines that are inserted through the new pipe. The bursting head consists of four or more interlocking segments, which are hinged at the ends and at the middle. An axially mounted hydraulic piston drives the lateral expansion and contraction of the bursting head.

16.2.3 Static Pipe Bursting (PB-ST)

In the static pull system, the force for breaking the existing pipe comes only from pulling the bursting head forward. Either a rod assembly or a winch cable pulls the head (see Fig. 16.5).The rod assembly or winch cable is inserted through the existing pipe and attached to the front

FIGURE 16.5 Static pipe bursting. (*Courtesy of TT Technologies.*)

of the bursting head at the insertion pit. The tensile force applied to the bursting head must overcome the frictional force and also breaking force of existing pipe and is therefore significant. The cone-shaped bursting head transfers this horizontal pulling force into a radial force, which breaks the existing pipe and provides a space for the new pipe.

If a rod assembly is used for pulling, the bursting process is done in consecutive sequences, rather than continuously. Prior to bursting, the segmented rods are inserted into the existing pipe from the reception pit. The rods are only a few feet long, and during insertion, they are threaded together to reach the bursting head at the insertion pit. At the insertion pit, they are attached to the front end of the bursting head, and the new pipe is connected to its rear end. In each sequence during the bursting, the hydraulic unit in the reception pit pulls the rods for the length of individual rods, and the rods are separated from the rest of the rod assembly as they reach the reception pit. If a winch cable is used instead of rods, the pulling process can be continuous. However, a typical cable system does not transmit as a large pulling force to the bursting head as a rod assembly.

16.2.4 Pipe Splitting (PB-PS)

Pipe splitting system (see Fig. 16.6) is used for pipes that are not brittle, like steel and ductile iron gas mains. Instead of the bursting head, the system uses a splitter, which cuts the existing pipe along one line on the bottom and opens it out, rather than fracturing it. The splitter is pulled through the existing pipe by either a wire rope or steel rods. The splitting system consists of one or more of three parts: (1) a pair of rotary slitter wheels, which make the first cut, (2) a hardened sail blade on the underside of the splitter, which follows, and (3) an expander, whose conical shape and off-centered alignment force the split pipe to expand and unwrap. The unwrapping of the pipe is smooth, without generating hoop stresses or longitudinal bending in the pipe walls that could cause high pulling forces and jamming. The splitting and unwrapping of the existing pipe creates a hole immediately behind the splitter and large enough to allow the new pipe to be pulled in. The existing pipe dislocates to a position above the hole and over the new pipe, thus protecting the new pipe from damage.

16.2.5 Pipe Insertion (PB-PI)

The pipe insertion system (Fig. 16.7) pushes or jacks a new pipe into the existing deteriorated pipe. This system uses the columnar strength of segmented *bell-less* jacking pipe to advance the *lead train* through the existing pipe. The lead train consists of five sections: the lead, a heavy steel guide pipe, which maintains the alignment within the center of the existing pipe; the cracker that fractures the existing pipe; the cone expander that radially expands the fractured line into the surrounding soil; the front jack, a hydraulic cylinder that

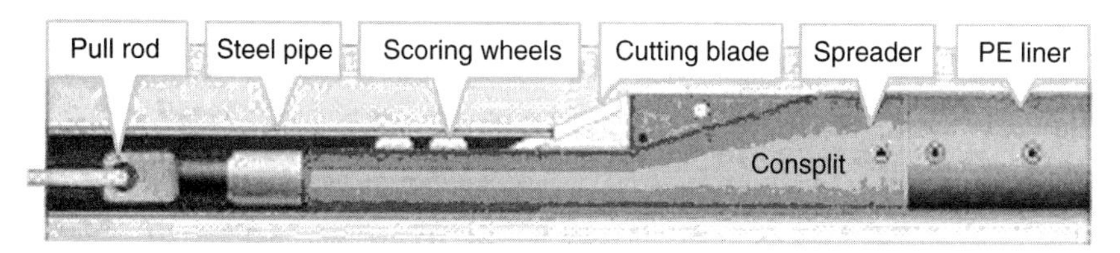

FIGURE 16.6 Pipe splitting system. (*Courtesy of ConSplit.*)

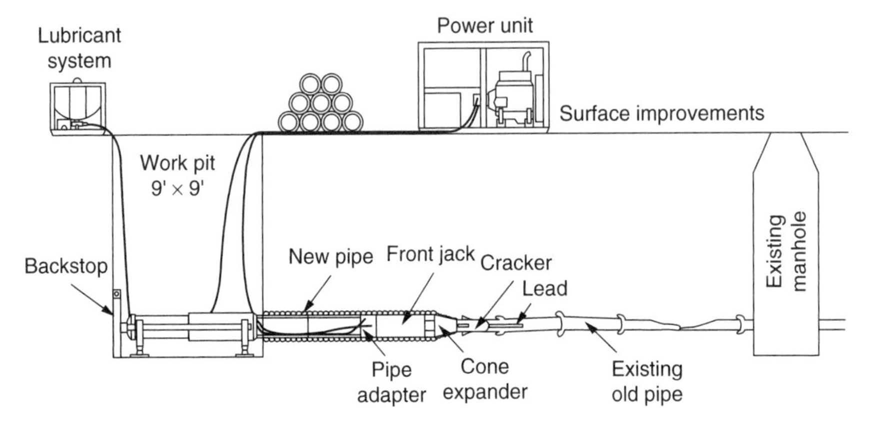

FIGURE 16.7 Pipe insertion. (*Courtesy of Tenbusch Insertion Method.*)

provides axial thrust to the (above) penetration or compaction pieces; and the pipe adapter that provides mating surfaces—linking the new pipe to the front jack.

The last section, the pipe adapter, is fitted with a lubricant injection port where the lubricant (polymer or bentonite) can be injected into the annular space surrounding the new replacement pipe. The introduction of a lubricant allows for the efficient replacement of the existing pipelines even in soft sticky clays or wet sands.

Dual flexible hose sections, which transport lubricant and hydraulic fluid to the front train, are fed through each new pipe section. Each new hose section is connected to the previous sections and to the operator's control panel with quick-disconnect couplings. Using the new pipe as a support column, the front jack advances the lead train into the existing pipe independent of the advance of the new pipe column. The new pipe is jacked behind the lead train, piece by piece, by the jacking frame (in the entry or insertion pit). The primary jacking frame applies the required thrust to advance the new pipe column (as the front jack is retracted). Instrumentation and controls at the operator's control panel (at the jacking frame) allow the operator to *feel* his way through the existing pipe as the new pipe column and front train are *inch-wormed* into the existing pipe. Upon completion of the pipe replacement, the lead train is disassembled easily inside a typical 4-ft diameter receiving manhole and the new pipe is jacked into its final position. Pipe insertion technique is illustrated in Fig. 16.7.

16.3 PIPE REMOVAL METHODS

16.3.1 Pipe Reaming (PR-PR)

Pipe reaming is a modified back reaming method used in directional drilling, which is specially adapted for pipe replacement (see Fig. 16.8). First, the pilot drill string is inserted through the existing pipe. Next, a specially designed reaming tool is attached to the drill string and pulled back through the existing pipe, while installing the new pipe. The reamer has cutting teeth, which grinds and pulverizes the existing pipe through a *cut and flow* process, rather than a compaction. The pipe fragments and the excess material from upsizing are carried with the drilling fluid to manholes or reception pits, and retrieved with a vacuum truck or slurry pump for disposal.

FIGURE 16.8 Pipe reaming process. (*Courtesy of Nowak Pipe Reaming.*)

The system provides for the removal of existing vitrified clay pipe (VCP), PVC, asbestos cement (AC), or nonreinforced concrete pipe and simultaneous replacement with a high density polyethylene (HDPE) or PVC pipe of equal or larger diameter. Usually double diameter replacement is the practical maximum upsize. Removal of the existing pipe is accomplished by *back reaming* using a directional drilling machine with a cutter head that breaks the existing pipe to small pieces, which are carried along with excavated soil by the drilling fluid to an extraction point.

HDPE, restrained joint PVC, and ductile iron pipes provide the new pipe that will satisfy performance as well as installation requirements. The HDD equipment requirements for the pipe reaming system generally serve two purposes for the contractors. The purchase of HDD equipment provides the contractor with not only the ability to enter the trenchless market with the pipe reaming system, but also to enter the HDD market using the same equipment. A modified backreaming tool is the only special tool required that may be modified at the contractor's shop or purchased from a manufacturer. Whether or not the prospective user owns a directional drill, the initial investment can be readily amortized. The directional drill units may also be rented.

16.3.2 Pipe Eating (PR-PE)

Pipe eating is a modified MT system specially adapted for pipe replacement (see Chap 11 for a detailed description of the MT method). The microtunnel boring machine (MTBM) crushes the existing defective pipeline and removes the particles by the circulating slurry system. A new pipe is simultaneously installed by jacking it behind the MT machine. The new pipe may follow the line of the old pipe on the entire length, or may cross the elevation of the old pipe on a limited segment only. The MT system is remotely controlled and is guided with a laser line from the drive pit. During the operation, the MTBM removes or *eats* whatever is in its way, the existing pipe or the ground itself. The MTBM has a cuttinghead and a shield section. The cuttinghead has cutting teeth and rollers that cut the pipe, and cutting arrangements close to the edge of the shield cut the ground to the required diameter to take the new pipe. The cuttinghead is cone-shaped, which puts the existing pipe material into tension and thus reduces the heavy wear of cutting teeth. The shield section carries the cuttinghead and its hydraulic motor system. The MTBM is launched from a drive pit, where a thrust frame is located. The jacking frame provides a thrust force that is applied to the new pipe to push the MTBM forward.

16.3.3 Pipe Ejection or Extraction (PR-PX)

Pipe ejection (a modified version of pipe jacking) and pipe extraction (modified static pipe bursting) are pipe removal systems, in which the unbroken existing pipe is removed from the ground, while the new pipe is simultaneously installed. The existing pipe is broken into pieces only as it completely exits out of the ground. These techniques are applicable only for existing pipes with sufficient remaining thrust capacity to withstand the push or pull forces. These methods are used in projects with shorter replacement drives to avoid high frictional resistance.

In pipe ejection, the jacking frame in the insertion pit pushes out the existing pipe by jacking the new pipe segments forward. The new pipe is placed against the existing pipe, and as the new pipe is jacked, the existing pipe is pushed toward a reception pit or manhole. At the reception pit, the existing pipe is broken into pieces and removed. The jacking frame and the insertion pit are sized to fit the length of individual new pipe sections.

In pipe extraction, the new pipe is pulled into place of the existing pipe. An extraction machine is placed into reception pit, and the replacement pipe is fed from the entry or insertion pit. A pulling device (a pulling rod assembly) is inserted through the existing pipe, and attached to the extraction machine at one end, and a tool assembly, which is connected to the new pipe, at the other end. The tool assembly consists of a centralizing device, pull plates, and a cylindrical expander or plug, which allows the method to handle size-for-size replacement or upsizing.

16.4 METHOD APPLICABILITY

It is important to choose the best ILR method for specific conditions of a project. Such parameters as the site space available, the chemical composition of the fluid carried by the pipeline, the pressure and the temperature in the pipeline, number of laterals, number of manholes, number of bends, type of deterioration, and so on, must be assessed from project to project. Table 16.1 provides an overview of the applications and possible limitations for ILR methods.

ILR techniques allow the existing pipeline to be upsized while removing displaced joints and realigning the pipeline. Moreover, the method does not require much of the preparatory work—pipeline cleaning, grinding of intruding lateral connections, root cutting,

TABLE 16.1 Overview of In-Line Replacement Applications and Possible Limitations

Pipeline type	Suitable	Notes
Sewers	Yes	
Gas pipelines	Yes	A
Potable water pipelines	Yes	
Chemical or Industrial pipelines	Yes	
Pipelines with bends	Yes	B
Circular pipes	Yes	
Noncircular pipes	Yes	C
Pipelines with varying cross sections	Yes	D
Pipelines with lateral connections	Yes	
Pipelines with deformations	Yes	E
Pressure pipelines	Yes	

Notes: A. Grooves in the pipe surface must be documented and ventilation must be assured to avoid gas pockets; B. Bends up to 20 degrees can be accommodated; C. The new pipe will have a circular shape no matter the shape of the existing pipe; D. The entire drive will have the same size after replacement; E. In-line replacement will remove deformities.

cutting of displaced joints, and so on—required by other renewal methods. The subsurface conditions must be thoroughly investigated for suitability of ILR methods. For example, when PB is used below the water table or when the existing pipeline is surrounded by sand, the pneumatic pipe burster can create a vacuum that prevents progress. This problem can be solved by lowering the water table or by using other pipe bursting methods, such as static heads which advance solely on high pulling forces without the use of pneumatic actions.

16.4.1 Range of Applications

ILR can be applied on a wide range of pipe sizes and types, in a variety of soil and site conditions. The maximum pipe size for PB is 48 in, although larger pipes have been replaced. It is expected that bursting of large pipes (e.g., 80 in) will also be possible with larger equipment in the future. The most common PB is size-for-size, or upsizing the diameter of the existing pipe up to three sizes (e.g., 8 to 12 in). Large upsizing requires more energy and causes more ground disturbance. Larger sizes can slow the replacement operation and therefore need more engineering evaluation for method and site applicability.

The existing pipes suitable for PB are typically brittle materials, such as vitrified clay, cast iron, plain concrete, asbestos, or plastics (PVC or PE). Reinforced concrete pipe (RCP) can also be successfully replaced. Ductile iron and steel pipes are not ideal for PB, but can be replaced using pipe reaming.

The bursting length is usually between 300 and 400 ft, which is a typical distance between sewer manholes, nut drive lengths of up to 1500 ft have been achieved. A long burst requires more powerful equipment. Pipe reaming replacement has been conducted for drive lengths of 1300 ft.

16.4.2 Unfavorable Conditions

Unfavorable site conditions for PB are (1) expansive soils, (2) obstructions in the form of completely collapsed pipe, (3) metallic point repairs that reinforce the existing pipe with ductile material, (4) concrete envelopment, and (5) adjacent pipes or utility lines that are very close to the pipe being replaced. These obstructions do not necessarily prevent the PB, and the problem is usually solved with localized excavations. Limitations for ILR and PB applicability are in three forms: (1) economical aspect (its comparative costs versus cost of open-trench replacement), (2) safety aspect (potential damage to the neighboring utilities and structures), or (3) technical aspect (the ability to provide sufficient energy to complete the operation).

PB operation may create outward ground displacements adjacent to the pipe alignment. The ground displacements dissipate rapidly away from the bursting operation. The bursting operation can cause ground heave or settlement above or at some distance from the pipe alignment. The most critical conditions for ground displacement are when: the pipe to be burst is shallow and ground displacements are primarily directed upward. Also when significant upsizing percentages for large-diameter pipes are used, and deteriorated existing utilities are present within a distance of twice or three times diameter of the existing pipe.

Pneumatic PB may create noticeable ground vibrations on the surface above bursting operation. Unless PB is carried out at very close distances, it is unlikely to damage existing utilities or nearby structures. As a rule of thumb, the head should not pass closer than 2.5 ft from buried pipes and 8 ft from sensitive surface structures. If distances are less than these, special measures must be taken to protect the existing structures, such as excavating the crossing point to relieve stress on the existing pipe surroundings.

For most soil conditions, it is simply necessary to provide the required power to burst pipe, displace the soil and pull the replacement pipe. Longer bursts can be accomplished more easily in favorable ground conditions. Bentonite or polymer lubrication may be injected into the annular space behind the bursting head to help keep the annular space open and to reduce the frictional drag on the replacement pipe.

16.5 CONSTRUCTION CONSIDERATIONS

16.5.1 Service Laterals

Service connections (sewer, gas, and water) to existing pipeline can create problems during bursting. This is regardless of whether these connections have been excavated prior to bursting or not. The services are usually excavated prior to bursting to provide temporary bypass service and to protect the services during the bursting operation. If the connections are not excavated prior to bursting, they can easily be damaged and the damage to the service lateral may happen at some distance from the connection. However, if the connections are excavated prior to bursting, a hump in the profile of the replacement pipe may be created due to reduced resistance to upward movement of the replacement pipe at lateral point. This problem could be reduced by excavating beneath the pipe, as well as above the pipe, at the connection. For the reaming or pipe eating process that have less impact on the service lines, a wye with a riser to the surface can be installed on the service line about 2 to 3 ft from the main, and a plug between the main and the wye inserted prior to bursting. During the bursting the riser can be monitored and after bursting the connection to the main can be excavated and reconnected.

16.5.2 Insertion and Reception Pits

For sewer applications, insertion and reception pits are usually excavated in place of manholes. For replacement of gas and water lines, service pits can be expanded and used for the insertion and reception pit. All pits must be sloped or shored in accordance with Occupational Safety and Health Administration (OSHA) standards. For all static rod and cable pull machines, the bursting machine should be properly braced to resist the horizontal force necessary for the bursting operation. This may require the pit or manhole wall to have a thrust block with proper structural capabilities. Inadequate thrust block structural capacity to resist the forces can cause wall deformation that may result in soil failure and surface heave near the wall.

Different PB systems have different requirements in terms of the space required in the reception pit. Some systems may be able to operate within the existing manholes and others may need to excavate a pit for the pulling frame. The insertion pit must be in sufficient size to allow the pipe to be inserted. For continuous HDPE pipes, this means that the pipe must be able to be fed from the surface into the existing pipe alignment without overstressing the pipe in bending. Pipe manufacturers' guidelines must be followed for minimum bending radius.

16.5.3 Bursting Operation

The bursting of the existing pipe can be performed as a continuous action if the replacement pipe is continuous and a winch with continuous cable is used. When rigid rods are used, the bursting operation temporarily halts, for each rod sections to be unthreaded and removed from the reception pit. In addition, when the pipe is installed in segments, the preparation of each successive pipe segment also interrupts the operation.

16.5.4 Reconnection of Service

The reconnection of service, sealing of the annular space at the manhole location, or backfilling of the insertion pit for new PE pipe must be delayed for the manufacturer's recommended time, but normally not less than 4 h. This period allows for PE pipe shrinkage due to cooling and pipe relaxation owing to the tensile stresses induced in the pipe during installation. Following the relaxation period, the annular space in the manhole wall may be sealed. Sealing is extended a minimum of 8 in into a manhole wall in such a manner as to form a smooth, watertight joint. Ensuring a proper bond between the PVC or PE replacement pipe and the new manhole wall joint is critical.

Service connections can be reconnected with specially designed fittings by various methods. The saddles, made of a material compatible with that of new pipe, are connected to create a leak-free joint. Different types of fused saddles (electrofusion saddles, conventional fusion saddles) are installed in accordance with manufacturer's recommended procedures. Connection of new service laterals to the pipe also can be accomplished by compression-fit service connections. After testing and inspection to ensure that the service meets all the required specifications of the service line, the pipeline returns to service.

16.6 ADVANTAGES

- With ILR methods, a wide range of existing pipe types and diameters are possible to replace.
- Increase in diameter is possible without disturbance to ground surface provided adequate cover is available.
- New pipe will follow alignment of the existing pipe.
- In PB, the existing pipe is left underground eliminating the need for its disposal.
- The main advantage of ILR over other trenchless renewal methods, such as cured-in-place pipe (CIPP), fold-and-form (FF) pipe, sliplining (SL), and so on, is the ability to upsize the existing pipes.

16.7 LIMITATIONS

- Usually drive and reception shafts are required.
- Working space is needed above ground for ancillary construction equipment.
- Reconnection of laterals by open-cut is required.
- Bypassing the flow is usually required.
- Ground movements, vibrations, and possibility of damaging nearby utilities and existing structures must be evaluated for specific conditions of each project.

REVIEW QUESTIONS

1. Describe ILR and name its different methods.

2. Describe the different variations of PB method.

3. Compare pipe eating and pipe reaming methods.
4. In what conditions would you recommend PB method?
5. In what conditions would you recommend pipe reaming method?
6. What are the construction considerations for PB method? Explain.
7. Describe the effects of ground conditions on PB.
8. Describe how surrounding utilities may affect PB process.
9. Discuss the applicability of existing pipe materials for ILR process.

REFERENCES

Atalah, A. (1998). *The Effect of Pipe Bursting on Nearby Utilities, Pavement, and Structures*, Technical Report TTC-98-01, Trenchless Technology Center, Louisiana Tech University, Ruston, La.

Brahler, C. (2000). How to use pipe bursting effectively, *Proceedings of UCT 2000 Conference*, Houston, Tex., January 25–27.

Campbell, D. B., and R. Prentice. (1992). Water main extraction: Technology advances and case study, *Proceedings of NASTT No-Dig International '92*, Washington, DC, April 16.

Falk, C., and D. Stein. (1994). Presentation of a soil-mechanical model of the dynamic pipe bursting system, *Proceedings of No-Dig International '94*, Copenhagen, Denmark, pp. D3-1–D3-10.

Glynn, J., and W. Simonds. (2000). Multiple diameter pipe bursting: Is it the right sewer rehabilitation solution for your project, *Proceedings of NASTT No-Dig 2000*, Anaheim, Ca., April 9–12, 2000, pp. 133–148.

Howell, N., and F. Gowdy. (1999). Improving pipe bursting capabilities, *No Dig International*, 10(10) 28–32.

Kramer, S. R., W. J. McDonald, and J. C. Thomson. (1992). *An Introduction to Trenchless Technology*, Van Nostrand Reinhold, New York, NY, p. 223.

Leach, G., and K. Reed. (1989). Observation and assessment of the disturbance caused by displacement methods of trenchless construction, *Proceedings of No-Dig International '89*, London, UK, pp. S2.4.2–12.

Marino, S., M. Rocco, and J. Pipkin. (2000). Conquering the curve in Omaha, Nebraska: Size for size replacement of three curved sanitary sewers using static head pipe bursting and 8-inch clay no-dig pipe, *Proceedings of NASTT No-Dig 2000*, Anaheim, Ca., April 9–12, 2000, pp. 149–165.

Meinolf, R. (2000). New developments in trenchless lateral connections, *Proceedings of 18th International Conference ISTT No-Dig 2000*, Perth, Australia, October 15–19, 2000, ASTT, Perth, Australia, pp. 249–258.

Najafi, M. (1999). Overview of pipeline renewal methods, *Proceedings of Trenchless Pipeline Renewal Design & Construction '99*, Kansas City, Missouri, November 1–2, 1999, p. 9.

Norris, C., and D. Holmberg. (1999). Case study: Design objectives and construction of sewer interceptor rehabilitation projects with pipe bursting method, *Proceedings of the WEFTEC '99 Conference: Collection Systems Rehabilitation and O&M*, Salt Lake City, Utah, August 1–4, 1999, Session 15: Trenchless technology.

Poole, A., R. Rosbrook, and J. Reynolds. (1985). Replacement of small-diameter pipes by pipe bursting, *Proceedings of 1st Inernational Conference on Trenchless Construction for Utilities: No-Dig '85*, London, UK.

Rogers, C., and D. Chapman. (1995). Ground movement caused by trenchless pipe installation techniques, *Proceedings of the Transportation Research Board*, 74th Annual Meeting, January 1995, Washington, DC.

Rogers, C. D. F. (1996). Ground displacements caused by trenching and pipe bursting, *No-Dig International*, February 1996, ISTT, London, UK.

Simicevic, J., and R. L. Sterling. (2000). *Survey of Bid Prices for Trenchless Rehabilitation and Replacement of Pipelines and Manholes*, TTC report, August 2000, p. 60.

Sims, W. (2000). Large diameter pipe bursting to upgrade a sanitary trunk sewer in an environmentally sensitive river valley: The city of Nanaimo's experience, *Proceedings of NASTT No-Dig 2000*, Anaheim, Calif., April 9–12, 2000, pp. 167–186.

Simicevic, J., and R. L. Sterling. (2004). *Pipe Bursting Guidelines, TTC Technical Report No. 2001.02*. Available at Trenhttp://www.latech.edu/tech/engr/ttc/publications/guidelines_pb_im_pr/bursting.pdf.

Sterling, R. L. and J. D. Thorne. (1999). *Trenchless Technology—Applications in Public Works,* prepared for *APWA*, Kansas City, Mo., p. 238.

Sterling, R. L., A. Atalah, and P. Hadala. (1999). Design and application issues for pipe bursting, *No-Dig Engineering*, 2nd, 3rd, and 4th Quarters 1999, 6(2):8–14.

CHAPTER 17
LATERAL RENEWAL

17.1 INTRODUCTION

Service laterals or service connections branch off the sewer mains and connect building sewers to the public sewer main lines (Fig. 17.1). Laterals may be as small as 4 in. in diameter, and normally ranging from 15 to 100 ft in length. Laterals are built with any of the pipe materials listed in Chap. 6, but most commonly they are plastic (such as PVC), concrete, vitrified clay, and ductile iron pipes. They are normally constructed at a minimum self-cleansing grade from the building to the main sewer. At the main sewer, the grade may change abruptly for the line to descend to the main sewer. Therefore, laterals may enter sewer mains at angles ranging from 30 to 90 degrees from the axial flow direction. In some housing developments, the same route is used to install both the potable water service connections and the sewer lateral. Any leaks in the potable water line can enter the lateral sewer line if it is not watertight.

To reduce infiltration/inflow (I/I) and to ensure a tight seal between the lateral and the main line, renewing laterals are the next step after renewing the main sewer lines. Infiltration is frequently found at the lateral connections between the main sewer pipeline and service laterals (Fig. 17.2). For many years the effect of leaking lateral sewers on the collection system and the treatment facilities was considered insignificant. Recent research studies sponsored by the U.S. Environmental Protection Agency (EPA) have indicated that in many cases a significant percentage of the extraneous flows is the result of the defects in sewer laterals. These defects include cracked, broken, or open-jointed pipes. Laterals historically have transported water from inflow sources such as roof drains, cellar and foundation drains, basement sump pumps, and clean water from commercial and industrial effluent lines. With new EPA regulations, such as CMOM, (see Chap. 3), there will be tighter controls of sanitary sewer overflows (SSOs), which means an even greater emphasis on stopping I/I through service laterals and main lines.

The construction and maintenance of service laterals is complicated because separate entities may have jurisdiction over different portions of the laterals. Usually the portion between the building's plumbing and drain system and the property line is the building owner's responsibility. This portion is considered an extension of the in-structure facilities, and therefore, it is ordinarily installed under plumbing or building codes and tested and approved by the building inspectors. The section of the sewer lateral between the property line and the main sewer line, including the sewer main connection, usually is installed under sewer pipe codes and local regulations and is a property of the city or municipality. The inspection and maintenance of this section of the lateral is a responsibility of the Public Works Departments and Engineering Division of the city, township, or county.

FIGURE 17.1 Service laterals connecting to main sewer line. (*Courtesy of Insituform Technologies.*)

17.2 SERVICE LATERAL RENEWAL METHODS

Some of the renewal methods used for larger main sewers are also applicable for service laterals. These methods include lateral cured-in-place pipe, in-line replacement methods (such as lateral pipe bursting), lateral thermoformed pipe, and lateral coatings and linings. Such trenchless renewal methods as sliplining may not be applicable to small-diameter laterals because of possible loss in hydraulic capacity. Unlike renewal of main sewers, access to laterals can often be limited, making renewal more difficult. This is combined with the additional problems of tree roots, existing surface and subsurface structures, and landscaping over laterals that make open-cut method and digging pits complicated. It should be noted that some of the trenchless construction methods, such as pilot-tube microtunneling (PTMT) and horizontal directional drilling (HDD) can be used to install a new lateral in place of the old one.

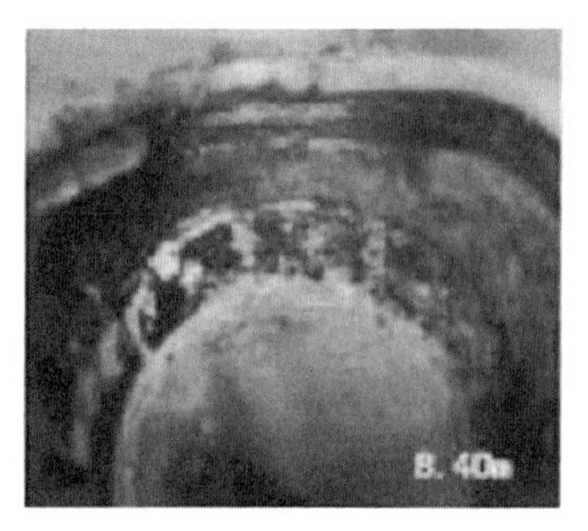

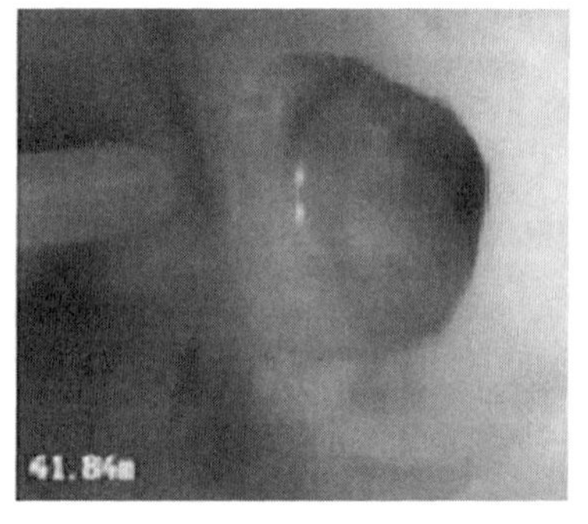

FIGURE 17.2 Defective (left) and renewed (right) lateral connection with the main sewer lines. (*Courtesy of Perma-Liner.*)

17.2.1 Method Installation

Before commencement of any sewer lateral renewal project, the following preparatory work is usually conducted:

- Preliminary inspection of all laterals (usually a few months prior to start of renewal project)
- Decision on renewal method (or methods) based on specific conditions of the existing laterals; note that lateral conditions can vary widely, even in a local area
- Mobilization
- If necessary, small pit excavation at the property line; note that this may not be required with the use of appropriate renewal method
- Lateral cleaning and reinspection (preinstallation inspection)
- Renewal installation (see the following sections for specific methods)
- If necessary, cleanout installation
- Postinspection
- Lateral reconnection
- Surface restoration and demobilization

Due to small size and length of individual installations, usually no air testing is conducted for quality control of a lateral renewal project. To measure effectiveness of I/I reduction, a continuous flow monitoring before and after lateral renewal can be conducted.

17.2.2 Lateral Cured-In-Place Pipe (L-CIPP)

The technique of renewing laterals with cured-in-place pipe (CIPP), also called short CIPP, is almost the same as that used to renew main pipelines (see Chap. 13). Using the same materials and methods as the original CIPP process, this method provides a structural, jointless, and nonleaking service lateral pipe. The new pipe itself can be manufactured in many ways, but in principle it consists of a short felt belt tube impregnated with thermal-cure or ambient-cure polyester or epoxy resin. The new pipe can be installed from cleanout or the main sewer. The polyester or epoxy resin is cured by heating (either physically or chemically) or by ultraviolet light. To consider applicability of CIPP for lateral renewal, such factors as number of bends (usually more than three bends makes lateral CIPP method impractical) and offset joints, collapsed pipe, and other obstructions should be evaluated.

Primary Characteristics. An important difference between installation from a cleanout and installation from a main sewer is that it is only possible to form a watertight seal between the lateral and the main sewer in one operation if the new pipe is installed from the main sewer. This is achieved by fitting the new pipe with a transition piece profile that bonds to the main sewer. When installing new pipes from a cleanout, the lateral connection can also be sealed by subsequently fitting a transition piece. Typically, the wall thickness of lateral new pipes is 0.18 in.

Depending on the condition and the design of the lateral, a new liner pipe can either reduce or increase the hydraulic capacity of the lateral as this is determined by the cross-sectional area of the pipe and the roughness of the renewed lateral. Reducing the diameter of laterals usually reduces their capacity, but this is counterbalanced by the fact that new pipes are considerably smoother than the existing laterals and a renewal method may improve the hydraulic capacity in the displaced joints, fractures, and the like. CIPP can usually be used

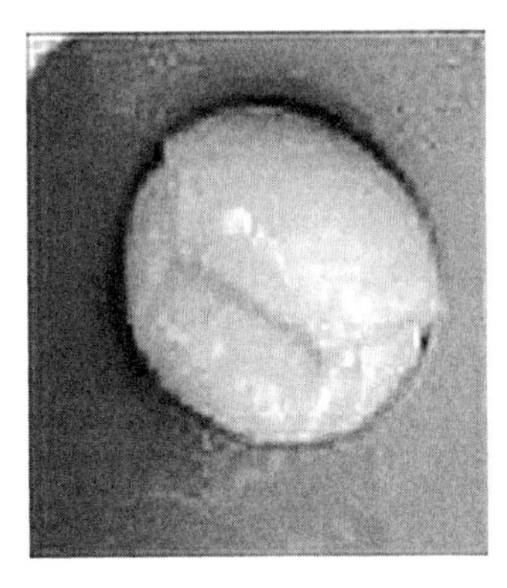

FIGURE 17.3 The inversion process. (*Courtesy of Performance Liner®.*)

for renewing most laterals where there have been no problems relating to self-cleaning ability or insufficient capacity.

Lateral CIPP with air inversion only requires one access point to the existing lateral and is more time efficient (Fig. 17.3). Digging one small hole (3 ft × 3 ft) at the property line would not only give access to the new pipe but also allow installation of a new inspection chamber (cleanout). By using this method, the construction work can commence within the city's right-of-way. The equipment required for CIPP air inversion include a trailer that contains the air-inversion machine, the two-part resin, liner material, two inversion heads, air compressors, generators, hand tools, sod cutter, power tamper (compactor), inspection chamber (cleanout), roller (to impregnate liner with resin), various fittings, and a backhoe.

Cleaning and Inspection. Before starting any renewal process, it is necessary to thoroughly clean the pipe. Typically a high-pressure water jet is used to free the service lateral of roots, mineral deposits, grease, sand, and sludge. The inspection process involves propelling a small-diameter camera into the service lateral from a robot positioned in the main line. This inspection provides valuable information regarding the condition of the lateral. The inspection is usually conducted by working remotely from the main line, thereby avoiding any digging or disruption.

Installation Process. An installation assembly unit is winched through the main line to the damaged service connection. The assembly unit is then used to invert the liner pipe into the service lateral. Then, a curing process is initiated, during which the liner is held in place by the assembly unit's inversion hose. Once the new lateral pipe has cured, the assembly unit is removed, leaving the service lateral renewed and ready for use. Figure 17.4 shows the renewed lateral from the main sewer line.

17.2.3 Lateral Pipe Bursting (L-PB)

Pipe bursting method for renewing service laterals is same as pipe bursting for renewing main pipelines (see Chap. 16). Pipe bursting is mainly used for laterals that have capacity and/or structural problems. Lateral pipe bursting is the only trenchless renewal method that can significantly increase the hydraulic capacity of laterals by installing a larger diameter

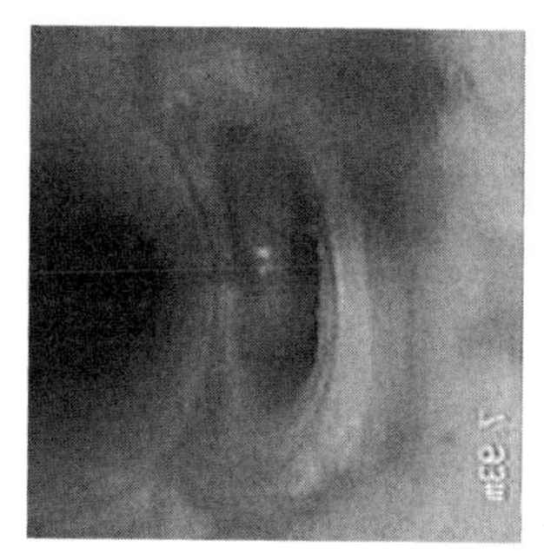

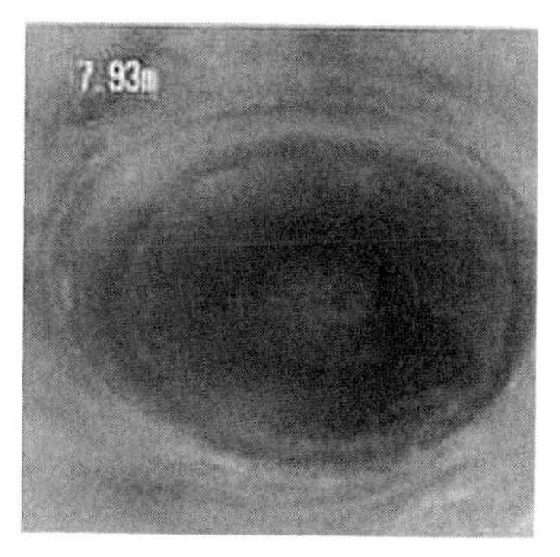

FIGURE 17.4 Two views of the renewed lateral from the main sewer line. (*Courtesy of Perma-Liner.*)

new pipe. Lateral pipe bursting usually requires excavation at two points, one at the private property (next to the house) and another at the main line. Thus, this method requires more preparation and renewal time.

Primary Characteristics. Pipe bursting lateral lines has increased dramatically over the past several years with advances in guide head technology and the development of smaller, space efficient tools. Pneumatic tools can be used to burst common lateral pipes (cast iron, clay, nonreinforced concrete, acrylonitrile-butadiene-styrene and other plastics) with diameters as small as 4 in. In the bursting procedure, a pneumatic tool is placed into the existing pipe at the entry pit or manhole. A winch line, based at the exit area pit and attached to the head of the tool, guides it through the existing pipe. Continuous percussion from the pneumatic tool fractures the existing pipe. The tool effectively hammers its way through, forcing the fragments into the surrounding soil, while pulling the new pipe into place behind it. Same size or larger pipe can be installed with the pneumatic pipe bursting system.

Several factors dictate whether pipe bursting is appropriate for the renewal of a failed service lateral. These considerations include: existing pipe material, diameter, condition, depth, and length; new pipe diameter requirements; soil conditions; and location of peripheral utilities. By using the pneumatic pipe bursting process, considerable lengths of trench work can be avoided. Small entry and exit pits are much less disruptive and costly than long deep trenches. Pipe bursting considerations and risks include depth of existing pipe, previous point repairs, pipe sags, turns and bends, collapsed lines, joint settlements and misalignments, pipes with various diameters, underground water, difficult access for the renewal process, and adjacent utilities.

17.2.4 Lateral Thermoformed Pipe (L-ThP)

Thermoformed pipe (ThP) can also be used for lateral renewal. This method slightly reduces the cross-sectional area of the lateral in the same way as CIPP does. However, slight reduction in the cross-sectional area does not result in reduction in hydraulic capacity as the new pipe will improve hydraulic capacity by providing a smooth surface. The lateral ThP method is often easier to install than lateral CIPP and is mainly used in laterals where project conditions, such as access points, bends, deformations, and so on, allow. For more information on ThP, refer to Chap. 21.

17.2.5 Lateral Underground Coatings and Linings (L-UCL)

Four chemical grouting and lining methods (see Chap. 20) are currently available for sealing lateral sewers for nonstructural purposes: pump full method, sewer sausage method (patented process), camera-packer method (patented process), and spray-on lining method.

Pump Full Method. This method involves injecting a chemical grout through a conventional sealing packer from the sewer main up the lateral sewer to an installed plug. As the grout is pumped under pressure, it is forced through the pipe faults into the surrounding soil where a seal is formed after the gel has set. After the sealing has been accomplished, excess grout is augured from the lateral sewer and it is returned to service.

Sewer Sausage Method. This method uses a camera-packer unit in the sewer main with the injection of grout from the sewer main up the lateral sewer to seal the pipeline. A tube is inserted into the lateral sewer before sealing to reduce the quantity of grout used and to minimize the amount of cleaning required after the sealing has been completed. The grout is pumped under pressure around the tube, up the lateral sewer, and through any pipe faults into the surrounding soil where the seals are formed after the gel sets.

Camera-packer Method. Unlike the other methods described, this method only repairs faults seen through an inspection camera. First, a miniature camera and a specialized sealing packer are inserted into the lateral sewer. Using a winch cable previously inserted from the lateral sewer access to the downstream manhole of the sewer main, the camera-packer unit is pulled into the lateral sewer. The camera-packer is then slowly pulled back out, repairing cracks that are seen through the camera. First, the deepest leaking joints are sealed. Joints and cracks are sealed in a manner similar to the conventional methods used for sealing joints in sewer mains. Once the repairs have been completed, the equipment is removed and the lateral sewer is returned to service. The costs for this method vary, depending on the number of cracks and the difficulties encountered. While estimating costs, allowances should be made for such factors as difficult site access and pit excavation.

Spray-on Linings Method. Recent developments in spray-on lining technology show significant promise for renewing small-diameter sewers such as service laterals for nonstructural purposes. The basic technology is the same as for larger sewers, although with miniaturized equipment. Difficulty may be experienced with bond in the areas of high infiltration.

17.2.6 Other Lateral Renewal Considerations

The above service lateral renewal methods will reduce inflow/infiltration (I/I) resulting from high groundwater levels and from rainstorms. Several of these methods can be used for structural renewal of sewer laterals. In addition to these renewal methods, efforts need to be made to remove sources of inflows that may be connected to the lateral sewer. Inflow sources connected are usually located on private property. In these situations, a public awareness or public relations program is often needed. Such programs are intended to persuade private property owners (such as home and business owners), without threat of legal consequences, to make the needed changes to help correct a community problem.

17.3 ADVANTAGES

Sewer lateral renewal methods provide tools for jointless, nonleaking, structural or non-structural, and tight-fitting solutions. These methods can be applied at congested areas where open-cut method is impractical.

17.4 LIMITATIONS

Such factors as number of bends (usually more than three bends makes lateral method impractical) and offset joints, collapsed pipe, previous point repairs, pipe sags, collapsed lines, joint settlements and misalignments, pipes with various diameters, difficult access, and other obstructions may limit application of lateral renewal methods.

REVIEW QUESTIONS

1. Why is lateral renewal important? Explain the main objectives of lateral renewal projects.
2. Describe the characteristics of service laterals (common lengths, diameters, slope, existing pipe types, and so on).
3. Describe lateral renewal methods.
4. In what conditions would you recommend lateral CIPP?
5. In what conditions would you recommend lateral pipe bursting method?
6. In what conditions would you recommend coatings and linings methods?
7. Explain why even after lateral renewal there might be other sources of inflow to the sanitary sewer system and how can it be resolved.
8. What are advantages and limitations of sewer lateral renewal methods? Explain.

REFERENCES

Atkinson, K. (2000). *Sewer Rehabilitation Techniques*, Subterra Systems, Dorset, UK.

Australian Society for Trenchless Technology. (2004). *ASTT – ISTT Trenchless Technology Guidelines Part 1: Spray Lining*. Available at http://www.astt.com.au/Sliplining.pdf.

Insituform Technologies. (2004). Available at www.insituform.com.

Iowa Statewide Urban Design and Specifications (SUDAS) (2004). *Design Manual: Chapter 14*, Ames, Iowa. Available at http://www.iowasudas.org.

Performance Liner. (2004). Available at http://www.performanceliner.com.

Perma-Liner. (2004). Available at http://www.perma-liner.com/innersealrehab.html.

Scandinavian Society for Trenchless Technology (SSTT) (2002). *No-Dig Handbook*, Copenhagen, Denmark.

Simicevic, J. (2004). *Currently Available Products and Techniques for Lateral Rehabilitation*, Trenchless Technology Center (TTC), Louisiana Tech University. Available at http://www.latech.edu/tech/engr/ttc/publications/other/rehab.pdf.

CHAPTER 18
LOCALIZED REPAIR

18.1 INTRODUCTION

In this book the term *renewal* has be used extensively to cover all methods of trenchless technology that *extend the design life* of a deteriorated existing pipe. Therefore, renewal methods cover such terms as rehabilitation, renovation, replacement, and similar terms that may be used in other publications. The term *repair* is used when the pipe defect is fixed without necessarily extending the design life of the existing pipeline. Local repair falls into this category under which a defect may be temporarily and/or locally fixed without renewing the whole section of the pipeline.

Local defects may be found in an otherwise sound pipeline, as a result of cracking or joint settlements or failure. Therefore, localized repair (LOR) or point source repair (PSR) methods are used to solve a number of problems such as cracks, broken pipe, hammer taps, root intrusion, infiltration, debris, soil erosion, exfiltration, and misaligned pipe sections. LOR methods cover a broad range of techniques such as robotic repair, grouting, cured-in-place pipe (CIPP), internal seal, shotcrete, and so on. Some of these LOR systems have been developed for sewer pipes and some others designed to seal joints in pressure pipelines. Many of these techniques are variations of full-length pipe renewal systems.

As for any construction project, the economics of LOR techniques should be considered case by case only, but clearly in many situations repair of individual defects may be more cost-effective than renewing the entire length of the pipeline from manhole to manhole where damage is localized. As a general rule, it has been suggested that LOR can be economically viable if less than 25 percent of the sewer length contains defects, although this rule will vary according to the individual circumstances. Just as for any renewal method, preinspection and cleaning are essential to commencing LOR systems. For those LOR methods, which require a bond with the existing pipe, all traces of grease and debris must be removed.

There are four broad categories of LOR system (Fig. 18.1): robotic repair, grouting, internal seal, and point CIPP.

18.2 PRIMARY CHARACTERISTICS

As stated previously, when local defects are found in otherwise sound pipeline, LOR are considered. Coatings are sprayed to the interior wall of the existing pipe and work by adhesion, or in the case of robotic repairs, the defect is filled with epoxy. Systems are available for remote-controlled resin injection to seal localized defects in the range of 4 to 30 in. in diameter.

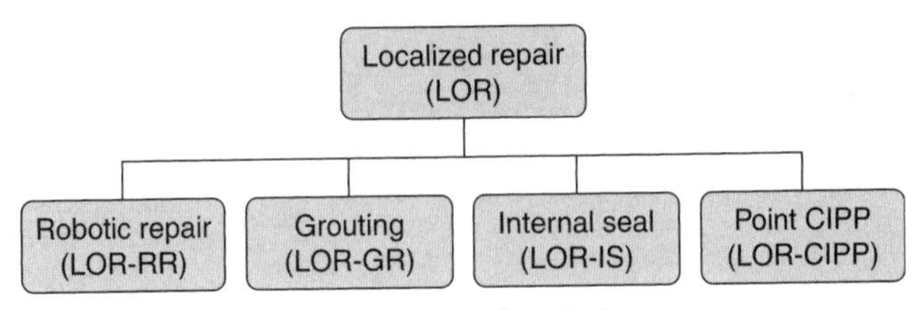

FIGURE 18.1 Categories of localized repair methods.

LOR devices may be used to provide an existing deteriorated pipe with additional load-bearing capacity that is equivalent to a masonry arch. Also they could provide added capacity or support to damaged pipes to sustain loads. Another use is to provide a seal against infiltration and exfiltration, corrosion protection, and in the cases of missing pipe sections, internal seal methods can be used to fill in these gaps.

All forms of LOR methods provide some structural enhancement. Point CIPP systems, including internal seal methods, are basically a new pipe inserted or cured within the old one. Although in practice the restraint offered by the existing pipe will enhance the *stand-alone* performance of the new pipe section significantly, for a conservative design, the structural properties of new pipe can be measured in isolation from the existing pipe. For example, the ring stiffness of a close-fit pipe or CIPP may be increased by a factor of up to 7 by restraint from the existing pipe considering a good support, although for design purposes a much lower factor is used.

Point CIPP and internal-seal methods are regarded as structural repairs. However, other methods such as resin or grout injection behave in a composite manner with the existing pipe and often the surrounding ground behaves differently. Because there is no separate element, and the strength of the repair is more difficult to calculate, these systems are often termed *nonstructural*. However, these methods can increase structural performance of the existing pipes by stabilizing and sealing the surrounding ground. The elimination of erosion resulting from the infiltration or exfiltration cycle is also important in extending the useful life of the pipeline.

18.3 ROBOTIC REPAIRS

18.3.1 Primary Characteristics

Robotic repair is one of the newest technologies (since 1990s) of trenchless pipeline repair. Robotic repair systems, largely developed in Switzerland, are used mainly in the gravity pipelines and comprise a grinding robot and a filler robot. The former removes intrusions, and also mills out cracks to provide a good surface key for the repair materials. The filler robot applies an epoxy mortar into the slot formed by the grinder, and trowels off the material to a smooth finish. Robots are available for use in pipes from 8 to 30-in diameter. Typically, the smaller versions will operate in diameters up to 8 to 16 in, and the larger ones from 12 in upward. Various axle and wheel sets are used to position the robot centrally within the pipe.

Robotic point repair is used either as stand-alone or as a precursor to the other renewal methods. As a stand-alone, robotics point repair is used to repair radial, longitudinal, and spider cracks. The process also lends itself to repairing broken joints, slip joints, open joints, protruding service connection (Fig. 18.2), recessed service connections, roots, and

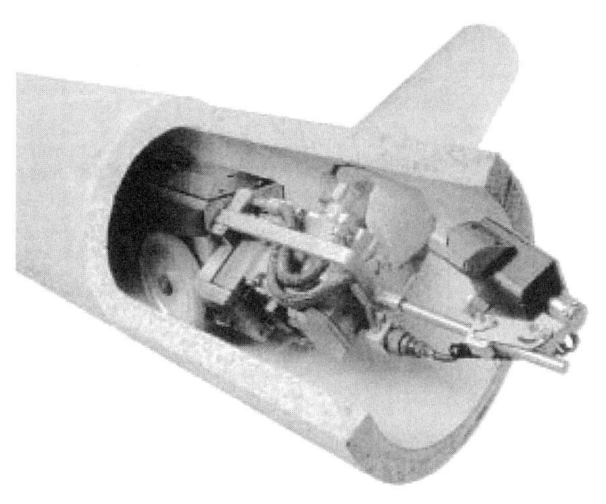

FIGURE 18.2 Cutting or grinding robot removing an intruding lateral connection. (*Courtesy of KA-TE System AG.*)

other foreign objects that are usually found in collection pipeline systems. The robotics process uses the epoxy resins for grouting. If chemical sealant was applied and all grease removed prior to the application of the epoxy, the epoxy will bond to the pipe medium and permanently seal the application location from further infiltration.

The grinding head is hydraulically driven, giving high torque at low speeds, and can be fitted with various shapes of diamond or carbide cutters suitable for clayware, concrete, polymeric materials, and even steel. Some of the more powerful grinding robots can cut through steel reinforcement. It is common for the cutters to be cooled by a water spray issuing from the central hub. Figure 18.2 shows a cutting or grinding robot removing an intruding lateral connection.

The wheels are usually driven by electric motors, as are the head rotation and extension functions. A CCTV camera attached to the head monitors the operation of the robot, and a further camera can usually be added for forward view. Some grinding robots have the facility to inject a sealing compound through a hollow drill, which prevent infiltration from affecting the mortar applied by the filler robot. They may also be fitted with a high-pressure water jet to remove the debris created by the grinding operation.

Typically, cracks are milled out to a width and depth of between 1 and 1.5 in. It is essential to clean the area of the repair thoroughly after grinding, because any dust, mud, or debris will prevent the mortar from adhering. Intruding laterals, grout deposits, and hard encrustation can also be removed.

The properties of the epoxy mortar are important, because the material is normally applied to wet surfaces. The two-part mortar may either be mixed prior to filling the canister on board the filler robot, or, in some designs, the components are loaded into the robot separately and mixed at the outlet as they are used. As with the grinding robots, the filler robots are self-propelled and have on-board cameras. The epoxy is applied by a system of remotely controlled nozzles and spatulas, the material being forced out of the canister by a piston driven by compressed air. Alternatively, the mortar can be injected through a flexible plate or former pressed against the pipe wall (Fig. 18.3).

In addition to the filling slots milled by the grinding robot, the filler robot can apply epoxy around poorly made connections, to seal the connection to the main pipe. Some systems allow the use of special formers or shields, which act as a temporary shutter and allow

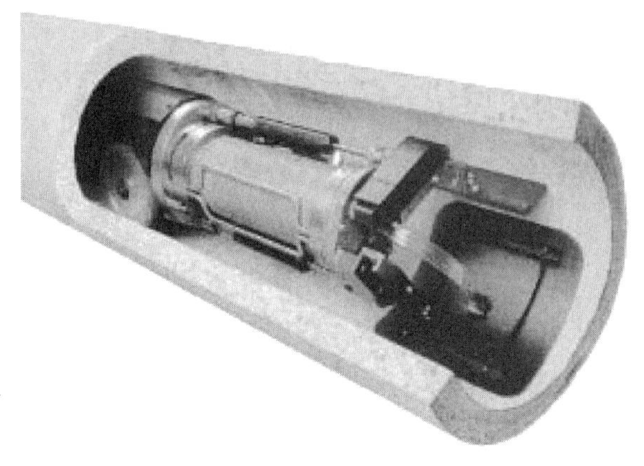

FIGURE 18.3 Filler robot injecting epoxy resin through a flexible plate pressed against the pipe wall. (*Courtesy of KA-TE System AG.*)

defective lateral connections to be remolded in epoxy mortar. They can also insert an inflatable stopper up a branch to assist in reforming a connection and to temporarily stop the flow.

All robot functions are controlled from a central console housed in a vehicle, which also contains hose reels, a compressor, a hydraulic power pack, and other ancillary equipment. There is also a hoist for lifting the robots into and out of manholes. The main source of power is a large generator, which is usually trailer-mounted. Robotic repair systems are versatile, but they need a consistent workload to be economically viable. They have found favor in regions, which offer a predictable workload for LORs, but have been less successful in commercial terms where the demand is only sporadic.

18.3.2 Method Description

Robotic point repairs are carried out by an operator manipulating the robotics functions by remote-controlled CCTV. As a first step, the robot is positioned at the defect area and is surveyed for the best starting position. Chemical grout is implemented if any infiltration of water is present. The operator then begins to grind out the crack(s). This accomplishes two goals: (1) the crack is cleared off all foreign material and by physical properties stopped from further cracking because of the groove cut and (2) the groove created, gives a larger surface area to inject the epoxy resin.

The second step is to fill the void area with epoxy. This is accomplished by making sure that the groove is fully filled and is flush with the pipe wall. Once the epoxy cures, (1 to 2 h, complete 8 days), the pipeline is restored and returned to service without reduction in flow.

Advantages. The robotic point repair is carried out at the defect location using a remote-controlled camera, with good precision. With this method, pipe structural integrity as well as the infiltration barrier is improved.

Limitations. Because of robotic design, elliptical pipelines cannot be repaired with this method. Also, small-diameter pipelines that have extensive defects, that is, crushed or caved-in walls, or slip joints that have fallen greater than 2 in, are not possible to repair. The robotic equipment is expensive and requires a significant investment from a contractor. Therefore, its efficient use requires a continuous amount of workload.

18.4 GROUTING

18.4.1 Primary Characteristics

Grouting is one of the oldest methods of pipeline repair. In recent years, there have been new advances in products and equipment for a wide variety of grouting techniques.

18.4.2 Chemical Grouting

Chemical grouting is normally used for sewer lines and is applied to leaking joints and circumferential cracks. Chemical grouting may also seal small holes and radial cracks. Chemical grouting can be applied to pipeline joints and manhole walls using special tools and techniques. Some chemical grouting is used in concrete, brick, clay sewers, and other pipe materials to fill voids in backfill outside the sewer wall. Such backfill voids, if not corrected, can reduce lateral support of the wall and allow outward movement resulting in the rapid deterioration of the integrity of the pipe. Chemical grouting does not provide an effective seal where joints or circumferential cracking problems are present because of the ongoing settlement or shifting of pipelines. Therefore, it is not effective to use chemical grouting to seal longitudinal cracks or to seal joints where the pipe near the joint is longitudinally cracked. Chemical grouting is normally undertaken to control groundwater infiltration in gravity pipeline caused by leaking pipe joints or circumferential cracking of pipe walls. Grouting mainly should be considered for pipelines in good structural conditions.

Chemical Grouting for External Repairs. External repair is performed by excavating from outside the pipe or from inside the pipe depending upon the pipe diameter. The common methods are variations of chemical and cement grouting. These methods are appropriate for solving problems of significant groundwater movement, washouts, soil settlement, and soil voids.

Chemical grouts for external repairs comprise a mixture of three or more water-soluble chemicals, which produce stiff gels from properly catalyzed solutions. The grouts produce a solid precipitate as opposed to cement; or cement or clay grouts that consist of suspensions of solid particles in a fluid. The reaction in the solution may be either chemical or physiochemical and may involve only the constituents of the solution or these constituents reacting with other substances encountered. In the latter type, the reaction causes a decrease in fluidity and a tendency to solidify and form occlusions in channels or fill voids in the material into which the grout has been injected.

Chemical Grouting for Internal Repairs. Internal repair methods are performed from within the pipeline, either remotely or through person entry. The various installation methods are discussed in this section.

Chemical grouting for internal repairs is most commonly used to reduce infiltration. It can be used to seal deteriorated and leaking pipe joints, service connections, open joints, maintenance holes, and structures. Grouting will not provide structural repair, and is inappropriate for longitudinal cracks and broken or crushed pipes. Although chemical grouting is normally used in small-diameter pipes (less than 24 in) using remote-controlled equipment, larger-diameter pipes can also be successfully grouted. Grouting of pipelines is generally accomplished using a sealing packer and CCTV camera. The packer is usually made of a hollow metal cylinder having an inflatable rubber sleeve on each end of a center band. The grout is pumped into the space created between the two inflated sleeves straddling the pipe joint. Depending upon the type of packer, the grout and the initiator solutions are mixed together, applied into this space and forced through the joint leak into the surrounding soil. The grout displaces the groundwater and fills the voids between the soil particles.

A remote CCTV is used to position the packer on the pipe joint and to inspect the joints before and after the sealing operation. Cables through a sewer section from maintenance hole to maintenance hole pull the sealing packer and CCTV. In addition, air or water testing equipment is used to determine the effectiveness of the sealing. On person-entry sewers, maintenance holes and structures, leaking joints and/or walls can be injected with chemical grouts using a nozzle-type applicator.

There are several different types of chemical grouts, which are categorized as either gel or foam. With each of the grouts there are a multitude of different types of additives, for example, initiators, activators, inhibitors, and various fillers. The general grout formulations are chemicals and water. When there is groundwater present, a higher concentration of chemicals are used because of the dilution potential. Because of soil and moisture variability, formulating the correct mixture is largely dependent on site and project conditions. Many parameters affect the grouting performance, such as, viscosity, gel time, temperature, pH, presence of entrained oxygen in the solution, contact with particular metals, influence of ultraviolet rays, presence of mineral salts, velocity of groundwater flows, and capabilities of placement equipment.

The properties of the grout vary in appearance, solubility, swelling and shrinkage, corrosiveness, stability, and strength. The various additives for grouts also affect, viscosity, density, color, strength, and shrinkage. For a proper formulation, the environmental conditions shall be taken into consideration and this must be done on a case-by-case basis. Another critical aspect of effective grouting is the proper operation of the equipment, that is, the grout packer, pumps, tanks, formulation, mixing, and application. The premixing of the final grout mixture is conducted in two separate tanks. One is the grout tank and the other is the catalyst tank. The grout tank contains water, grout, and buffer, whereas the catalyst tank contains water, oxidizer catalyst, and at times fillers.

Chemical Gel Grouts. The most commonly used gel grouts are of the acrylamide, acrylic, acrylate, and urethane-base types. All the gels are resistant to most chemicals found in sewer pipelines. All produce a gel-soil mixture that is susceptible to shrinkage cracking. All except the urethane-base gels are susceptible to dehydration. Using chemical additives can reduce these deficiencies. In addition to the chemical differences in composition, discussed under materials, there are the following important differences:

Acrylamide-base gel is significantly more toxic than the others (grout toxicities are of concern only during handling and placement or installation). The nontoxic urethane-base gels are EPA approved for potable water pipelines. Urethane gels use water as the catalyst; whereas the other gels use other chemicals. Therefore, the urethane gels require avoiding additional water contamination for proper curing.

Acrylamide-Base Gel. Acrylamide grout is mixed in proportions that produce stiff gels from dilute water solutions when properly reacted. Before a successful application with acrylamide grout, desired result, nature of the grout zone, application equipment, alternative procedures, and the plan of injection must be evaluated.

The grout injection plan shall include gel times to be used, quantities of acrylamide, and the probable layout for grout injection points. After work starts, the work plan shall be modified based on the additional information that is developed during the job. No natural soil or rock formations has yet been found in which a gel will not form. However, the injected solution must remain in the grout zone until gelation occurs. In dry soils and in flowing groundwater, there is often a tendency for the grout to disperse. Gravity and capillary forces in dry soils disperse the grout solution and may cause the gel to be ineffective. As in pipe joint grouting, dispersion can be avoided by saturating the soil prior to grouting and by the use of very short gel times. In soil void work, however, a dry soil mass cannot be as efficiently stabilized as a soil mass below the water table.

Dilution around the outer edges of the grout bulb may occur in wet soils when long gel times are used. The flowing groundwater distorts the normal shape of the stabilized mass and can displace it in the direction of flow. In turbulent flow conditions, dilution can be minimized by short gel times. In open formations or in fissures, solids such as bentonite or cement may be added to the grout solution to help produce a more complete block to flowing water. Acrylamide application is most successful in saturated or partially saturated soils.

Acrylamide is used primarily to reduce leakage rather than increase strength. It does, however, improve the structural integrity indirectly by stabilizing the surrounding soil mass. Acrylamide is a toxic chemical that can be absorbed into the body through broken skin, inhalation, and swallowing. Because of this toxicity, potential hazards in handling and usage can occur if not supervised by technically qualified personnel.

Acrylic-and Acrylate Base Gel. The acrylic grouts are water solutions of acrylic resins with several types for different applications. The acrylate grouts are very similar to acrylic grouts. After being mixed with catalysts a cohesive gel is formed. The set time can be closely controlled from a few seconds for flowing water condition to several hours for normal conditions. These grouts are good for use in sewer pipe joints, maintenance holes, and structures. The acrylic grouts have a viscosity similar to water, prior to gelation. These grouts have a tendency to swell in water allowing a watertight sealing effect.

Urethane-Base Gel. Urethane grout is a solution of a prepolymer, which cures upon reaction with water. During the reaction the gel remains hydrophilic, that is, it absorbs water plus holds it within a cured gel mass. Upon being cured, the resultant gel is resistant to the passage of water. As the prepolymer is cured by water, premature contamination by other water must be avoided during application. The formulation obviously is water sensitive and various water-to-grout ratios provide various strengths. The ratio needed to provide a strong gel ranges from 5:1 to 15:1, by volume. Ratio less than this will produce a foam reaction, where greater ratios will produce a weak gel. An example formulation would involve an 8:1 ratio of compounded material for pipe joint grouting.

Urethane-Base Foam. Urethane foam is used primarily to stop infiltration into the pipe. These leaks occur through cracks in the foundation or base (or wall), through joints formed by the base, corbel or upper frame interfaces, or through pipe penetrations into the maintenance hole wall. Grout injections are placed into predrilled holes and the grout is injected under pressure. When cured it forms a flexible gasket or plug in the leakage path. When mixed with an equal amount of water, the grout expands and quickly cures to a tough, flexible closed-cell rubber. In some applications the material is used without premixing with water, however, it will eventually require an equal amount of water for total cure.

18.4.3 Resin Injection System

Another method of grouting is resin injection using epoxy resin. A tube is lowered into the pipe and moved by either pulling or pushing it to the location of damaged area. The tube is then inflated to pipe diameter to fit tightly against the walls of the pipe. The resin is released around the tube as well as around the damaged area. Depending on the size of the damaged area, the excess resin is forced out through the pipe into the soil, which creates a seal on the outside of the pipe. The repair head remains in place for approximately 90 to 120 min to allow the resin to harden enough to remove the head cleanly. The entire process is performed by remote-control with the aid of a CCTV. The curing process takes about 24 to 36 h and is not affected by lack of applied pressure or by the presence of water or air.

Resin injection systems fall into two categories: those whose principal function is to seal the pipeline against infiltration and exfiltration, and those that aim to restore the structure of a damaged pipe. A common method of sealing leaking joints in gravity pipelines is to use a special packer, which combines the functions of leakage testing and grout injection. Joint testing and sealing may or may not be *localized*, depending on how many joints fail, but can be included under this heading as it aims to identify and cure a specific defect. A packer with inflatable end elements is positioned across a pipe joint and pressurized to isolate the joint. Air or water pressure is then applied to the center section of the packer, and the rate of pressure loss through the joint is measured. If the loss exceeds a specified limit, a sealing gel is injected into the joint through the packer, and the joint is retested. Refer to references at the end of this chapter for more information on resin injection systems.

18.4.4 Fill-and-Drain Systems

A different approach to leak sealing is taken by *fill-and-drain* techniques, which repairs the main sewer, branches, and manholes in one operation. The section to be sealed is first isolated and then filled from a manhole with a chemical solution (usually sodium silicate). After a predetermined interval to allow the chemical to permeate through leaking joints and cracks, the solution is pumped out. The section is then filled with a second proprietary chemical solution, and this reacts with the residue of the first chemical to form a waterproof gel. The second chemical is then pumped out and the pipe is cleaned to remove any residue. As with packer-injected sealants, the effect of the system is to turn the manhole and pipe surround into an impermeable mass around the points of leakage. Because of plant requirements and the volumes of materials, fill-and-drain systems are aimed more at large-scale leakage control projects than at the treatment of isolated lengths. They have the advantage of treating leaks throughout the whole system in a single operation.

18.4.5 Cement Grouting

Cement grout consists of slurry (particulate suspension) of cement and water. Materials such as sand, bentonite, or accelerators may be added. Because of its particulate nature the use of cement grout is normally restricted to fractured rock and large-grained soils, where the voids are large enough to facilitate penetration and permeation.

Cement. Portland cement grout can be used to form impermeable subsurface barriers but is restricted in application to medium sands or coarser materials because of the larger size of cement particles (see microfine cement). For purposes of filling voids and washouts adjacent to sewer pipelines, various Portland cements have been used successfully. Type III cement is often selected because of its small particle size.

A range of water cement ratios can be used depending on subsurface conditions. Strength characteristics are generally not important where grouting is primarily intended to fill voids surrounding a buried sewer pipeline. Proper design of the water to cement ratio results in grout mixtures, which are easy to mix and inject, and have strengths of 500 to 1000 psi. Clays can be added to cement for grouting, to form gels, and to prevent settlement of the cement from suspension. They do, however, have the problem of no well-defined setting time and they have a slow strength development. Because of this, they are normally not used in sewer void grouting when groundwater is present.

Portland cement grouts may also be used as a filler and accelerator in silicate grouts and may be compatibly mixed with acrylamide to improve water shutoff capabilities and injection characteristics. For extremely large void filling applications, other cements such as pozzolan and fly ash mixes can be used and are more economical than straight Portland cement and soil-cement mixtures.

Micro-Fine Cement. Microfine cement consists of finely ground cement that allows it to penetrate fine sands that ordinary Portland cement cannot penetrate. Beside excellent permeability, it provides the needed strength and durability with a set time of 4 to 5 h. When combined with sodium silicate, with no organics, a 1 to 3 min gel time can be attained for underground water control.

Compaction Grouting. Compaction grouting is the injection of very stiff, low-slump, mortar-type grout under relatively high pressure to displace and compact soils in place. Compaction grouting acts as a radial hydraulic jack, physically displacing the soil particles and moving them closer together. The technique is used to strengthen loose, disturbed, or soft soils or for the control of settlement. Compaction grouting is used primarily on large pipelines applied through the pipe wall into the surrounding soil.

Reinforced Shotcrete Placement. Shotcrete is the application of concrete or mortar conveyed through a hose and pneumatically projected at high velocity onto a surface. Shotcrete refers to wet process, whereas the dry mix process refers to *gunite*. Shotcrete and gunite linings can provide structural strength and improved hydraulic performance, with improvement in corrosion resistance.

Shotcrete and gunite are used in large-diameter (usually 42 in and larger) sewers. These linings can also be used in sewers and other structures. Various segmented repairs can be incorporated onto the finished cement for total corrosion protection. The shotcrete processes, are similar in material use and generally use steel mesh for reinforcing. Also, various latex polymers can be added for improved bond strength, reduced absorption and permeability, and increased chemical resistance. For more information on coatings and linings including shotcrete and gunite, see Chap. 20.

18.5 INTERNAL SEAL

Internal seals can be used for structural repair pipe joints and missing pipe sections. It can be used in both worker-entry and nonworker-entry pipes.

18.5.1 Primary Characteristics

Internal seals can be conducted on pipe diameters ranging from 6 to 110 in. For diameters between 6 and 24 in, a stainless steel sleeve, which is wrapped in polyethylene foam, is used. This sleeve and an inflatable sewer plug lock are placed. The plug is then deflated and a visual inspection takes place.

For pipe diameters between 24 and 110 in, a polyvinyl chloride (PVC) sleeve is used. It consists of six longitudinal segments. The edges are grooved to lock each segment together. The sleeve is wrapped in a polyethylene foam gasket. This gasket provides the radial force to secure the sleeve in place. If grouting is required, small portholes are drilled in the sleeve and filled with grout upon completion. This system can be used to repair any length. The sleeves come in lengths of 12, 24, and 35 in, and any number of them can be used to repair a pipe.

FIGURE 18.4 Stainless steel and rubber internal repair sleeve for gravity pipes. (*Courtesy of Toa Grout Kogyo Co.*)

18.5.2 Mechanical Joint Sealing

Another method of joint sealing involves installing, across the joint, a metal band or clip faced with an elastomeric material, which forms a seal with the inner surface of the pipe.

Systems of this type are available for both gravity and pressure pipes. Mechanical sealing systems have the advantage of not relying on in situ chemical reactions, and can also be installed quickly. The cost of the materials is, however, higher than for cured in place methods. Mechanical systems are available for sealing joints in man-entry pressure pipes rated at up to 300 psi. Only 30 to 45 psi pressure is needed to tighten the seal, so low-modulus pipe materials such as polyethylene (PE) and PVC are not overstressed. Repair modules are available for pipes from 25 to 120 in diameter, and the seal can be made of nitrile butadiene rubber (NBR) rubber for gas applications or ethylene propylene diene monomer (EPDM, a thermoset material) rubber for potable water. Tapered versions are manufactured for use between pipes of varying diameter, and to seal the annulus at the ends of sliplined pipes (Fig. 18.4).

Mechanical repair modules are also available for nonworker-entry gravity pipes from 8 to 25 in diameter. The stainless steel inner sleeve is in the form of a scrolled clip, which can expand in diameter, and an integral ratchet mechanism prevents the unit from contracting again. The outer rubber sleeve also has bands of hydrophilic rubber to give a watertight seal with the existing pipe. The repair modules are installed by means of an inflatable packer, which expands the clip and presses the rubber against the pipe wall. The packer can then be deflated and withdrawn. Variations are available for sealing around connections, and for structural renovation in conjunction with a cured-in-place sleeve.

Pipe Rerounding. Rerounding is not a stand-alone technique, but is intended to reshape a deformed pipe prior to patch repair or relining. An expander unit is used to reround the pipe and installs a metal or plastic clip, which holds the pipe fragments in position until a patch or liner is installed. The expander can be made from an elastomeric material inflated with hydraulic pressure, or it may be a variant of a hydraulic mole with steel *petals* that are forced outward by hydraulic rams. The plastic or metal clip is scrolled around the expander prior to insertion, and held in place with bands or tape. The clip usually has some form of ratchet or locking arrangement to keep it at its expanded size. After positioning under CCTV control, the clip is expanded with sufficient pressure to reround the pipe.

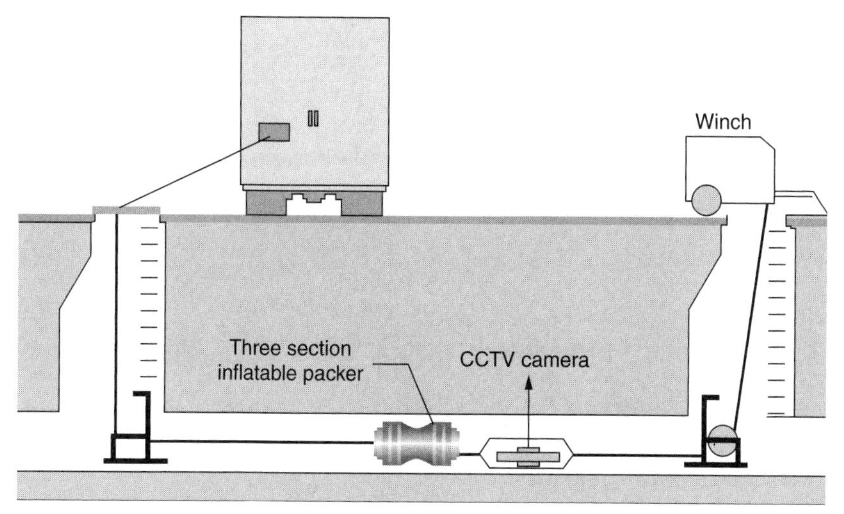

FIGURE 18.5 Joint sealing using an inflatable packer.

Although a useful trenchless technology technique, pipe rerounding is not always successful. During expansion, the unit will take the path of least resistance. For example, if there is a void beneath the pipe invert, the clip may be forced down when expanded, rather than reshaping the top of the pipe. After the rerounding operation, it is important to follow with relining or patch repair as soon as possible, because the clip has limited structural strength and may soon become deformed if left without support.

18.5.3 Method Description

There are different propriety techniques for this method. Prior to the operation, the pipe must be thoroughly cleaned and inspected to identify all the obstructions such as displaced joints, crushed pipes, and protruding service laterals. The operation then continues according to instruction from the system manufacturer and type of application. Figure 18.5 illustrates an internal seal process using stainless steel internal sleeve. For example, link-pipe, a propriety technique, involves covering the damaged area with an internally applied prefabricated stainless steel sleeve. As these repair sleeves can be installed through manhole no excavation is needed. The sleeves are used to repair structurally damaged pipes. They can also be used to restore missing pipe sections without excavation. Infiltration and exfiltration sealing is made recreating a sound pipe. Not counting the setup time, an individual repair takes approximately 20 min with a crew of two. From one setup 25 repairs have been made in a 5-h work period. Figure 18.6 presents an innovative application of link seal process where it is installed under water.

Advantages

- This method can be used for structural LORs.
- Such problems as pipe breakage, joint settlement, and long cracks can be fixed with this method.

FIGURE 18.6 South Florida water management district Everglades culvert repair using link seal. (*Courtesy of Link Pipe.*)

- Link seal can be used for variety of applications, such as gravity sewers, pressure pipes, culverts, and gas mains.
- Usually no excavation is required.

Limitations

- An internal pipe repair will leave a ridge on the inside circumference of the pipe, decreasing the overall diameter as much as 1 in at point of repair, which may decrease hydraulic capacity of the existing pipe.
- The existing pipe must be circular.

18.6 POINT CIPP

The applications of point CIPP are for pipelines that are sound but may contain isolated pipe lengths that have failed. The materials used in point CIPP repair are the same as the regular CIPP methods, which have more than three decades of proven performance.

18.6.1 Primary Characteristics

The point CIPP includes a liner reinforcement, which is made up of two layers of fiberglass woven roving with a layer of polyester sandwiched between them. The three layers are sewn together with polyester thread at approximately 1 in spacing using a zigzag pattern. The liner is cut and impregnated on site with epoxy resin to suit the particular requirements of the host pipe. The liner material holds extra resin, and when compressed gives up it to fill the cracks and surface roughness, thereby providing a better seal between the liner and

FIGURE 18.7 Preparation and installation of a point CIPP. (*Courtesy of Perma-Liner.*)

the host pipe. The usual diameter range is 2 to 24 in and the maximum length is up to 50 ft long with most liners being between the ranges of 3 to 15 ft in length. The minimum thickness is usually $^1/_8$ in and the maximum thickness is normally $^3/_8$ in with thicker liners available for specific conditions. The curing time is normally 1 to 4 h depending on the type of resin used and pipe diameter. Most CIPP contractors provide LOR systems using standard CIPP tube and resin materials to form short length of CIPP. After curing, the liner withstands external hydrostatic pressure of 30 psi or more according to design considerations. Figure 18.7 illustrates preparation and installation of a point CIPP.

18.6.2 Method Description

Most point CIPP techniques entail impregnating a fabric with a suitable resin, pulling this into place within the sewer around an inflatable packer or mandrel, and then filling the packer with water, steam, or air under pressure to press the patch against the existing sewer wall while the resin cures. Both thermal-cure and ambient-cure systems are available. Although point CIPP methods are short versions of cured-in-place liners, often the fabrics and resins are stronger because material economy is less significant in the overall installation cost. The fabric is commonly polyester needle-felt (nonwoven), either on its own or in combination with glassfiber. Some systems use a multilayer sandwich, the glassfiber providing strength while the felt acts as a resin carrier.

Although polyester resin may be used as in the full-length liners, epoxy resin is a common alternative for LORs. Epoxies can be formulated to be water-immiscible, whereas polyester resins are adversely affected by water prior to cure. This may be particularly relevant in techniques, which are designed for installation without diverting the flows in the pipeline. The drawback of epoxy resins, apart from their higher cost, is that the cure regime is more critical. Most epoxy-based systems require thermal cure, and ambient cure is generally restricted to polyester resins.

The patch may be preformed into a tube of the correct circumference, or it may simply be a rectangle of material, which is wrapped around the packer and scrolls out to the wall of the pipe when the packer is inflated. In the latter case an overlap will be visible in the finished tube, but this is of little consequence to the performance of the repair (Fig. 18.8).

Impregnation of the fabric is either carried out on site or preimpregnated at the shop and refrigerated for delivery to site. While onsite impregnation is generally acceptable, care is needed to avoid health risks and the spillage of chemicals, some of which are toxic in their unreacted state. Refer to Chap. 13 for more information on CIPP process. During mixing and impregnation, it is important to exclude air from resin as far as possible. Entrained air

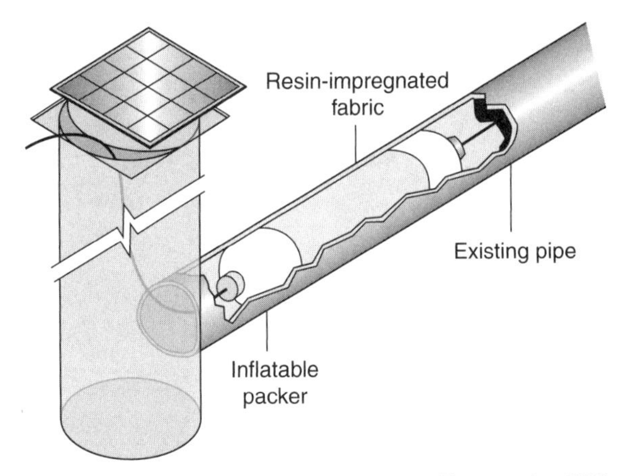

FIGURE 18.8 Typical arrangement for installing a point CIPP inflatable packer. (*Courtesy of Subterra.*)

weakens the material and, in severe cases, can result in porosity. It may be impossible to avoid air entrainment, especially with viscous resins, and some systems overcome this with vacuum impregnation.

With both ambient and thermal cured systems, it is essential to limit the rise in temperature of the materials until the patch is inflated within the pipe. One of the most common causes of failure is premature cure, where the patch has started to harden before it is in position. Exothermic cure starts as soon as the components of the resin are mixed, and the rate of temperature rise increases with the volume of resin. Mixed resin should therefore be applied to the fabric immediately and not left in the mixing vessel. Due regard should also be paid to the temperature of the surface on which the fabric is placed during impregnation and, once impregnated; the patch should be positioned and inflated without delay.

Packers are generally made from an elastomer such as rubber, and internal pressure first expands the packer and then presses the patch against the pipe wall. Compressed air is used for inflation with most ambient-cure systems. Thermal cure methods may use a mixture of compressed air and steam, or hot water, which is recirculated between the packer and a boiler unit on the surface. Care is needed to avoid overpressurization, especially with water-inflated systems where the packer is subjected to both a static head and an additional pumping head, which is related to fluid friction in the return hose.

The curing time depends on the resin formulation, the thickness of the patch, the temperature within the packer (in thermal-cure systems), and the temperature of the existing pipe wall. A high groundwater table will tend to act as a heat sink, cooling the outer surface of the patch, and additional curing time should be allowed. After curing, the packer is deflated and removed. The patch should then be inspected, and any lateral connections reopened using the same techniques available for full-length liners.

Advantages

- The point CIPP repair is installed according to reliable ASTM 1216 design standard
- Diameters up to 60 in possible
- Lengths from 3 to 50 ft possible
- Point CIPP bonds to the existing pipe structure providing a tight fit

- Point CIPP eliminates infiltration and prevents root growth at the location of application
- Usually bypass pumping is not required
- Point CIPP provides a structural repair at the location of application
- The point CIPP provides tapered smooth ends

Limitations

- While point CIPP provides a structural solution at the location of application, it may not enhance the overall structural capability of pipe from manhole-to-manhole.
- The price of point CIPP against other LOR needs to be evaluated.
- Any possible reduction in hydraulics capacity of the existing pipe must be evaluated.

ACKNOWLEDGMENTS

Sections of this chapter have been excerpted from the references listed. The reader is referred to these references for more information.

REVIEW QUESTIONS

1. In what project conditions are LORs applicable? Explain.
2. In what project conditions would you recommend grouting?
3. In what project conditions would you recommend internal seals?
4. In what project conditions would you recommend point CIPP?
5. In what project conditions would you recommend robotic repairs?
6. What are the advantages and limitations of LORs?
7. What is meant by *renewal* versus *repair*?
8. Check with your city or community's public works departments to see if they have used any LORs. Obtain full details of the project and prepare a report.
9. Check on the Web or your telephone directory to see which contractors provide LORs in your area. Interview one of these contractors to get a case study of a LOR job and prepare a report.

REFERENCES

American Logiball Inc. Product Catalogue, Jackson, Main. Available at http://www.logiball.com.

Bureau of Engineering Technical Document Center *Sewer Design Manual: Part F*, Bureau of Engineering, Los Angeles, Calif. Available at http://eng.lacity.org/techdocs/sewer-ma/.

Iseley D. T., and M. Najafi, (1995). *Trenchless Pipeline Renewal*, The National Utility Contractors Association (NUCA), Arlington, Va.

ISTT-ASTT *Guide Part 1: Localized Repairs and Sealing,* Australian Society for Trenchless Technology. Available at http://www.astt.com.au/Localised.pdf.

Link-Pipe Product Catalogue, Richmond Hill, Ontario, Canada. Available at http://www.linkpipe.com.

Najafi, M. (1994) *Trenchless Pipeline Renewal: State-of-the-Art Review*. Trenchless Technology Center (TTC), Louisiana Tech University, Ruston, La.

North American Grout Marketing Association. *Guide to Successful Chemical Grouting*, Brentwood, Tn.

Scandinavian Society for Trenchless Technology (SSTT) (2002). *No-Dig Handbook*. Copenhagen, Denmark.

CHAPTER 19
MODIFIED SLIPLINING

19.1 INTRODUCTION

This chapter presents those sliplining methods where the new pipe fits closely with the shape of the existing pipe and the diameter reduction is minimized. These methods include spiral wound pipe (SWP), panel lining (PL), and formed-in-place pipe (FIPP). The PL and FIPP methods are specifically designed for larger diameter gravity sewers, such as more than 48 in. These large-diameter pipelines are usually located at greater depths than smaller diameter pipes. Open-cut replacement can involve insurmountable social costs because of the enormous impact such work has on local residents, businesses, and traffic. Additionally, modified sliplining methods fill the gap among trenchless renewal techniques to target noncircular pipe shapes.

19.2 SPIRAL WOUND PIPE (SWP)

This trenchless renewal method is generally used for gravity sewers only. In this process, a new pipe is installed inside the existing pipe from a continuous strip of polyvinyl chloride (PVC) 8 to 12 in. in width. The strip has tongue-and-groove castings on its edges. It is fed to a special winding machine placed in a manhole, which creates a continuous helically wound liner that proceeds through the existing pipe. The continuous spiral joint, which runs the length of the pipe, is watertight. Upon completion of the lining process, grouting of the annulus space between the lining and the existing pipe wall is usually required. If the lining pipe is fitted closely to the existing pipe wall, it may remove the need for grouting.

This method can be used for both circular and egg-shaped pipes. Further, as the new pipe is formed directly against the wall of the existing pipe, this method can be used to renew an existing pipe with a minimal loss of pipe cross-sectional area as compared to conventional sliplining methods (see Chap. 14 for conventional sliplining methods). As for any trenchless renewal method, there is a possibility of actually increasing hydraulic capacity of pipes lined in this manner because of reduction in the friction factor.

Accurate determination or prediction of structural integrity of a renewed pipe is not a simple task. For cases in which the existing pipe is structurally sound, the new pipe liner is designed to carry only the hydrostatic load owing to the groundwater above the pipe. For cases in which the existing pipe is failing structurally, the renewal method may be designed as a stand-alone structural pipe, but analysis must involve the composite action of the thin PVC shell, the surrounding grout, and the remaining existing pipe wall. This design process should be considered as structural renewal only if appropriate analysis can show the composite liner pipe (including grout and existing pipe) to be stiff enough to carry the sustained dead and live loads as well as the hydrostatic loads.

TABLE 19.1 Overview of Spiral Wound Applications for Different Conditions

Pipeline type	Suitable	Notes
Sewers	Yes	
Gas pipelines	No	
Potable water pipelines	No	
Chemical/industrial pipelines	Yes	A
Pipelines with bends	Yes	
Circular pipes	Yes	
Noncircular pipes	Yes	
Pipelines with varying cross sections	Yes	B
Pipelines with lateral connections	Yes	
Pipelines with deformations	Yes	B
Pressure pipelines	No	

Notes: A. Application for chemical or industrial pipelines is dependent on the project conditions and type of flow. B. There may be a significant diameter reduction for pipes with varying cross sections and deformations.

19.2.1 Primary Characteristics

The SWP technique forms a new pipe in situ by using PVC-ribbed profiles with interlocking edges. This method can be used for either structural or nonstructural purposes, depending on the grout and design factors. There are two variations of this method. One fabricates a sewer pipe in the manhole by helically winding a continuous PVC fabric. The second method, which is used for larger diameters (up to about 30 in), uses preformed panels inserted in-place in the existing sewer. Excavation usually is not required for this process. Laterals can be reconnected by local excavations or by a remote-controlled cutter. Table 19.1 presents an overview of applications and possible limitations for the spiral wound method.

19.2.2 Method Description

The PVC liner is fed through a machine to form a smooth-bore SWP. The winding machine can work from the bottom of a manhole and can directly feed a liner into an existing sewer pipe. The seams in the spirally wound liner are mechanically sealed using patented joints specific to each manufacturer. The diameter of the new sewer pipe is usually slightly less than the existing pipe. It is therefore necessary to grout the annular space between the two pipes. However, recent developments have led to a process in which the spiral liner is expanded inside the existing pipe until it is flush with the pipe wall thus eliminating or minimizing the need for grout. Figure 19.1 illustrates this process.

19.2.3 Advantages

The lining pipe may be formed on-site by spirally winding a strip. Access and installation via manholes is possible. The lining is capable of accommodating large radius bends. No pipe storage on-site is necessary. Any diameter within the winding machine's range can be selected. This method has low mobilization costs and low-to-moderate overall costs.

(a)

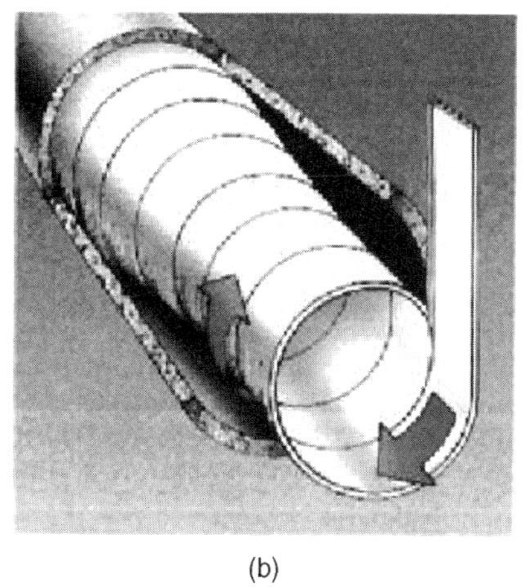

(b)

FIGURE 19.1 (*a*) Spiral wound installation machine, (*b*) expanding spiral winding. (*Courtesy of Danby of North America, Inc.*)

19.2.4 Limitations

This method requires some skill and needs trained personnel to operate the winding equipment. Continuous fusion, solvent-welded or mechanical joints are required. Reduction in capacity might be significant. This method leaves an annular space, which must be grouted. The resulting PVC pipe has a continuous, closely-spaced joint.

19.3 PANEL LINING (PL)

19.3.1 Background

In 1979, Water Research Center (WRc) in the United Kingdom was commissioned by the government and several other companies to undertake a 5-year research in methods and materials for structural renewal of large-diameter noncircular sewers. The US$18 million contract was completed in 1984 and the WRc's manual on sewerage rehabilitation design was published. This manual has provisions for three design types I, II, and III. Type I design assumes bonding of new pipe with existing pipe. This design type relies on the composite action created between the existing sewer pipe, the new sewer lining pipe, and the bonding agent that holds them together (normally a structural grout). Type I design takes into account vertical pressures resulting from traffic loadings and soil overburden to depths of up to approximately 26 ft. This design type does not consider the effects of external hydraulic head or groundwater.

Type I design relies on the efficient transfer of loads from the existing pipe to the sewer lining through the grout. Therefore, it is of primary importance that the grout be of sufficient strength, and have sufficient resistance to shear at the liner pipe and the existing sewer pipe wall interfaces. Foam or low strength grouts, therefore, are not suitable for type I designs.

Type II design, like type I, is intended for renewal applications that use some form of internal lining. In this type of design, the liner pipe thickness is calculated such that it is capable of withstanding external hydraulic loading conditions. Type II design does not take into account the effects of soil overburden or traffic loading. Type II liner pipes are designed as *stand-alone* systems using formulae derived from Timoshenko's equations for flexible pipes. The composite action created by bonding of the liner pipe and the existing sewer pipe, therefore, is not considered in this design type.

Type III renewal systems are intended for use where external ground water load is not present (or likely to be built in future) and where corrosion resistance or improvement in flow capacity is the main renewal objective. Type III designs are therefore considered nonstructural.

Where a combination of type I and type II designs are used, the design is referred to as a type II design with a type I check. In these cases, the required liner pipe thickness is calculated using type II design formulae, with this thickness used in type I design calculations. The liner pipe is then modified until it meets the factor of safety requirements for both design types. Refer to WRc publications or references at the end of this chapter for more information on these design types.

19.3.2 Primary Characteristics

Many systems for renewing large-diameter pipes are currently available. One of these methods is fiberglass liner, also called fiberglass reinforced polyester panel (FRPP), manufactured as pipe segments or panels. Glassfiber liners can be manufactured to fit any size or shape. The main structure of the liner and its make-up wall incorporates a corrosion barrier on the inside surface followed by a layer of preformed, powder bound, chopped strand glass mat consolidated with isophthalic resin. Figure 19.2 presents an egg-shaped fiberglass liner.

The liner pipes are made of various combinations of fiberglass mats, glass fibers, polyester, vinylester, epoxy resin, and sometimes silica sand. When the liner pipes are machine-made, a 0.1 in (approximately 2 to 3-mm) thick fiberglass mat is laid first and then impregnated with polyester to form a corrosion-resistant layer. Subsequently, the mould is rotated while polyester-soaked fibers are spun on. Fine silica sand is sometimes applied at the same time as the fibers. The process is continued until the required wall thickness is achieved.

Channeline is a brand name, which is hand-made in moulds and can therefore be produced either as complete pipes or as pipe segments. This system also allows the production of asymmetrical profiles or profiles that have negative curvature on some portions of the existing pipe length—as in the case of profiles with floor channels, for example. As in other liner systems, the type of resin and fiberglass used is chosen on the basis of requirements for strength and resistance to chemicals and extreme temperatures.

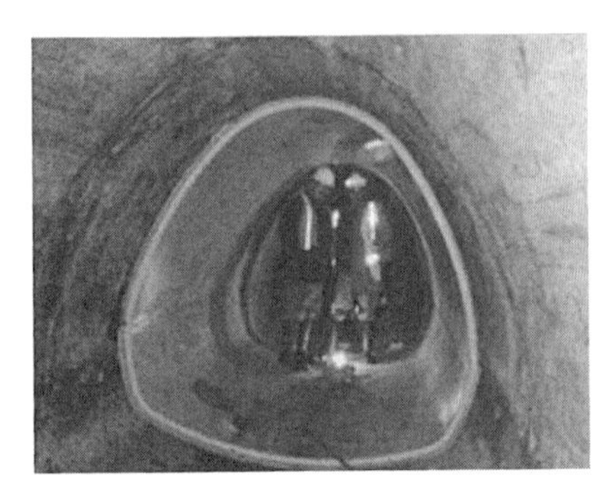

FIGURE 19.2 Channeline fiberglass panels. (*Courtesy of Kenny Construction Company.*)

Panel liners can be designed and constructed as self-supporting, flexible pipes. Alternatively, strength can be achieved by sandwiching together the existing pipe, the liner, and the concrete filling the annular space between the two. Different design criteria are used in different parts of the world. Where there are no national standards, guidelines from the WRc in the United Kingdom as described previously are often used.

In addition to static loading calculations, the hydraulic capacity of the new pipeline should be investigated and calculated. Despite the fact that PL renewal method certainly reduces the existing pipe diameter, the resulting smooth surface of the new pipe would usually increase hydraulic capacity in relation to the existing pipeline.

19.3.3 Panel Lining Applications

As for other renewal projects, it is important to choose the best method for each individual project. The available space, the chemical composition of the fluid carried by the pipeline, the pressure and temperature in the pipeline, the number of laterals, the number of manholes, and so on must be assessed form project to project. Table 19.2 presents an overview of applications and possible limitations for the PL method.

19.3.4 Method Description

Preparatory Work. Renewing large-diameter pipelines requires detailed planning and close collaboration between the pipeline owner, the contractor, and other parties involved in the project. In the case of storm sewers, the PL method has the advantage that flow bypassing is not always necessary. Flooding usually does not damage work in progress, however, safety of workers inside the pipe must be considered. Before starting any renewal project, it is important to carry out a detailed preliminary survey of the project site and inspection and cleaning of existing pipe. Other preparatory planning include:

- Any planned overpumping, including its duration and the number of consumers affected
- Planned traffic diversion and their advertising
- Pipeline shutoff (timing, duration, number of laterals affected)
- Grinding of intruding service line connections
- Measurement of the existing pipe internal diameter at different locations
- Grinding or cutting away of deposits
- Root cutting if necessary
- Closing or blocking off unused service lines
- Restoration of manholes
- Sealing leaks to prevent groundwater infiltration

Safety Issues. During the installation, certain safety and work environment requirements according to local and national regulations must be met. Most important of all, OSHA's safety regulations for confined-space entry must be addressed by a competent person. The existing pipeline must be effectively closed off upstream (and possibly also downstream) to ensure that storm or wastewater flow cannot enter during the construction. If this is accomplished with inflatable sealing plugs, the plugs must be securely held in place, with possible chains, and the flow pressure should be constantly monitored. Two plugs should be positioned upstream as an extra precaution. The bypass pumps must be constantly

TABLE 19.2 Overview of Panel Lining Applications for Different Conditions

Pipeline type	Suitable	Notes
Sewers	Yes	A
Gas pipelines	No	
Potable water pipelines	No	
Chemical/industrial pipelines	Yes	
Pipelines with bends	Yes	
Circular pipes	Yes	
Noncircular pipes	Yes	
Pipelines with varying cross sections	Yes	
Pipelines with lateral connections	Yes	
Pipelines with deformations	Yes	
Pressure pipelines	No	

Notes: A. The method is suitable for manhole-to-manhole renewal of sewer lines with diameters larger than 48 in.

monitored to ensure that the level of storm or wastewater behind the barricade does not become so high to subject the plugs to excessive loads.

Adequate light must be supplied and ventilation must be provided. A gas detector must be used continuously to monitor oxygen levels and to warn personnel of the presence of poisonous or explosive gasses.

Escape routes are necessary in both directions and a watchman in the launch pit must continuously be in touch with personnel in the pipeline. There must be a telephone and emergency equipment on-site, and the workers must be instructed in the emergency procedures. As is the case with all other sewer work, workers must, of course, be vaccinated.

Installation Process. Workers need to manually position each prefabricated fiberglass pipe segment. The pipe sections are inserted into the existing pipeline through manholes and access pits if necessary. Point or localized repairs are sometimes needed before installation of PLs to take care of such problems as excessive infiltration/inflow (I/I) and misalignments. Most panel liners are machine-made as centrifugal-cast or spun sections. Panel liners usually have tongue and groove joints that are sealed with either rubber sealing rings or polyurethane or epoxy filler. Liner segments have specially developed longitudinal joints that are sealed with epoxy filler.

Drive lengths of up to 3000 ft can be installed from a manhole or a shaft (Fig. 19.3). However, ventilation and escape routes must be provided at short intervals. Over the manholes or access pits, work sites should be established and cordoned off to ensure the safety of workers and passers-by. At these work sites, there should be enough room for necessary installation equipment. A crane should be positioned above the launch pit to lower panels to the pipeline entrance. There should also be a compressor, a generator as well as tool storage and personnel facilities.

Kenny Construction Company reported the construction of a renewal project involving over 3 mi of a 7 and 8 ft horseshoe sewer that served many of Chicago's western suburbs. The existing pipeline had been originally constructed in the 1920s and never been renewed. Much of the concrete lining had shown decay from sewage gases that corrode the concrete. Over a mile of the sewer was under active railway lines where shafts could not be installed unless the railroad was temporarily diverted. So Kenny chose to renew this sewer pipe with *Channeline*™ glass reinforced panels. The panels were two pieces per section and were fabricated to varying shapes.

FIGURE 19.3 An entry shaft used for panel lining project in Chicago. (*Courtesy of Kenny Construction Company.*)

The panels were hauled into the sewer for about 3000 ft before they were installed. Then, the annular space between the existing sewer and the PL was grouted.

Panel Installation. The panels are transported to their position in the pipeline. Panel installation usually starts at the opposite end of the existing pipeline so as to finish at the entry-pit. Packing pieces are wedged between the panel liner and the existing pipe as installation progresses. If there is room, structural irregularities can be evened out to a certain extent if the dimensions of the panels and the existing pipe are large enough to allow adjustment in the positioning of individual panels.

Once the panels have been positioned throughout the length of the pipeline the annular space is filled with structural grout (Fig. 19.4). Care must be taken during the pumping process to prevent the panels from being subjected to excessive external pressure. This is accomplished by checking the pump pressure and the floatation forces that the panels are subjected to by the fluid concrete. In this respect, it should be noted that concrete creates greater floatation than water because of its higher density. Using foamed concrete, which has a lower density than regular structural concrete, can restrict floatation forces but foaming reduces the compressive strength of concrete. Another possibility for reducing floatation forces is to inject the concrete in two or three stages while allowing it to set between successive injections.

Changes in pipe diameter can be negotiated with prefabricated transition panels. Such panels are often expensive because a mould must be made for each individual irregularity. Another possibility is to cut the panels to be installed on either side of the irregularity so that they fit and then connect the two with a matrix made by hand on site. Bends are similarly negotiated using prefabricated segments or panels cut to fit.

As for other renewal methods, service line connections and laterals are reopened once the panels are in place. Watertight connections can be established with filler, fiberglass matrixes or transition profiles—or a combination of these methods. Depending on various factors, including the diameter and the length of the pipeline, the weight of the panels and the number of connections, advance rates of 20 to 100 ft/day can be expected. Figure 19.5 illustrates

FIGURE 19.4 Filling the annular space with grout. (*Courtesy of Channeline.*)

the PL installation process. PL may be advantageous to segmental sliplining (where a whole pipe section is pushed inside the existing pipe) for the following reasons:

- Existing pipeline access may be limited to the existing manholes where there is only room for inserting panels.
- Pipe transport from the factory to the renewal site can be expensive and problematic where large-diameter pipelines are used. PL allows delivery in segments and subsequent assembly either in the pipeline itself or—if the access conditions to the existing pipeline allow—above ground.

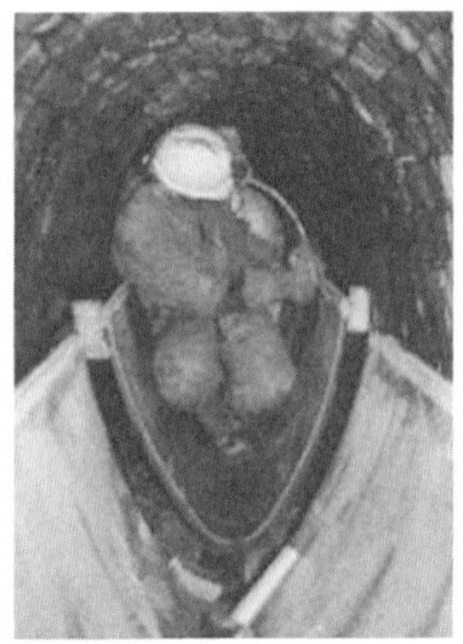

FIGURE 19.5 Panel lining installation process. (*Courtesy of Channeline.*)

Project Delivery. After installation it is often necessary to carry out the completion work. This may, for example, take the following steps:

- Renewing manholes and end seals
- Reinstating service lines
- Conducting a final inspection and leak testing
- Notifying residents and businesses about work completion and conducting a public survey for improvements in future projects

19.3.5 Advantages

- PL can be manufactured in any worker-entry diameter and any shape (circular, oval, vertically sided, symmetrical, or nonsymmetrical)
- The PL can serve for both corrosion control and structural purposes
- PL can be installed in restricted flow conditions, which in many cases eliminates the need for bypassing the flow
- PL can be used for renewal of culverts
- PL can be used for long-drive length where access is limited
- Nonstandard shapes up to 21 ft in diameter are possible
- There might be an improved hydraulic capacity by providing a lower coefficient of friction
- The panel liner is abrasive resistant and chemical resistant

19.3.6 Limitations

- PL can be used *only* in worker-entry pipes
- Significant reduction in pipe cross-sectional area is possible
- Access shafts may be required
- The annular space must be grouted

19.4 FORMED-IN-PLACE PIPE (FIPP)

The FIPP application includes renewal of wastewater, stormwater, and culverts for worker-entry diameters up to 12 ft regardless of the shape and the material of the existing pipe.

19.4.1 Primary Characteristics

The FIPP technique comprises a uniform concentric ring of two or more thin sheets of high-density polyethylene (HDPE), in which the outer sheet or ring (the *preliner*) is smooth and the second or inner (the *inliner*) is studded. The studs provide an annular space (between the preliner and the inliner) into which a high-strength, nonshrink, proprietary grout (the *injector*) is injected that subsequently hardens. The result is a continuous rigid liner that tightly fits and conforms to the existing pipe. The liner system is unique because a new pipe

TABLE 19.3 Overview of Formed-In-Place Pipe Applications for Different Conditions

Pipeline type	Suitable	Notes
Sewers	Yes	A
Gas pipelines	No	
Potable water pipelines	No	
Chemical/industrial pipelines	Yes	
Pipelines with bends	Yes	
Circular pipes	Yes	
Noncircular pipes	Yes	
Pipelines with varying cross sections	Yes	
Pipelines with lateral connections	Yes	
Pipelines with deformations	Yes	
Pressure pipelines	No	

Notes: A. The method is suitable for manhole-to-manhole renewal of sewer lines with diameters larger than 48 in.

is literally *formed in place* as the grout hydrates inside a plastic tube and mechanically locks around the plastic anchors or studs.

The FIPP technology successfully combines the features and advantages of several very different sewer renewal methods, but the finished product is a unique composite pipe that can support fully deteriorated pipes and accommodate the odd shapes and offsets of the existing pipe. Table 19.3 presents an overview of applications and possible limitations for the PL method.

Basic Options. The FIPP technology provides a variety of systems for renewal of sewers, namely:

- *The basic system.* The basic system (Fig. 19.6) consists of a single, thin HDPE sheet (the inliner) that is smooth on its inner circumference and has integral, extruded anchors on its outer circumference. At the factory, it is cut to a custom length to fit the existing sewer and formed into a ring (i.e., a tube), by wedge-welding its edges so that it will fit snugly into the existing sewer. The height of the anchors, which typically range from 0.40 to 0.63 in, determines the thickness of the ring-shaped annulus between the existing pipe and the inliner. The annulus is filled with *injector* grout to fill cracks and to surround the voids and the missing segments of the existing pipe.
- *The preliner system.* The preliner system (Fig. 19.7) consists of a studded inliner and an additional, smooth HDPE liner (the *preliner*). Like the inliner, the preliner sheets are similarly formed into hollow tubes by wedge-welding their edges along their axial lengths. An inliner can thus be pulled through a preliner to create a tube within a tube, with a defined annular space being created by the integral anchors on the inliner. This defined annular space allows for the accurate determination of the amount of *injector* grout that will be used to form the new, renewed sewer. This system, where the injector grout is contained between the inliner and the preliner, is recommended for use in areas where grout migration outside of the existing pipe cannot be tolerated, in areas where a high groundwater table exists, and where predetermined pipe stiffness is required to renew the failing sewer.
- *The leak detection system.* This system provides a leak detection annulus by inserting a third liner (called a *spacer sheet*) around the outside circumference of the preliner.

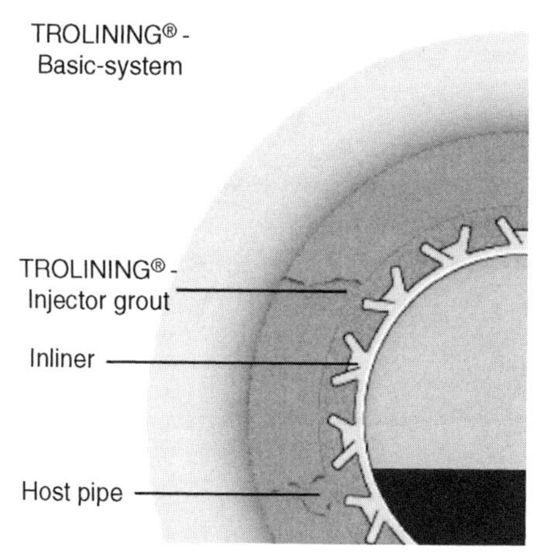

FIGURE 19.6 Formed-in-place pipe (FIPP) basic system. (*Courtesy of Trolining.*)

The spacer sheet has rounded hemispherical spacer studs extruded on its inside surface to create an additional void space, for leak detection, by monitoring for liquid between the preliner and the spacer sheet. Like the inliner and preliner, the spacer sheet is formed into a tube and axially welded, at the factory, to fit snugly into the existing sewer. Thus, the completed system features three concentric HDPE liners with a captive ring of high-strength grout. This is effectively a double-wall containment system—a pipe within a pipe. This system allows full 360-degree leak detection with minimal reduction of the

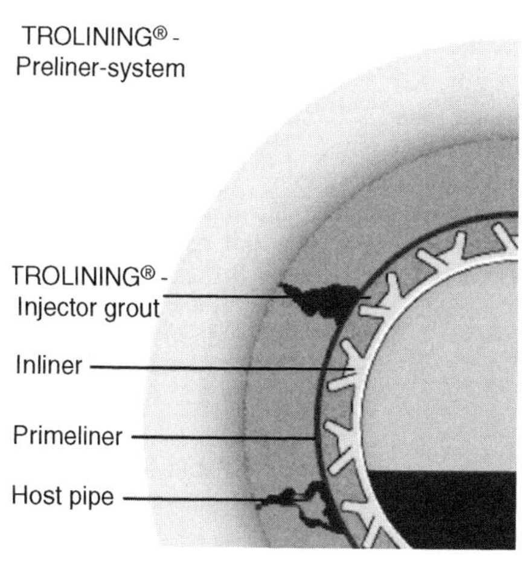

FIGURE 19.7 Formed-in-place pipe (FIPP) preliner system. (*Courtesy of Trolining.*)

existing pipe and is thus suited for the environmentally sensitive waste streams as an alternative to above-ground piping where leaks are more easily discovered.

- *The double-liner system.* This system consists of two studded inliners, installed concentrically within a preliner thus resulting in increased stiffness. It is used for renewal of large-diameter sewers as well as fully deteriorated sewers.
- *The fiber optic option.* Fiber optic cables, electric cables, and the like can be installed in the wall of the liner eliminating the need of for open-cut operations when the cables are needed later. In municipal systems, the renewed sewer becomes a right-of-way that can earn a fee from the cable users, thereby assisting in cost recovery for sewer renewal. Although this option is unlikely to be a revenue source in a private establishment such as sewers in an industrial facility, it could save money on future communication systems within the facility.

19.4.2 Method Description

- *Prefabrication.* The HDPE liners are prefabricated in the manufacturing plant using the dimensions of the existing pipe. The liner sheets are mechanically formed into hollow tubes and lap-welded along the edges of their axial length to form tubular liner sections designed to fit snugly into the existing pipe. The finished double-seam weld is then tested for strength and continuity with 90 psi air pressure. All results are documented in a test certificate for each liner section prior to leaving the factory. The finished liner tubes are rolled, wrapped, and shipped to the job site.
- *Installation.* Following the completion of all preparatory procedures, the rolls of the custom-made liner sections are prepositioned near their respective manholes. A towing bridle for the first liner is affixed to the end of the liner roll. A down sheave is positioned at the top of the manhole to fold and guide the liner for winching through the manhole opening or entry pit, and transporting inside the existing sewer. At the base of the manhole shaft where the liner first enters the sewer, a protective HDPE bend fitting is positioned to protect the liner as it bends through 90 degrees for horizontal travel through the existing pipe. A winch at the next downstream manhole pulls the liner through the sewer.

  The folded liner tube allows for unconfined pulling through the existing pipe. A specially selected winch is used through which the pulling force can be controlled so that the allowable material tensile capacity is not exceeded, with a safety factor of two.

  After the first liner tube has been pulled into position and secured, inflatable sewer plugs are inserted into the ends of the liner and inflated to an air pressure of 5 psi. This opens the liner and facilitates a preliminary leak test of the tube.

  Once all the liner tubes are concentrically inserted, positioned, and air-tested, their ends are sealed using extrusion-welding technology. Certified FIPP welding technicians are used to perform this task. Grout ports and vent tubes are also welded onto the liner ends to facilitate the grouting operations. All finished welds are visually inspected and spark-tested in accordance with the FIPP manufacturer's procedures.

  Flow-through sewer plugs are then inserted in both liner ends, and the interior of the liner tube is filled with water. A rising water column at the upstream manhole pressurizes the liner so that it expands to fill and conform to the existing pipe. For a successful installation test, a pressure head of not less than 16 ft is maintained for 2 h during which there should be no leakage or loss of water column. To prevent liner collapse, the water column is maintained throughout the grouting operations and for at least 12 h thereafter.
- *Grouting.* Grouting operations take place manhole-to-manhole from downstream to upstream manhole to ensure that air is forced out of the annulus between the inliner and the preliner.

A quality control station is setup adjacent to the downstream manhole, complete with a flow cone, weight scales, stopwatch, thermometer, grout bottles, and documentation sheets. Certified FIPP technicians are assigned to perform continuous quality control duties throughout grouting operations.

Quality control samples of grout are taken at a minimum of every 10 min to ensure that grout parameters are kept within the design specifications. The interactive nature of the grouting process allows for adjustments to accommodate any small change in water to cement ratios or viscosity.

If desired, compression testing of hardened grout can be completed in accordance with ASTM C109 to ascertain a 28-day compressive strength equal to or greater than 10,000 psi. Other tests can include tensile properties (ASTM C496), and early volume change (ASTM C 827) with a trigger level of 0.5 percent or greater change in volume.

Following the completion of the grouting operations, the grout and water columns are secured, and the grout is allowed to harden for a minimum of 12 h prior to release of the water column and the internal head pressure.

Finally, the water is drained and the liner is robotically inspected for grout voids or imperfections. The translucent nature of the liner facilitates the rapid detection and location of any faults in the installation. Reinstatement of laterals can be performed from the surface or by using robotic cutting tools and epoxy shoe technology.

19.4.3 Advantages

- FIPP can be used in any shape (circular, oval, vertically sided, symmetrical, or nonsymmetrical)
- FIPP can serve for both corrosion control and structural purposes
- FIPP can be used for renewal of culverts
- FIPP can be used for long-drive length where access is limited
- Nonstandard shapes up to 12 ft in diameter are possible
- There might be an improved hydraulic capacity by providing a lower coefficient of friction
- FIPP is abrasive resistant and chemical resistant

19.4.4 Limitations

- Significant reduction in pipe cross-sectional area is possible
- Access shafts may be required
- The annular space must be grouted

REVIEW QUESTIONS

1. Compare SWP, PL, and FIPP.
2. In what project conditions would you recommend each of the above methods?
3. What are the safety measures taken before starting PL installation?
4. What are the advantages and the limitations of each of the method described in this chapter?

5. Describe the three design principles for PL process.
6. Describe the preparatory work required for PL process.
7. Describe the installation process for the three methods described in this chapter.
8. Why is flow bypassing not required for PL? Explain.

REFERENCES

Channeline (2004). Available at http://www.swpipeline.com/Segmental_GRPx.html.

Kenny Construction Company. (2004). Available at http://www.kennyconstruction.com/underground/channeline.htm.

Pearson, C., and B. Khamanian. (2004). *Rehabilitation of Non-Circular Shaped Tunnels Using Conformal Glassfiber Reinforced Polyester Panel (GRPP) System—A Case Study in Successful Design and Execution of Project and Lessons Learned*, No-Dig 04 New Orleans, North American Society for Trenchless Technology (NASTT), Arlington, Va.

Scandinavian Society for Trenchless Technology (SSTT). (2002). *No-Dig Handbook*, Copenhagen, Denmark.

Water Research Centre. (1990). *Sewerage Rehabilitation Manual*, Swindon, UK.

CHAPTER 20
UNDERGROUND COATINGS AND LININGS

20.1 INTRODUCTION

Coatings and linings[1] have been used to protect and renew water and sewer infrastructure (pipelines, tanks, reservoirs, clarifiers, primary and secondary retention and treatment basins, pump stations, diversion boxes, manholes, and other structures) for decades. Shotcrete, an air-assisted spray-on lining method for cementitious products, was developed at the dawn of the 20th century in Allentown, Pennsylvania, and became accepted as a construction method in 1910. Today, high-tech polymer coatings and composite lining methods are used to restore, protect, repair, and renew a wide range of concrete, masonry, and steel structures.

The principal objective of a coating or lining is to apply a monolithic layer that inhibits further deterioration. Type of deterioration is dependent upon underground infrastructure under consideration. In water mains, it is characterized by tuberculation, scale buildup, and corrosion that can choke off a pipe's flow capability. Coatings are effective at eliminating infiltration while providing containment in both sewage collection systems and water distribution pipelines. In corrosive sewer environments, pipe crowns, pump stations, and manholes can lose an inch or more of concrete in less than a year. Coatings can mitigate further degradation and, if needed, can structurally enhance and renew severely damaged structures. The most common materials used for these diverse purposes are cement mortar, 100-percent solid thermoset polymers, sheet linings, and cured-in-place liners.

20.2 MATERIAL TYPE AND PURPOSE

Coatings and linings may be applied to renew and protect aging underground infrastructure or to protect new structures to increase their service life. The primary materials used for coatings and linings fall into four broad categories of cementitious, polymers, sheet liners, and cured-in-place liners. These methods are sometimes used in conjunction with one another.

[1]The distinction between coatings and linings is not clear so the terms are used interchangeably in practice.

20.2.1 Ideal Properties for Liner Materials

In conjunction with chemical resistance and monolithic coverage, adhesion is generally regarded as a required attribute of coatings. Other attributes vary greatly between polymer and cementitious coatings, the most prevalent types used in underground infrastructure protection and renewal. Some may be excellent for bridging moving cracks in concrete structures but may have low chemical resistance owing to inherently higher porosity; others may exhibit excellent long-term strength but poor adhesion in damp environments. As for any renewal method, true project needs should be evaluated and matched with proven product attributes.

Moisture can weaken a coating's curing process as well as its ability to bond to the existing structure. Although moisture is relatively easy to mitigate in above-surface structures, it cannot be completely avoided below grade, especially in concrete and masonry structures. Therefore, a coating with high moisture tolerance offers an adhesion advantage for below-grade coating projects. Epoxies can generally be formulated to offer the best moisture tolerance, although some urethanes and urea also offer moderate tolerance or require the use of an epoxy primer. Other attributes to analyze include structural enhancement, permeability and chemical resistance, quick return-to-service, maintenance, and visibility.

20.2.2 Cementitious

Cementitious materials are used in both water and sewer applications primarily for its most basic characteristic—cost-effectiveness in corrosion protection. In water applications, cement mortar lining is most common and serves two main functions: the alkalinity of the cement inhibits corrosion (especially in case of an iron pipe), and the relatively smooth internal surface reduces hydraulic roughness and improves flow characteristics. It should be noted that cement mortar lining is also applied to many new cast iron and ductile iron water pipes to inhibit corrosion. In this application, the lining does not fulfill a structural function; rather it reduces the rate at which the pipe will deteriorate. This lining is not appropriate for pipes, that leak, or where corrosion has reduced the wall thickness significantly. It is also important to apply sufficient thickness of mortar to create the alkaline environment at the mortar or iron interface for iron water pipes. As with steel reinforcement in concrete structures, inadequate cover to the metal will allow the onset of corrosion, which will cause the mortar to crack and spall.

In sewer applications, cementitious linings comprise various base materials such as Portland cement and calcium aluminate. These materials are used to line underground structures to prevent infiltration and rebuild structures deteriorated because of age and erosion from infiltration inflow (I/I). Calcium aluminate cements are frequently used to provide some sulfide resistance in renewing sewer structures where mild corrosion is evident. Cementitious materials are also frequently used as a basecoat material for polymer linings in more corrosive sewer environments.

20.2.3 Polymers

The Clean Water Act in the United States introduced efforts to reduce unwanted inflow and infiltration into the collection systems. Therefore, more concentrated sewage without heavy metal and industrial pollutants results in much more aggressive microbiologically induced corrosion (MIC) problems. Accelerated corrosion problems in today's infrastructure, as well as greater awareness of corrosion issues, have led to the development and the use of coatings to increase infrastructure life expectancies. As a result of inherent safety, performance, and quick return-to-service attributes, solvent-free 100-percent solids epoxy, urethane, and urea coatings have proven to have unique advantages for successful renewal and corrosion protection. These thermoset coatings are essentially inert plastics when cured and therefore resistant to most corrosive elements found in these environments. They have longer life expectancies when properly applied by professionally trained applicators using specialized spray equipment to achieve optimal performance as defined by the product manufacturers. Because these products contain no solvents, they require no evaporative process to cure and do not emit destructive volatile organic compounds (VOCs) found in many coatings used in above-grade coatings. Polymer formulations with 100 percent solids are safer in confined spaces and can also be formulated for thicker, structure enhancing applications. Such products now dominate the underground infrastructure protection industry because many can cure and bond to concrete, brick, steel, and cast iron, in damp underground environments while providing necessary protection from the aggressive sewer environment.

Compared to cementitious products, most polymers deliver superior chemical resistance in the most severe corrosive environments, although manufactures should be consulted for specific recommendations. Polymers can be formulated to be structural or nonstructural, including extraordinary long-term strengths, flexibility, and elongation. Most polymers used in underground applications have been formulated for *ultra high-build* thicknesses of 40 to 250 mils single-coat capability towing to the jagged and deep profiles found on new and deteriorated concrete surfaces. Polymer linings are often used to topcoat cement mortar linings in severe duty environments.

Epoxies. Epoxies are considered quick setting (2 to 6 h) and are known to have the best moisture tolerance making them ideal for penetrating into porous substrates in damp underground environments. Epoxies are generally rigid and can be formulated with high tensile and flexural strengths for use in structural renewal or with lower strengths but excellent chemical resistance to be used as a protective (nonstructural) barrier. Unlike other polymers, epoxies generally do not contain toxins; important when considering extensive underground applications.

Urethanes and Ureas. Urethanes and ureas, and hybrid combinations thereof, are usually rapid setting (30 sec) and exhibit moisture sensitivity sometimes requiring a moisture tolerant epoxy primer. They use an isocyanate catalyst, which can require more extensive ventilation procedures in confined spaces. They can be formulated to be very flexible or rigid. The flexible formulations can be used in nonstructural (protective) applications (Fig. 20.1), whereas the rigid products supply structural enhancement to deteriorated structures and corrosion protection.

Polyesters. Polyesters use styrenated resins, which are rigid, moisture sensitive, and can require more extensive ventilation procedures in confined spaces. The resin must provide the same expansion and contraction rates as the substrate after curing. There are very few polyester formulations that are used as independent coatings in underground application. Those available require intense field chemistry and trained installers. Polyesters are frequently used in cured-in-place-pipe (CIPP) lining systems.

20.2.4 Sheet Linings

PVC or polyethylene sheet liners have been used to protect underground infrastructure for the past 50 years. The progression of polymer linings now provides technological and cost-effective alternatives to sheet liners in the existing structures. Today, these systems are

FIGURE 20.1 Flexible polyurethane liquid formulation applied to chimney seal product. (*Courtesy of Raven Lining Systems.*)

more typical in new construction where the sheets can be anchored into poured concrete by placing the sheets inside concrete forms (Fig. 20.2). Joints must be welded after the forms are released. Sheet liners are used in renewal, generally in conjunction with the application of a poured-in-place concrete system. In recent years, systems have been developed whereby the sheets are applied atop urethane mastic coatings providing a composite system of a polymer coating and sheet lining.

FIGURE 20.2 Sheet lining in new construction. (*Courtesy of Raven Lining Systems.*)

20.2.5 Cured-in-Place Linings

Cured-in-place liners are polymer epoxy impregnated fiberglass or polyester sheets that are placed over corroded surfaces. They are essentially a hybrid polymer system with most of the advantages therein and the added advantage of high mechanical properties.

20.3 INSTALLATION PROCESS

20.3.1 Surface Preparation

The application of any coating or lining requires surface preparation. It is necessary to clean and profile the substrate prior to application. Also, most protective materials require that all active infiltration be stopped prior to application and some require a dry surface to attain long-term adhesion to the existing structure. Coating systems can be either machine or hand applied. Small-diameter (<36 in ID) pipe applications are generally carried out using spin-cast equipment, a rotating head that is winched through the pipeline. Large-diameter applications can also be robotically applied or, more likely, installed with hand-held spray equipment similar to what is used for lining diversion boxes, pump station wet wells, manholes, and similar underground structures where worker-entry is possible.

The objective of surface preparation is to provide a clean, sound surface with adequate profile and porosity that is essential to ensure proper adhesion. Proper site inspection is important to be able to specify the right coating product and appropriate thickness. Because of the "out-of-sight, out-of-mind" nature of underground infrastructure, there are always unanticipated difficulties, ranging from active I/I to the presence of various contaminants, which makes the task of coating application rather difficult hence the strong need for professional, experienced applicators. There are, however, additional resources available to assist both specifier and trained coating applicator in selecting the right surface preparation method(s). In particular, the National Association of Corrosion Engineers (NACE), Society for Protective Coatings (SSPC), and the International Concrete Repair Institute (ICRI) have prepared extensive guidelines including visual guides that aid in proper selection of coatings, surface preparation, and application methods.

Cleaning is the first and most commonly inadequately performed task before coating application. For cleaning in worker-entry structures, mechanical abrasion is preferred whenever practical, but rust, latent concrete, and other surface contaminants can generally be removed by low-to-high pressure water cleaning, abrasive blasting, shot blasting, and power or hand tooling. For small or hard-to-reach places, hand grinders and wire brushing may be used. If oil, grease, or other hydrocarbon deposits have contaminated the surface, steam, hot water blasting, or other methods may be necessary to clean and decontaminate the surface chemically (Fig. 20.3).

Nonworker-entry pipe cleaning techniques can include robotic low-and high-pressure water jetting, scraping, pigging, rack-feed borers, and mechanically driven devices such as cutters and chain flails. There is often a balance to be drawn between removing all traces of corrosion and avoiding damage to the pipe wall, and some of the more aggressive techniques should be used with caution. Pipe scrapers are designed to remove hard deposits and nodules when winched through a pipe, and consist of a number of spring steel blades mounted on a central shaft. A towing eye is fitted to each end of the shaft, allowing the tool to be pulled back if necessary. Wire brush pigs comprise two circular wire brushes on a central shaft with a towing eye at each end, and are used to remove loose deposits and dust prior to lining. They may also be used to remove debris loosened by a pipe scraper. Cleaning pigs are available in a wide range of types, and are usually molded from hard resin with an

FIGURE 20.3 Hydrocarbon removal prior to coating. (*Courtesy of Raven Lining Systems.*)

abrasive outer layer. Some have carbide studs around the barrel to remove hard deposits. They are normally driven through the main by water pressure, and can travel distances of several kilometers in continuous pipelines. In a heavily encrusted pipe, pigging is carried out in stages using pigs of increasing size. Foam pigs are generally pushed through a pipeline by air or water pressure, but versions are available that can be pulled through with a towing rope. They are generally used to remove dust or fluids from pipes of any material, and are also suitable for line drying.

Surface preparation procedures strongly depend on the type of surface being prepared and the profile necessary for the coating being applied. Some recommended surface preparation procedures for different materials are discussed below. It is very important to understand that the performance standard for cleaning is paramount, how that is accomplished will be up to the contractor who may have to use a myriad of different techniques to achieve the desired standard.

Concrete and Masonry. All concrete and masonry surfaces must be sound, clean, and dust and contaminant free. Existing coatings that have failed because of incompatibility or debonding must be completely removed. Any form of release, curing compounds, toppings, waxes, oils, greases, and the like must be removed prior to application.

Cleaning for concrete or masonry materials can generally be accomplished with low-pressure water cleaning using equipment capable of 5,000 psi at 4 gal/min according to the NACE No. 5/SSPC-SP 12 and NACE No. 6/SSPC-SP 13. Mechanical methods such as high-pressure water cleaning, water jetting, abrasive blasting, shotblasting, grinding, or scarifying may be used to remove previous coatings, laitance, and disintegrated material. Detergent cleaning and hot water blasting may be necessary to remove oil and grease from the substrate. Chemical cleaning such as acid etching with muriatic acid can be used in certain situations. However, care must be taken to remove all the residual acid prior to application of any coating. Whatever method(s) are selected for use, it must be ensured that a uniform, sound, and clean surface without excessive damage is obtained.

Steel. Steel structures being coated should be prepared to a white or near white metal according to the NACE standard (NACE No. 1/SSPC SP-5 or NACE No. 2/SSPC SP-10) based upon its intended service environment and manufacturer recommendations for that specific coating product. Prior to blasting, all surfaces should be decontaminated should oils, grease, soluble salts, or other contaminants be present. All scale, deposits, weld splatter, and soluble salts should be removed, and all rough and sharp edges rounded off prior to coating. Surface preparation is generally accomplished by dry abrasive blast or power tool cleaning. Vacuum sweep can be used to remove the remaining dust and debris. Coating must be applied quickly to avoid flash rust or contamination of prepared surfaces.

20.3.2 Repair Materials

Repair and patching is necessary for final surface preparation, especially on concrete and masonry substrates. The procedures used for this are also dependent on the type of substrate and profile created during the cleaning process.

Concrete and Masonry. Where necessary, voids should be filled and jagged surfaces profiled with repair materials compatible with the coating. Any area exhibiting movement or cracking owing to expansion and contraction should be grouted and patched according to appropriate crack repair or expansion joint procedure provided by the manufacturer. All surfaces that show exposed structural steel, spalling greater than $^3/_4$ in deep or cracks greater than $^3/_8$ in wide, should be patched using a quick-setting, high-strength cement mortar or high-build, nonsagging epoxy grout after initial cleaning. All concrete that is not sound or has been damaged by chemical exposure should be removed. If, in areas to be patched, reinforcement is missing and radial cracking from a spall site exists, steel replacement should be considered by the project engineer.

In masonry structures where loss of mortar has created gaps greater than $^1/_4$ in. in diameter between the bricks or blocks, the voids can generally be filled using a compatible quick-setting, high-strength cement mortar (see Fig. 20.4). Whenever structural integrity is questioned, a basecoat ($^1/_2$ in or greater) of high-strength cement mortar or additional epoxy coating or grout should be considered.

Steel. Steel surfaces should be thoroughly inspected, and when necessary, ultrasonically tested to detect thin spots in the structure, which may need reinforcement. A fiberglass fabric patch can be applied whenever corrosion or erosion has removed the safety factor of steel (generally greater than 50 percent of the original thickness). Fiberglass fabric may be

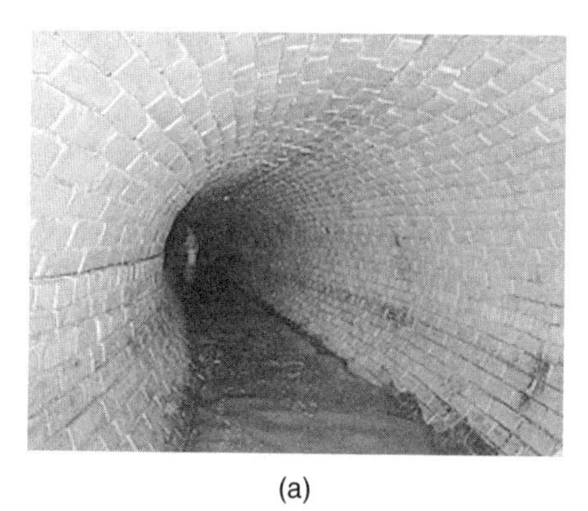

(a)

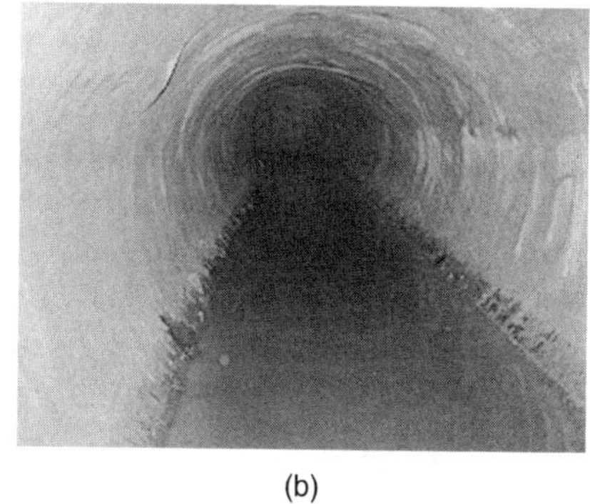

(b)

FIGURE 20.4 (*a*) Brick interceptor to be coated; (*b*) brick interceptor after cementitious profiling. (*Courtesy of Raven Lining Systems.*)

rolled into applied resin or chopped glass spray applied with the resin. Additional polymer coats and fiberglass layers may be applied to achieve greater thickness, fill remaining voids, cover exposed fibers, or more strength.

20.3.3 Application Equipment and Process

Coating systems range widely in their applications. In each application, the material is applied in one or more layers to the interior of a structure that has been adequately cleaned and prepared. Most coating systems provide for both mechanical and chemical adhesion. The system can be used to coat the entire or a portion of the structure.

Coating systems are applied both by machine and by hand. With water pipelines (nonworker-entry) the principal objective is to remove scale and corrosion, and then apply a coating that further inhibits deterioration and seals minor leaks. As most mains are structurally sound, the most common techniques to improve flow and prolong pipe life is to renew the existing pipes with cement mortar or polymeric resins, applied by a robotic spraying machine that is winched through the pipe at a constant predetermined rate. The American Water Works Association (AWWA) standard *C602-00: Cement-Mortar Lining of Water Pipelines in Place—4 in (100 mm) and larger* and the AWWA Manual for Rehabilitation of Water Mains (M28) provide detailed procedures.

Application equipment for sewers are different from water pipelines. Development work has been carried out on spray-on lining techniques for nonworker-entry sewers, but so far no such method has achieved commercial prominence. This may be partly because of the different requirements of sewer renewal, where the aim is usually to increase the pipe's resistance to the external loading rather than to prevent corrosion, and partly because of the practical difficulties of ensuring that I/I to the sewer is completely stopped while the material is being applied and cured. In the United States spray-on lining is seldom used in gas mains, although in some countries it is used extensively in gas service pipes. The following sections cover various application methods as related to the coating types used in underground applications.

Cementitious

NonWorker-Entry Pipes. Cement mortars are applied to the inside wall of nonworker-entry sewers to form linings with limited structural integrity. Application is generally carried out by a spraying machine, which is either fed through hoses from the surface, or, particularly in larger pipes, may have its own hopper containing premixed mortar (see Fig. 20.5). Forward speed control of the machine is important to produce a consistent thickness of mortar. Trowelling follows spray application, which may be carried out by rotating spatulas fitted to the spraying machine, or sometimes by a simple tubular shield of the required internal diameter, which is pulled through behind the machine. Whatever system is used, it is essential to centralize the equipment within the existing pipe so that the coating is of constant thickness around the whole perimeter.

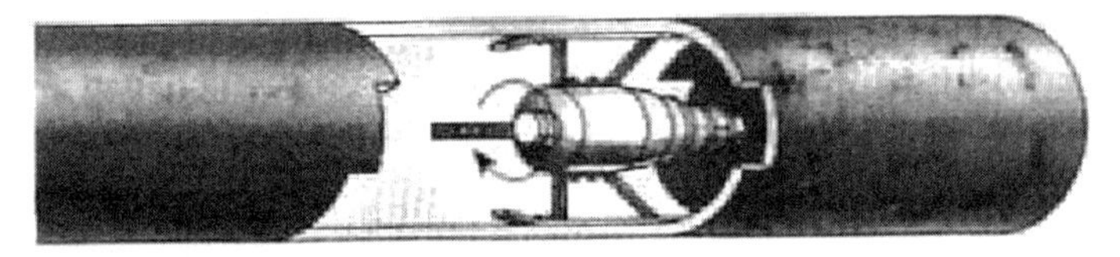

FIGURE 20.5 Typical cement mortar spray head. (*Courtesy of Subterra.*)

Process for NonWorker-Entry Pipelines

- The pipeline is cleaned to remove encrustation and protrusions
- Remote inspection of the main is carried out to ensure no infiltration and a relatively clean surface for adhesion of the coating
- The cement is mixed on-site and pumped though lines to a spinner head
- The spinner head is winched through the main at a required speed
- The cement is allowed to set
- A uniform 0.15-in coating of cement is achieved
- Ordinary Portland cement or a blend containing blast furnace slag is used for lining

Worker Entry Pipes and Other Underground Structures. For worker-entry pipes, spraying of gunite (dry concrete) or shotcrete (wet concrete) is possible. It is essentially a stabilization technique for old brickwork and masonry pipelines, the sprayed concrete forms a new inner skin which enhances strength, reduces leakage, and prevents further loss of mortar. The concrete is trowelled after application to produce a smooth surface. Reinforcement may be fixed to the existing pipe wall prior to applying the concrete, adding to the structural strength and creating a new reinforced concrete pipe with the existing pipeline. The design of the concrete mix is important, to ensure full penetration and encapsulation of the reinforcement. Precast concrete units are commonly used to line the invert, especially if the flow in the pipe cannot be stopped off completely.

American Concrete Institute (ACI) and American Society for Testing and Materials (ASTM) define shotcrete as the method of placing concrete or mortar onto a surface at high velocity. ACI 506 adds "at sufficient high velocity to achieve full compaction." Shotcrete is a method of placing concrete upside down and vertically without the aid of forming. Table 20.1 presents a typical shotcrete mix design.

In Gunite, a trademark for more than 50 years, sand cement mix of specific aggregate gradation and cement content is used. For spraying operation, a pressure vessel machine is used that conveys and applies the material at high velocity. Aggregate contains 3 to 6 percent moisture providing for a short period of prehydration. Balance of water is added at the nozzle. If the water to cement ratio is approximately 0.35 to 0.40 the resulting concrete will have superior compaction and bonding properties. During the laboratory testing at Lehigh University, it was indicated that a high impact velocity, high fine content, low water-cement ratio will achieve a greater density and durability, low permeability, and higher resistance to attack by carbonates and sulfates (Stier, 2000). These tests further indicated that higher velocity would provide a better bonding.

Since the 1980s, with development of concrete pumps and better understanding of concrete mix, the application of shotcrete and gunite has increased. These developments

TABLE 20.1 Typical Shotcrete Mix Design

Material	Quantity
Cement (can be type I, II, or III)	700 lb (7.5 sacks)
Concrete sand (SSD)	2790 lb
Water (added at predampener and nozzle)	320 lb (38 gal)
Total weight	3810 lb
Unit weight	144 pcf
Volume	1 yd^3

include longer stroke for pumping, better nozzles for injection of air and acceleration, and updated material technology for pumpability, high range of water reducers to decrease water-cement ratio and microsilica to increase density. The same admixtures used in structural concrete are used in shotcrete with same results. For example, adding silica fumes improves adhesion, increases cohesive properties of mix, reduces rebound (which is substantial for dry-mix), and improves washout resistance. For reinforcement, wire mesh, reinforcing bars, or steel or synthetic fibers are used.

In Situ Reinforced Concrete. Continuous in situ reinforced concrete lining systems for worker-entry pipelines from 35 in to over 220 in diameter use reinforcing steel attached to the sewer wall. Lightweight formwork is then installed, usually in 6.5 ft lengths, which are locked together with pins. Concrete is pumped under pressure behind the formwork to create a monolithic lining. The process accommodates bends and slight deformation, and produces a smooth finish with low hydraulic roughness. Connections can be formed through the use of adapted shutters, and flows can be maintained without overpumping.

Process for Worker-Entry Pipelines

- The sewer section is cleaned to remove encrustation and protrusions
- No high levels of infiltration are allowed and a relatively clean surface for adhesion, of the coating is provided
- Coating is pumped through a hand-held hose and sprayed onto the section to be coated
- Mortar is allowed to set

Major Advantages

- Lateral connections are easily handled
- No limitation on sewer size
- Variations in cross section can be readily accommodated
- Low cost

Major Limitations

- Relatively slow installation
- Requires safe conditions for worker entry
- High-level operator skill is required
- Potential for lateral restriction
- Water pH sensitive

Polymers

NonWorker-Entry Pipelines. Epoxy lining function is to provide corrosion protection and a smooth bore. Epoxy or polyurethane materials are centrifugally sprayed to the inside walls of nonworker-entry circular sewers. The objective is to make the resin bond with the prepared internal surface of the pipe, forming a monolithic coating that protects against corrosion and enhances flow. Epoxy coatings are generally much thinner than cement mortar linings, and therefore, do not cause significant bore reduction. In comparison with cement, epoxies also cure more quickly, do not affect the pH of the water, do not wear, and can significantly enhance flow. However, any defect in the epoxy coating may allow corrosion to start and, unlike cement mortar, there is then no alkalinity to inhibit deterioration chemically. Epoxy resins are also slightly more expensive.

FIGURE 20.6 Epoxy resin lining. (*Courtesy of Subterra.*)

Epoxy resins that have been approved for the lining of water mains by relevant authorities such as the National Sanitary Foundation (NSF) and Underwriters Laboratories (UL) do not impair the quality of the conveyed water, provided they are mixed, cured, and applied properly. AWWA Manual 28 requires only the use of approved linings for water mains.

The resin is applied by a spraying machine, which usually has a rotating nozzle (see Fig. 20.6). The typical 40 mils thickness of the coating is controlled by the flow rate and the forward speed of the machine. In most systems, the resin base and hardener are fed through separate hoses and are combined by a static mixer just behind the spray nozzle. Ideally, the cure time should be as short as possible to minimize the period during which the main is out of service, and also to reduce the risk of contamination of the resin prior to cure. However, too rapid a cure carries the risk of blocking the static mixer or the nozzle. Unlike with cement mortar lining, the resin is not smoothed or trowelled after spraying, and the surface quality depends on the application technique and the properties of the material.

Installation Process

- The pipeline is cleaned to remove encrustation and protrusions
- Inspection of the main is carried out to ensure no infiltration and a relatively clean surface for adhesion of the coating
- A two-component resin is mixed and pumped though lines to a spinner head
- The spinner head is winched through the main at a required speed
- The resin is cured for a specified time
- A uniform, smooth coating, typically 40 mils thick is achieved

Major Advantages

- No pH effect
- Higher flow

- No wear for longer life
- Laterals do not need reconnection
- Variations in cross section can be readily accommodated
- Relatively quick installation or high production

Major Limitations

- Control of infiltration is required to prevent precure lining disbondment or collapse
- No structural integrity
- Slightly higher cost

Worker-Entry Pipelines and Other Underground Structures. Lining is normally carried out using specialized spray equipment with a hand spray gun in worker-entry pipelines, lift stations, junction boxes, clarifiers, tanks, and other underground structures. Most polymer systems use plural component spray equipment customized to the characteristics of the particular polymer to ensure properly mixed and metered output. Such equipment can typically pump material over 500 ft to access difficult jobsites, or the equipment can sometimes be dropped into larger structures to accommodate extended application needs. Figure 20.7 illustrates spray application of epoxy and the lift stations used to lower the workers in manhole or other kinds of worker-entry underground structures.

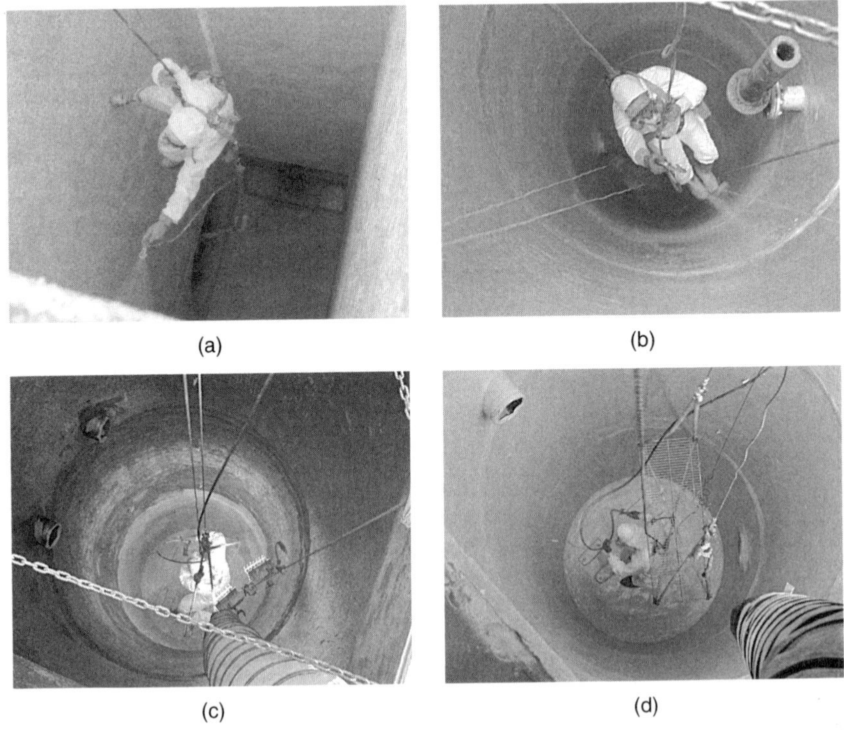

(a) (b) (c) (d)

FIGURE 20.7 (*a*) Epoxy spray application; (*b*) epoxy spray manhole; (*c*) lift station during application of epoxy; (*d*) lift station after application of epoxy. (*Courtesy of Raven Lining Systems.*)

Composite Systems. The use of high-strength cementitious materials topcoated with polymer coatings is a cost-effective solution to renewal of severely deteriorated structures. The products must be compatible and installed in accordance with manufacturer recommendations.

20.4 INSPECTION AND TESTING

Inspection and testing methods during installation of coatings and linings contained within this section are available through the product manufacturers and associations such as the NACE, ASTM, and SSPC. In general, some common procedures include the following.

20.4.1 Visual Inspection

Visual inspection and project documentation should be carried out at a minimum by the project inspector. A trained inspector who is knowledgeable in the products used and their installation process should conduct a final visual inspection. Any deficiencies in the finished coating should be marked and repaired according to the coating or lining manufacturer's recommendations.

20.4.2 Coating Thickness

For cementitious and epoxy coatings, thickness of the applied coating should be randomly checked prior to the set of the coating. A wet film thickness gauge, such as those meeting ASTM D-4414, *Standard Practice for Measurement of Wet Film Thickness of Organic Coatings by Notched Gages*, can be used too with epoxies to ensure uniform thickness during application (Fig. 20.8). Cementitious products generally will use a depth gauge to measure the applied thickness. This is not practical for the rapid setting urethanes and ureas. The use of an ultrasonic thickness gauge can be used after polymer coatings have set or cured. This

FIGURE 20.8 Wet film thickness test. (*Courtesy of Raven Lining Systems.*)

testing should only be performed by personnel trained in ultrasonically evaluating the thickness of thick-film coatings (>100 mils) on nonferrous substrates. The equipment should follow the standard for *Measurement of Dry Coating Thickness With Magnetic Gages* (SSPC PA-2) by the SSPC. The equipment should have the capability to calculate the average of a set of readings to determine specification compliance, such as a PosiTector 100D.

20.4.3 Pinholes

Spark testing or holiday detection can be used to detect holidays or pinholes in a polymer coating and inadequately welded seams in sheet liners. After the protective coating has set hard to the touch or after welding has been completed the system can be inspected with high-voltage holiday detection equipment. Surfaces should first be dried, an induced holiday is then be made on to the coated concrete surface and serves to determine the minimum or maximum voltage to be used to test the coating for holidays at that particular area. The spark tester is initially set at 100 V per 1 mil (25 μ) of film thickness applied but may be adjusted as necessary to detect the induced holiday (refer to NACE RPO188-99). All detected holidays are marked and repaired following coating or lining manufacturer's recommendations.

20.4.4 Adhesion

Bond strength of polymer coatings can be measured in accordance with ASTM D4541. Any areas detected to have inadequate bond strength should be evaluated by the project engineer. Further bond tests may be performed in that area to determine the extent of potentially deficient bonded area and repairs made in accordance with the manufacturer's recommendations. An understanding between the owner and installer should be made prior to testing taking place regarding *inadequate adhesion*. There are many variables that play a part in analyzing the results of an adhesion test. Adhesion testing is a destructive test and consideration should be made as to the extent of testing in a structure in which

FIGURE 20.9 Adhesion testing. (*Courtesy of Raven Lining Systems.*)

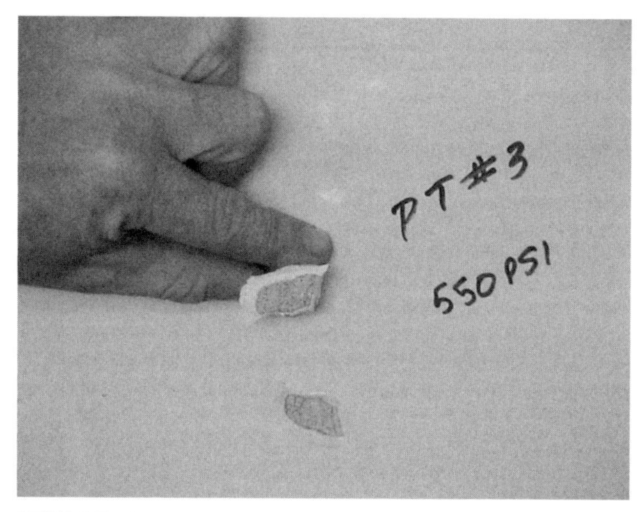

FIGURE 20.10 Adhesion testing results of pull-off. (*Courtesy of Raven Lining Systems.*)

the objective is to provide a monolithic coating. Figures 20.9 and 20.10 illustrate the pull-off adhesion test.

20.4.5 Vacuum and Exfiltration Testing

Manholes and pipelines can be tested for sealing integrity in accordance with ASTM standards set for vacuum and exfiltration testing. For more information on these testing procedures, refer to ASTM standards.

REVIEW QUESTIONS

1. Describe the various types of coatings and linings available and their applications.
2. Under which project conditions would you recommend coatings and linings?
3. Compare cementitious and polymer coatings and linings.
4. Describe the gunite and shotcrete process. In what project conditions are they applicable?
5. Describe the importance of pipe cleaning process for coating and lining applications.
6. What are the advantages and the limitations of coatings and lining processes?
7. Check with your city's public works department to see if they have used linings and coatings on their pipeline systems. If yes then obtain and report project specifics.
8. Check the Web or your phone directory to see if there are contractors in your area who provide coating services. Obtain one or two case studies of their projects and report all the specifics.

REFERENCES

American Shotcrete Association (ASA). (2004). Farmington Hills, Mich. Available at http://www.shotcrete.org.

ASTM. International The published standards of the American Society for Testing and Materials, West Conshohocken, Pa. Available at www.astm.org.

Atkinson, K. (2000). *Sewer Rehabilitation Techniques*, Subterra Systems, Dorset, UK.

Australian Society for Trenchless Technology (2004). *ASTT—ISTT Trenchless Technology Guidelines Part 1: Spray Lining.* Available at http://www.astt.com.au/Sliplining.pdf.

Australian Society for Trenchless Technology (2004), *ASTT—ISTT Trenchless Technology Guidelines Part 2: Renovation of Large Diameter Pipes and Chambers.* Available at http://www.astt.com.au/Sliplining.pdf.

City of Los Angeles Bureau of Engineering Technical Document Center (1992). *Sewer Design Manual: Part F*, Los Angeles, Calif. Available at http://eng.lacity.org/techdocs/sewer-ma/.

NACE International (2004). The published standards of National Association of Corrosion Engineers Houston, Tex. Available at http://www.nace.org.

Raven Lining Systems. (2004). *System Selection Guide*, Raven Lining Systems, Tulsa, Okla.

SSPC (2004). The published standards of the Society of Protective Coatings, Pittsburgh, Pa. Available at www.sspc.org.

Stier, B. (2000). Shotcrete for sewer applications. Presentation at the Midwest Society for Trenchless Technology (MSTT) Symposium, University of Missouri, St. Louis, Miss. November 1–3.

CHAPTER 21
THERMOFORMED PIPE

21.1 INTRODUCTION

Since 1988, thermoformed pipe (ThP) has been used extensively in the United States with over 21 million ft installed by 2003. All design properties of ThPs are established at the manufacturing facility under ASTM prescribed quality assurance protocols and can only negligibly be influenced by field construction crews. This technique can be used for sewers systems, potable water and gas supply lines, and industrial applications. ThP can be used for structural (including renewal of severely distressed pipelines) or nonstructural purposes and for pipelines from 4 in to greater than 30 in diameter and for lengths up to 1500 ft. This technology can negotiate bends in the existing pipeline and generally provides a small footprint, minimal community disruption, and very brief service disruption.

21.1.1 Primary Characteristics

This type of trenchless pipeline renewal uses a new polyvinyl chloride (PVC) or polyethylene (PE) pipe that is expanded by thermoforming to fit tightly inside the host pipe. Both PVC and PE have a long performance history in pipe applications that verifies not only their structural capacity, but also other important long-term performance characteristics such as chemical and abrasion resistance. There are three methods of ThP as shown in Fig. 21.1.

In the first method, called fold and formed (F&F), PVC pipes are flattened in the factory during production, then wound onto large reels, and folded during insertion. F&F methods can be used for gravity and/or pressure pipelines, including sanitary sewer, storm sewer and culvert, and potable water pipes and can be designed to provide full, independent structural integrity. Following the delivery to the renewal site, the new PVC pipe is heated with steam until it becomes flexible, allowing it to be pulled from manhole-to-manhole with a winch cable through the existing pipe. Once in place, the new pipe is forced against the inside surface of the existing pipe using steam and air pressure to form a new PVC pipeline tightly inside the old pipe. After installation is completed, lateral connections are reopened using a remote-controlled robotic cutter. Maximum diameter varies across vendors with sizes available up to 24 in. Ability to negotiate bends varies across vendors from 60 to 90 degrees depending upon the extruded diameter of the PVC pipe. Figure 21.2 presents an F&F process.

The second ThP method is deformed and reformed (D&R), where high-density polyethylene (HDPE) pipe is deformed into a U shape in the factory and wound into large coils. This method is used for gravity and/or pressure pipelines and for structural purposes. The new HDPE pipe is pulled at ambient temperature from manhole-to-manhole with a winch

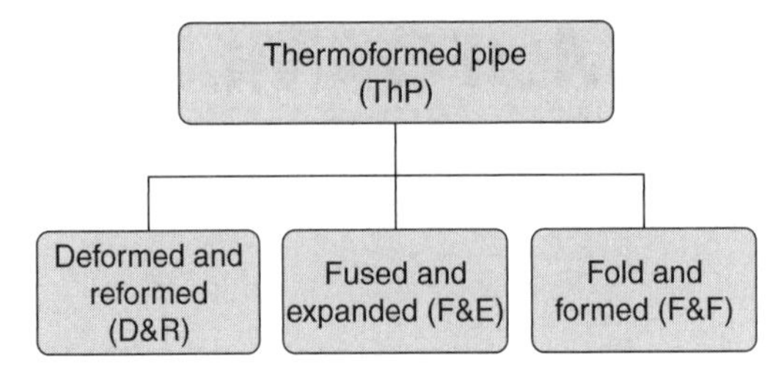

FIGURE 21.1 Three main variations of ThP process.

cable through the existing pipe. After the new pipe is inserted into position, it is heated with steam to revert it to its round memory and pressurized to push it out against the host pipe. As for the F&F method, lateral connections are reopened using a remote-controlled robotic cutter. Maximum diameter is 24 in with sizes above 18 in butt-fused and deformed in the field immediately prior to installation. Bends up to 22.5 degrees can be negotiated with the D&R method (Fig. 21.3).

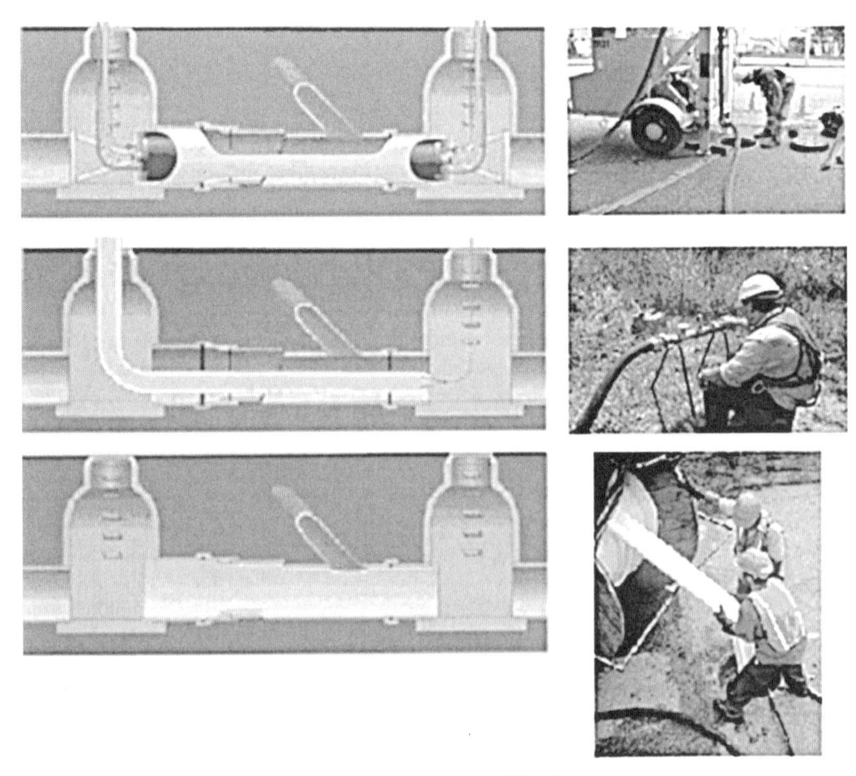

FIGURE 21.2 Fold and form insertion process. (*Courtesy of Ultraliner.*)

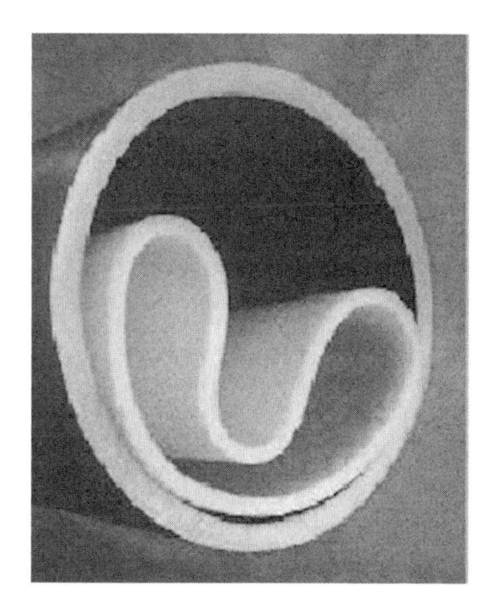

FIGURE 21.3 Deformed and reformed pipe. (*Courtesy of U-Liner.*)

With the third method, called fused and expanded (F&E), PVC pipes are fused in the field prior to insertion (Fig. 21.4). F&E pipes can be used for high-pressure pipelines exceeding 150 psi, including potable water pipes. Following delivery to the renewal site, the new PVC pipe is butt-fused and inserted through access pits as would be typical of sliplining. Once in place, the new pipe is heated with a hot liquid and highly pressurized to thermoform the new pipe tightly against the inside surface of the existing pipe. As this technology has been relatively recently introduced to the marketplace, the maximum available diameter continues to expand with sizes exceeding 30 in having been installed.

21.2 THERMOFORMED PIPE APPLICATIONS

As for any other renewal method, it is important to choose the best renewal method for each individual project. The type of application, working space available, chemical composition of the fluid carried by the pipeline, the pressure and temperature in the pipeline, number of

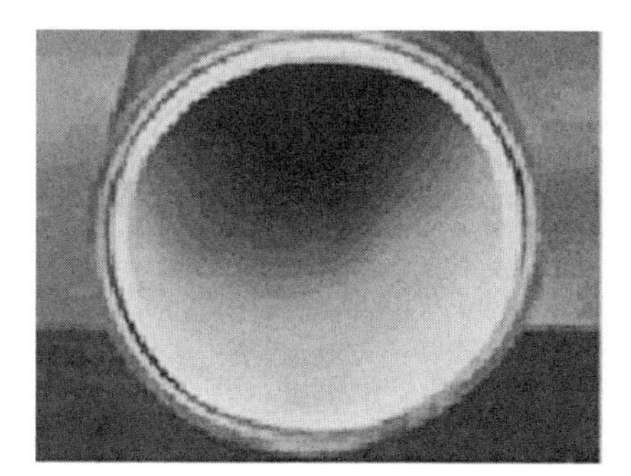

FIGURE 21.4 Fused and expanded (F&E) method. (*Courtesy of Underground Solutions.*)

TABLE 21.1 Overview of Thermoformed Pipe Applications and Possible Limitations

Pipeline type	Suitable	Notes
Sewers	Yes	
Gas pipelines	Yes	A
Potable water pipelines	Yes	B
Chemical or industrial pipelines	Yes	C
Straight pipelines	Yes	
Pipelines with bends	Yes	D
Circular pipes	Yes	
Noncircular pipes	Yes	E
Pipelines with varying cross sections	Yes	F
Pipelines with lateral connections	Yes	G
Pipelines with deformations and misalignments	Yes	
Pressure pipelines	Yes	

Note: A. Primarily PE pipes; B. Check all local and national codes for all materials that come into contact with potable water. Both PVC and PE are routinely suitable for use in potable water applications. Pipe pressure ratings need to be taken into consideration. Some methods will require the excavation of entry and reception pits. Valves, service connections, and sharp bends must be dug free before renewal work is begun; C. Temperature exposure must be considered in the design and often requires a significant reduction in the design modulus. Chemical exposure limits vary across vendors; D. To avoid folds or wrinkles at bends, the new pipe must be extruded substantially smaller than the inside diameter of the existing pipeline; E. ThPs are suitable for use with oval and elliptical pipes with minor design adjustments; F. The new pipe must be extruded smaller than the smallest inside diameter of the existing pipeline; G. For sewers, it is not necessary to excavate lateral connections, as they can be reopened from inside the new pipe using robotic cutters. It is always necessary to excavate service line connections on potable water pipelines.

laterals, depth, diameter, number of bends, any misalignments or joint settlements, the number of manholes, and so on must be assessed form project to project. ThP can be used for structural or nonstructural purposes to renew both gravity and pressure pipelines. Table 21.1 provides an overview of the Thp applications and possible limitations.

21.3 CONSTRUCTION CONSIDERATIONS

21.3.1 General

As with any construction project, it is the responsibility of the individual contractors to comply with current work safety and environmental regulations. Product and technical specifications should be used to ensure that the chosen product is suitable for the renewal in question. For example for waterlines, it is important to ensure that all product information and approvals according to local and state laws are available. Any surplus liner pipe must be safely disposed off. Information on how the material used can be disposed off or recycled can be obtained from the material supplier or the manufacturer. Well-known products like PE, PVC, and modified PVC that are currently used to manufacture Thps and can be delivered to approved waste collection points or returned to the manufacturer for reuse or recycling.

21.3.2 Preliminary Evaluation and Inspection

To achieve the best possible result of pipeline renewal using thermoformed method, it is important to carry out detailed preliminary surveying to determine the following:

- The conditions of the existing pipeline, using appropriate inspection method (see Chap. 3).
- The need for cleaning, based on inspection results
- Existing pipeline material, length, and depth
- Pipeline dimensions, deformations, and displaced joints
- Watertable depth and its fluctuations
- Location of laterals
- Traffic loads
- Service line data, including dimensions, unused service lines, service lines, and intruding service line connections
- Manhole condition
- Location and size of access pits

21.3.3 Preparatory Work

Once necessary preliminary surveying has been completed and renewal is about to start, the following preparatory work is recommended:

- Any planned overpumping, its duration, and the number of consumers affected
- Planned traffic diversions and their advertising
- Pipeline shutoff (timing, duration, number of laterals affected)
- Location and size of pits
- Inspection of bends in the existing pipeline

Usually, the preparatory work required would depend on what the pipeline is designed to carry, and whether it is a sewer or pressure line. For a sewer, for example, preparatory work would include:

- Inspection and possible grinding of intruding service line connections
- Inspection and possible repair of severely displaced joints on the existing pipeline
- Inspection and possible grinding or cutting away of scattered deposits
- Water jetting or root cutting
- Closing or blocking off unused service lines
- Restoration of the existing pipelines or parts of it when badly deformed
- Prevention of groundwater infiltration (temporary sealing to simplify renewal)
- General grinding and cleaning of internal corrosion throughout the pipeline

21.3.4 Installation Phase

As for any renewal project, installation is an important phase of the ThP process. It is therefore important for both the pipeline owner and the contractor to ensure that both the production of the new pipe and its installation are carried out under controlled conditions that can subsequently be documented in writing. Vendors should be requested to provide detailed installation guidelines.

As some of the ThP are frequently used for potable water and gas supply lines, it is necessary to excavate valves, service connections, and sharp bends before installing the line. F&F and D&R pipes for sewers can usually be installed from manhole-to-manhole with no excavations required.

Insertion. There are significant variations on the insertion process across vendors. F&F pipes are generally, preheated, whereas D&R and F&E pipes are inserted at ambient temperatures. F&F and D&R pipes are inserted through the manholes, whereas F&E pipes require access pits. In every case, the pipes are inserted with the aid of a winch cable, which is pulled through the pipeline to be renewed. For ambient insertion, it must be ensured that the new pipe is not subjected to excessive bending. The new pipe must also be protected from damage caused by rubbing against the sides of the manhole or the entry and reception points. The force used during insertion must never exceed the permissible pulling force for the new pipe, which varies across vendors with their differing insertion practices.

The Thermoforming Process. After insertion, it is generally advisable to position a hydrophilic gasket between the renewal pipe and the existing pipe to provide a pressure-tight seal at the manhole connections. A flow-through plug or end cap is fitted over the end of the new pipe in the manhole so the new pipe can be heated by steam or by circulating a hot liquid. The new pipe is then heated to the temperature recommended by the manufacturer. The processing temperature used will vary across vendors and is influenced by the maximum allowable inside skin temperature, the minimum required outside skin temperature, and the thermal conductivity of the pipe material. When the recommended temperature has been reached, the new pipe is pressurized in accordance with manufacturer recommendations. This pressure is usually between 5 and 10 psi for F&F, 12 and 15 psi for D&R, and substantially higher pressures for F&E. The new pipe must then cool while still under pressure (e.g., using compressed air) to ensure that it remains tightly pressed against the existing pipe. After a complete cooling, the pipe can be depressurized and the end-seals can be removed.

Completion Work. In a gravity sewer application, the renewal pipe ends should be trimmed leaving a few inches flared into the manholes to lock the renewal pipe tightly against the existing pipeline. The renewal pipe ends should be properly shaped to ensure continued smooth flow. If a hydrophilic gasket was not used, end-sealing with epoxy or grout at the manhole terminations is recommended. Lateral connections are generally reinstated with robotic cutters. It is also advisable to take measures to seal the lateral connections by rehabilitating the laterals.

In pressure applications, pipe-end terminations can be accomplished with fittings or by flanging the renewal pipe and bolting the connections. Tees or branches must also be excavated and can be sealed with fittings.

21.3.5 Project Inspection and Delivery

After installation it is often necessary to carry out completion work. This may, for example, take the following steps:

- Leak testing
- Consumer information following service line reinstatement
- Final inspection to document the completed renewal

The main purpose of the above tasks is to document the renewal process in accordance with the contractor's own quality control and assurance standards and according to the technical specifications of the contract documents.

21.4 QUALITY ASSURANCE AND DOCUMENTATION

A final quality review compares the completed renewal work with the requirements and descriptions of technical specifications and contract documents. A final review would expose any variations or changes between the contract documents and the work of the contractor.

When a renewal project is completed, the contractor provides the pipeline owner with documentation of the quality of the work performed. In this respect, it is important that during the bidding phase the pipeline owner ensures that such documentation can be prepared and submitted on completion of the renewal work. The documentation should include the followings:

- Information on the products and installation method used
- Pre- and postinspection results
- Test results
- Any changes or deviations from contract documents and possible corrective work
- Information on lateral reinstatements and pipe ends or manhole connection and possible grouting process
- Recommendations for future inspections, cleaning, and other renewal works to be completed as part of a comprehensive inflow/infiltration or structural plan, or both

21.5 ADVANTAGES

The ThP method has the following advantages over comparable renewal technologies:

- The product pipe is manufactured at the factory; therefore, the installation process in the field is faster.
- For the same reason above, quality assurance process is more efficient. Usually an original PVC or PE with recognized physical and chemical properties is installed. All design properties are established at the factory and can not be influenced by the construction crews.
- There are no noxious or caustic chemicals to impact the environment or disrupt the community.
- Many of the ThP methods have start/stop capability, reducing the risk of excavation.
- Being a Thp, the reduction in the existing pipe's cross-sectional area is minimal and flow rates are not adversely impacted.
- The ThP method can solve pipeline internal corrosion problems, water quality problems associated with pipeline internal corrosion, leakage from corrosion holes, cracks, and failed pipeline joints, and flow capacity problems arising from pipe tuberculation and deposits.

- The ThP method controls infiltration, eliminates root intrusion at pipe joints and cracks, eliminates exfiltration.
- The ThP method can be designed to provide full, independent structural integrity with up to a 100-year design life.
- The new pipe can be installed via a manhole or insertion pit.
- The new pipe is capable of accommodating large radius bends.
- The new pipe can be installed up to 1500 ft; therefore, it will have a smooth and jointless pipe between manholes with better flow characteristics than the excising pipe.
- Internal lateral connections are possible.

21.6 LIMITATIONS

ThP has the following limitations:

- Limited diameter range is available, and longer drive lengths are often limited in the larger diameters.
- For fused and expanded (F&E) pipe, a large working space is required to lay the butt-fused string of PVC pipe before insertion into the existing pipe.
- In many cases bypassing the flow during the installation is required.
- For water mains, locations of valves and connections generally have to be excavated.

REVIEW QUESTIONS

1. What are the primary characteristics of a ThP?
2. In what conditions would you recommend a ThP?
3. How does ThP differ from other renewal methods?
4. Describe the three main variations of the ThP method.
5. What type of preparatory work is usually performed for ThP?
6. What are construction considerations of ThP?
7. What are the environmental considerations of ThP?
8. Describe in detail the ThP installation process.
9. Check your local city phone directory to see which companies or contractors offer ThP method. Interview one of these companies to see how many feet of pipe they have installed and which variation of ThP they have used. Check the prices they offer and compare it with other renewal methods.
10. Check your local city public works department of public works to see if they have used ThP method. Obtain specific project conditions (location, diameter, existing pipe type, length, application, and so on) and their decision criteria to select this method over other methods.
11. List the main advantages and limitations of ThP.

REFERENCES

Iseley D. T., and M. Najafi. (1995). T*renchless Pipeline Renewal*, The National Utility Contractors Association (NUCA), Arlington, Va.

Najafi, M. (1994). *Trenchless Pipeline Renewal: State-of-the-Art Review*, Trenchless Technology Center (TTC), Louisiana Tech University, Ruston, La.

Scandinavian Society for Trenchless Technology (SSTT). (2002). *No-Dig Handbook*, Copenhagen, Denmark.

CHAPTER 22
SEWER MANHOLE RENEWAL

22.1 INTRODUCTION

Manholes are windows to the sewer system, as they are the most visible point in identifying the condition of the underground sewer infrastructure. Without a complete manhole renewal plan, corrosion and infiltration/inflow (I/I) problems will not be solved in these crucial structures.

Although manholes are easily accessible, they offer a variety of challenging problems that are frequently misunderstood and overlooked. Like pipelines, manholes come in a variety of materials and sizes. They are commonly made of brick-mortar, precast concrete, concrete block, and cast-in-place concrete. At each major change of direction or grade in the sewer system, a manhole is usually constructed to facilitate flow, cleaning, and inspection. Open-curved channels in the base of the manhole allow sewage to change direction with a minimum amount of friction from one pipe segment into another.

There are currently more than 20 million manholes in the United States with over half installed prior to 1960. As we drive through communities, one can often determine the age of a system by looking at the manhole lid. In the older systems, the lids may be 18 in. in diameter and have numerous vent holes. The manhole chamber below the lid of older systems is generally constructed of brick and mortar. Newer systems will generally have a lid of 24 or 30 in with only one or two pick holes and is constructed of precast concrete. These early observations tell us a little about the potential problems we may encounter as old manholes suffer from a variety of serious problems. They are subject to erosion from surface runoff and groundwater intrusion, corrosion from liquids and gases, wearing from dynamic traffic loads, and general deterioration from age. Out of sight, their degradation is not easily monitored. When neglected, their complete collapse is likely to result as shown in Fig. 22.1.

Irrespective of the age, the location of a manhole within a system often dictates the type of problems manholes will develop. These problems include erosion, corrosion, and wear from dynamic traffic loads, any of which will create structural fatigue and/or serious I/I. Other influences that can contribute to the structural decline of a manhole include soil movement and water table fluctuation (hydrostatic loading). Soil movement can be because of natural occurring events such as earthquakes or freeze-thaw cycles or induced by nearby construction activities.

Old concrete is relatively porous and very rigid. Once a rigid structure has experienced movement, there is a good possibility of foundation damage. A fracture in the wall of a manhole can also open up the possibility for erosion. Approximately half of the manholes that have been installed in the United States were never pressure tested to ensure water or air tightness. This constant water intrusion takes a wearing toll on the porous concrete over

FIGURE 22.1 Manhole collapse. (*Courtesy of Action Products Marketing Corp.*)

time. As shown in Fig. 22.2, for the case of a brick manhole, water movement alone can remove mortar between bricks, and aid in the loss of bricks and the potential for failure of the structure over time. If we add the corrosive nature of modern sewers and its no wonder concrete fails, often very quickly.

22.2 MANHOLE COMPONENTS AND TYPICAL PROBLEMS

Manholes comprise two main sections: the floor and the shaft. The problems encountered in the two are often of different types (see Table 22.1). Corrosion, for example, typically attacks

FIGURE 22.2 Structural failure. (*Courtesy of Action Products Marketing Corp.*)

TABLE 22.1 Different Manhole Components and Typical Defects

Components	Definition	Typical defects
Cover	The lid that provides access to the interior of manhole	• Open vent or pick holes with ponding • Bearing surface worn or corroded • No gasket or bolt for gasketed and bolted covers • Poor fitting; loose or tight • Cracked or broken
Frame	The cast or ductile ring that supports the lid	• Bearing surface worn or corroded • No gasket for gasketed frames • Cracked or broken • Frame offset from chimney
Frame seal	Material or device that prevents intrusion of water at joints between frame and chimney; It includes a frame and cone, or frame and flat top slab	• No seal • Leaking at frame and/or chimney joint • Cracked or missing seal
Clear opening	The smallest entry-way to the manhole	
Chimney	The narrow vertical section built from brick or concrete adjusting ring that extends from top of the cone to the frame and cover	• Cracked, broken • Deteriorated
Joint seal	Material or device to prevent intrusion of water at the joint between precast wall sections or cone and wall section	• Missing or leaking joint seal
Cone	The reducing section, which tapers concentrically or eccentrically from the top wall joint to the chimney or the frame and cover (sometimes referred as corbel when made of brick)	• Leaking because of poor fit, multiple vent holes, damaged adjusting rings or shifted frames • Deteriorated
Wall	The vertical barrel portion extending just above the bench joint to the cone	• Unplugged leaks resulting in voids, missing bricks, and mortar joints • Deteriorated
Pipe seal	The material or device at the pipe and wall or cone interface for preventing entry of water	• Corrosion caused by chemical and/or bacteriological actions • Missing or leaking pipe seal

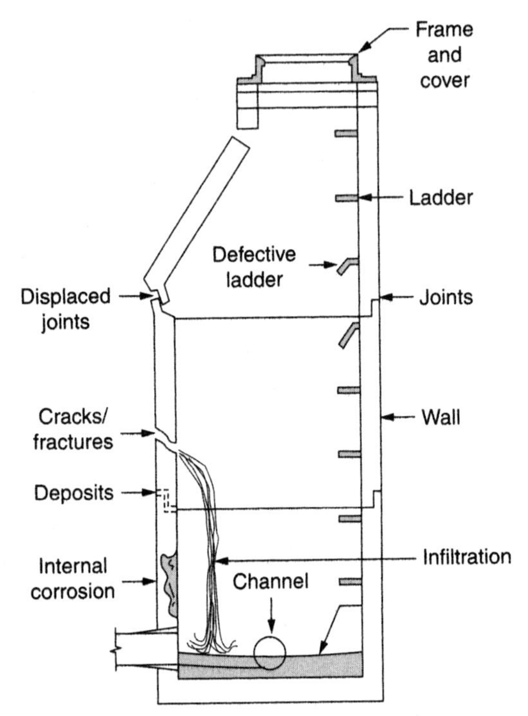

FIGURE 22.3 Cross-section of a defective manhole.

TABLE 22.2 Common Causes of Manhole Problems

Problem	Description
Corrosion	Hydrogen sulfide attacks above the water level. The floor may therefore remain intact while remainder of the manhole is subject to corrosion.
Cracks or fractures	Cracks and fractures typically occur as the result of poor construction, inferior materials, and/or outside influences. They result in leaks. Depending on their extent and location, cracks and fractures will reduce manhole strength and impair function.
Displaced joints	Displaced or open joints occur for the same reason as cracks and fractures. Manhole strength may remain unaffected, but leaks result.
Deposits	Deposits often result from the action of aggressive groundwater. Lime is leached from the concrete, resulting in leaks and manhole weakening.
Infiltration	Infiltration is caused by groundwater leaks. All the above-mentioned problems can cause infiltration. Typically, leaks result in greater quantities of water being delivered to water treatment plants, thus reducing their efficiency. Exfiltration of wastewater may occur if the groundwater table is below the manhole.

the part of the manhole above the water level. Other problems are typically found in limited areas in the manhole. Leaks, for example, are most often found in joints and cracks. Subsequently, renewal work can be prioritized on the basis of the extent of the problem and its location in the manhole. Typical problems that may occur in a manhole include inflow, infiltration, corrosion, displaced joints, cracks and fractures, defective or no floor, no berms, poor flow owing to deposits, defective steps, and defective cover. Table 22.1 presents the different manhole components and typical defects associated with these components. A cross section of a defective manhole is shown in Fig. 22.3. The most common causes of manhole problems are described in Table 22.2.

22.3 MANHOLE INSPECTION AND SAFETY ISSUES

In manhole inspections, the following areas should be addressed:

- Location
- Clear opening
- Cover
- Ponding
- Structural and leakage condition of manhole components
- Construction material used to build the manhole (brick, concrete, precast, and so on)
- Rim to invert measurements
- Shape of manhole and its size
- Evidence of surcharge
- Sizes of incoming and outgoing pipes

Among the above problems, structural problems (risk of collapse), capacity (flow), corrosion problems, and top rings or covers should be prioritized. Both I/I conditions and structural conditions should be quantified according to standard guidelines prepared by American Society of Civil Engineers (ASCE) and Water Environment Federation (WEF). During the inspection, safety precautions such as traffic control, worksite safety, confined space-entry, equipment use, and maintenance should be taken. Atmospheric conditions in manholes should be determined before entry and proper equipment and procedures should be used and followed for entry. Also, adequate ventilation should be provided. Also all debris and unsafe steps should be thoroughly removed. Before application of any renewal method, the manhole must be cleaned thoroughly.

Problems differ greatly from manhole to manhole and the renewal method used must be adapted to suit the manhole in question. The following sections provide details of manhole problems and some recommendations for selecting appropriate methods.

22.4 CLASSIFICATION OF MANHOLE PROBLEMS AND SELECTION OF PROPER METHOD

With the advent of effective pipe repair technologies, manholes are receiving increased attention. Leaks, sealed out when the pipes are lined, are likely to travel to the next point of least resistance that are very often the manholes. Engineers have learned that neglecting such an integral part of the system only shifts the problem from the pipe to the manholes. When any part fails, the whole system fails. Thankfully, manholes are the easiest segment to investigate and the *most effective and least costly* to repair. Selecting the proper method is critical to a successful renewal program. As each type of problem is different, each demands an engineered solution. Regardless of the renewal mechanism, bonding or adhering to the original structure is critical for success. Understanding the environment, the steps needed to prepare the surfaces, and a good understanding of the products being used are all steps toward achieving a successful project.

Being aware of these various types of problems will aid in the investigation of manholes in a system. Once manhole defects have been identified, they can be classified into different types. Each type of defect can then be related back to an effective repair method.

22.4.1 Type I Problems: Frame and Cover

These are single repairable defects that can be easily and usually inexpensively repaired, such as frames and covers that leak because of poor fit multiple vent holes, damaged adjusting rings, or shifted frames. They can be replaced or fitted with under the cover inflow protectors.

Sewer dish inserts (Fig. 22.4) are installed under the cover creating a waterproof seal between the frame and underground structure. Inserts prevent the intrusion of rainwater through cover vent holes, damaged frames or, degraded covers. Effluent gases are vented through a nonmechanical device located in the insert.

The grade adjustment area is prone to develop I/I problems owing to traffic loading and weather conditions. Damaged adjusting rings can receive an internal or external chimney seal (Fig. 22.7). Internal seals are available in mechanical or liquid hand applied versions made of a flexible membrane (Figs. 22.5 to 22.8). Elastomeric urethane and epoxies are hand applied to seal the frame and grade adjustment area with enough flexibility to withstand traffic loading and weather conditions. Mechanical seals are also available creating a watertight seal to the frame and another watertight seal on the top of the cone section. If the exterior of the structure is exposed during the renewal process, sealing the frame and grade adjustment area eliminates I/I. Protective rubber jackets seal the frame and grade adjustment area to the cone section.

FIGURE 22.4 Stainless steel sewer dish insert. (*Courtesy of The Rainstopper.*)

FIGURE 22.5 Mechanical seal. (*Courtesy of Cretex Specialty Products.*)

Replacing degraded adjusting rings, frames, and covers reduce I/I. Standard products include precast concrete adjusting rings, cast iron frames, and covers. Newer adjusting ring and frame technologies include high-density polyethylene and rubber products that are lightweight and impervious to structural erosion and effluent gases (Figs. 22.9 and 22.10).

Individual leaks that have not yet resulted in structural erosion can be sealed with water plug or chemical grouts (Fig. 22.11). Chemical grouts can eliminate inflow and infiltration that develop in underground structures. Gel-type chemical grouts are injected from the interior of the structure to the exterior to form a gelatinous curtain around the structure (Fig. 22.12). The gelatinous curtain prevents groundwater from entering into the structure and stabilizes the soil. In precast structures, urethane foam grouts are used to rebuild the joint gaskets.

FIGURE 22.6 Installed mechanical seal. (*Courtesy of Cretex Specialty Products.*)

FIGURE 22.7 Installed external seal. (*Courtesy of Sealing Systems, Inc.*)

Eroded benches and leaking channels can likewise be patched and sealed. Bench and channel repair can be accomplished using cementitious repair mortars, epoxies, and urethanes that are sprayed or troweled on, or both. These processes structurally enhance the existing bench extending its life expectancy. Total replacement of the existing bench includes the complete removal of the old bench and channel material. New concrete is used to rebuild the bench while hand-forming the new channel (Fig. 22.13). New products on the market today include preformed fiberglass channels used when replacing the existing bench (Fig. 22.14). The new preformed fiberglass channel is set in place and the concrete is poured around the new channel.

Damaged steps create a serious hazard and should be removed. The trend within the U.S. cities is not to replace them to reduce liability for unauthorized entry and to force workers to use required safety equipment for all entry. These individual type I repairs generally range from US$50 to $500.

FIGURE 22.8 Flexible internal seal. (*Courtesy of United States Concrete Products.*)

FIGURE 22.9 HDPE frame and risers. (*Courtesy of United States Concrete Products.*)

FIGURE 22.10 Rubber frame. (*Courtesy of LifeSpan.*)

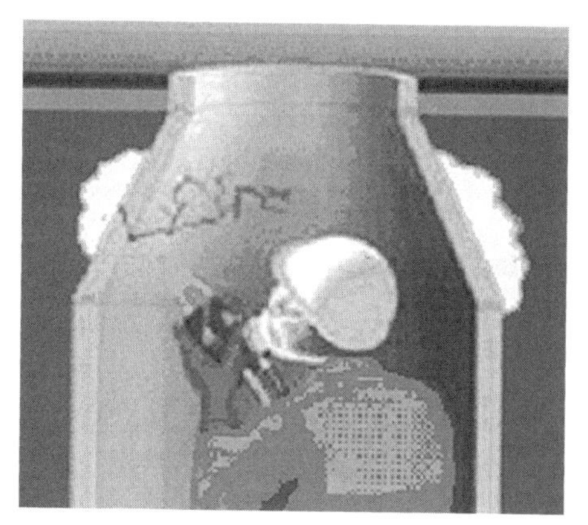

FIGURE 22.11 Chemical injection grout. (*Courtesy of Avanti International.*)

FIGURE 22.12 Completed grout injection. (*Courtesy of Avanti International.*)

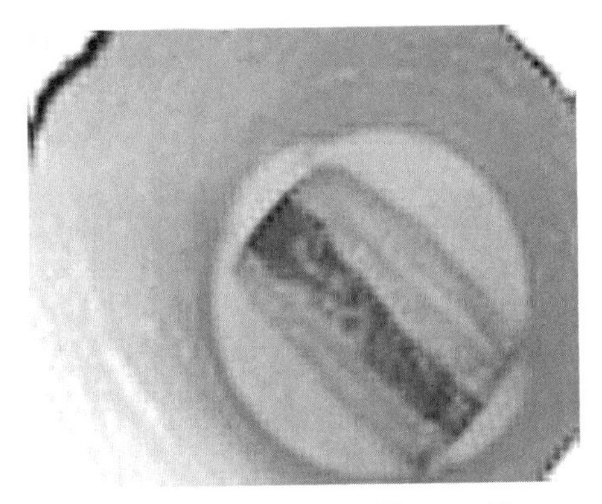

FIGURE 22.13 Bench repaired with mortar. (*Courtesy of Action Products Marketing Corp.*)

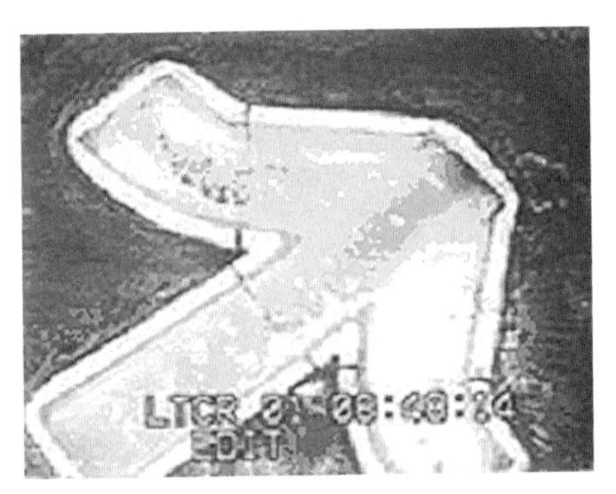

FIGURE 22.14 Preformed fiberglass channel. (*Courtesy of Reliner/Duran Inc.*)

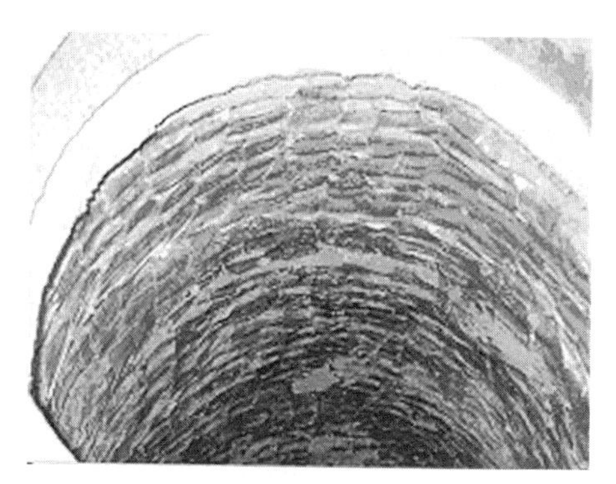

FIGURE 22.15 Brick manhole after leak plugging. (*Courtesy of Action Products Marketing Corp.*)

22.4.2 Type II Problems: Structural Defects

Type II problems are more serious where corrosion is not evident but some structural damage has occurred from continuous leaks resulting in voids, missing bricks, and mortar joints or when multiple type I problems are evident. In these cases, a structural liner that reinforces and seals the damaged structure is warranted. Cementitious, epoxy, and urethane (*polymer*) liners can be hand sprayed, troweled, or centrifugally cast onto the prepared interior of the old manhole to create a structural shell of sufficient strength to withstand the groundwater pressure and dynamic traffic loads (Figs. 22.15 and 22.16). The thickness of the liners is determined by the depth, diameter, and the condition of manhole. Similar to Standard dimension ratio (SDR) ratings of pipe, large-diameter manholes require correspondingly greater thickness for the same strength value, that is, manholes that are 72 in. in diameter require the liner to have a cementitous thickness of 1.5 in. to attain a value equivalent to a $^1/_2$ in thickness in a 48 in.-diameter manhole. Likewise, manholes that are deeper and have greater groundwater pressures require a greater thickness than the shallower portion of the same manhole. Structural reinforcement and sealing with a cementitious lining costs about US$8 to $12 per ft^2 of coverage or US$10 to $15 per square foot with a polymer.

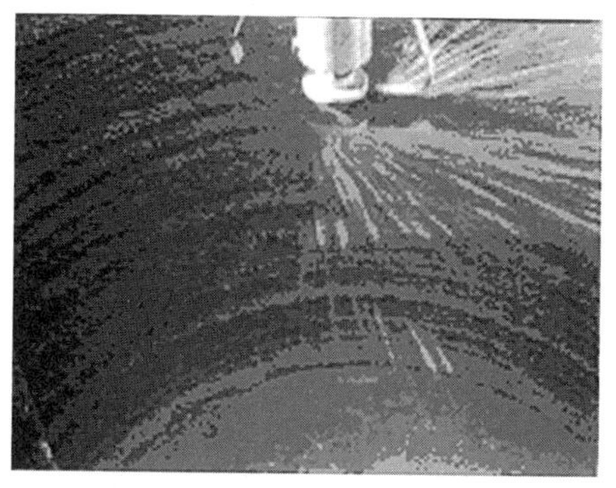

FIGURE 22.16 Centrifugal casting of cementitious lining. (*Courtesy of Action Products Marketing Corp.*)

22.4.3 Type III Problems: Corrosion

Type III problems relate to corrosion, the most common cause of severe deterioration in concrete sanitary sewer structures. Industry studies estimate that corrosion is the primary source of structural damage in three-fourths of all precast concrete manholes and lift stations installed in the last 30 years. The damage and the corresponding repair and replacement costs are counted in billions of dollars.

The source of this corrosion can be either chemical or bacteriological. When corrosion occurs in the areas contacted by the flow, then chemicals are present in sufficient concentrations and temperatures to corrode the lower portions of the manholes. This type of corrosion is common in industrial sewers. The most common corrosion in municipal sewerage is caused by acid generating bacteria, *microbiologically induced corrosion* (MIC). Thiobacillus bacteria thrive in the slime layer above the flows and secrete sulfuric acid. Wherever hydrogen sulfide gas is produced, these bacteria will grow rapidly because the gas is their source of nourishment. Their rate of growth and subsequent damage is affected by the amount of nutrients *biological oxygen demand* (BOD) *levels*; long retainage times that allow the sewerage to go septic; very warm temperature; and turbulence that releases hydrogen sulfide gas. This is why it is most severe in coastal plains and warm climates.

The corrective action for this type of manhole and lift station renewal is to place a protective barrier between the corrosive bacteria and the structural substrate. The most appropriate solution depends upon the severity of the deterioration.

Type III-A: Light Corrosion. This is when the corrosion has done little damage or before it has begun on new structures. In this case, a polymer coating may be applied to the properly prepared surface for protection. Thorough preparation and expert application is key to success of this method. Low-pressure water cleaning (NACE No. 5) and abrasive blasting are generally utilized to accomplish proper decontamination and cleanliness necessary to expose a sound, clean substrate. The industry offers a variety of coating materials such as epoxies, polyurethanes, poly-ureas, and blended hybrid polymers that can be brushed, roller-applied, spray applied (Figs. 22.17 and 22.18), or centrifugally cast. The thickness (commonly 65 to 125 mils) must be sufficient to prevent vapor penetration and it must be free of any holidays or pinholes that would permit corrosion on concrete behind the coating. Prices for coatings used for type III-A defects are commonly US$10 to $15 per ft^2 of coverage.

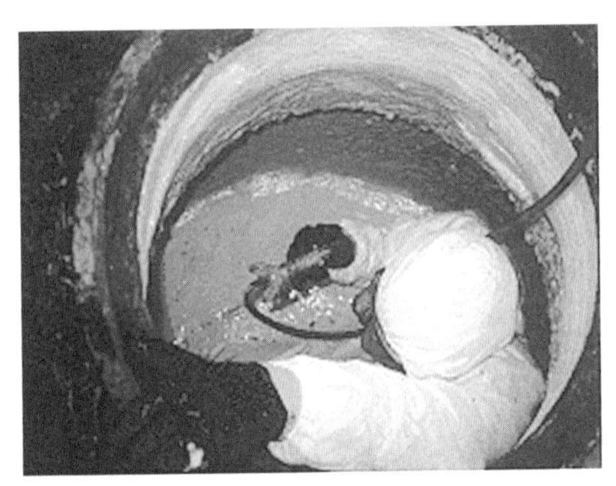

FIGURE 22.17 Spray applied epoxy. (*Courtesy of Raven Lining Systems.*)

FIGURE 22.18 Finished spray applied urethane. (*Courtesy of SprayRoq, Inc.*)

Type III-B: Extensive Corrosion. These defects occur when the manhole has deteriorated to the point of becoming structurally deficient. The coating in this case must be applied at a thickness that will compensate for the structural value of the lost wall. This can be done by using a rigid polymer coating of sufficient strength and applied at a thickness equal to 125–375 mils ($^1/_4$ to $^3/_8$ in.) or by using a composite or laminated liner that combines structural reinforcement and corrosion protection at similar combined thickness and strength values. Common sources for the latter methods are cementatious liner or polymer (CLP) composites (Fig. 22.19) and fiberglass or polymer (FGP) composites. The FGP may be hand-laid (Fig. 22.20) or it may be a custom fitted bag (Fig. 22.21) that is expanded and cured in place with steam. Costs for these methods range from US\$20 to \$30 per ft^2 of coverage.

1. Liners are prefabricated to structure size. On-site, structures are pressure-cleaned and grouted where necessary, while steps and obstructions are removed.
2. Epoxy-coated liner inserted into manhole interior.

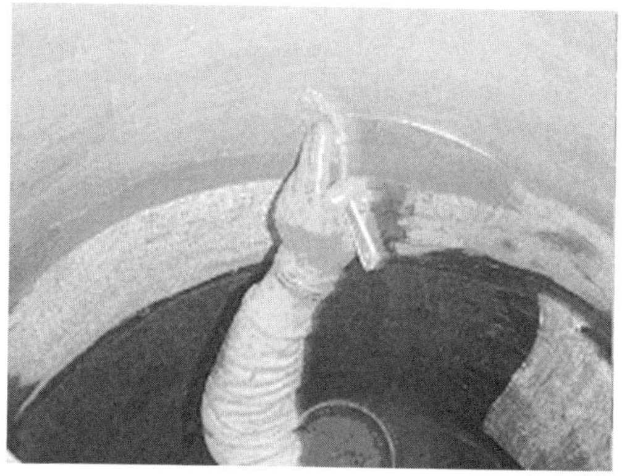

FIGURE 22.19 CLP. (*Courtesy of Action Products Marketing Corp.*)

FIGURE 22.20 Hand-applied FGP. (*Courtesy of Protective Liner Systems, Inc.*)

3. Flexible diaphragm is inflated. Heat introduced within the system cures the epoxy, forming a protective barrier within the structure. The bond is enhanced because of inflation pressure during the cure.
4. Structures are virtually renewed, sealed from further corrosion and infiltration.

Separate and distinct alternatives to type III-B defects are cementitious liners that retard or prevent production of acid-producing bacteria. Calcium aluminate cement generally retards the growth of bacteria by making surface conditions less favorable to their development than Portland cements. Calcium aluminates have been successfully used for more than 78 years in extreme wastewater applications worldwide. The first U.S. application was in 1959 located at the Hyperion Treatment Plant in Southern California. Over the past 6 years, a certain EPA registered antibacterial additive has proven to entirely prevent growth of the bacteria. In either case, the applied thickness is the same as for type II repairs with correspondingly lower costs.

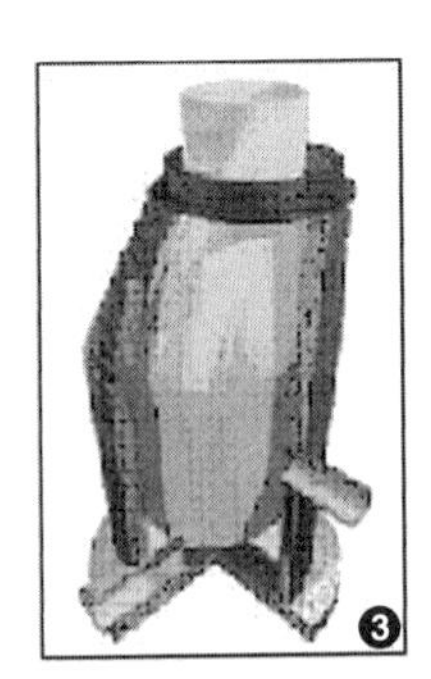

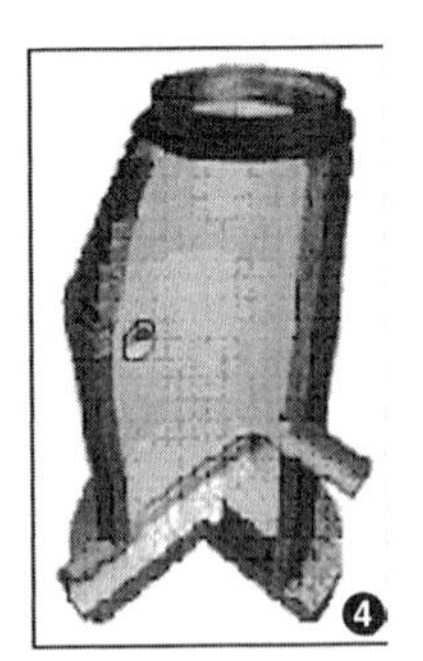

1. Liners are prefabricated to structure size. On-site, structures are pressure-cleaned and grouted where necessary, while steps and obstructions are removed.
2. Epoxy-coated liner inserted into manhole interior.
3. Flexible diaphragm is inflated. Heat introduced within the system cures the epoxy, forming a protective barrier within the structure. The bond is enhanced due to inflation pressure during the cure.
4. Structures are virtually renewed, sealed from further corrosion and infiltration. Manholes and pump stations, for example, can be completely rehabilitated in 4 to 8 hours, with little or no downtime.

FIGURE 22.21 Custom fabricated FGP. (*Courtesy of Terre Hill Concrete Products.*)

22.4.4 Type IV Problems: Severe Structural Damage

The type IV category defines severe structural damage that requires replacement. These are easily determined because field staff can recognize structures in this category that are near collapse. Replacement of some type is the only remaining option for this category. Excavation, removal, and resetting of a new structure is the most costly option because it requires bypassing of flows, traffic interruption, allowance for other buried utilities, and general social disruption. Depending upon the location, size of the pipeline and its depth, and diameter, costs can range from US$120 to $300 per square foot.

No-dig and partial dig solutions are available for replacement. No-dig methods use forms that can be passed through the existing opening and assembled entirely within the old manhole. An annulus space between the forms and the old wall is filled with standard DOT ready-mix concrete and vibrated into place creating a new, joint-free wall about 3 in. thick. In corrosive environments, a plastic liner with interlocking protrusions is fitted onto the form before placement of the concrete so that when the forms are removed, a fully plastic-lined concrete manhole remains in place. Partial excavation is required to sleeve the old structure. In this option, the entire cone is removed and a smaller-diameter fiberglass or plastic pipe is sleeved inside (Figs. 22.22 and 22.23). The annulus between the sleeve and the old wall is similarly filled with a DOT ready-mix concrete and then a new cone is installed and the pavement replaced. These options offer substantial savings with costs in the range of US$25 to $80 per square foot.

The most important part of a manhole renewal is at the end of the project. When plastic linings, polymer coatings, or polymer composites are used, the project is not finished until the final repair product has been thoroughly inspected and tested to verify the installation integrity. A visual inspection will give the owner's inspector the opportunity to verify that the manhole is leak free and coated as specified. Spark testing or the use of a holiday detector verifies that the new polymer coating or plastic lining is free from pinholes or voids and is uniform in thickness. The National Association of Corrosion Engineers' publication, *NACE RP0188-Standard Recommended Practice Discontinuity (Holiday) Testing of New Coatings on Conductive Substrates*, and *ASTM D4787 Standard Practice for Continuity Verification of Liquid or Sheet Linings Applied to Concrete Substrates*, are references for inspectors. Verification of bond/adhesion is another test that can be conducted on polymer coatings after the installed products have cured. The ASTM D-4541 *Pull-off strength of coatings using a portable adhesion tester* uses a portable adhesion testing device to verify the bond of the polymer coating to the prepared substrate.

FIGURE 22.22 Fiberglass sleeve. (*Courtesy of L.F. Manufacturing, Inc.*)

FIGURE 22.23 Poured-in-place manhole with embedded plastic lining. (*Courtesy of Action Products Marketing Corp.*)

The bond test is conducted to failure and values are recorded for pull-off strength and mode of failure (adhesive, cohesive, substrate). These results are compared to predeter mined expectations set forth in project specifications. Identification of repair process such as failure at the coating and less than 100 psi pulloff strength, will be regarded as adhesion failure. The contractor must remove the coating to soundly adhered edges, re-perform surface preparation procedures, and recoat the failed surfaces. The inspecting engineer must have knowledge on how to conduct this test, utilize consistent equipment for the evaluation and understand how the two types of data are used in the evaluation of adhesion. A drawback to this method is that it is destructive and will require that the test locations be repaired although this can be done relatively quickly with good access.

22.5 SUMMARY OF MANHOLE RENEWAL METHODS

Table 22.3 presents a summary of renewal methods with applicable products available in the market. Table 22.4 presents a decision matrix for renewal methods.

22.6 ADVANTAGES AND LIMITATIONS OF MANHOLE RENEWAL METHODS

Table 22.5 presents the advantages and limitations of different manhole renewal methods.

REVIEW QUESTIONS

1. Why are manholes so important in a comprehensive gravity pipeline renewal system? Explain.
2. What are the main manhole components?

TABLE 22.3 Summary of Manhole Renewal Methods with Applicable Products

Method	Product	Primary use	Corrosion resistance	Training and equipment	Cost (US$)
Specialty products	Patching and plugging compounds	I/I control	Poor	Minimal	Varies with use
	Lid seals, hole plugs, inserts, dishes	I/I control	Poor	Minimal	20–130 ea.
	Chimney seals	I/I control	Poor to good	Minimal	275–415 ea.
	Flexible frame/ grade ring sealant	I/I control	Poor to good	Minimal	250–415 ea.
Structural replacement	Cast-in-place liners Precast cement Composite (polymer topcoat of cementitious)	Structural	Poor to excellent	Requires training and specialized equipment	20–50 per sq. ft
Cured-in-place liners	Resin-impregnated bag liner	Structural and corrosion	Excellent	Requires training and specialized equipment–labor intensive	High–20–54 per sq. ft
Pressure grout	Avanti	I/I control, void stabilization	Excellent	Minimal	400–850 per 9-ft, 48 in. diameter manhole
Coatings and linings—cementitious	Shotcrete	I/I control and structural (for 0.5–4 in thickness)	Good	Specialized equipment with moderate training	6–15 per sq. ft
Coatings and linings–polymers	Epoxy, urethanes, and ureas	Corrosion, I/I and structural	Good to excellent	Specialized equipment and training	8–20 per sq. ft
Thermoplastic liners	Sheet liner	Corrosion protection (used in conjunction with cast-in-place liners)	Good to excellent	Moderate–depends on the product	25–50 per sq. ft

3. Explain the microbiologically induced corrosion (MIC) process and its importance in manhole renewal.
4. Explain the four types of manhole problems.
5. What method(s) would you recommend for structural problems? Explain why.

TABLE 22.4 Decision Matrix for Manhole Renewal Methods

Manhole component	Grouting	Coating	Structural renewal	Corrosion	Frame/cover chimney
Cover	Not applicable	Not applicable	Not applicable	Asphalt coating	Replace seal with gasket inserts
Frame	Not applicable	Not applicable	Not applicable	Asphalt coating	Replace seal with gasket inserts
Frame or chimney joint	Not rec	Not rec	Not rec	Flexible seal	Replace
Cone	For active I/I Structurally sound brick MH Precast without corrosion	No evidence or low potential of movement Deteriorated mortar joints Observed I/I	Structurally degraded with minimum 36 in. diameter	Sulfide resistance coatings and linings	Gasket material seal
Wall	For active I/I Structurally sound brick MH Precast without corrosion	No evidence or low potential of movement Deteriorated mortar joints Observed I/I	Structurally degraded with minimum 36 in. diameter	Sulfide resistance coatings and linings	Not applicable

TABLE 22.5 Advantages and Limitations of Different Manhole Renewal Methods

Renewal method	Advantages	Limitations
Replace cover only	Proper fit, eliminates holes and cover leakage	May not get good seal with the existing frame or bearing surface
Seal cover with gasket Seal cover with asphalt mastic	Eliminates inflow inexpensively Eliminates inflow and it is inexpensive	Gaskets may become loose and leak, and can drop into sewer; difficult to install properly prevents easy access to manholes
Replacement of manhole frame and cover	Improves service life and alignment, adjusts grade and eliminates leakage	Excavation is required, so it may involve additional cost for pavement replacement
Seal existing cover or install insert	Eliminates inflow and stops rattle	Raises cover slightly
Sealing of frame-chimney joint and chimney above cone	Eliminates inflow while allowing movement of frame	Reduced access in chimney and more cost

6. What method (s) would you recommend for corrosion problems? Explain why.
7. Describe common causes of manhole problems.
8. What safety precautions are used for manhole inspections and repairs?
9. Check with your city or township public works department to see what types of manhole problems they are facing and what solutions they have used.
10. Check your telephone directory or the Internet to see which manhole contractors provide renewal services in your city.

REFERENCES

Committee on Manhole Rehabilitation of the Pipeline Division of the American Society of Civil Engineers, Richard E. Nelson, Chairman. (1997) "Manual and Report No. 92, Manhole Inspection and Rehabilitation," ASCE.

Habibian, A. (2000). *Design Principles for Manhole Renewal Technologies*, Presentation at the Midwest Society for Trenchless Technology Symposium, University of Missouri, St. Louis, Miss.

Strong-Seal Systems. (2000). *Manhole Maintenance Programs*, Presentation at the Midwest Society for Trenchless Technology Symposium, University of Missouri, St. Louis, Miss.

Lafarge Calcium Aluminates, Inc. Company Catalogue, Chesapeake, Va.

Madewell Company Catalogue, Alpharetta, Ga.

GLOSSARY OF TERMS

A

ABS: Acrylonitrile-butadiene-styrene; a form of thermoplastic.

AC: Asbestos cement; a composite material used in pipe construction.

Adapter ring: In microtunneling, a fabricated ring usually made from steel that serves to mate the microtunneling machine to the first pipe section. This ring is intended to create a waterproof seal between the machine and the spigot of the first joint.

Advance rate: Speed of advance of a TBM or other trenchless construction through the ground, generally expressed as feet per day (meters per day).

AMP: Asset management plan; a structured approach for utilities to achieve long-term defined service standards or an external bearing used to isolate the final drive from the thrusting force of the machine.

Anaerobic: Able to live and grow where there is no oxygen.

Angle of repose: The angle, which the sloping face of a bank of loose earth, gravel, or other material, makes with the horizontal.

Annular filler: Material for grouting the annular space between the existing pipeline and the new lining pipe.

Annulus: Free space between the existing pipe and the lining pipe.

Anode: The electrode at which oxidation or corrosion occurs (opposite of cathode).

Apparent tensile strength: A value of tensile strength used for comparative purpose that is determined by tensile testing pipe rings in accordance with ASTM D2290. This differs from true tensile strength of the material because of a bending moment induced by a change in the contour of the ring as it is tested. Apparent tensile strength may be at yield, rupture, or both.

Aqueduct: Large pipe or conduit to convey water from distant source.

Auger: A flighted drive tube having hex couplings at each end to transmit torque to the cutting head and transfer the spoil back to the machine.

Auger boring: Also horizontal auger boring, A technique for forming a bore from a drive pit to a reception pit, by means of a rotating cutting head. Spoil is removed back to the drive shaft by helically wound auger flights rotating in a steel casing. The equipment may have limited steering capability. See guided auger boring.

Auger machine: A machine to drill earth horizontally by means of a cutting head and auger or other functionally similar device. The machine may be either cradle or track type.

Auger MTBM: A type of microtunnel boring machine, which uses auger flights to remove the spoil through a separate casing placed through the product pipeline.

Auger TBM: Tunnel boring machine in which the excavated soil is removed to the drive shaft by auger flights passing through the product pipeline pushed in behind the TBM.

B

Back reamer: A cutting head attached to the leading end of a drill string to enlarge the pilot bore during a pullback operation to enable the carrier, sleeve, or casing to be installed.

Backstop: Also thrust block, a reinforced area of the entrance pit wall directly behind the track or where the jacking loads will be resisted.

Band: A ring of steel welded at or near the front of the lead section of casing to cut relief and strengthen the casing (used in horizontal auger boring).

Barrel: The vertical section in a manhole between the cone and the benchwall.

Base: The slab structure, which supports a manhole.

Base resin: Plastic materials prior to compounding with other additives or pigments.

Base track: See master track.

Bedding: A prepared layer of material below a pipeline to ensure uniform support.

Benching or bench: The floor of a manhole into which the channel is set; the bench is raised so that it can drain to the channel.

Benchwall: The horizontal bottom of the manhole normally out of the flow path.

Bent sub: An offset section of drill stem close behind the drill head that allows steering corrections to be made by rotation of the drill string to orientate the cutting head (used in horizontal directional drilling).

Bentonite: Colloidal clay sold under various trade names that form a slick slurry or gel when water is added; also known as driller's mud (see drilling fluids).

Biological corrosion: Corrosion that results from a reaction between the pipe material and organisms such as bacteria, algae, and fungi.

Bits: Replaceable cutting tools on the cutting head or drill string.

Bore: A generally horizontal hole produced underground, primarily for the purpose of installing services.

Boring: (1) The dislodging or displacement of spoil by a TBM, a rotating auger or drill string to produce a hole called a bore. (2) An earth-drilling process used for installing conduits or pipelines. (3) Obtaining soil samples for evaluation and testing.

Boring machine: An automated mechanism to drill earth.

Boring pit: Also entry pit or drive pit, an excavation in the earth of specified length, depth, and width for placing the boring machine on required line and grade.

Breakout: Controls the joint make and/or break mechanism.

Building combined sewers: A small-diameter pipe that conveys both wastewater and drain-stormwater from a single property to a combined sewer.

Building sanitary drain: Also service lateral or service connection, a small-diameter pipe that conveys wastewater from a single property (e.g., domestic home) to a sanitary sewer. See building sewer.

Building sewer: The conduit that connects building wastewater sources, to the public or street sewer, including lines serving homes, public buildings, commercial establishments, and industry structures. In this specification, the building sewer is referred to in two sections. (1) the section between the building line and the property line, frequently specified and supervised by plumbing or housing officials; and (2) the section between the property line and the street sewer, including the connection thereto frequently specified and supervised by sewer, public works, or engineering officials (also referred to as *house sewer, building connection, service connection*, or *lateral connection*).

Burst strength: The internal pressure required to cause a pipe or fitting to fail within a specified time period.

Butt-fusion: A method of joining polyethylene and PVC pipe where two pipe ends and are rapidly brought together under pressure to form a homogeneous bond.

Bypass pumping: The transportation of sewage that flows around a specific sewer pipe or line section or sections via any conduit for the purpose of controlling sewage flows in the specified section or sections without flowing or discharging onto public or private property.

Bypass: An arrangement of pipes and valves whereby the flow may be passed around a hydraulic structure or appurtenance. Also, a temporary setup to route flows around a part of a sewer system.

C

Can: A principal module, which is part of a shield machine as in microtunneling or TBMs. Two or more may be used, depending on the installation dimensions required and the presence of an articulated joint to facilitate steering. May also be referred to as a trailing tube.

Carriage: The mechanical part of a nonsplit boring machine that includes the engine or drives motor, the drive train, thrust block, and hydraulic cylinders.

Carrier pipe: The tube, which carries the product being transported, and which may go through casings at highway and railroad crossings. It may be made of steel, concrete, clay, plastic, ductile iron, or other materials. On occasion it may be bored direct under the highways and railroads.

Cased bore: A bore in which a pipe, usually a steel sleeve, is inserted simultaneously with the boring operation. Usually associated with horizontal auger boring or pipe ramming.

Casing: A pipe used to line bore holes through which a pipe(s) called carrier pipes or ducts are installed. Usually not a product pipe.

Casing adapter: A circular mechanism to provide axial and lateral support of a smaller-diameter casing than that of the casing pusher.

Casing pipe method: Method in which a casing, generally steel, is pipe jacked into place, within which a product pipe is inserted later.

Casing pipe: A pipe installed as external protection to a product pipe.

Casing pusher: The front section of a boring machine that distributes the thrusting force of the hydraulic cylinders to the casing and forms the outside of the spoil ejector system.

Catch basin: A small buried structure to divert overland stormwater flow into sewer flows.

Catchment: A geographical area draining to a sewer or receiving water.

Cathode: The electrode of an electrolytic cell at which reduction is the principal reaction (electrons flow toward the cathode in the external circuit). Typical cathodic processes are cations taking up electron and being discharged, oxygen being reduced, and the reduction of an element or group of elements from a higher to a lower valence state.

Cathodic: A process by which the corrosion of a metal pipeline may be protected by the use of an electrical current.

Cathodic corrosion: An unusual condition (especially with Al, Zn, Pb) in which corrosion is accelerated at the cathode because the cathodic reaction creates an alkaline condition, which is corrosive to certain metals.

Cathodic protection: Preventing corrosion of a pipeline by using special cathodes (and anodes) to circumvent corrosive damage by electric current. Also a function of zinc coatings on iron and steel drainage products is galvanic action.

Caulking: General term which, in trenchless technology, refers to methods by which joints may be closed within a pipeline.

CCTV: Closed-circuit television used to carry out internal inspection and survey of pipelines.

Cell classification: Method of identifying plastic materials, such as polyethylene, as specified by ASTM D3350, where the cell classification is based on these six properties: (1) density of base resin, (2) melt index, (3) flexural modulus, (4) tensile strength at yield, (5) ESCR and (6) hydrostatic design basis and color.

Cellar drain: A pipe or a series of pipe that collect wastewater which leaks, seeps, or flows into subgrade parts of structures and discharge them into building sewers or by other means dispose of such wastewaters into sanitary, combined, or storm sewers (also referred to as *basement drain*).

CFM: Cubic feet per minute; a measure of flow volume. One CFM equals 0.472 liters per second.

Channel: A prepared flow route within the bench of a manhole that conveys the incoming flow to the downstream pipe.

Chemical grouting: Method for the treatment of the ground around a shaft or pipeline, using noncementitious compounds, to facilitate or make possible the installation of an underground structure.

Chemical resistance: Ability to render service in the transport of a specific chemical for a useful period of time at a specific concentration and temperature.

Chemical stabilization: Renovation method in which a length of pipeline between two access points is sealed by the introduction of one or more compounds in solution into the pipe and the surrounding ground and, where appropriate, producing a chemical reaction. Such systems may perform a variety of functions such as the sealing of cracks and cavities, the provision of a new wall surface with improved hydraulic characteristics or ground stabilization.

Chimney: The small vertical section between a manhole frame and cone, which is built from brick, masonry, or concrete adjusting rings.

Chippers: See bits.

CIPP: Cured-in-place pipe; a renewal technique whereby a flexible resin-impregnated tube is installed into an existing pipe and then cured to a hard finish, usually assuming the shape of the existing pipe.

Circumferential coefficient of expansion and contraction: The fractional change in circumference of a material for a unit change in temperature; expressed as inches of expansion or contraction per inch of original circumference.

Circumferential: The perimeter around the inner surface of a circular pipe cross section.

Closed face: The ability of a tunnel boring machine to close or seal the facial opening of the machine to prevent, control, or slow the entering of soils into the machine. Also may be the bulk heading of a hand-dug tunnel to slow or stop the inflow of material.

Closed-circuit television inspection (CCTV): Inspection method using a closed-circuit television camera system with appropriate transport and lighting mechanisms to view the interior surface of sewer pipes and structures.

Close-fit: Description of a lining system in which the new pipe makes close contact with the existing defective pipe at normal or minimum diameter. An annulus may occur in sections where the diameter of the defective pipe is in excess of this.

Coefficient of thermal expansion and contraction: The fractional change in length of a material for a unit change in temperature.

Cohesionless: A material that does not readily bond with other particles.

Cohesive soil: A soil that when unconfined has considerable strength when dried, and that has significant cohesion when saturated.

Cold bend: To force the pipe into a curvature without damage, using no special tools, equipment, or elevated temperatures.

Collapse: Critical failure of a pipeline when its structural fabric disintegrates.

Collaring: The initial entry of casing or a cutting head into the earth.

Collection system: A network of sewers that serves one or more catchment areas.

Collector sewer: A sewer located in the public way collects the wastewaters discharged through building sewers and conducts such flows into larger interceptor sewers and pumping and treatment works (referred to also as *street sewer*).

Combined sewer system: A single network of sewers designed to convey stormwater as well as sanitary flows.

Compressed air method: General term in trenchless technology that refers to the use of compressed air within a tunnel or shaft to balance groundwater and prevent ingress into an open excavation.

Compression gasket: A device that can be made of several materials in a variety of cross sections and that serves to secure a tight seal between two pipe sections (e.g., "O" rings).

Compression ring: A ring fitted between the end bearing area of the bell and spigot to help distribute applied loads more uniformly. The compression ring is attached to the trailing end of each pipe and is compressed between the pipe sections during jacking. The compression rings compensate for slight misalignment, pipe ends that are not perfectly square, gradual steering corrections, and other pipe irregularities. Compression rings are also referred to as spacers.

Conduit: A broad term that can include pipe, casing, tunnels, ducts, or channels. The term is so broad that it should not be used as a technical term in boring or tunneling.

Cone: The section between the top of a manhole wall and chimney or the frame. The diameter of the manhole is reduced over the cone section to receive the frame. The cone section may be concentric or eccentric.

Continuous pipe: A single continuous pipe lining or jointed sections to form a continuous lining.

Continuous sliplining: See sliplining or lining with continuous pipe.

Control console: An electronic unit inside a container located on the ground surface, which controls the operation of the microtunneling machine. The machine operator drives the tunnel from the control console. Electronic information is transmitted to the control console from the heading of the machine. This information includes head position, steering angle, jacking force, progression rates, machine face torque, slurry and feed line pressures, and laser position. Some control consoles are equipped with a computer that tracks the data for a real-time analysis of the tunnel drive.

Control lever: A handle that activates or deactivates a boring machine function.

Conventional trenching: Method in which access is gained by excavation from ground level to the required level underground for the installation, maintenance, or inspection of a pipe, conduit, or cable. The excavation is then backfilled and the surface reinstated.

Conventional tunneling: Methods of tunnel construction ranging from manual excavation to the use of self-propelled tunnel boring machines. Where a lining is required bolted segmental rings are frequently used.

Corbel: For brick manholes this term is sometimes used in place of cone, and indicates the gradual reduction in diameter by indenting brick

Core area: That part of a sewer network containing the critical sewers, and other sewers where hydraulic problems are severe and merit detailed investigation.

Corrosion: The destruction of a material or its properties because of a reaction with its (environment) surroundings.

Corrosion fatigue: Fatigue-type cracking of metal caused by repeated or fluctuating stresses in a corrosive environment characterized by shorter life than would be encountered as a result of either the repeated or fluctuating stress alone or the corrosive environment alone.

Corrosion rate: The speed (usually an average) with which corrosion progresses (it may be linear for a while); often expressed as though it were linear, in units of mdd (milligrams per square decimeter per day) for weight change, or mpy (milligrams per year) for thickness changes.

Corrosion resistance: Ability of a material to withstand corrosion in a given corrosion system.

Corrugated pipe: Pipe with ridges (corrugations) going around it to make it stiffer and stronger. The corrugations are usually in the form of a sine wave, and the pipes are usually made of galvanized steel or aluminum.

Cover: The lid at the top of the manhole, which can be removed when access to the interior of the manhole is required.

Cracks: Fracture lines visible around the circumference, or along the length of a pipe, or both.

Cradle machine: A boring machine typically carried by another machine that uses winches to advance the casing.

Cradle: A structure constructed from concrete or masonry that provides structural support to a pipe. It typically surrounds the bottom and the sides of a pipe up to the springing line.

Creep: The dimensional change, with time, of a material, such as plastic, under continuously applied stress after the initial elastic deformation.

Crossing: Pipeline installation in which the primary purpose is to provide one or more passages beneath a surface obstruction.

Crown: (1) Top of pipe segment, or (2) the highest elevation within a pipe.

Crush lining: See pipe eating.

CSO: Combined sewer overflows; a hydraulic relief point within combined sewer systems designed to discharge excess flows during wet weather to receiving waters.

Cured-in-place pipe (CIPP): A lining system in which a thin flexible tube of polymer or glass fiber fabric is impregnated with thermoset resin and expanded by means of fluid pressure into position on the inner wall of a defective pipeline before curing the resin to harden the material. The uncured material may be installed by winch or inverted by water or air pressure, with or without the aid of a turning belt.

Cut and cover: See open cut.

Cutterhead: Any rotating tool or system of tools on a common support that excavates at the face of a bore, usually applies to the mechanical methods of excavation.

Cutting bit (cutter head): The actual teeth and supporting structure that is attached to the front of the lead auger, drill stem, or front face of the TBM. It is used to reduce the material that is being drilled or bored to sand or loose dirt so that it can be conveyed out of the hole. Usually applies to mechanical methods of excavation, but may also include fluid jet cutting.

D

Dead man: A fixed anchor point used in advancing a saddle or cradle-type boring machine.

Deck assembly: Drive train assembly for a split design boring machine.

Deformed and reshaped: See modified sliplining.

Dereliction: The gradual decay of a sewer or a pipe network accelerated by the negligence of effective maintenance.

Diameter of reamer: Largest diameter of reamer (in horizontal directional drilling).

Diameter of standard bit: Maximum outside diameter of standard bit.

Die draw: Site cold deformed close-fit PE liner.

Dimension ratio (DR): See standard dimension ratio (SDR).

Dimple: A term used in tight fitting pipeline reconstruction, where the new plastic pipe forms an external departure or a point of expansion slightly beyond the underlying pipe wall where unsupported at side connections. The dimples are used for location and reinstatement of lateral sewer service.

Directional drilling: A steerable system for the installation of pipes, conduits, and cables in a shallow arc using a surface launched drilling rig. Traditionally the term applies to large-scale crossings in which a fluid-filled pilot bore is drilled using a fluid-driven motor at the end of a bent sub, and is then enlarged by a washover pipe and back reamer to the size required for the product pipe. The positioning of a bent sub provides the required deviation during pilot boring. Tracking of the drill string is achieved by the use of a downhole survey tool.

Discrete pipe: Sections of pipe to be joined.

Discrete sliplining: See segmental sliplining or lining with discrete pipes.

Diverting: Modifying the normal wastewater flow to allow access to some specific sewer structure; often includes bypass pumping.

Dog plate: See thrust block.

Dogs: Moveable protrusions in the thrust block that engage holes or blocks in the track.

Drill bit: A tool that cuts the ground at the head of a drill string, usually by mechanical means but may include fluid jet cutting.

Drill string: (1) The total length of drill rods or pipe, bit, swivel joint, and so on in a drill borehole. (2) System of rods used with cutting bit or compaction bit attached to the drive chuck.

Drilling fluid or mud: A mixture of water and usually bentonite and/or polymer continuously pumped to the cutting head to facilitate cutting, reduce required torque, facilitate the removal of cuttings, stabilize the borehole, cool the head, and lubricate the installation of the product pipe. In suitable soil conditions water alone may be used.

Drive chuck: The female hex connector located within the casing pusher.

Drive or entry or jacking shaft or pit: Excavation from which trenchless technology equipment is launched for the installation of a pipeline, conduit, or cable. It may incorporate a thrust wall to spread reaction loads to the soil.

Drop connection manhole: A manhole in which the influent pipe enters from above the effluent invert depth. If the drop occurs in the manhole itself, it is called an *internal* drop manhole. If the drop occurs a few feet upstream of the manhole, it is referred to as an *external* drop.

Drop manhole: If the upstream pipe is at a greater elevation than the manhole invert then two inlet connections to the manhole are made. One is through the wall at the same grade as the upstream pipe, the other is at the invert so as to direct flows through the channel. Incorporating a pipe drop in the upstream pipe makes the invert connection. The pipe drop may be outside or inside the manhole.

Dry bore: Any drilling or rod pushing system not employing drilling fluid in the process. Usually associated with guided impact moling, but also some rotary methods.

Duct: (1) In many instances, a term interchangeable with pipe. (2) In the boring industry, it is usually used for small plastic or steel pipes that enclose wires or cables for electrical or communication usage. (3) Conduit inside which a utility service is installed.

DWF: Dry weather flow; is the base flow in a sewer prior to rain-induced flows.

DWI: Dry weather inflow; is the result of flow entering the collection system from connected sources, which are not caused by rainfall. Typically, this could include water from fire fighting, hydrant abuse, street washing, sump pumps, and so on.

E

Earth piercing: (1) Term commonly used in North America as an alternative to impact moling. (2) The use of a tool, which comprises a percussive hammer within a suitable casing, generally of torpedo shape. The hammer may be pneumatic or hydraulic. The term is usually associated with non-steered devices without rigid attachment to the launch pit, relying upon the resistance (friction) of the ground for forward movement. During operation the soil is displaced, not removed. An unsupported bore may be formed in suitable ground, or a pipe drawn in, or pushed in, behind the tool. Cables may also be drawn in.

Earth pressure balance (EPB) machine: Type of microtunneling or tunneling machine in which mechanical pressure is applied to the material at the face and controlled to provide the correct counterbalance to earth pressures to prevent heave or subsidence. The term is usually not applied to those machines where the pressure originates from the main pipe jacking rig in the drive shaft/pit or to systems in which the primary counterbalance of earth pressures is supplied by pressurized drilling fluid.

Earth pressure balance shield: Mechanical tunneling shield that uses a full face to support the ground in front of the shield and usually employs an auger flight to extract the material in a controlled manner.

Effluent: A generic term used to indicate the relative strength of sewer flows; from stormwater to wastewater to industrial effluent, for example.

Emergency repair: An unscheduled repair that must be made during a pipe failure or collapse. This type of repairs may cost many times more than the planned repair costs and may not be as effective and/or permanent.

Emergency stop: A red, manually operated push button that, when activated, stops all functions of the machine.

Entrance pit: (1) See boring pit or dive shaft.

Entry/exit angle: Angle to horizontal (the ground surface) at which the drill string enters or exits in forming the pilot bore in a horizontal directional drilling operation.

EPDM (ethylene-propylene diene): Type of rubber that has excellent resistance to ozone, sunlight, and oxygen. It also has excellent resistance to acids, alkalis, and ketones. Plus it has excellent heat resistance and aging. However it has poor resistance to fuels and oils.

Epoxy: Resin formed by the reaction of bisphenol and Epichlorohydrin.

Epoxy lining: A curable resin system based on epoxy resins.

Estimated remaining life (ERL): A value determined by the inspectors based on experience, judgment, and guidelines within their manual that predicts the amount of time that a sewer structure will remain in a *fit-for-use* condition.

Exfiltration: The leakage or discharge of flows being carried by pipes or sewers out into the ground through leaks in pipes, joints, manholes, or other sewer system structures; the reverse of *infiltrations*.

Exit pit: See reception shaft.

Exit shaft: See reception shaft.

Expander: A tool, which enlarges a bore during a pullback operation by compression of the surrounding ground rather than by excavation. Sometimes used during a pipe bursting process as well as during horizontal directional drilling.

F

Face stability: Stability of the excavated face of a tunnel or pipe jack operation.

Face: Wall of the entrance pit into which the bore is made.

Fairings: Molding features at the ends of pipes, usually of varying dimensions to the main pipe, to facilitate easy jointing.

Flexural modulus: The slope of the curve defined by flexural load versus resultant strain. A high flexural modulus indicates a stiffer material.

Flexural strength: The strength of a material in bending expressed as the tensile stress of the outermost fibers at the instant of failure.

Flight: The spiral plates surrounding the tube of an auger.

Fluid cutting: (1) An old trenchless method where pressurized fluid jets are mainly used to provide the soil cutting action. (2) A process using high-pressure fluid to wash out the face of a utility crossing without any mechanical or hand excavation of the soils in the face.

Fluid-assisted boring or drilling: A type of horizontal directional drilling technique using a combination of mechanical drilling and pressurized fluid jets to provide the soil cutting action.

Fold and form lining: Method of pipeline renewal in which a liner is folded to reduce its size before insertion and reversion to its original shape by the application of pressure, or heat, or both. See also lining with close-fit pipes.

Fold and form pipe: A pipe renewal method where a plastic pipe manufactured in a folded shape of reduced cross-sectional area is pulled into an existing conduit and subsequently expanded with pressure and heat. The reformed plastic pipe fits snugly and takes the shape of the ID of the host pipe.

Force-main: A pipeline that conveys sanitary, combined, or stormwater flow under pressure from a pumping (or lift) station to a discharge point (treatment plant).

Forward rotation: The clockwise rotation of the auger as viewed from the machine end.

Frame ring: In a manhole, it is the metal frame, which supports the cover.

Frame: A cast iron unit at the ground surface that provides entry into the manhole.

Free boring: A horizontal auger boring method without the use of casing installed at the same time as the hole is cut. This method should be used with caution as it is appropriate for certain project and site conditions.

G

Gallons per minute (GPM): The U.S. customary unit to measure liquid volume discharge rate. One gallon per minute equals 0.063 liters per second.

Geographical information system (GIS): A computer software system designed to store, manipulate, analyze, and print geographically referenced information.

Gravity sewer: A sewer that is designed to operate under open channel conditions (below pipe full capacity) up to a maximum design flow at which point it will become surcharged.

GRC: Glassfiber reinforced concrete; a renewal lining material.

Ground mat: Usually used in horizontal directional drilling, metal mats rolled out on either side of drill rack for operators and crew to stand on during operation to give grounding protection in case of electrical strike.

Ground rod: This is a copper or brass rod that is hand driven into the ground and is connected to the drill rack and mats to provide adequate grounding of unit and personnel.

Groundwater table (or level): Upper surface of the zone of saturation in permeable rock or soil (when the upper surface is confined by impermeable rock, the water table is absent).

Grout: (1) Material used to seal pipeline and manhole cracks; also used to seal connections within pipe or sewer structures. (2) A material, usually cement or polymer based, used to fill the annulus between the existing pipe and the lining; and also to fill voids outside the existing pipeline. (3) A material such as cement slurry, sand, or pea gravel that is pumped into voids.

Grouting: (1) Filling of the annular space between the host pipe and the carrier pipe. Grouting is also used to fill the space around laterals and between the new pipe and manholes. Other uses of grouting are for localized repairs of defective pipes and ground improvement prior to excavation during new installations. (2) The process of filling voids, or modifying or improving ground conditions. Grouting materials may be cementitious, chemical, or other mixtures. In trenchless technology, grouting may be used for filling voids around the pipe or shaft, or for improving ground conditions. (3) A method of filling voids with cementitous or polymer grout.

GRP: Glass reinforced plastic, a family of renewal linings. Often generically known as reinforced plastic mortar (RPM) and reinforced thermosetting resin (RTR).

Guidance system: The guidance system continuously confirms the position of the TBM.

Guide rail: Device used to support or guide, first the shield and then the pipe within the drive shaft during a pipe jacking or utility tunneling operation.

Guided auger boring: A modified version of horizontal auger boring method. See Chap. 7 on horizontal auger boring method.

Guided boring: This term is used in Europe for small-diameter horizontal directional drilling method. See Chap. 10.

Guided drilling: See guided boring.

Gunite: A renewal technique that employs steel reinforcement fixed to the inside surface of an existing sewer line, which is sprayed with dry concrete. See Chap. 20.

GWI: Groundwater infiltration; results from the movement of groundwater into sections of a gravity system below the water table through defects, cracks, and voids.

H

HDPE: High density polyethylene, see polyethylene.

Head (static): The height of water above any plane or point of references. (The energy possessed by each unit of weight of a liquid, expressed as the vertical height through which a unit of weight would have to fall to release the average energy posed.) Standard unit of measure shall be the foot. Head in feet for water is 1 psi = 2.310 ft.

Heaving: A process in which the ground in front of a tunneling or pipe jacking operation may be displaced forward and upward, causing an uplifting of the ground surface.

Height of cover (HC): Distance from crown of a pipe or conduit to the finished road surface, or ground surface, or the base of the railway.

Helicoid: In horizontal auger boring, a section of auger flight.

High density polyethylene (HDPE): A plastic resin made by the copolymerization of ethylene and a small amount of another hydrocarbon. The resulting base resin density, before additives or pigments, is greater than 0.941 g/cc.

Holiday: Any discontinuity or bare spot in a coated surface.

Hoop stress: The circumferential force per unit areas, psi, in the pipe wall owing to internal pressure.

Horizontal directional drilling (HDD): See directional drilling.

Horizontal earth boring machine: A machine used to bore horizontally through the earth by means of a rotating tool, or nonrotating pushing or piercing tool.

Horizontal earth boring (HEB): The use of auger boring machines to prepare holes by the installation of a casing whereby the spoil is removed by the use of augers (see Chap. 7 on horizontal auger boring).

Horizontal rotary drilling: The mechanical installation of pipe or casing by rotating methods, which do not use augers for the removal of spoil. Usually uses a mixture of water and bentonite to remove spoil.

Host pipe: In trenchless renewal methods, it is the existing or deteriorated pipe.

Hydraulic cleaning: Techniques and methods used to clean sewer lines with water, for example, water pumped in the form of a high velocity spray and water flowing by gravity or head pressure. Devices include high velocity jet cleaners, cleaning balls, and hinged disc cleaners.

Hydraulic gradient line (HGL): An imaginary line through the points to which water would rise in a series of vertical tubes connected to the pipe. In an open channel, the water surface itself is the hydraulic grade line.

Hydrogen sulfide: An odorous gas found in sewer systems with chemical formula of H_2S.

I

I/I: Infiltration or inflow; this is, the sum of all the extraneous contributions to the collection system.

ICG: Internal condition grade; numeric criteria applied to visual images of sewers to develop a priority rating based on defects observed.

Impact moling: Method of creating a bore using a pneumatic or hydraulic hammer within a casing, generally of torpedo shape. The term is usually associated with nonsteered or limited steering devices without rigid attachment to the launch pit, relying upon the resistance of the ground for forward movement. During the operation the soil is displaced, not removed. An unsupported bore may be formed in suitable ground, or a pipe drawn in, or pushed in, behind the impact moling tool.

Impact ramming: See pipe ramming.

In-line replacement: The process of breaking out of an existing pipeline and the installation of a new service on the same line. See Chap. 16.

Infiltration or inflow (I/I): The total quantity of water from both infiltration and inflow without distinguishing the source.

Infiltration: Penetration of groundwater into the sewer system through cracks and defective joints in the pipeline, or through lateral connections, or manholes.

Inflow: Storm water discharged into a sewer system and service connections from sources on the surface.

Interceptor sewer: A sewer that conveys flow from a trunk sanitary sewer or dry weather flow plus a small volume of storm water from a trunk-combined sewer to a wastewater treatment plant.

Interjack pipes: Pipes specially designed for use with an intermediate jacking station used in pipe jacking and microtunneling operations.

Interjack station: See intermediate jacking station.

Intermediate jacking method: Pipe jacking or microtunneling method to redistribute the jacking force by the use of intermediate jacking stations.

Intermediate jacking station (IJS): A fabricated steel cylinder fitted with hydraulic jacks that are incorporated into jacking pipes between two pipe segments. Its function is to distribute the jacking load over the pipe string on long drives and thereby decrease the total jacking forces exerted on the pipe string.

Inversion: The process of turning a fabric tube inside out with water or air pressure as is done at installation of a cured-in-place pipe.

Invert: (1) The lowest point on the pipe circumference; also the defined channel in the manhole platform that directs flow from inlet pipe to outlet pipe. (2) The inside bottom, lowest elevation, of a pipe.

J

Jacking force: Force applied to pipes in a pipe jacking operation.

Jacking frame: A structural component that houses the hydraulic cylinders used to propel the microtunneling machine and pipeline. The jacking frame serves to distribute the thrust load to the pipeline and the reaction load to the shaft wall or thrust wall.

Jacking pipes: Pipes sections with smooth outside joints designed to be installed using pipe jacking techniques.

Jacking pit: See jacking shaft.

Jacking shaft: Excavation from which trenchless technology equipment is launched for the installation or renewal of a pipeline.

Jacking shield: A steel cylinder from within which the excavation is carried out either by hand or machine. Incorporated within the shield are facilities to allow it to be adjusted to control line and grade.

Jacking station (JS): See immediate jacking station (IJS).

Jacking: The actual pushing of pipe or casing in an excavated hole. This is usually done with hydraulic cylinders (jacks), but has been done with mechanical jacks and air jacks.

Jet cutting (jetting): See fluid cutting.

Joint sealing: Method in which an inflatable packer is inserted into a pipeline to span a leaking joint, resin, or grout being injected until the joint is sealed and the packer then removed.

Joints: The means of connecting sectional lengths of sewer or pipeline system into a continuous line using various types of jointing materials. The number of joints depends on the pipe section lengths used in the specific pipeline project. See Chap. 6.

L

Lateral: A service line that transports wastewater from individual buildings to a main sewer line.

Lateral connection: The point at which the downstream end of a building drain or sewer connects into a larger-diameter sewer.

Launch pit: Also known as drive pit, but more usually associated with *launching* an impact moling tool.

Launch seal: A mechanical seal; usually comprises a rubber flange that is mounted to the wall of the drive shaft. The flange seal is distended by the TBM as it passes through creating a seal to prevent water or lubrication inflow into the shaft during tunneling operations.

lb/ft: The US customary units for weight per unit length. One lb/ft equals 1.488 kilograms per meter.

Lead pipe: The leading pipe designed to fit the rear of a jacking shield and over which the trailing end of the shield is fitted.

Liner plate: A proprietary product, used to line tunnels instead of casing, and comes in formed steel segments. When these segments are bolted together they form a structural tube to protect the tunnel from collapsing. The segments are made so that they may be bolted together from inside the tunnel.

Lining: A renewal process where a new pipe material is inserted or cured in place to give an existing pipe a new design life.

Live insertion: Installation of a liner while the existing pipe remains in service.

Localized (spot) repair: Repair work on an existing pipe, to an extent less than the run between two access points or manholes.

Locator: An electronic instrument used to determine the position and strength of electro-magnetic signals emitted from a transmitter (sonde) in a directional drilling operation, or from existing underground services, which have been energized, thereby identifies its location. In horizontal directional drilling it is referred to as a walkover system.

M

Man-entry: Also worker-entry, describes any inspection, construction, renewal, or repair process, which requires an operator to enter a pipe, duct, or bore. OSHA currently has no minimum size limit for worker-entry operations; however, they address a much broader concept of *confined space* in Title 29 Code of Federal Regulations Part 1910.146. The minimum size for which this is currently permissible in the United Kingdom is 900 mm (approximately, 36 in). Many trenchless technologies do not require worker entry inside the pipe.

Manhole: A structure that allows access to the sewer system.

Manual inspection: Method of sewer inspection that usually involves physical entry and hands-on examination.

Measurement while drilling (MWD): Borehole survey instrumentation that provides continuous information simultaneously with drilling operations, usually transmitting to a display at or near the horizontal directional drilling rig.

Mechanical cleaning: Methods used to clean sewer lines of debris mechanically with devices such as rodding machines; bucket machines winch-pulled brushes, and so on.

Microtunnel boring machine (MTBM): See microtunneling.

Microtunneling: A trenchless construction method for installing pipelines. Microtunneling uses all of the following features during construction: (1) Remote controlled—The microtunneling boring machine (MTBM) is operated from a control panel, normally located on the surface. The system simultaneously installs pipe as spoil is excavated and removed. Personnel entry is not required for routine operation. (2) Guided—The guidance system usually references a laser beam projected onto a target in the MTBM, capable of installing gravity sewers or other types of pipelines to the required tolerance, for line and grade. (3) Pipe jacked—The process of constructing a pipeline by consecutively pushing pipes and MTBM through the ground using a jacking system for thrust. (4) Continuously supported—Continuous pressure is provided to the face of the excavation to balance groundwater and earth pressures.

Midi-rig: Steerable surface-launched horizontal directional drilling equipment for installation of pipes, conduits, and cables. Applied to intermediate sized drilling rigs used as either a small directional drilling machine or a large guided boring machine. Tracking of the drill string may be achieved by either a downhole survey tool or a walk-over locator.

Million gallons per day (mgd): The U.S. customary unit for flow measurement. One MGD equals 43.81 liters per second.

Mini-horizontal directional drilling (Mini-HDD): Small-diameter horizontal directional drilling. In Europe, it is called guided boring.

Mixed face: A soil condition that presents two or more different types of material in the path of the bore.

Modified sliplining (MSL): A range of techniques in which the liner is reduced in cross-sectional diameter before insertion into the carrier pipe. It is subsequently restored close to its original diameter, generally forming a close-fit with the original pipe. There are different methods of cross-sectional area reduction. See Chap. 15.

N

Nonworker-entry: Size of pipe, duct, or bore less than that for worker-entry.

O

Obstruction: Any natural or man-made object that lies on the path of bore or tunnel and has a potential to stop the boring or tunneling operation.

Open-cut: The method by which access is gained to the required level underground for the installation, repair, or replacement of a pipe, conduit, or cable. The excavation is then backfilled and the surface restored. See conventional trenching.

Open face shield: Shield in which manual excavation is carried out from within a steel tube at the front of a pipe jacking operation.

Outfall: An outlet to a sewer system.

Ovality: There are two definitions: (a) the difference between the maximum and mean diameter divided by the mean diameter, and (b) the difference of the mean and minimum divided by the mean, at any one cross section of a pipe, generally expressed as a percentage.

Overcut: The annular space between the excavated borehole and the outside diameter of the pipe.

Overflow: The excess water that flows over the ordinary limits of a sewer, manhole, or containment structure.

P

Packer: See compression ring.

PE: Polyethylene; a form of thermoplastic pipe.

pH: A measure of the acidity or alkalinity of a solution. A value of seven is neutral; lower numbers indicate more acidity.

Physical pipe inspection: The crawling or walking through manually accessible pipe lines. The logs for physical pipe inspection record information of the kind detailed under television inspection. Manual inspection is only undertaken when field conditions permit this to be done safely. Precautions are necessary.

Piercing tool: Similar to closed-face pipe ramming but for small-diameter (2 to 6 in) boring used for cable installations under roadways.

Pilot bore: The action of creating the first (usually steerable) pass of any boring process, which later requires back reaming or similar enlarging process to install the product pipe. Most commonly applied to horizontal directional drilling but also used in pilot tube microtunneling and guided auger boring systems.

Pilot tube method: A multistage method of accurately installing a product pipe by use of a guided pilot tube and followed by upsizing to install the product pipe.

Pipe bursting: A pipe replacement method for breaking the existing pipe by brittle fracture, using force from within, applied mechanically, the remains being forced into the surrounding ground. At the same time a new pipe, of the same or larger diameter, is drawn in behind the bursting tool. The pipe-bursting device may be based on an impact moling tool to exert diverted forward thrust to the radial bursting effect required, or by a hydraulic device inserted into the pipe and expanded to exert direct radial force or a static hammer. For new pipe, generally a HDPE pipe is used, but currently PVC, ductile iron, clay, and GRP is also used. Also known as pipe cracking and pipe splitting.

Pipe displacement: See pipe bursting.

Pipe eating: A pipe replacement technique, usually conducted by use of a horizontal directional drilling rig, in which a defective pipe is excavated during the backreaming operation. Also a microtunneling machines can be used where the existing pipe is excavated together with the surrounding soil as for a new installation. The microtunneling shield machine will usually need some crushing capability to perform effectively. The defective pipe may be filled with grout to improve steering performance.

Pipe jacking: A system of directly installing pipes behind a shield machine by hydraulic jacking from a drive shaft such that the pipes form a continuous string in the ground.

Pipe joint sealing: A method of sealing leaking or defective pipe joints that permit infiltration of groundwater into sewers by means of injecting chemical grout into and/or through the joints from within the pipe.

Pipe lubricant: See lubrication.

Pipe pusher: A machine that pushes or pulls a rod or pipe to produce a bore by means of compaction without rotation or impact.

Pipe ramming: A nonsteerable system of forming a bore by driving an open-ended steel casing using a percussive hammer from a drive pit. The soil may be removed from the casing by augering, jetting, or compressed air.

Pipe segment: A specific portion of the sewer or pipeline system; which usually runs between two structures (e.g., manhole, trap tanks, sumps); identified with unique sewer or pipe structure ID number.

Pipe splitting: Replacement method for breaking an existing pipe by longitudinal slitting. At the same time a new pipe of the same or larger diameter may be drawn in behind the splitting tool. See also pipe bursting.

Pipeline rehabilitation (PSR): See pipeline renewal.

Pipeline renewal: The in situ renewal of an existing pipeline, which has become deteriorated. The selection of appropriate renewal method is dependent on type of application and characteristics and types of defects of the existing pipe. See Chap. 5 for method selection criteria.

Plastic: Any of a variety of thermoplastic and thermoset material used in pipeline construction and renewal (e.g., polypropylene, PVC, fiberglass reinforced plastics, polyester felt reinforced pipe, epoxy and polyester mortars, and so on).

Point source repair (PSR): See localized repair.

Pointing: Method of repairing a brick sewer or manhole by the application of cement mortar where mortar loss has occurred.

Polyester: Resin formed by condensation of polybasic and monobasic acids with polyhydric alcohols.

Polyethylene (PE): A ductile, durable, virtually inert thermoplastic composed by polymers of ethylene. It is normally a translucent, tough solid. In pipe grade resins, ethylene-hexene copolymers are usually specified with carbon black pigment for weatherability.

Polyolefin: A family of plastic material used to make pipes.

Polypropylene (PP): A type of plastic pipe from the polyolefin family.

Potholing: Digging a vertical hole to visually locate a utility.

Preparatory cleaning: Internal cleaning of pipelines, particularly sewers, prior to inspection, usually with water jetting and removal of material where appropriate.

Preventative maintenance: Routine maintenance designed to prevent pipeline system failures and resulting emergency repairs.

Product pipe: Permanent pipeline for operational use. Pipe for conveyance for water, gas, sewage, and other products.

Protruding: To be projecting outward.

psi: Pounds per square inch. The U.S. customary unit for pressure. One psi equals 6.895 kiloNewtons per square meter.

Pull-back force: The tensile load applied to a drill string during the pull back process. Horizontal directional drilling rigs are generally rated by their maximum pull-back force.

PVC: Polyvinyl chloride; a form of thermoplastic pipe.

R

Radian: An arc of a circle equal in length to its radius; or the angle at the center measured by the arc.

Ramming: A percussive hammer is attached to an open-end casing, which is driven through the ground. See pipe ramming.

Receiving pit: (1) See exit pit. (2) An opening in the earth located at the expected exit of the cutting head or TBM. (3) The pit that is dug at the end of the bore, opposite the jacking pit.

Receiving shaft: See reception shaft.

Reception or exit shaft or pit: Excavation into which trenchless technology equipment is driven and recovered following the installation of the product pipe, conduit, or cable. See receiving pit.

Rehabilitation: See renewal.

Reinstatement: Method of backfilling, compaction, and resurfacing of any excavation to restore the surface and underlying structure to enable it to perform its original function.

Remote-control system (microtunneling): The remote-control system monitors and controls the MBTM, the automated transport system, and the guidance system from a location not in the MTBM.

Renewal: All aspects of upgrading with a new design life for the performance of existing pipeline systems. Includes rehabilitation, renovation, and replacement.

Renovation: See renewal.

Repair: Reconstruction of short pipe lengths, but not the reconstruction of a whole pipeline. Therefore, a new design life is not provided. In contrast, in pipeline renewal, a new design life is provided to existing pipeline system.

Replacement: See renewal.

Rerounding: A preparatory process, which involves the insertion of an expansion device into a distorted pipe to return it to a circular cross section. This is usually carried out prior to the insertion of a permanent liner or supporting band.

Resin impregnation (wet-out): A process used in cured-in-place pipe installation process where a plastic coated fabric tube is uniformly saturated with a liquid thermosetting resin while air is removed from the coated tube by means of vacuum suction.

Resin injection: The localized repair of pipes, usually sewers, by injection of a resin formulation into cracks or cavities, which subsequently cures to prevent leakage and further deterioration. It may also increase the structural strength of the pipeline.

Resins: An organic polymer, solid or liquid; usually thermoplastic or thermosetting.

Restoration: The backfilling, compaction, and resurfacing of any excavation to restore the surface and underlying structure to enable it to perform its original function.

Retract: The motion retracting TBM away from the cutting face.

Reverse: In horizontal auger boring, the counterclockwise rotation of the auger as viewed from the machine end.

RII: Rainfall induced infiltration; is a particular form of infiltration, which is similar to stormwater inflow. RII generally occurs during and immediately after rainfall events.

Ring compression: The principal stress in a confined thin circular ring subjected to external pressure.

Riser: A thin ring located between the frame and cone of a manhole. Used to bring the frame and cover final grade. Commonly, several risers are present in a manhole.

Robot: Remote-control device with CCTV monitoring, used mainly in localized repair work, such as cutting away obstructions, reopening lateral connections, grinding and refilling defective areas, and injecting resin into cracks and cavities.

Rod pushing: Method of forming a pilot bore by driving a closed pipe head with rigid attachment from a launch pit into the soil, which is displaced. See thrust boring.

Roller cone bit or reamer: A bit or reamer in which the teeth rotate on separate, internal shafts that are usually aligned perpendicular to line. Used for boring rock.

Rotary rod machine: A machine used to drill earth horizontally by means of a cutting head attached to a rotating rod (not an auger). Such drilling may include fluid injected to the cutting head through a hollow rod.

RPM: Reinforced plastic mortar; a form of thermoset plastic pipe.

RTR: Reinforced thermosetting resin; a form of thermoset plastic pipe within the GRP family.

S

Saddle: In horizontal auger boring, a vertical support mechanism to hold the casing in position while starting (collaring) the bore.

SBR (styrene butadiene): Type of rubber that has good abrasion resistance and excellent impact and cut-and-gouge resistance. Used as gasket material.

SDR: See standard dimension ratio.

Seal: A watertight bond.

Sediment: Particles that settles on the pipe invert, causing a reduction in cross-sectional area.

Seepage: Water escaping through or emerging from the ground along rather extensive line or surface, as contrasted with a spring, the water of which emerges from a single spot.

Segmental concrete tunnel liner: Used the same way as liner plate except that they are tunnel liners made of concrete.

Segmental lining: See segmental sliplining.

Segmental sliplining: See sliplining.

Self-cleansing: A consequence of good hydraulic design when the pipe invert is kept relatively free of sediments by ensuring adequate flow velocities.

Semi-structural liner: A liner that in its own entity does not have the required strength to withstand internal, external, or both types of loading from soil column, traffic, and groundwater pressure for the design life of the product, but will offer some level of structural support against internal pressure.

Separate system: A system that uses sanitary sewers to convey the wastewater and stormwater sewers to carry the stormwater.

Sewage: Wastewater transported in a sewer.

Sewer: An underground pipe or conduit for transporting stormwater, or wastewater, or both.

Sewer cleaning: The use of mechanical or hydraulic equipment to dislodge, transport, and remove debris from sewer lines.

Sewer interceptor: A sewer, which receives the flow from collector sewers and conveys the wastewaters to treatment facilities.

Sewer lateral: A building sewer (sometimes referred to as a sewer lateral or house lateral) is the pipeline between the public sanitary sewer line, which is usually located in the street, and the indoor plumbing. See building sewer.

Sewer pipe: A length of conduit, manufactured from various materials and in various lengths, that when joined together can be used to transport wastewaters from the points of origin to a treatment facility. Types of pipe are acrylonitrile-butadiene-styrene (ABS); asbestos-cement (AC); brick pipe (BP); concrete pipe (CP); cast iron pipe (CIP); polyethylene (PE); polyvinylchloride (PVC); and vitrified clay pipe (VCP).

Skin friction: Resistance to advancement caused by soil pressure around the pipe or casing.

Sleeve pipe: A pipe installed as external protection to a product pipe.

Slipline: A renewal technique covering the insertion of one pipe inside an existing pipe.

Sliplining (SL): (1) General term used to describe methods of lining with continuous pipes and lining with discrete pipes. (2) Insertion of a new pipe by pulling or pushing it into the existing pipe and grouting the annular space. The pipe used may be continuous or a string of discrete pipes. The latter is also referred to as segmental sliplining.

Slurry chamber: Located behind the cutting head of a slurry microtunneling machine. Excavated material is mixed with slurry in the chamber for transport to the surface.

Slurry line: A series of hoses or pipes that transport tunnel muck and slurry from the face of a slurry microtunneling machine to the ground surface for separation.

Slurry separation: A process where excavated material is separated from the circulation slurry.

Slurry shield method: Method using a mechanical tunneling shield with closed face, which employs hydraulic means for removing the excavated material and balances the groundwater pressure. See also earth pressure balance machine.

Slurry: A fluid, mainly water mixed with bentonite and sometimes polymers, used in a closed loop system for the removal of spoil and for the balance of groundwater pressure during tunneling and microtunneling operations.

Social costs: Costs incurred by society as a result of underground pipeline construction and renewal. These include but not limited to traffic disruptions, environmental damages, safety hazards, inconvenience to general public, and business losses owing to road closures. See Chap. 2.

Soft lining: See cured-in-place pipe (CIPP) or lining with CIPP.

Sonde housing: The horizontal directional drilling head, which houses the transmitter (sonde) radio-sending unit.

Specific gravity: The density of a material divided by the density of water usually at 4°C. As the density of water is nearly 1 g/cm^3, density in g/cm^3 and specific gravity are numerically nearly equal.

Spiral lining: A technique in which a ribbed plastic strip is spirally wound by a winding machine to form a liner, which is inserted into a defective pipeline. The annular space may be grouted or the spiral liner expanded to reduce the annulus and form a close-fit liner. In larger diameters, the strips are sometimes formed into panels and installed by hand. Grouting the annular space after installation is recommended.

Spiral weld pipe (casing): Pipe made from coils of steel plate by wrapping around a mandrail in such a manner that the welds are a spiral helix.

Split design: A boring machine having the capability of being broken down into two or more elements to reduce the lifting weight.

Spoil (muck): Earth, rock, and other materials displaced by a tunnel, pipe or casing, and removed as the tunnel, pipe, or casing is installed. In some cases, it is used to mean only the material that has no further use.

Spot repair: See localized repair.

Spray lining: A technique for applying a lining of cement mortar or resin by rotating a spray head, which is winched through the existing pipeline.

Springline: (1) An imaginary horizontal line across the pipe that passes between the points where the pipe has its greatest cross-sectional width. (2) Midpoint of a pipe cross section (equal vertical distance between the crown and the invert of the pipe).

SSES Sewer system evaluation survey: Mainly for I/I surveys to determine the degree and location of flows entering the collection system.

Stabilization: See chemical stabilization.

Standard dimension ratio (SDR): Defined as the ratio of the outside pipe diameter to wall thickness. Same as DR.

Steerable moling: Method similar to impact moling with a limited steering capability

Steering head: In horizontal auger boring, a moveable lead section of casing that can be adjusted to steer the bore.

Subsidence: The settlement of the ground, pipeline, or other structure. The effects may not be evenly distributed and/or immediately noticeable. Differential settlement may occur.

Sump: A depression usually in the drive pit to allow the collection of water and the installation of a sump pump for water removal.

Swageing: The reduction in diameter of a polyethylene pipe by passing it through one or more dies. The die may be heated if necessary.

Swagelining: A method of sliplining whereby the diameter of the PE pipe is temporarily reduced by swageing prior to insertion in the defective pipe. After insertion, the pipe is expanded by means of steam or a rerounding device.

Swivel: In horizontal directional drilling, it is used to attach product pipe (to be pulled into drilled hole) to drill pipe to prevent it from rotating.

T

Target shaft or/pit: See reception or exit shaft or pit.

TBM: See tunnel boring machine.

Teeth: See bits.

Televise: Process by which a sewer or pipeline or lateral is inspected with a closed-circuit television camera.

Thermoplastic (TP): A polymer material, such as polyethylene, that will repeatedly soften when heated and harden and reformed when cooled. TPs are generally much easier to recycle than their thermoset (see below) counterparts.

Thermoset (TS): A polymer material, such as rubber, that does not melt when reheated. TS polymers can be formed initially into almost any desired shape, but they cannot be reformed at a later time.

Thrust block: See backstop.

Thrust boring: A method of forming a pilot bore by driving a closed pipe or head from a thrust pit into the soil which is displaced. Some small-diameter models have steering capability achieved by a slanted pilot-head face and electronic monitoring. Back reaming may be used to enlarge the pilot bore. Also loosely applied to various trenchless installations methods. See rod pushing.

Thrust jacking method: Method in which a pipe is jacked through the ground without mechanical excavation of material from the front of the pipeline.

Thrust pit: See drive pit.

Thrust ring: A fabricated ring that is mounted on the face of the jacking frame. It is intended to transfer the jacking load from the jacking frame to the thrust bearing area of the pipe section being jacked.

Thrust: Force applied to a pipeline or drill string to propel it through the ground.

Torque: The rotary force available at the drive chuck.

Track: A set of longitudinal rails mounted on cross members that support and guide a horizontal auger boring machine.

Trenching: See open-cut or conventional trenching.

Trenchless technology (TT): Also no-dig, techniques for underground pipeline and utility construction and replacement, rehabilitation, renovation (collectively called renewal), repair, inspection, leak detection, and so on, with minimum or no excavation from the ground surface.

Trunk sewer: A sewer into which at least two-branch sanitary, combined, or storm water sewers connect. It conveys the flow to the interceptor sewer. The trunk sewer is the longest connection sewer in any drainage basin. Also sometimes known as a *main* sewer.

Tuberculation: Localized corrosion at scattered locations resulting in knob-like mounds.

Tunnel: An underground conduit, often deep and expensive to construct, which provides conveyance and/or storage volumes for wastewater, often involving minimal surface disruption.

Tunnel boring machine (TBM): (1) A full-face circular mechanized shield machine, usually of worker-entry diameter, steerable, and with a rotary cutting head. For pipe jacking installation it leads a string of pipes. It may be controlled from within the shield or remotely such as in microtunneling. (2) A mechanical excavator used in a tunnel to excavate the front face of the tunnel (mole, tunneling head).

U

Uncased bore: Any bore without a lining or pipe inserted, that is, self-supporting, whether temporary or permanent. Not recommended except in special conditions.

Underground utility: Active or inactive services or utilities below ground level.

Upset: The inadvertent action of a horizontal auger boring machine that rotates the machine and track from its normal and upright position to another position.

Upsizing: Any method such as pipe replacement or pipe bursting that increases the cross-sectional area of an existing pipeline by replacing with a larger-diameter pipe.

Utility corridor: Duct in which two or more different utility services are installed with access for maintenance.

V

VCP: Vitrified clay pipe.

VCT: Vitrified clay tile or vitrified clay tile pipe.

Velocity head: For water moving at a given velocity, the equivalent head through which it would have to fall by gravity to acquire the same velocity.

Voids: (1) Holes on the outside of the pipe in the surrounding soil or material. (2) A term generally applied to paints to describe holidays, holes, and skips in the film. Also used to describe shrinkage in castings or welds.

W

Walkover system: See locator.

Washover pipe: In horizontal directional drilling, a rotating drill pipe of larger diameter than the pilot drill pipe and placed around it with its leading edge less far advanced. Its purpose is to provide stiffness to the drilling pipe to maintain steering control over long bores, to reduce friction between the drill string and the soil and to facilitate mud circulation. See Chap. 10 for horizontal directional drilling method.

Water jetting: (1) Method for the internal cleansing of pipelines using high-pressure water jets. (2) An obsolete method for cutting earth with water jetting.

Water table: (1) The depth of the groundwater. (2) The upper limit of the portion of ground wholly saturated with water.

Waterline: Maximum liquid level in a sewer pipe or structure during normal operating cycles.

Weatherability: The properties of a plastic material that allows it to withstand natural weathering; hot and cold temperatures, wind, rain, and ultraviolet rays.

Wet-out: The process of injecting resin into, and distributing it throughout, a hose or tube, which will then be installed into the pipeline and cured in place.

Winch: Mechanical device used to pull the CCTV cameras or cleaning tools through a pipe.

Wing cutters: In horizontal auger boring, appendages on cutting heads that will open to increase the cutting diameter of the head when turned in a forward direction, and close when turned in a reverse direction. They are used to cut clearance for the casing pipe.

Wrapped casing: A coating on pipe for protection from corrosion usually composed of asphalt and asphalt-coated paper. Some coatings may contain plastic, fiberglass, coal tar, or other materials.

ACRONYMS AND ABBREVIATIONS

Acronym	Description
ABS	Acrylonitrile-butadiene-styrene
AC	Asbestos cement
AMP	Asset management planning
CBS	Controlled boring system
CCFRPM	Centrifugally cast fiberglass reinforced polymer
CCTV	Closed-circuit television
CFM	Cubic feet per minute
CFP	Close-fit pipe
CI	Cast iron
CIP	Cast iron pipe
CIPP	Cured-in-place pipe
CM	Compaction methods
CMOM	Capacity, management, operation, and maintenance
DIP	Ductile iron pipe
D&R	Deformed and reformed
DWF	Dry weather flow
DWI	Dry weather inflow
EPB	Earth pressure balance
EPDM	Ethylene propylene diene monomer
ER	Epoxy resin
ERL	Estimated remaining life
ERW	Electrical resistance welding
F&E	Fused and expanded
FF	Fold and formed pipe
F&F	Fold and formed
FIPP	Formed-in-place pipe
FLCS	Full line chemical stabilization
FRPP	Fiberglass reinforced polyester
GBM	Guided boring method (Horizontal Directional Drilling)
GBR	(ASCE) Geotechnical baseline report

GIS	Geographical information system
GN	Gunite
GPM	Gallons per minute
GPR	Ground penetrating radar
GRC	Glassfiber reinforced cement
GRP	Glassfiber reinforced polyester
GWI	Groundwater infiltration
HAB	Horizontal auger boring
HDD	Horizontal directional drilling
HDPE	High density polyethylene
HEB	Horizontal earth boring
I&I	Infiltration and inflow
I/I	Infiltration/inflow
IJS	Intermediate jacking station
ILR	In-line replacement
J/CS	Joint/crack sealing
L-CIPP	Lateral-cured-in-place pipe
LCP	Lining with continuous pipe
LOR	Localized repair
LOR-CIPP	Localized repair–cured-in-place pipe
LOR-GR	Localized repair–grouting
LOR-IS	Localized repair–internal seal
LOR-RR	Localized repair–robotic repair
L-PB	Lateral-pipe bursting
LR	Lateral renewal
LSP	Lining (with) segmental pipe
LSWP	Lining with spirally wound pipe
L-ThP	Lateral-thermoformed pipe
L-UCL	Lateral-underground coatings and lining
MDPE	Medium density polyethylene
MFP	Mechanically folded pipe
MS	Mechanical sleeves
MSL	Modified sliplining
MTBM	Microtunnel boring machines
MTM	Manhole-to-manhole
NBR	Nitrile butadiene rubber
OC	Open-cut
ORM	Operation, reliability, and maintenance
PB	Pipe bursting
PB-HY	Pipe bursting-hydraulic

PB-PE	Pipe bursting–pneumatic
PB-PS	Pipe bursting–pipe splitting
PB-ST	Pipe bursting–static
PCCP	Prestressed concrete cylinder pipe
PCP	Polymer concrete pipe
PE	Polyethylene
PFA	Pulverized fuel ash
PJ	Pipe jacking
PL	Panel lining
PP	Polypropylene
PR	Pipe removal
PRC	Polyester resin concrete
PR-HAB	Pipe removal–horizontal auger boring
PR-MT	Pipe removal–microtunneling
PR-PR	Pipe removal–pipe reaming
PR-T	Pipe removal–tunneling
PR-PX	Pipe removal–pipe extraction
PSR	Point source repair (same as localized repair)
PTMT	Pilot tube microtunneling
PVC	Poly-vinyl-chloride
PVDF	Poly-vinylidene chloride
PVDM	Polyvinylidene difluoride membranes
QA/QC	Quality assurance/quality control
RCP	Reinforced concrete pipe
RDP	Reduced diameter pipe
RII	Rainfall induced infiltration
RPM	Reinforced plastic mortar
RR	Robotic repairs (or robotic renewal)
RTR	Reinforced thermosetting resin
S/S	Stabilization/sealing
SBR	Styrene-butadiene rubber
SCS	Stress corrosion cracking
SDR	Standard dimension ratio
SH	Shotcrete
SL	Sliplining
SMR	Sewer manhole renewal
SOL	Spray on lining
SP	Steel pipe
SPT	Standard penetration test
SR	Spot repair (same as localized repair)

SRPC	Sulfate resisting portland cement
SRw/WL	Steel ribs with wooden lagging
SSBS	Sanitary sewer bypasses
SSC	Sulfide stress cracking
SSES	Sewer system evaluation survey
SSO	Sanitary sewer overflows
SUE	Subsurface utility engineering
SWP	Spiral wound pipe
TBM	Tunnel boring machine
TCM	Trenchless construction methods
ThP	Thermoformed pipe
TLP	Tunnel liner plates
TRM	Trenchless renewal methods
TT	Trenchless technology
UCL	Underground coatings and linings
UT	Utility tunneling
VOC	Volatile organic compounds
VCP	Vitrified clay pipe
WQS	Water quality standard
WWTP	Wastewater treatment plant

Organizations Related to Trenchless Technology

Acronym	Description	Website
AASHTO	American Association of State Highway Transportation Officials	http://transportation1.org/aashtonew
AEM	Association of Equipment Manufacturers	http://www.aem.org/
AGA	American Gas Association	http://www.aga.org
ANSI	American National Standards	http://www.ansi.org
APWA	American Public Works Association	http://www.apwa.net
ASA	American Shotcrete Association	http://www.shotcrete.org
ASCE	American Society of Civil Engineers	http://www.asce.org
ASTM	American Society of Testing and Materials	http://www.astm.org

ASTT	Australian Society of Trenchless Technology	http://www.astt.com.au
AWWA	American Water Works Association	http://www.awwa.org
BAMI	Buried Asset Management Institute	—
CATT	Center for Advancement of Trenchless Technologies, University of Waterloo, Canada	http://www.civil.uwaterloo.ca/catt
CIGMAT	Center for Innovative Grouting Materials and Technology	http://geml.cive.uh.edu/
CUIRE	Center for Underground Infrastructure Research and Education	http://www.msucuire.org
DCCA	Directional Crossing Contractors Association	http://www.dcca.org
DIPRA	Ductile Iron Pipe Research Association	http://www.dipra.org
EPA	Environmental Protection Agency	http://www.epa.gov
GRI	Gas Research Institute	http://www.gri.org
ICRI	International Concrete Repair Institute	http://www.icri.org/
ISTT	International Society of Trenchless Technology	http://www.istt.com
MSTT	Midwest Society for Trenchless Technology	http://www.mstt.org
NACE	National Association of Corrosion Engineers	http://www.nace.org/nace/index.asp
NASSCO	National Association of Sewer Service Companies	http://www.nassco.org
NASTT	North American Society for Trenchless Technology	http://www.nastt.org
NSF	National Science Foundation	http://www.nsf.gov/
NUCA	National Utility Contractors Association	http://www.nuca.com
OSHA	Occupational Safety and Health Administration	http://www.osha.gov
PCCA	Power and Communication Contractors Association	http://www.pccaweb.org
PPI	Plastic Pipe Institute	http://www.plasticpipe.org

PRC	Pipeline Research Council	http://www.prci.com
SESTT	Southeast Society for Trenchless Technology	http://www.sestt.org
SSPC	The Society of Protective Coatings	http://www.sspc.org/
TRB	Transportation Research Board	http://gulliver.trb.org
TTC	Trenchless Technology Center, Louisiana Tech University	http://www.latech.edu/tech/engr/ttc
WEF	Water Environment Federation	http://www.wef.org
WRC	Water Research Center	http://www.wrcplc.co.uk

CONVERSION TABLE

From	To	Multiply by
Linear		
mil	in	0.001
in	mm	25.4
ft	m	0.3048
yards	m	0.9144
ft	in	12
mile	km	1.609
mile	ft	5280
Area		
in^2	mm^2	645.16
ft^2	m^2	0.0929
sq yards	m^2	0.8361
sq mi	km^2	2.5889
acres	km^2	4.0469×10^{-3}
ha	km^2	0.01
acres	ft^2	43,560
m^2	ha	10^4
Volume		
in^3	cm^3	16.39
ft^3	m^3	0.02832
cu yards	m^3	0.7646
in^3	L	0.01639
Capacity		
British pt	L	0.568
U.S. pt	L	0.4731
British qt	L	1.136
U.S. qt	L	0.9463
British gal	L	4.546
U.S. gal	L	3.785
Mass and Weight		
oz	g	28.349
lb	kg	0.45359
tons	kg	1016.05
tons	t	1.016
t	tons	0.9842
Density (Mass/Unit Volume)		
g/cm^3	kg/m^3	1000
lb/ft^3	kg/m^3	16.018

From	To	Multiply by
Specific Weight (Weight/Unit Volume)		
lb/ft^3	N/m^3	157.1
lb/in^3	lb/ft^3	1728
Force		
lb	kN	0.004448
tons	kN	9.96401
kN	kg (f)	102.0
Velocity		
mi/h	km/h	1.609
ft/s	m/s	0.3048
ft/min	m/s	0.00508
ft/min	m/min	0.305
Volume Flow Rate		
ft^3/s	gal/min	449
m^3/s	ft^3/s	35.3
m^3/s	gal/min	15,850
gal/min	L/min	3.785
m^3/s	L/min	60,000
m^3/s	ft^3/min	2120
m^3/hr	L/min	16.67
ft^3/s	m^3/hr	101.9
Pressure		
psi	kN/m^2	6.895
psi	atm	0.0680
psi	kg/m^2	9.80665
psi	Pa	6894.757
psi	lb/ft^2	144
N/m^2	Pa	1
atm	Psi	14.696
atm	Pa	101.325
bar	Pa	10^5
bar	psi	14.5

Temperature Conversion

C to F first deduct 32, multiply by 5 then divide by 9

F to C multiply by 9, divide by 5, add 32

APPLICABLE STANDARDS

STANDARD SPECIFICATIONS FOR TCM

Standard Pipe Specifications

Plastic Pipe

- ASTM D1248 Polyethylene Plastics Molding and Extrusion Materials
- ASTM D3350 Polyethylene Plastics Pipe and Fittings Materials
- ASTM D2657 Heat-Joining Polyolefin Pipe and Fittings
- ASTM D2683 Socket-Type Polyethylene Fittings for Outside Diameter Controlled
- ASTM D3034-00 Standard Specification for Type PSM Poly(Vinyl Chloride) (PVC) Sewer Pipe and Fittings
- ASTM F1290 Electrofusion Joining Polyolefin Pipe and Fittings
- ASTM F1336-02 Standard Specification for Poly(Vinyl Chloride) (PVC) Gasketed Sewer Fittings
- ASTM F1732-96 Standard Specification for Poly(Vinyl Chloride) (PVC) Sewer and Drain Pipe Containing Recycled PVC Material
- ASTM F1901 Polyethylene (PE) Pipe and Fittings for Roof Drain Systems
- ASTM F412 Terminology Relating to Plastic Piping Systems
- ASTM F480 Thermoplastic Well Casing Pipe and Couplings Made in Standard
- ASTM F679-01 Standard Specification for Poly(Vinyl Chloride) (PVC) Large-Diameter Plastic Gravity Sewer Pipe and Fittings
- ASTM F789-95a Standard Specification for Type PS-46 and Type PS-115 Poly(Vinyl Chloride) (PVC) Plastic Gravity Flow Sewer Pipe and Fittings
- ASTM F794-01 Standard Specification for Poly(Vinyl Chloride) (PVC) Profile Gravity Sewer Pipe and Fittings Based on Controlled Inside Diameter
- ASTM F949-01a Standard Specification for Poly(Vinyl Chloride) (PVC) Corrugated Sewer Pipe With a Smooth Interior and Fittings
- AWWA C900-97 Polyvinyl Chloride (PVC) Pressure Pipe, and Fabricated Fittings, 4–12 in (100–300 mm), for Water Dist
- AWWA C903-02 Polyethylene-Aluminum-Polyethylene and Cross-Linked Polyethylene Composite Pressure

- AWWA C905-97 Polyvinyl Chloride (PVC) Pressure Pipe and Fabricated Fittings, 14–48 in (350–1,200 mm)
- AWWA C906-99 Polyethylene (PE) Pressure Pipe and Fittings, 4–63 in (100–1,575 mm), for Water Dist and Trans
- AWWA C907-91 Polyvinyl Chloride (PVC) Pressure Fittings for Water—4–8 in (100–200 mm)
- AWWA C908-01 PVC Self-Tapping Saddle Tees for Use on PVC Pipe
- AWWA C909-02 Molecularly Oriented Polyvinyl Chloride (PVCO) Pressure Pipe, 4–12 in (100–300 mm), for Water Dist
- AWWA C950-01 Fiberglass Pressure Pipe
- ASTM F1962-99 Standard Guide for Use of Maxi-Horizontal Directional Drilling for Placement of Polyethylene Pipe or Conduit under Obstacles, Including River Crossings
- ASTM D2321-00 Standard Practice for Underground Installation of Thermoplastic Pipe for Sewers and Other Gravity-Flow Applications

Ductile-Iron Pipe and Fittings

- AWWA C150/A21.50-02 ANSI Standard for Thickness Design of Ductile-Iron Pipe
- AWWA C110/A21.10-98 ANSI Standard for Ductile-Iron and Gray-Iron Fittings, 3–48 in (76–1, 219 mm), for Water
- AWWA C111/A21.11-00 ANSI Standard for Rubber-Gasket Joints for Ductile-Iron Pressure Pipe and Fittings
- AWWA C115/A21.15-99 ANSI Standard for Flanged Ductile-Iron Pipe with Ductile-Iron or Gray-Iron Threaded Flanges
- AWWA C151/A21.51-02 ANSI Standard for Ductile-Iron Pipe, Centrifugally Cast, for Water
- AWWA C200-97 Steel Water Pipe 6 in (150 mm) and Larger
- AWWA C207-01 Steel Pipe Flanges for Waterworks Service-Sizes 4–144 in (100–3,600 mm)
- AWWA C208-01 Dimensions for Fabricated Steel Water Pipe Fittings
- AWWA C219-01 Bolted, Sleeve-Type Couplings for Plain-End Pipe
- AWWA C220-98 Stainless-Steel Pipe, 4 in (100 mm) and Larger (Includes addendum C220a-99)
- AWWA C221-01 Fabricated Steel Mechanical Slip-Type Expansion Joints
- ASTM A53-97 Standard Specification for Pipe, Steel, Black and Hot-Dipped, Zinc-Coated, Welded and Seamless (1997)
- ASTM A106 Standard Specification for Seamless Carbon Steel Pipe for High-Temperature Service
- ASTM A139-96 Standard Specifications for Electric-Fusion (Arc)-Welded Steel Pipe (NPS 4 and over)
- ASTM A252 Standard Specification for Welded and Seamless Steel Pipe Pile
- ASTM A500-96 Standard Specification for Cold-Formed Welded and Seamless Carbon Steel Structural Tubing in Rounds and Shapes
- ASTM A716-95 Standard Specification for Ductile Iron Culvert Pipe
- ASTM A746-95 Standard Specification for Ductile Iron Gravity Sewer Pipe

Concrete Pipe

- ASTM C14/C14M-95: Standard Specification for Concrete Sewer, Storm Drain, and Culvert Pipe. ASTM Vol. 4.05
- ASTM C76-03 Standard Specification for Reinforced Concrete Culvert, Storm Drain, and Sewer Pipe
- ASTM C361-03ae1 Standard Specification for Reinforced Concrete Low-Head Pressure Pipe
- ASTM C443-03 Standard Specification for Joints for Concrete Pipe and Manholes, Using Rubber Gaskets
- ASTM C478-03a Standard Specification for Precast Reinforced Concrete Manhole Sections
- ASTM C497-03a Standard Test Methods for Concrete Pipe, Manhole Sections, or Tile
- ASTM C655-02 Standard Specification for Reinforced Concrete D-Load Culvert, Storm Drain, and Sewer Pipe
- ASTM C822-03 Standard Terminology Relating to Concrete Pipe and Related Products
- ASTM C923-02 Standard Specification for Resilient Connectors between Reinforced Concrete Manhole Structures, Pipes, and Laterals
- ASTM C924-02 Standard Practice for Testing Concrete Pipe Sewer Lines by Low-Pressure Air Test Method
- ASTM C969-02 Standard Practice for Infiltration and Exfiltration Acceptance Testing of Installed Precast Concrete Pipe Sewer Lines
- ASTM C1103-03 Standard Practice for Joint Acceptance Testing of Installed Precast Concrete Pipe Sewer Lines
- AWWA C300-97 Reinforced Concrete Pressure Pipe, Steel-Cylinder Type
- AWWA C301-99 Prestressed Concrete Pressure Pipe, Steel-Cylinder Type
- AWWA C302-95 Reinforced Concrete Pressure Pipe, Noncylinder Type
- AWWA C303-02 Concrete Pressure Pipe, Bar-Wrapped, Steel-Cylinder Type
- AWWA C304-99 Design of Prestressed Concrete Cylinder Pipe

Asbestos-Cement Pipe

- AWWA C400-93 (R98) Asbestos-Cement Pressure Pipe, 4–6 in (100–400 mm), for Water Dist and Trans
- AWWA C401-93 (R98) Selection of Asbestos-Cement Pressure Pipe, 4–16 in (100–400 mm), for Water Dist Sys
- AWWA C402-00 Asbestos-Cement Transmission Pipe, 18–42 in (450–1,050 mm), for Water Supply Service
- AWWA C403-00 The Selection of Asbestos-Cement Transmission Pipe, Sizes 18–42 in (450–1,050 mm)
- ASTM C301-98 (2003) Standard Test Methods for Vitrified Clay Pipe
- ASTM C425-04 Standard Specification for Compression Joints for Vitrified Clay Pipe and Fittings
- ASTM C828-03 Standard Test Method for Low-Pressure Air Test of Vitrified Clay Pipe Lines

- ASTM C1091-03a Standard Test Method for Hydrostatic Infiltration Testing of Vitrified Clay Pipe Lines

Standard Test Methods for Pipes

- ASTM D1598-02 Standard Test Method for Time-to-Failure of Plastic Pipe under Constant Internal Pressure
- ASTM D1599-99 Standard Test Method for Resistance to Short-Time Hydraulic Failure Pressure of Plastic Pipe, Tubing, and Fittings
- ASTM D2122-98 Standard Test Method for Determining Dimensions of Thermoplastic Pipe and Fittings
- ASTM D2290-00 Standard Test Method for Apparent Hoop Tensile Strength of Plastic or Reinforced Plastic Pipe by Split Disk Method
- ASTM D2412-02 Standard Test Method for Determination of External Loading Characteristics of Plastic Pipe by Parallel-Plate Loading
- ASTM D2837-01ae1 Standard Test Method for Obtaining Hydrostatic Design Basis for Thermoplastic Pipe Materials
- ASTM F1417-92 (1998) Standard Test Method for Installation Acceptance of Plastic Gravity Sewer Lines Using Low-Pressure Air
- ASTM F1473 Notch Tensile Test to Measure the Resistance to Slow Crack Growth in PE pipes and resins
- ASTM F948 Time-to-Failure of Plastic Piping systems and Components Under constant Internal Pressure with Flow
- ASTM F1804-97 Practice for Determining Allowable Tensile Load for Polyethylene (PE) Gas Pipe during Pull-In Installation

Pipe Installation

- AWWA C600-99 Installation of Ductile-Iron Water Mains and Their Appurtenances
- AWWA C602-00 Cement-Mortar Lining of Water Pipelines in Place 4 in (100 mm) and Larger
- AWWA C603-96 (R00) Installation of Asbestos Cement Pressure Pipe
- AWWA C605-94 Underground Installation of Polyvinyl Chloride (PVC) Pressure Pipe and Fittings for Water
- AWWA C606-97 Grooved and Shouldered Joints
- AWWA C206-97 Field Welding of Steel Water Pipe

Valves and Hydrants

- AWWA C500-02 Metal-Seated Gate Valves for Water Supply Service (Includes addendum C500a-95)
- AWWA C502-94 Dry-Barrel Fire Hydrants
- AWWA C503-97 Wet-Barrel Fire Hydrants
- AWWA C504-00 Rubber-Seated Butterfly Valves
- AWWA C507-99 Ball Valves, 6–48 in (150–1,200 mm)
- AWWA C508-01 Swing-Check Valves for Waterworks Service, 2–24 in (50–600 mm) NPS
- AWWA C509-01 Resilient-Seated Gate Valves for Water Supply Service

- AWWA C510-97 Double Check Valve Backflow Prevention Assembly
- AWWA C511-97 Reduced-Pressure Principle Backflow Prevention Assembly
- AWWA C512-99 Air Release, Air/Vacuum, and Combination Air Valves for Waterworks Service
- AWWA C513-97 Open-Channel, Fabricated-Metal Slide Gates
- AWWA C515-01 Reduced-Wall, Resilient-Seated Gate Valves for Water Supply Service
- AWWA C540-02 Power-Actuating Devices for Valves and Slide Gates
- AWWA C550-01 Protective Epoxy Interior Coatings for Valves and Hydrants
- AWWA C560-00 Cast-Iron Slide Gates

Service Lines

- AWWA C800-01 Underground Service Line Valves and Fittings

STANDARD SPECIFICATIONS FOR TRM

Cured-In-Place Pipe (CIPP)

- ASTM F1216-98 Standard Practice for Rehabilitation of Existing Pipelines and Conduits by the Inversion and Curing of a Resin
- ASTM F1743-96 Standard Practice for Rehabilitation of Existing Pipelines and Conduits by Pulled-In-Place Installation of Cured-In-Place Thermosetting Resin Pipe (CIPP)
- ASTM D5813-95 Standard Specification for Cured-In-Place Thermosetting Resin Sewer Pipe
- ASTM F2207-02 Standard Specification for Cured-In-Place Pipe Lining System for Rehabilitation of Metallic Gas Pipe
- ASTM F2019-00 Standard Practice for Rehabilitation of Existing Pipelines and Conduits by the Pulled in Place Installation of Glass Reinforced Plastic (GRP) Cured-In-Place Thermosetting Resin Pipe (CIPP)

Plastic Pipe

- ASTM F1504-02 Standard Specification for Folded Poly(Vinyl Chloride) (PVC) Pipe for Existing Sewer and Conduit Rehabilitation
- ASTM F1533-97 Standard Specification for Deformed Polyethylene (PE) Liner
- ASTM F1697-02 Standard Specification for Poly(Vinyl Chloride) (PVC) Profile Strip for Machine Spiral-Wound Liner Pipe Rehabilitation of Existing Sewers and Conduits
- ASTM F1735-02 Standard Specification for Poly(Vinyl Chloride) (PVC) Profile Strip for PVC Liners for Rehabilitation of Existing Man-Entry Sewers and Conduits
- ASTM F1803-97 Standard Specification for Poly(Vinyl Chloride)(PVC) Closed Profile Gravity Pipe and Fittings Based on Controlled Inside Diameter
- ASTM F1606-95 Standard Practice for Rehabilitation of Existing Sewers and Conduits with Deformed Polyethylene (PE) Liner
- ASTM F1698-02 Standard Practice for Installation of Poly(Vinyl Chloride)(PVC) Profile Strip Liner and Cementitious Grout for Rehabilitation of Existing Man-Entry Sewers and Conduits

- ASTM F1741-01e1 Standard Practice for Installation of Machine Spiral Wound Poly(Vinyl Chloride) (PVC) Liner Pipe for Rehabilitation of Existing Sewers and Conduits
- ASTM F1741-02a Standard Practice for Installation of Machine Spiral Wound Poly(Vinyl Chloride) (PVC) Liner Pipe for Rehabilitation of Existing Sewers and Conduits
- ASTM F1867-98 Standard Practice for Installation of Folded/Formed Poly Vinyl Chloride (PVC) Pipe Type A for Existing Sewer and Conduit Rehabilitation
- ASTM F1871-02 Standard Specification for Folded/Formed Poly Vinyl Chloride Pipe Type A for Existing Sewer and Conduit Rehabilitation
- ASTM F1871-98 Standard Specification for Folder/Formed Poly Vinyl Chloride Pipe Type A for Existing Sewer and Conduit
- ASTM F1947-98 Standard Practice for Installation of Folded Poly Vinyl Chloride (PVC) Pipe into Existing Sewers and Conduits
- ASTM F585-94 (2000) Standard Practice for Insertion of Flexible Polyethylene Pipe into Existing Sewers
- ASTM D1785-04 Standard Specification for Poly(Vinyl Chloride) (PVC) Plastic Pipe, Schedules 40, 80, and 120
- ASTM D2104-03 Standard Specification for Polyethylene (PE) Plastic Pipe, Schedule 40
- ASTM D2239-03 Standard Specification for Polyethylene (PE) Plastic Pipe (SIDR-PR) Based on Controlled Inside Diameter
- ASTM D2241-04a Standard Specification for Poly(Vinyl Chloride) (PVC) Pressure-Rated Pipe (SDR Series)
- ASTM D2321-00 Standard Practice for Underground Installation of Thermoplastic Pipe for Sewers and Other Gravity-Flow Applications
- ASTM D2412-02 Standard Test Method for Determination of External Loading Characteristics of Plastic Pipe by Parallel-Plate Loading
- ASTM D2447-03 Standard Specification for Polyethylene (PE) Plastic Pipe, Schedules 40 and 80, Based on Outside Diameter
- ASTM D2683-98 Standard Specification for Socket-Type Polyethylene Fittings for Outside Diameter-Controlled Polyethylene Pipe and Tubing
- ASTM D2729-03 Standard Specification for Poly(Vinyl Chloride) (PVC) Sewer Pipe and Fittings
- ASTM D2774-04 Standard Practice for Underground Installation of Thermoplastic Pressure Piping
- ASTM D2846/D2846M-99 Standard Specification for Chlorinated Poly(Vinyl Chloride) (CPVC) Plastic Hot- and Cold-Water Distribution Systems
- ASTM D2992-01 Standard Practice for Obtaining Hydrostatic or Pressure Design Basis for Fiberglass (Glass-Fiber-Reinforced Thermosetting-Resin) Pipe and Fittings
- ASTM D3034-00 Standard Specification for Type PSM Poly(Vinyl Chloride) (PVC) Sewer Pipe and Fittings
- ASTM D3035-03a Standard Specification for Polyethylene (PE) Plastic Pipe (DR-PR) Based on Controlled Outside Diameter
- ASTM D3261-03 Standard Specification for Butt Heat Fusion Polyethylene (PE) Plastic Fittings for Polyethylene (PE) Plastic Pipe and Tubing
- ASTM D3262-04 Standard Specification for Fiberglass (Glass-Fiber-Reinforced Thermosetting-Resin) Sewer Pipe

- ASTM D3350-02a Standard Specification for Polyethylene Plastics Pipe and Fittings Materials
- ASTM D3517-04 Standard Specification for Fiberglass (Glass-Fiber-Reinforced Thermosetting-Resin) Pressure Pipe
- ASTM D3567-97(2002) Standard Practice for Determining Dimensions of Fiberglass (Glass-Fiber-Reinforced Thermosetting Resin) Pipe and Fittings
- ASTM D3681-01 Standard Test Method for Chemical Resistance of Fiberglass (Glass-Fiber-Reinforced Thermosetting-Resin) Pipe in a Deflected Condition
- ASTM D3754-04 Standard Specification for Fiberglass (Glass-Fiber-Reinforced Thermosetting-Resin) Sewer and Industrial Pressure Pipe
- ASTM D4161-01 Standard Specification for Fiberglass (Glass-Fiber-Reinforced Thermosetting-Resin) Pipe Joints Using Flexible Elastomeric Seals
- ASTM D4396-99ae1 Standard Specification for Rigid Poly(Vinyl Chloride) (PVC) and Chlorinated Poly(Vinyl Chloride) (CPVC) Compounds for Plastic Pipe and Fittings Used in Nonpressure Applications
- ASTM D5365-01 Standard Test Method for Long-Term Ring-Bending Strain of Fiberglass (Glass-Fiber-Reinforced Thermosetting-Resin) Pipe
- ASTM D5813-04 Standard Specification for Cured-In-Place Thermosetting Resin Sewer Piping Systems
- ASTM F441/F441M-02 Standard Specification for Chlorinated Poly(Vinyl Chloride) (CPVC) Plastic Pipe, Schedules 40 and 80
- ASTM F442/F442M-99 Standard Specification for Chlorinated Poly(Vinyl Chloride) (CPVC) Plastic Pipe (SDR-PR)
- ASTM F477-02e1 Standard Specification for Elastomeric Seals (Gaskets) for Joining Plastic Pipe
- ASTM F512-95(2001)e1 Standard Specification for Smooth-Wall Poly(Vinyl Chloride) (PVC) Conduit and Fittings for Underground Installation
- ASTM F585-94(2000) Standard Practice for Insertion of Flexible Polyethylene Pipe into Existing Sewers
- ASTM F679-03 Standard Specification for Poly(Vinyl Chloride) (PVC) Large-Diameter Plastic Gravity Sewer Pipe and Fittings
- ASTM F714-03 Standard Specification for Polyethylene (PE) Plastic Pipe (SDR-PR) Based on Outside Diameter
- ASTM F758-95(2000) Standard Specification for Smooth-Wall Poly(Vinyl Chloride) (PVC) Plastic Underdrain Systems for Highway, Airport, and Similar Drainage
- ASTM F789-95a Standard Specification for Type PS-46 and Type PS-115 Poly(Vinyl Chloride) (PVC) Plastic Gravity Flow Sewer Pipe and Fittings
- ASTM F794-03 Standard Specification for Poly(Vinyl Chloride) (PVC) Profile Gravity Sewer Pipe and Fittings Based on Controlled Inside Diameter
- ASTM F810-01 Standard Specification for Smoothwall Polyethylene (PE) Pipe for Use in Drainage and Waste Disposal Absorption Fields
- ASTM F894-98a Standard Specification for Polyethylene (PE) Large Diameter Profile Wall Sewer and Drain Pipe
- ASTM F949-03 Standard Specification for Poly(Vinyl Chloride) (PVC) Corrugated Sewer Pipe with a Smooth Interior and Fittings
- ASTM F1055-98e1 Standard Specification for Electrofusion Type Polyethylene Fittings for Outside Diameter Controlled Polyethylene Pipe and Tubing

- ASTM F1336-02 Standard Specification for Poly(Vinyl Chloride) (PVC) Gasketed Sewer Fittings
- ASTM F1533-01 Standard Specification for Deformed Polyethylene (PE) Liner
- ASTM F1606-95 Standard Practice for Rehabilitation of Existing Sewers and Conduits with Deformed Polyethylene (PE) Liner
- ASTM F1668-96(2002) Standard Guide for Construction Procedures for Buried Plastic Pipe
- ASTM F1697-02 Standard Specification for Poly(Vinyl Chloride) (PVC) Profile Strip for Machine Spiral-Wound Liner Pipe Rehabilitation of Existing Sewers and Conduits
- ASTM F1698-02 Standard Practice for Installation of Poly(Vinyl Chloride)(PVC) Profile Strip Liner and Cementitious Grout for Rehabilitation of Existing Man-Entry Sewers and Conduits
- ASTM F1735-02e1 Standard Specification for Poly(Vinyl Chloride) (PVC) Profile Strip for PVC Liners for Rehabilitation of Existing Man-Entry Sewers and Conduits
- ASTM F1741-02a Standard Practice for Installation of Machine Spiral Wound Poly(Vinyl Chloride) (PVC) Liner Pipe for Rehabilitation of Existing Sewers and Conduits
- ASTM F1743-96(2003) Standard Practice for Rehabilitation of Existing Pipelines and Conduits by Pulled-In-Place Installation of Cured-In-Place Thermosetting Resin Pipe (CIPP)
- AWWA C105/A21.5-99 ANSI Standard for Polyethylene Encasement for Ductile-Iron Pipe Systems
- AWWA C222-99 Polyurethane Coatings for the Interior and Exterior of Steel Water Pipe and Fittings
- AWWA C224-01 Two-layer Nylon-11 Based Polyamide Coating System for Interior and Exterior of Steel Water Pipe and Fittings
- AWWA C209-00 Cold-Applied Tape Coatings for the Exterior of Special Sections, Connections, and Fittings for Steel Water Pipe
- AWWA C210-97 Liquid-Epoxy Coating Systems for the Interior and Exterior of Steel Water Pipelines
- AWWA C213-01 Fusion-Bonded Epoxy Coating for the Interior and Exterior of Steel Water Pipelines
- AWWA C214-00 Tape Coating Systems for the Exterior of Steel Water Pipelines
- AWWA C215-99 Extruded Polyolefin Coatings for the Exterior of Steel Water Pipelines
- AWWA C216-00 Heat-Shrinkable Cross-Linked Polyolefin Coatings for the Exterior of Special Sections, Connections, and Fitting
- AWWA C203-02 Coal-Tar Protective Coatings and Linings for Steel Water Pipelines, Enamel and Tape, Hot-App. (Incl. add. C203a-99)

Concrete Pipe

- AWWA C205-00 Cement-Mortar Protective Lining and Coating for Steel Water Pipe 4 in (100 mm) and Larger-Shop Application
- AWWA C104/A21.4-95 ANSI Standard for Cement-Mortar Lining for Ductile-Iron Pipe and Fittings for Water

Ductile-Iron Pipe and Fittings

- AWWA C153/A21.53-00 ANSI Standard for Ductile-Iron Compact Fittings for Water Service
- AWWA C116/A21.16-98 ANSI Std. for Protective Fusion-Bonded Epoxy Coatings Interior and Exterior Surface Ductile-Iron/Gray-Iron Fittings
- AWWA C218-02 Coating the Exterior of Aboveground Steel Water Pipelines and Fittings
- AWWA C217-99 Cold-Applied Tape Coatings for Ext. Spec. Sect., Conn., and Fittings for Buried/Submerged Steel Water Pipelines

PLASTIC PIPE INSTITUTE PUBLICATIONS

Technical Reports

- PPI TR-3 Policies and Procedures for Developing Hydrostatic Design Basis (HDB), Pressure Design Basis (PDB), Strength Design Basis (SDB), and Minimum Required Strengths (MRS) Ratings for Thermoplastic Piping Materials or Pipe (2003)
- PPI TR-4 PPI Listing of Hydrostatic Design Basis (HDB), Strength Design Basis (SDB), Pressure Design Basis (PDB) and Minimum Required Strength (MRS) Ratings for Thermoplastic Piping Materials or Pipe (2003)
- PPI TR-5 A listing of standards for the various types of plastics piping are promulgated by numerous standards-making organizations (2001)
- PPI TR-7 Recommended Methods for Calculation of Nominal Weight of Solid Wall Plastic Pipe (2000)
- PPI TR-9 Recommended Design Factors for Pressure Applications of Thermoplastic Pipe Materials. (2000)
- PPI TR-11 Resistance of Thermoplastic Piping Materials to Micro- and Macro-Biological Attack (2000)
- PPI TR-14 Water Flow Characteristics of Thermoplastic Pipe (2001)
- PPI TR-18 Weatherability of Thermoplastic Piping Systems (1999)
- PPI TR-19 Thermoplastic Piping for the Transport of Chemicals (2000)
- PPI TR-21 Thermal Expansion and Contraction in Plastics Piping Systems (2001)
- PPI TR-22 Polyethylene Plastic Piping Distribution Systems for Components of Liquid Petroleum Gases (2000)
- PPI TR-30 Investigation of Maximum Temperatures Attained by Plastic Fuel Gas Pipe Inside Service Risers (2000)
- PPI TR-33 Generic Butt Fusion Joining Procedure for Polyethylene Gas Pipe (2001)
- PPI TR-34 Disinfection of Newly Constructed Polyethylene Water Mains. (2001)
- PPI TR-35 Chemical and Abrasion Resistance of Corrugated Polyethylene Pipe (1999)
- PPI TR-36 Hydraulic Considerations for Corrugated Polyethylene Pipe (1998)
- PPI TR-37 CPPA Standard Specification (100-99) for Corrugated Polyethylene (PE) Pipe for Storm Sewer Applications (1999)
- PPI TR-38 Structural Design Method for Corrugated Polyethylene Pipe (1996)

- PPI TR-39 Structural Integrity of Non-Pressure Corrugated Polyethylene Pipe (1997)
- PPI TR-40 Evaluation of Fire Risk Related to Corrugated Polyethylene Pipe (1996)
- PPI TR-41 Generic Saddle Fusion Joining Procedure for Polyethylene Gas Piping (2000)
- PPI TR-42 Agriculture and Drainage—Inseparable Science (2002)
- PPI TR-43 Design Service Life of Corrugated High Density Polyehtylene (HDPE) Pipe (2003)

Technical Notes

- PPI TN-5 Equipment used in the Testing of Plastic piping Components and Materials. (1998)
- PPI TN-6 Polyethylene (PE) Coil Dimensions (2001)
- PPI TN-7 Nature of Hydrostatic Stress Rupture Curves (2000)
- PPI TN-11 Suggested Temperature Limits for the Operation and Installation of Thermoplastic Piping in Non-Pressure Applications (1999)
- PPI TN-13 General Guidelines for Butt, Saddle, and Socket Fusion of Unlike Polyethylene Pipes and Fittings (2000)
- PPI TN-14 Plastic Pipe in Solar Heating Systems (2001)
- PPI TN-15 Resistance of Solid Wall Polyethylene Pipe to a Sanitary Sewage Environment (2000)
- PPI TN-16 Rate Process Method for Projecting Performance of Polyethylene Piping Components (1999)
- PPI TN-17 Cross-linked Polyethylene (PEX) Tubing (2001)
- PPI TN-18 Long-Term Strength (LTHS) by Temperature Interpolation (1998)
- PPI TN-19 Pipe Stiffness for Buried Gravity Flow Pipes (2000)
- PPI TN-20 Special Precautions for Fusing Saddle Fittings to Live PE Fuel Gas Mains Pressurized on the Basis of a 0.40 Design Factor (2000)
- PPI TN-21 PPI PENT test investigation Phase1.xls (2000)
- PPI TN-23 Guidelines for Establishing the Pressure Rating for Multilayer and Coextruded Plastic Pipes (2002)
- PPI TN-26 Erosion Study on Brass Insert Fittings used in PEX Piping Systems (2002)
- PPI TN-27 Commonly Asked Questions about HDPE Pipe for Water Applications (2003)

Reference Documents

- An Introduction to Trenchless Technology by S.R. Kramer, W.J. McDonald and J.C. Thomson, Chapman & Hall, 1992.
- Buried Pipe Design by A.P. Moser, McGraw-Hill, 2001.
- Clay Pipe Engineering Manual, National Clay Pipe Institute (NCPI), 1995.
- Concrete Pipe Design Manual, American Concrete Pipe Association, 2000.
- Geotechnical Baseline Reports for Underground Construction, American Society of Civil Engineers (ASCE), 1997.

- Handbook of PVC Pipe Design and Construction, Uni-Bell PVC Pipe Association, 1991.
- Horizontal Directional Drilling Good Practices Guidelines, HDD Consortium, North American Society for Trenchless Technology (NASTT), 2001.
- Horizontal Auger Boring Projects, American Society of Civil Engineers (ASCE), 2004.
- No-Dig Handbook, Scandinavian Society for Trenchless Technology (SSTT), 2002.
- Pipeline Crossings, American Society of Civil Engineers (ASCE), 1996.
- Pipeline Design for Installation by Horizontal Directional Drilling, American Society of Civil Engineers (ASCE), Scheduled Publication Date 2005.
- Pipeline Engineering by H. Liu, Lewis Publishers, 2003.
- Rehabilitation and Maintenance of Drains and Sewers by D. Stein, Ernst & Sohn, 2001.
- Selected ASTM Standards on Concrete Pipe, American Concrete Pipe Association 2001.
- Soil Mechanics in Engineering Practice by K. Terzaghi, R.B Peck and G. Mesri, John Wiley and Sons, 1996.
- Standard Construction Guidelines for Microtunneling, CI/ASCE 36-01, American Society of Civil Engineers (ASCE), 2001.
- Standard Guideline for the Collection and Depiction of Existing Subsurface Utility Data, CI/ASCE 36-01, American Society of Civil Engineers (ASCE), 2003.
- Standard Practice for Direct Design of Precast Concrete Pipe for Jacking in Trenchless Construction, American Society of Civil Engineers (ASCE), 2000.
- Structural Mechanics of Buried Pipes by R.K. Watkins and L.R. Anderson, CRC Press, 2000.
- Steel Pipe – A Guide for Design and Installation, American Water Works Association, 1989.
- Trenchless Construction Methods and Soil Compatibility Manual by D.T. Iseley, M. Najafi and R. Tanwani, National Utility Contractors Association (NUCA), 1999.
- Trenchless Installation of Conduits Beneath Roadways, Synthesis of Highway Practice 242, Transportation Research Board (TRB), 1997.
- Trenchless Pipeline Rehabilitation by D. T. Iseley and M. Najafi, National Utility Contractors Association (NUCA), 1995.
- Trenchless Pipeline Rehabilitation by M. Najafi, Trenchless Technology Center (TTC), Louisiana Tech University, 1994.

INDEX

ABOUT THE AUTHOR

Dr. Mohammad Najafi, P.E., is Assistant Professor of Construction Management and Director of the Center for Underground Infrastructure Research and Education (CUIRE) at Michigan State University. From 1994 through 2002, he was the Technical Editor of *No-Dig Engineering Journal* and currently is the Pipeline Editor of *ASCE Journal of Transportation Engineering*. He lives in East Lansing, Michigan.